Simulation Foundations, Methods and Applications

Series Editor

Louis G. Birta, University of Ottawa, Ottawa, ON, Canada

Advisory Editors

Roy E. Crosbie, California State University, Chico, CA, USA
Tony Jakeman, Australian National University, Canberra, ACT, Australia
Axel Lehmann, Universität der Bundeswehr München, Neubiberg, Germany
Stewart Robinson, Loughborough University, Loughborough, Leicestershire, UK
Andreas Tolk, Old Dominion University, Norfolk, VA, USA
Bernard P. Zeigler, University of Arizona, Tucson, AZ, USA

More information about this series at http://www.springer.com/series/10128

Louis G. Birta • Gilbert Arbez

Modelling and Simulation

Exploring Dynamic System Behaviour

Third Edition

 Springer

Louis G. Birta
School of Electrical Engineering
and Computer Science
University of Ottawa
Ottawa, ON, Canada

Gilbert Arbez
School of Electrical Engineering
and Computer Science
University of Ottawa
Ottawa, ON, Canada

ISSN 2195-2817   ISSN 2195-2825   (electronic)
Simulation Foundations, Methods and Applications
ISBN 978-3-030-18871-9   ISBN 978-3-030-18869-6   (eBook)
https://doi.org/10.1007/978-3-030-18869-6

This Springer imprint is published by the registered company Springer Nature Switzerland AG
The registered company address is: Gewerbestrasse 11, 6330 Cham, Switzerland

To our wives, Suzanne and Monique, and to the next and future generations: Christine, Jennifer, Alison, Amanda, Julia, Jamie, Aidan and Mika

Preface

Overview

Modelling and simulation is a tool that provides support for the planning, design and evaluation of dynamic systems as well as the evaluation of strategies for system transformation and change. Its importance continues to grow at a remarkable rate, in part because its application is not constrained by discipline boundaries. This growth is also the consequence of the opportunities provided by the ever- widening availability of significant computing resources and the expanding pool of human skill that can effectively harness this computational power. However, the effective use of any tool and especially a multi-faceted tool such as modelling and simulation involves a learning curve. This book addresses some of the challenges that lie on the path that ascends that curve.

Consistent with good design practice, the development of this book began with several clearly defined objectives. Perhaps the most fundamental was the intent that the final product provides a practical (i.e. useful) introduction to the many facets of a typical modelling and simulation project. The important differences in his regard between the discrete-event and the continuous-time domains would need to be highlighted. As well, the importance of illustrative projects from both the discrete and continuous domains would need to be recognized as a necessary part of providing practical insights. To a large extent, these objectives were the product of insights acquired by the authors over the course of several decades of teaching a wide range of modelling and simulation topics. Our view is that we have been successful in achieving these objectives.

Features

We have taken a project-oriented perspective of the modelling and simulation enterprise. The implication here is that modelling and simulation is, in fact, a collection of activities that are all focused on one particular objective; namely, providing a credible resolution to a clearly stated goal, a goal that is formulated within a specific system context. There can be no project unless there is a goal. All the constituent sub-activities work in concert to achieve the goal. Furthermore the 'big picture' must always be clearly in focus when dealing with any of the sub-activities. We have strived to reflect this perspective throughout our presentation.

The notion of a conceptual model plays a central role in our presentation. While this is not especially significant for projects within the continuous time domain inasmuch as the differential equations that define the system dynamics can be correctly regarded as the conceptual model, it is very significant in the case of projects from the discrete-event domain. This is because there is no generally accepted view of what actually constitutes a conceptual model in that context. This invariably poses a significant hurdle from a pedagogical point of view because there is no abstract framework in which to discuss the structural and behavioural features of the system under investigation. The inevitable (and unfortunate) result is a migration to the semantic and syntactic formalisms of some computer programming environment.

We have addressed this issue by presenting a conceptual modelling framework for discrete-event dynamic systems, which we call the ABCmod framework (Activity-Based Conceptual modelling). While this book makes no pretence at being a research monograph, it is nevertheless appropriate to emphasize that the ABCmod conceptual modelling framework presented in Chap. 4 is the creation of the authors. This framework has continued to evolve from versions presented in previous editions of this book.

The basis for the ABCmod framework is the identification of relevant 'units of behaviour' within the system under investigation and their subsequent synthesis into individual behavioural components called 'activities'. The identification of activities as a means for organizing a computer program that captures the time evolution of a discrete-event dynamic system is not new. However, in our ABCmod framework, the underlying notions are elevated from the programming level to an abstract and hence conceptual level. A number of examples are presented to illustrate conceptual model development using the ABCmod framework. Furthermore, we demonstrate the utility of the ABCmod framework by showing how its constructs conveniently map onto those that are required in program development perspectives (world views) that appear in the modelling and simulation literature.

The Activity-Object world view presented in Chap. 6 is a recent outgrowth of the continuing development of the ABCmod concept. This world view which has an object-oriented basis preserves the activity perspective of the ABCmod framework and makes the translation of the conceptual model into a simulation model entirely transparent. This, in turn, simplifies the verification task.

Audience

This book is intended for students (and indeed, anyone else) interested in learning about the problem-solving methodology called modelling and simulation. A meaningful presentation of the topics involved does necessarily require a certain level of technical maturity on the part of the reader. An approximate measure in this regard would correspond to a science or engineering background at the senior undergraduate or the junior graduate level.

More specifically our readers are assumed to have a reasonable comfort level with standard mathematical notation, which we frequently use to concisely express relationships. There are no particular topics from mathematics that are essential to the discussion but some familiarity with the basic notions of probability and statistics play a role in the material in Chaps. 3 and 7. (In this regard, a Probability and Statistics Primer is provided in Annex 2). Some appreciation for the notions relating to ordinary differential equations is necessary for the discussions in Chaps. 8, 9 and 11. A reasonable level of computer programming skills is assumed in the discussions of Chaps. 5, 6 and 9. We use Java as our programming environment of choice in developing simulation models based on the Three-Phase world view and the Activity-Object world view. The GPSS programming environment is used to illustrate the process-oriented approach to developing simulation models. (We provide a GPSS Primer in Annex 3). Our discussion of the modelling and simulation enterprise in the continuous time domain is illustrated using the numerical toolbox provided in MATLAB. A brief overview of relevant MATLAB features is provided in Annex 4.

Organization

This book is organized into four Parts. Part I has two Chapters that present an overview of the modelling and simulation discipline. In particular, they provide a context for the subsequent discussions and, as well, the process that is involved in carrying out a modelling and simulation study. Important notions such as quality assurance are also discussed.

The five Chapters of Part II explore the various facets of a modelling and simulation project within the realm of discrete-event dynamic systems (DEDS). We begin by pointing out the key role of random (stochastic) phenomena in modelling and simulation studies in the DEDS realm. This, in particular, introduces the need to deal with data models as an integral part of the modelling phase. Furthermore, there are significant issues that must be recognized when handling the output data resulting from experiments with DEDS models. These topics are explored in some detail in the discussions of Part II.

As noted earlier, we introduce in this book an activity-based conceptual modelling framework (the ABCmod framework) that provides a means for formulating a description of the structure and behaviour of a model that originates within the

DEDS domain. An outline of this framework is provided in Part II. A conceptual model is a first step in the development of a computer program that will serve as the 'solution engine' for the project. The next step in creating simulation program is to transform the conceptual model into a simulation model. Chapter 5 introduces traditional 'world views' used in creating simulation models and outlines the transition from an ABCmod conceptual model to simulation models based on two of these world views, the Three-Phase and the process-oriented. The presentation in Chap. 6 shows how the transition task to a simulation model based on the newly proposed Activity-Object world view is considerably more straightforward and transparent. Various key aspects of the Java library (called ABSmod/J) that supports the Activity-Object world view, are presented.

Chapter 7 examines the many important facets of the experimentation process with DEDS simulation models. New to this third edition is a discussion of the area of study called 'design of experiments', whose concern is with assessing the relative importance of the many parameters that are frequently embedded in a conceptual model.

There are two Chapters in Part III of the book and these are devoted to an examination of various important aspects of the modelling and simulation activity within the continuous time dynamic system (CTDS) domain. We begin in Chap. 8 by showing how conceptual models for a variety of relatively simple systems can be formulated. Most of these originate in the physical world that is governed by familiar laws of physics. However, we also show how intuitive arguments can be used to formulate credible models of systems that fall outside the realm of classical physics.

Inasmuch as a conceptual model in the CTDS realm is predominantly a set of differential equations, the 'solution engine' is a numerical procedure. In Chap. 9, we explore several options that exist in this regard in the numerical mathematics literature and provide some insights into important features of the solution alternatives. Several properties of CTDS models that can cause numerical difficulty are also identified.

Part IV of this third edition is essentially new. It explores the increasingly important interface between optimization and the modelling and simulation enterprise. The development of numerical optimization procedures has been a topic of interest in the numerical mathematics literature for many years. A brief overview of some of the underlying concepts is provided in Chap. 10. In particular, the main dichotomy between gradient-dependent and gradient-independent (heuristic) procedures is introduced. In Chap. 11, the simulation-optimization problem is examined within the CTDS context and in Chap. 12 the same exploration is undertaken within the DEDS context. Chapter 12, in addition, includes a case study that illustrates many of the challenges that can arise.

Overview of New Content in This Third Edition

Several references to the new content in this third edition have been made above. A summary is provided below:

1. In Chap. 4, the presentation of events within a DEDS model has been extended. As well, two new behavioural constructs called the 'scheduled sequel activity' and its natural counterpart called the 'scheduled sequel action' have been introduced into the ABCmod conceptual modelling framework.
2. The designations of 'Class' and 'Set[N]' assignable to the *scope* property of an entity category has been renamed 'Transient' and 'Many[N]', respectively. Additional content relating to the use of entity categories that contain many entities (i.e. *scope* = Many[N]) has been incorporated. It is stressed that the use of this feature enables the 'parameterization' of activities, which can significantly simplify model development. This applies both to ABCmod conceptual models and their simulation model counterparts.
3. Several relatively minor but nevertheless significant improvements have been incorporated into the simulation model development process based on ABSmod/J.
4. A new section that introduces the area of study called Design of Experiments has been added to Chap. 7. Its application in experimentation with DEDS simulation models is illustrated.
5. In response to the increasing interest with embedding optimization sturdies in modelling and simulation projects, a new Part IV has been added to this third edition. It explores this topic in both the DEDS and the CTDS domains and illustrative examples are included.
6. A new Annex 1 has been added. Its purpose is to consolidate in one place complete versions of the various ABCmod conceptual models that are introduced as examples at various places in the book.

Web Resources

A website has been established to provide access to a variety of supplementary material that accompanies this book. The following are included:

1. A set of PowerPoint slides from which presentation material can be developed.
2. The ABSmod/J Java package (jar file), that can be used to develop Activity-Object simulation programs.
3. DEDS Activity-Object simulation programs using ABSmod/J for the ABCmod conceptual modules presented in Annex 1.
4. A methodology for organizing student projects.

Acknowledgements

We would, first of all, like to acknowledge the privilege we have enjoyed in recent years in having had the opportunity to introduce so many students to the fascinating world of modelling and simulation. In many respects, the material in this book reflects much of what we have learned in terms of 'best practice' in presenting the essential ideas that are the foundation of the modelling and simulation discipline.

We would also like to acknowledge the contribution made by the student project group called Luminosoft, whose members (Mathieu Jacques Bertrand, Benoit Lajeunesse, Amélie Lamothe and Marc-André Lavigne) worked diligently and capable in developing the initial version of a software tool that supports the ABCmod conceptual modelling framework that is presented in this book. Our many discussions with the members of this group fostered numerous enhancements and refinements that would otherwise not have taken place. Their initial work has necessarily evolved to accommodate the continuing evolution of the ABCmod framework itself.

The extensions and refinements that are embedded in this third edition of our book have consumed substantial amounts of time. To a large extent this has been at the expense of time we would otherwise have shared with our families. We would, therefore, like to express our gratitude to our families for their patience and their accommodation of the disruptions that our preoccupation with this book project has caused on so many occasions. Thank you all!

Finally, we would like to express our appreciation for the help and encouragement provided by Mr. Wayne Wheeler, Senior Editor, (Computer Science) for Springer. His enthusiasm for this project fuelled our determination to maintain a high level of quality in the presentation of our perspectives on the topics covered in this book. Thanks also to Mr. Simon Rees (Associate Editor, Computer Science) who always provided quick responses to our concerns and maintained a firm, but nevertheless accommodating oversight over the timely completion of this book project.

Ottawa, Canada   Louis G. Birta
                                                                 Gilbert Arbez

Contents

Part I
Fundamentals

Part I of this book establishes the foundations for the subsequent discussions about our topic of interest, namely, modelling and simulation. It consists of two chapters, namely, Chaps. 1 and 2.

In Chap. 1, we briefly consider a variety of topics that can be reasonably regarded as background material. A natural beginning is a brief look at a spectrum of reasons why a modelling and simulation study might be undertaken. Inasmuch as the notion of a model is fundamental to our discussions, some preliminary ideas that relate to this notion are presented. A generic 'full-service' gas station is used to illustrate some of the key ideas. We then acknowledge that modelling and simulation projects can fail and suggest a number of reasons why this might occur.

Monte Carlo simulation and simulators are two topics which fall within a broadly interpreted perspective of modelling and simulation. In the interests of completeness, both of these topics are briefly reviewed. We conclude Chap. 1 with a brief look at some of the historical roots of the modelling and simulation discipline.

Modelling and simulation is a multi-faceted, goal-oriented activity and each of the steps involved must be duly recognized and carefully carried out. Chapter 2 is concerned with outlining these steps and providing an appreciation for the modelling and simulation process. The discussion begins with an examination of the essential features of a dynamic model and with the abstraction process that is inherent in its construction. The basic element in this abstraction process is the introduction of variables. These provide the means for carrying out a meaningful dialogue about the model and its behaviour properties, which must necessarily be consistent with the goal of the study. Variables fall into three broad categories, namely, input variables, output variables and state variables. The distinctive features of each of these categories are outlined.

The modelling and simulation process gives rise to a number of artefacts and these emerge in a natural way as the underlying process evolves. These various artefacts are outlined together with their inter-relationships. The credibility of the results flowing from a modelling and simulation project is clearly of fundamental importance. This gives rise to the topic of quality assurance and we conclude Part I of the book by exploring various facets of this important topic. In particular, we examine the central role of verification and validation as it relates to the phases of the modelling and simulation activity.

Introduction

1

1.1 Opening Perspectives

This book explores the use of modelling and simulation as a problem-solving tool. We undertake this discussion within the framework of a modelling and simulation *project*. This project framework embraces two key notions; first, there is the notion of a 'system context', i.e. there is a system that has been identified for investigation, and second, there is a problem relating to the identified system that needs to be solved. Evaluating a set of solution options to a stated problem within the context of a dynamic system that is being studied is the goal of the modelling and simulation project. We use the term 'system' in its broadest possible sense; it could, for example, include the notions of a process or a phenomenon. Furthermore, the physical existence of the system is not a prerequisite; the system in question may simply be a concept, idea or proposal. What is a prerequisite, however, is the requirement that the system in question exhibit 'behaviour over time'—in other words, that it be a *dynamic* system.

Systems, or more specifically dynamic systems, are one of the most pervasive notions of our contemporary world. Broadly speaking, a dynamic system is a collection of interacting entities that produce some form of behaviour that can be observed over an interval of time. There are, for example, physical systems such as transportation systems, power-generating systems, manufacturing systems and data communications systems. On the other hand, in a less tangible form, we have healthcare systems, social systems and economic systems. Systems are inherently complex, and tools such as modelling and simulation are needed to provide the means for gaining insight into aspects of their behaviour. Such insight may simply serve to provide the intellectual satisfaction of deeper understanding or, on the other hand, may be motivated by a variety of more practical and specific reasons such as providing a basis for decisions relating to the control, management, acquisition or transformation of the system under investigation (the SUI).

© Springer Nature Switzerland AG 2019
L. G. Birta and G. Arbez, *Modelling and Simulation*, Simulation Foundations,
Methods and Applications, https://doi.org/10.1007/978-3-030-18869-6_1

The defining feature of the modelling and simulation approach is that it is founded on a computer-based experimental investigation that utilizes an *appropriate* model for the SUI. The model is a representation or abstraction of the system. The use of models (in particular, mathematical models) as a basis for analysis and reasoning is well established in disciplines such as engineering and science. It is the emergence and widespread availability of computing power that has made possible the new dimension of experimentation with complex models and, hence, the emergence of the modelling and simulation discipline. Furthermore, the models in question need not be constrained by the requirement of yielding to classical mathematical formalism in the sense generally associated with the models in science and engineering.

It must be emphasized, furthermore, that there is an intimate connection between the model that is 'appropriate' for the study and the nature of the problem that is to be solved. The important corollary here is that there rarely exists a 'universal' model that will support all modelling and simulation projects that have a common system context. This is especially true when the system has some reasonable level of complexity. Consider, for example, the difference in the nature of a model for an airliner first in the case where the model is intended for use in evaluating aerodynamic properties versus the case where it is simply a revenue-generating object within a business plan. Identification of the most appropriate model for the project is possibly the most challenging aspect of the modelling and simulation approach to problem-solving.

While the word 'modelling' has a meaning that is reasonably confined in its general usage, the same cannot be said for the word 'simulation'. Nevertheless, the phrase 'modelling and simulation' does have a generally accepted meaning and implies two distinct activities. The modelling activity creates an object (i.e. a model) that is subsequently used as a vehicle for experimentation. This experimentation with the model is the simulation activity, and in the view that prevails in this book, the experimentation is carried out by a computer program.

The word 'simulation' is frequently used alone in a variety of contexts. For example, it is sometimes used as a noun to imply a specialized computer program (as in, 'A simulation has been developed for the proposed system'.). It is also used frequently as an adjective (as in, 'The simulation results indicate that the risk of failure is minimal' or 'Several extensions have been introduced into the language to increase its effectiveness for simulation programming'). These wide-ranging and essentially inconsistent usages of the word 'simulation' can cause regrettable confusion for neophytes to the discipline. As a rule, we will avoid such multiplicity of uses of this word, but as will become apparent, we do use the word as an adjective in two specific contexts where the implication is particularly suggestive and natural.

1.2 Role of Modelling and Simulation

There is a wide range of possible reasons for undertaking a modelling and simulation study. Some of the most common are listed below (the order is alphabetical and hence should not be interpreted as a reflection of importance):

1. Comparison of control policy options;
2. Education and training;
3. Engineering design;
4. Evaluation of decision or action alternatives;
5. Evaluation of strategies for transformation or change;
6. Forecasting;
7. Performance evaluation;
8. Prototyping and concept evaluation;
9. Risk/safety assessment;
10. Sensitivity analysis;
11. Support for acquisition/procurement decisions;
12. Uncertainty reduction in decision-making.

It was noted earlier that the goal of a simulation project has a major impact on the nature of the model that evolves. However, it is also important to observe that the goal itself may be bounded by constraints. These typically are a consequence of limitations on the level of knowledge that is available about the SUI. The unavoidable reality is that the available knowledge about the underlying dynamics of systems varies considerably from system to system. There are systems whose dynamics can be confidently characterized in considerable detail and, in contrast, there are systems whose dynamics are known only in an extremely tentative fashion. An integrated circuit developed for a telecommunications application would fall into the first category, while the operation of the stock market would reasonably fall into the second. This inherent range of knowledge level is sometimes reflected in the terminology used to describe the associated models. For example, an integrated circuit model might be referred to as a 'deep' model while a model of the stock market would be a 'shallow' model. The goal of a modelling and simulation project is necessarily restricted to being relatively qualitative when only a shallow model is feasible. A quantitative goal is feasible only for those situations where deep knowledge is available. In other words, the available knowledge level significantly influences the nature of the goal that can be realistically formulated for a modelling and simulation study.

The centrality of the goal of a modelling and simulation project has been recognized in terms of a notion called the *experimental frame* (see [9] and [15]). This notion is rather broadly based and implies a mechanism to ensure appropriate compatibility between the SUI, the model and the project goal. This usually includes such fundamental issues as the proper identification of the data the model must deliver, identification and representation of pertinent features of the environment in which the model functions, its inputs, its parameters and its granularity.

A model plays the role of a surrogate for the system it represents, and its purpose (at least from the perspective of this book) is to replace the system in experimental studies. When the underlying system (i.e. SUI) does not exist (e.g. it may merely be an idea, concept or proposal), then the model is the only option for experimentation. But even when the SUI does exist, there are a variety of reasons why experimentation directly with it could be inappropriate. For example, such experimentation might be:

- Too costly (determining the performance benefit likely to be achieved by upgrading the hardware at all the switch nodes of a large data communications network),
- Too dangerous (exploring alternate strategies for controlling a nuclear reactor),
- Too time-consuming (determining the ecological impact of an extended deer hunting season, implemented over several consecutive years, on the excessive deer population in a particular geographical region),
- Too disruptive (evaluating the effectiveness of a proposed grid of one-way streets within the downtown core of an urban area),
- Morally/ethically unacceptable (assessing the extent of radiation dispersion following a particular catastrophic failure at some nuclear power-generating facility),
- Irreversible (investigating the impact of a fiscal policy change on the economy of a country).

Behavioural data is almost always easier to acquire from a model than from the system itself and this is another important reason for favouring experimentation with a model. Consider, for example, the challenges inherent in monitoring the force exerted on the blades of a rotating turbine by escaping combustion gases. Furthermore, the fact that the platform for the experimentation is a computer (or, more correctly, a computer program) ensures reproducibility of results which is an essential requirement for establishing the credibility of experimental results. Such reproducibility can generally only be approximated when experiments are carried out directly with an existing system.

1.3 The Nature of a Model

A particular perspective prevails throughout this book, namely, the perspective that a model is a representation of the SUI's behaviour and the modelling process is concerned with the development of this representation. It is often suggested that the task is to ensure that the behaviour of the model is as indistinguishable as possible from the behaviour of the SUI. This assertion is only partially correct. A more appropriate statement of the task at hand is to develop the model so that it captures the behaviour properties at a level of granularity that is appropriate to the goal of the study. The challenge is to capture all relevant detail and to avoid superfluous features

(One might recall here a particularly memorable quote from Albert Einstein: 'Everything should be made as simple as possible, but not simpler'.). For example, consider a project concerned with evaluating strategies for improving the operating efficiency of a fast-food restaurant. Within this context, it would likely be meaningless (and, indeed, nonsensical) to incorporate into the model, information about the sequence in which a server prepares the hot and cold drinks when both are included in a customer's order.

The notion of 'behaviour' is clearly one that is fundamental to our discussions, and in particular, we have implied that there is usually a need to evaluate behaviour. But what does this mean and how is it done? At this point, we have to defer addressing these important questions until a more detailed exploration of the features of models has been completed.

Modelling is a constructive activity, and this raises the natural question of whether the product (i.e. the model) is 'good enough'. This question can be answered only if there is an identified context, and as we shall see in the discussions to follow, there are many facets to this key issue. One context that is most certainly fundamental is the goal of the project. In other words, a key question is always whether the model is good enough from the point of view of the requirements implied by the project goal. The corollary of this assertion is that it is never meaningful to undertake a modelling study without a clear understanding of the purpose for which the model will be used. Perhaps the most fundamental implication of the above discussion is that it is never meaningful to undertake a study whose goal is simply 'to develop a model of XYZ'.

There are a variety of ways in which the representation of behaviour can be formulated. Included here are: natural language, mathematical formalisms, rule-based formalisms, symbolic/graphical descriptions and combinations of these. It is typical for several distinct formulations of the model (or perhaps only portions of it) to evolve over the course of the study. These alternatives are generally created in formats that are best suited to capturing subtleties or providing clarification.

But there is one format that plays a very special role. This is a representation that has been formulated in the format of a computer program. The importance of such a representation arises because that computer program provides the means for actually carrying out the experiments that are central to the modelling and simulation approach. The development of this essential representation is a task that follows a well-defined process; in fact, discussion of this process is one of the main themes of our presentation in this book.

We conclude this section by emphasizing one particularly important point, namely, that models are representations of the SUI's behaviour, and only in that special case when this representation is in the form of a computer program can a model's behaviour be observed. Nevertheless, in the discussions that follow we will frequently use the phrase 'the model's behaviour'. This must always be carefully assessed by the reader since in most cases the phrase is used as a compact substitution for the correct phrase, which is 'the behaviour implied by the model'. We will, from time to time, remind the reader of our simplified (but misleading) usage.

1.4   An Example Project (Full-Service Gas Station)

To illustrate some facets of the discussion above, we consider a modelling and simulation project whose system context (SUI) is a 'full-service' gas station with two islands and four service lanes (see Fig. 1.1). A significant portion of the customers at this station drive small trucks and vans which typically have gas tank capacities that are larger than those of most passenger vehicles. Often, the drivers of passenger cars find themselves queued behind these 'large-tank' vehicles which introduce substantially longer wait times when they arrive at the gas station. This can cause aggravation and complaints. The station management is proposing as a possible solution the restriction of these large-tank vehicles to two designated lanes. The goal of the modelling and simulation project would be to determine if this strategy would improve the flow of vehicles through the station.

Vehicles are obliged (via appropriate signage) to access the pumps from the east side, and after their respective gas purchases, they exit on the west side. Upon arrival, drivers always choose the shortest queue. In the case where two or more queues have the same shortest length, a random choice is made. An exception is when it is observed that a customer in one of the 'shortest queues' is in the payment phase of the transaction in which case that queue is selected by the arriving driver.

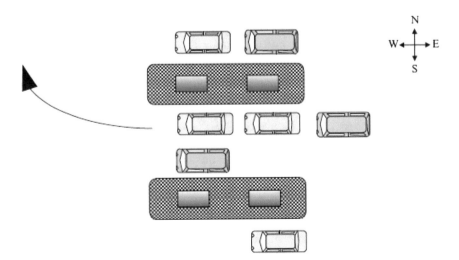

Fig. 1.1 Gas station project

Depending on the time of day, one or two attendants are available to serve the customers. The service activity has three phases. During the first, the attendant determines the customer's requirement and begins the pumping of gas (the pumps have a preset delivery amount and automatically shut off when the preset amount has been delivered or the tank is full). In addition, any peripheral service such as cleaning of windshields is carried out during this first phase. Phase two is the delivery phase during which gas continues to be pumped until delivery is completed. Phase three is the payment phase; the attendant accepts payment either in the form of cash or credit card. The duration of phase two is reasonably long, and an attendant typically has sufficient time either to begin a new phase one (serving a newly arrived customer) or to return to handle the phase three (payment) activity for a customer whose gas delivery is complete. The protocol is to give priority to a payment function before serving a newly arrived customer. It is standard practice for the payment function to be carried out by the same attendant who initiated the transaction.

The above text can be regarded as the initial phase in the modelling and simulation project. It corresponds to the notion of a *problem description*, which we examine more carefully in Chap. 2. Notice, however, that refinements to this description are almost certain to be necessary; these may simply provide clarification (e.g. what are the conditions that govern the attendant's options during phase two) or may introduce additional detail; e.g. what happens when a pump becomes defective or under what conditions does an arriving customer 'balk', i.e. decides the queues are too long and leaves. There is also the case, where a customer who has been waiting in a queue for an unacceptably long time leaves the gas station in frustration (reneging). Or, in fact, are balking and reneging even relevant occurrences? What about accommodating the possibility that drivers (or passengers) may need to use the washroom facilities and thereby 'hold' the pump position longer than is otherwise necessary?

The reader may, at this point, wonder when refinements to the problem description can be judged to be sufficient. (It may be useful for the reader to dwell on other possible refinements). This is by no means an inconsequential concern and a simple answer is elusive. It should be noted that refinements are, in fact, frequently incorporated in the problem description during the modelling phase, because missing information or information that lacks clarity is often discovered during model development. It needs to be stressed, however, that the main intent of the problem description is clarity in the documentation of both the problem to be solved and the relevant details of the SUI. Modelling consideration must be avoided. If the problem to be solved is significantly different, e.g. an environmental assessment of the gas station's operation or indeed in an analysis of its financial viability, the problem description would have to be adapted to the alternate perspective.

With the problem description at hand, model building can start. The full range of matters that are part of the model development phase (e.g. data modelling, identification of SUI performance data, simplifications and assumptions, etc.) are addressed in the discussions that follow in Chapters 2 through 5. Data modelling in the gas station context could, for example, focus on the determination of the arrival rate of vehicles, the proportion of vehicles that fall into the small truck/van category and service times for each of the three service phases.

1.5   Is There a Downside to the Modelling and Simulation Paradigm?

The implicit thrust of the presentation in this book is that of promoting the strengths of the modelling and simulation paradigm as a problem-solving methodology. However, one might reasonably wonder whether there exist inherent dangers or pitfalls. And the simple answer is that these do indeed exist! Like most tools (both technological and otherwise), modelling and simulation must be used with a good measure of care and wisdom. An appreciation for the limitations and dangers of any tool is a fundamental prerequisite for its proper use. We examine this issue within the modelling and simulation context somewhat indirectly by examining some reasons why modelling and simulation projects can fail.

(a)   Inappropriate statement of a goal.
       No project can ever be successful unless its objectives are clearly articulated and fully understood by all the stakeholders. This most certainly applies to any modelling and simulation project. The goal effectively drives all stages of the development process. Ambiguity in the statement of the goal can lead to much-wasted effort or yield conclusions that are unrelated to the expectations of the 'client' responsible for the initiation of the project.
       A second but no less important goal-related issue relates to the feasibility of achieving the stated goal. As suggested earlier, the project goal has to be consistent with the realities of the depth of knowledge that characterizes the SUI. Any attempt to extract precise knowledge from a shallow model will most certainly fail. There are other feasibility issues as well. For example, the available level of essential resources may simply not be adequate to achieve the goal. Such resources include time (to complete the project), talent (skill set; see (d) below), funding and data/information (relating to the SUI).

(b)   Inappropriate granularity of the model
       The granularity of the model refers to the level of detail with which it attempts to replicate features of the SUI. The level of granularity is necessarily bounded by the goal of the project, and care must always be taken to ensure that the correct level has been achieved (e.g. in the gas station example, is the treatment of reneging relevant?). Excessive detail increases complexity, and this can lead to cost overruns and/or completion delays that usually translate into project failure. Too little detail, on the other hand, can mask the very effects that have substantial relevance to the behaviour that is of critical interest. This is particularly serious because the failure of the project generally becomes apparent only when undesired consequences begin to flow from the implementation of incorrect decisions based on the study.

(c)   Ignoring unexpected behaviour.
       Although a validation process is recognized to be an essential stage in any modelling and simulation project, its main thrust generally is to confirm that expected behaviour does occur. On the other hand, testing for unexpected (counter-intuitive) behaviour is never possible. Nevertheless, such behaviour

can occur, and when it is observed, there often is a tendency to dismiss it, particularly when validation tests have provided satisfactory results. Ignoring such counter-intuitive, or unexpected, observations can lay the foundation for failure.

(d) Inappropriate mix of essential skills.

A modelling and simulation project of even modest size can have substantial requirements in terms of both the range of skills and the effort needed for its completion. A team environment is therefore common; team members contribute complementary expertise to the intrinsically multifaceted requirements of the project. The range of skills that needs to be represented among the team members can include project management, documentation, transforming domain knowledge into the format consistent with a credible dynamic model, development of data modules as identified in the data requirements, experiment design, software development and analysis of results. The intensity of coverage of these various areas is very much dependent on the specific nature of the project. Nevertheless, an inappropriate mix of skills can seriously impede progress and can ultimately result in project failure.

(e) Inadequate flow of information to the client.

The team that carries out a modelling and simulation project often does so on behalf of a 'client', who is not a member of the team. In such cases, care must be taken to ensure that the client is fully aware of how the project is unfolding in order to avoid the occurrence of a 'disconnect' that results in the delivery of a product that falls short of expectations. For example, a minor misinterpretation of requirements, if left uncorrected, can have consequences that escalate to the point of jeopardizing the project's success.

1.6   Monte Carlo Simulation

References to Monte Carlo simulation are often encountered in the modelling and simulation literature. This somewhat fanciful label refers to a problem-solving methodology that is loosely related to, but is very different from, the topic that we explore in this book. The term refers to a family of techniques that are used to find solutions to numerical problems. The distinctive feature of these techniques is that they proceed by constructing a stochastic (probabilistic) system, whose properties contain the solution of the underlying problem. The origins of the approach can be traced back to Lord Rayleigh, who used it to develop approximate solutions to simple partial differential equations. The power of the methodology was exploited by von Neumann and colleagues in solving complex problems relating to their work in developing a nuclear arsenal in the latter years of the Second World War. The Monte Carlo label for the methodology is, in fact, attributed to this group.

Perhaps, the simplest example of the method is its application to the evaluation of the definite integral:

$$I = \int_a^b f(x)dx \tag{1.1}$$

for the special case where $f(x) \geq 0$. The value of I is the area under $f(x)$ between $x = a$ and $x = b$. Consider now a horizontal line drawn at $y = K$ such that $f(x) \leq K$ for $a \geq x \geq b$ (see Fig. 1.2). The rectangle, R, enclosed by $x = a$, $x = b$, $y = 0$ and $y = K$ has the area $K(b - a)$ and furthermore $I \leq K(b - a)$. Suppose a sequence of points (x_i, y_i) is chosen at random within the rectangle R such that all points within R are equally likely to be chosen (e.g. by choosing from two uniform distributions oriented along the length and width of R). It can then be easily appreciated that the ratio of the number of points that fall either on the curve or under it (say, n) to the total number of points chosen (say, N) is an approximation of the ratio of I to the area of the rectangle, R. In other words,

$$n/N \approx I/[K(b - a)] \tag{1.2}$$

or

$$I \approx nK(b - a)/N \tag{1.3}$$

In the procedure, a point (x_i, y_i) is included in the count, n, if $y_i \leq f(x_i)$. The accuracy of the approximation improves as N increases.

The interesting feature in this example is that the original problem is entirely deterministic, and yet the introduction of probabilistic notions can yield an approximation to its solution.

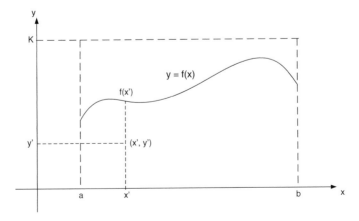

Fig. 1.2 Example of Monte Carlo simulation

The class of problems that can be effectively investigated by Monte Carlo simulation generally falls within the domain of numerical analysis. The approach provides an alternate, and often very effective, solution option for these problems. However, these problems do not fall within the scope of the modelling and simulation methodology because they lack the prerequisite of 'behaviour', i.e. an evolution over time. The reference to 'simulation' in the label for the approach could be regarded as a reflection of the dissimilarity between the solution mechanism and the inherent nature of the problem.

1.7 Simulators

There is frequent reference in the modelling and simulation literature to the notion of simulators. Most commonly, the notion is a reference to a training device or platform, and it is from that perspective that we explore the topic in the discussion that follows. Within the training context, a simulator can be viewed as a device that replicates those operational features of some particular system that are deemed to be important for the training of operators of that system. A characteristic feature of any simulator is the incorporation of some physical parts of the system itself as a means of enhancing the realism of the training environment; e.g. an actual control panel layout. Beginning with the development of flight simulators for training pilots (see Sect. 1.8 which follows), the use of simulators has spread into a wide range of domains, for example, there exist power plant simulators (both nuclear and fossil), battlefield simulators, air traffic control simulators and (human) patient simulators. An interesting presentation of recent applications of simulators in the training of health science professionals can be found in the various papers of the special journal issue [2].

The fundamental requirement of any simulator is the replication of system behaviour within a physical environment that is as realistic as possible from the perspective of an operator. Although the simulator incorporates some physical features of the system, substantial components of the system necessarily exist only in the form of models. In early simulators, these models were themselves physical in nature, but with the emergence of computing technology, the modelled portions of the system have increasingly exploited the modelling power of this technology.

The development of a simulator can be viewed as a modelling and simulation project, whose goal is to achieve an effective training environment. This, in particular, implies that the device must operate in 'real-time'; i.e. behaviour, as experienced by the operator, must evolve at a rate that corresponds exactly to that of the real system. This introduces additional important constraints on the models for those portions of the system that are being emulated, e.g. synchronization of 'virtual' time within the model with 'real', (i.e. clock) time.

Simulators can also contribute in a variety of ways to enhancing the educational experience of students especially in circumstances where alternatives are precluded (e.g. by budgetary constraints) or alternatively when the devices being examined are either 'hypothetical' or are no longer available in the marketplace. The areas of computer organization and operating systems are particularly well suited to this application, and relevant discussions can be found in Refs. [13] and [14].

Apart from their importance as training platforms and educational tools, it is interesting to observe that simulators also have a lighter side in their role within the entertainment industry. A multibillion-dollar 'computer-game' industry has emerged in recent years essentially built upon simulators for a vast array of vehicles, both imagined and unimagined.

Simulators represent an application area of the modelling and simulation paradigm. Inasmuch as our intent in this book is to avoid a focus on any particular application area, the topic of simulators will not be explicitly addressed in the sequel, apart from one exception. This is a brief examination of their important role in the history of modelling and simulation as discussed in the following section.

1.8 Historical Overview

The birth of the modelling and simulation discipline can reasonably be associated with the development of the Link Trainer by Edward Link in the late 1920s. The Link Trainer is generally regarded as the first successful device designed specifically for the training of pilots and represents the beginning of an extensive array of training tools called flight simulators. The initial Link Trainer clearly predates the arrival of the modern computer and its behaviour generating features were produced instead using pneumatic/hydraulic technology. As might be expected, flight simulators quickly embraced computer technology as it developed in the 1950s. The sophistication of flight simulators has continuously expanded, and they have become indispensable platforms for training not only aircraft pilots (both commercial and military) but also the members of the Apollo Missions and, as well, the various space shuttle teams. In fact, the technology and methodology demands made by the developers of flight simulators have had a great impact on the evolution of the modelling and simulation discipline itself.

While the development and evolution of simulators, in general, represent the initial (and probably pivotal) application area for modelling and simulation, it is the analog computer that represents the initial computing platform for the discipline. The commercial availability of these computers began in the early 1950s. The principles upon which this computing machine was based were originally formulated by Lord Kelvin in the latter part of the nineteenth century. The electronic realization of the concept was developed by Vannevar Bush in the 1930s.

The analog computer was primarily a tool for the solution of differential equations. The solution of such equations was obtained by direct manipulation of voltage signals using active elements called operational amplifiers. The computing environment was highly interactive and provided a convenient graphical output. While primarily electronic in nature, the early machines nevertheless relied on electromechanical devices to carry out basic non-linear operations such as multiplication and division. This often introduced serious constraints in terms of both solution speed and accuracy.

Programming the analog computer was a relatively complex and highly error-prone process since it involved physically interconnecting the collection of processing elements that were required for solving the problem at hand. The range of processing element types was relatively narrow but did include one that directly carried out an integration operation (not surprisingly called the 'integrator'). It was this device that provided the basis for the straightforward solution of differential equations. As a result, the problems that were best suited for solution emerged from the realm of engineering (e.g. aerospace flight dynamics and control system design).

By the mid-1960s, the speed of general-purpose digital computers and the software support for their use had improved to a point where it was apparent that they were going to make important contributions to the evolution of the modelling and simulation discipline. For example, their capabilities showed promise in providing an alternate solution tool for the same problem class that had previously fallen exclusively into the domain of the analog computer (thereby setting the stage for the demise of the 'worthy predecessor'). But perhaps even more significantly, it was clear that computing power was now becoming available to support a class of modelling and simulation projects that had been beyond the capabilities of the analog computer, namely, the class of discrete event problems that incorporate stochastic phenomena.

Over the next two decades of the 1970s and the 1980s, a wide variety of software support for modelling and simulation applications was developed. This made possible the initiation of increasingly more ambitious projects which, by and large, fall into two distinct realms, namely, the continuous (typically engineering design problems formulated around differential equation models) and the discrete event (typically process design problems incorporating stochastic phenomenon and queuing models).

Some interesting perspectives on the development of the modelling and simulation paradigm can be found in Nance and Sargent [6]. The evolution of specialized programming languages for modelling and simulation is an integral part of the history of the discipline, and a comprehensive examination of such developments within the discrete event context (up to the year 1986) can be found in Nance [4, 5]. The overview given by Bowden [1] and the survey results presented by Swain [12] provide some insight into the wide spectrum of commercially available software products in this domain. A comprehensive collection of contemporary simulation software is listed and summarized in [11]. Making correct choices can be a challenging task, and the work of Nikoukaran et al. [7] can provide useful guidance in this regard.

In spite of the relatively short time span of its history, the modelling and simulation discipline has given birth to a remarkably large number of professional organizations and associations that are committed to its advancement (see http://www.site.uottawa. ca/~oren/links-MS-AG.pdf, where a comprehensive listing is maintained). These span a broad range of specific areas of application, which is not unexpected because the concepts involved are not constrained by discipline boundaries.

The Society for Modeling and Simulation International (SCS) is one of many well-established professional organizations devoted to the modelling and simulation discipline (in fact, the SCS celebrated its 60th anniversary in 2012). Its scholarly publication history begins with topics that explored early applications of analog computing. An interesting article that examines topics covered in its more recent publication history (between 2000 and 2010) has been prepared by Mustafee et al. [3]. This comprehensive presentation contributes an insightful perspective of the evolution of the modelling and simulation discipline over the decade considered.

The impact of decisions that are made on the basis of results that flow from a modelling and simulation study is often very significant and far-reaching. In such circumstances, it is critical that a high degree of confidence can be placed on the credibility of the results, and this, in turn, depends on the expertise of the team that carries out the study. This has given rise to an accreditation process for those individuals who wish to participate in the discipline at a professional level. The accreditation option has obvious benefits for both the providers and the consumers of professional modelling and simulation services which are now widely available in the marketplace. This accreditation process has been developed under the auspices of the Modeling and Simulation Professional Certification Commission (see www.simprofessional.org).

An integral part of professionalism is ethical behaviour, and this has been addressed by Oren et al. [10] and Oren [8] who have proposed a code of ethics for modelling and simulation professionals (see www.scs.org/ethics/). This proposed code of ethics has already been adopted by numerous modelling and simulation professional organizations.

1.9 Exercises and Projects

(1.1)   A new apartment building is being designed. It will have ten floors and will have six apartments on each floor. There will, in addition, be two underground parking levels. The developer needs to make a decision about the elevator system that is to be installed in the building. The choice has been reduced to two alternatives: either two elevators each with a capacity of 15 or three smaller elevators each with a capacity of 10. A modelling and simulation study is to be undertaken to provide a basis for making the choice between the two alternatives.

(a)   Develop a list of possible performance criteria that could be used to evaluate the relative merits of the two alternative designs.

(b) Develop a list of behavioural rules that would be incorporated in the formulation of the model of the elevator system (e.g. when does an elevator change its direction of motion, which of the elevators responds to a particular call for service, where does an idle elevator 'park'?).

(c) A model's behaviour can 'unfold' only as a consequence of input data. Develop a list of input data requirements that would necessarily become an integral part of a study of the elevator system, e.g. arrival rates of tenants at each of the floors and velocity of the elevators.

(1.2) Consider the intersection of two roads in a rapidly expanding suburban area. Both of these roads have two traffic lanes. Because of the development of the area, the volume of traffic flow at this intersection has dramatically increased and so have the number of accidents. A large proportion of the accidents involve vehicles making a left turn, which suggests that the 'simple' traffic lights at the intersection are no longer adequate because they do not provide a left turn priority interval. The city's traffic department is evaluating alternatives for upgrading these traffic signals so that such priority intervals are provided to assist left-turning drivers. The need is especially urgent for traffic that is flowing in the north and south directions. If the upgrade is implemented, then a particular parameter value that will need to be determined is the proportion of time allocated to the priority interval. The planning department has decided to explore solution alternatives for the problem by undertaking a modelling and simulation study.

(a) Develop a list of possible performance criteria that would be used for evaluating the traffic flow consequences of the upgraded traffic signals.

(b) Formulate the behaviour rules that would be needed in the construction of the dynamic model (e.g. how do the queues of cars begin to empty when the light turns from red to green, how will right turning cars be handled).

(c) Develop a list of input data requirements for the behaviour generating a process for the model.

References

1. Bowden R (1998) The spectrum of simulation software. IEE Solut 30(May):44–54
2. Cohen J (ed) (2006) Endoscopy simulators for training and assessment skills. Gastrointest Endosc Clin N Am 16(3):389–610. See also: www.theclinics.com
3. Mustafee N, Katsaliaki K, Fishwick P, Williams MD (2012) SCS: 60 years and counting! A time to reflect on the society's scholarly contribution to M&S from the turn of the millennium. Simul Trans Soc Model Simul Int 88:1047–1071
4. Nance RE (1993) A history of discrete event simulation programming languages. ACM SIGPLAN Notices 28(3):149–175

5. Nance RE (1995) Simulation programming languages: an abridged history. In: Proceedings of the 1995 winter simulation conference, Arlington VA, IEEE, pp 1307–1313
6. Nance RE, Sargent RG (2002) Perspectives on the evolution of simulation. Oper Res 50:161–172
7. Nikoukaran J, Hlupic V, Paul RJ (1999) A hierarchical framework for evaluating simulation software. Simul Pract Theory 7:219–231
8. Ören TI (2002) Rational for a code of professional ethics for simulationists. In: Proceedings of the 2002 summer computer simulation conference, San Diego, CA, pp 428–433
9. Ören TI, Zeigler BP (1979) Concepts for advanced simulation methodologies. Simulation 32:69–82
10. Ören TI, Elzas MS, Smit I, Birta LG (2002) A code of professional ethics for simulationists. In: Proceedings of the 2002 summer computer simulation conference, San Diego, CA, pp 434–435
11. Rizzoli AE (2005) A collection of modelling and simulation resources on the internet. http:// www.masters.donntu.org/2006/kita/kondrakhin/library/art4.htm
12. Swain JJ (2017) Simulation software survey: simulated worlds. OR/MS Today 44(5):38–49
13. Wolffe G, Yurcik W, Osborne H, Holloday M (2002) Teaching computer organization/architecture with limited resources using simulators. In: Proceedings of the 33rd technical symposium on computer science education. ACM Press, New York
14. Yurcik W (2002) Computer architecture simulators are not as good as the real thing—they are better! ACM J Educ Resour Comput 1(4), (guest editorial, special issue on General Computer Architecture Simulators)
15. Zeigler BP, Praehofer H, Kim TG (2000) Theory of modeling and simulation: integrating discrete event and continuous complex dynamic systems, 2nd edn. Academic Press, San Diego

Modelling and Simulation Fundamentals

2

2.1 Some Reflections on Models

The use of models as a means of obtaining insight or understanding is by no means novel. One could reasonably claim, for example, that the pivotal studies in geometry carried out by Euclid were motivated by the desire to construct models that would assist in better understanding important aspects of his physical environment. It could also be observed that it is rare indeed for the construction of even the most modest of structures to be undertaken without some documented perspective (i.e. an architectural plan or drawing) of the intended form. Such a document represents a legitimate model for the structure and serves the important purpose of providing guidance for its construction. Many definitions of a model can be found in the literature. One that we feel is especially noteworthy was suggested by Shannon [23]: 'A model is a representation of an object, system or idea in some form other than itself'.

Although outside the scope of our considerations, it is important to recognize a particular and distinctive class of models called physical models. These provide the basis for experimentation activity within an environment that mimics the physical environment in which the problem originates. An example here is the use of scale models of aircraft or ships within a wind tunnel to evaluate aerodynamic properties; another is the use of 'crash-test dummies' in the evaluation of automobile safety characteristics. A noteworthy feature of physical models is that they can, at least in principle, provide the means for the direct acquisition of relevant experimental data. However, the necessary instrumentation may be exceedingly difficult to implement.

A fundamental dichotomy among models can be formulated on the basis of the role of time; more specifically, we note that some models are dynamic while others are static. A linear programming model for establishing the best operating point for some enterprise under a prescribed set of conditions is a static model because there is no notion of time dependence embedded in such a model formulation. Likewise, the use of tax software to establish the amount of income tax payable by an individual to the government can be regarded as the process of developing a (static)

© Springer Nature Switzerland AG 2019
L. G. Birta and G. Arbez, *Modelling and Simulation*, Simulation Foundations,
Methods and Applications, https://doi.org/10.1007/978-3-030-18869-6_2

model of one aspect of that individual's financial affairs for the particular taxation year in question. The essential extension in the case of a dynamic model is the fact that it incorporates the notion of 'evolution over time'. The difference between static and dynamic models can be likened to the difference between viewing a photograph and viewing a video clip. Our considerations throughout this book are concerned exclusively with dynamic models.

Another important attribute of any model is the collection of simplifications and assumptions that are incorporated into its formulation. We note first that these two notions are used with a variety of meanings in the modelling and simulation literature (in fact, they are sometimes used together as in 'simplifying assumptions'). Our perspective is as follows: a simplification is a choice among alternative ways of dealing with an aspect of the SUI and is focused on reducing complexity, while an assumption is an unsubstantiated choice to fill a 'knowledge gap' that is preventing progress at either the modelling or the simulation phases of the project.

Making the most appropriate simplification choices can be one of the most difficult aspects of model development. It is a non-trivial but essential task and has been the topic of considerable research, e.g. [3, 11, 24]. Simplification is sometimes associated with abstraction or granularity assessment. The underlying issue here is identifying the correct balance between complexity and credibility, where credibility must always be interpreted in the context of the goal of the project. An example would be the deliberate decision not to include the various rest breaks taken by the members of the maintenance team in a manufacturing system model.

It's worth observing that an extreme but not unreasonable view in this regard is that the development of any model is simply a matter of making the correct selection of simplifications from among the available options (often a collection of substantial size). Also, one might reasonably ask the question: when has a model become 'over-simplified'? The most straightforward answer is simply: when it fails the validation criteria that have been formulated.

A degree of information/knowledge deficiency is generally present in any model development activity. It gives rise to two distinct outcomes based on the modeller's awareness. When the modeller is aware of the deficiency but nevertheless needs to move forward with the model development, then an assumption is made about the missing (or imperfect) information. Typical examples here can be found in the often assumed stochastic data models that are common requirements in Discrete Event Dynamic System (DEDS)[1] models (in particular, in those circumstances where the SUI does not yet exist) or in the values of physical parameters such as the friction coefficient that impacts the flow of water in a river bed.

---

[1]By way of defining a Discrete Event Dynamic System, we adapt, with minor modification, the definition of an equivalent notion taken from [20]: *A discrete event dynamic system is a system that evolves over time in accordance with the abrupt occurrence, at possibly unknown irregular intervals, of events that may or may not have a physical nature.*

In the absence of awareness about knowledge deficiency, there is potential for serious shortcomings in the modelling process. One can only hope that such omission(s) from the model's formulation will become apparent during validation efforts or that the omitted detail will have no consequence upon the conclusions drawn from the study.

The simplifications and assumptions embedded in a model place boundaries around its domain of applicability and hence upon its relevance not only to the project for which it is being developed but also to any other project for which its reuse is being considered. They are rarely transparent. It is, therefore, of paramount importance to ensure, via the documentation for the project, that all users of the model are aware of its limitations as reflected in the simplifications and assumptions that underlie its development.

As might be expected, the inherently restricted applicability of any particular model as suggested above has direct and significant consequences upon the simulation activity. The implication is simply that restrictions necessarily emerge upon the scope of the experiments that can be meaningfully carried out with the model. This is not to suggest that certain experiments are impossible to carry out but rather that the value of the results that are generated is questionable. The phenomenon at play here parallels the extrapolation of a linear approximation of a complex function beyond a limited region of validity. The need to incorporate into simulation software environments a means for ensuring that experimentation remains within the model's range of credibility has been observed but the realization of this desirable objective has, however, proven to be elusive.

2.2 Exploring the Foundations

The discussion in this section provides some essential background for the development of the modelling and simulation process that is explored in Sect. 2.3.

2.2.1 The Observation Interval

In Chap. 1, we indicated that a modelling and simulation project has two main constituents. The most fundamental is the underlying 'system context', namely, the dynamic system whose behaviour is to be investigated (i.e. the SUI). The second essential constituent is the goal for the project; this has a significant impact on the manner in which the model is formulated and the experiments carried out with it. A subordinate, but nonetheless important, third constituent is the *observation interval* which is the interval of (simulated) time over which the behaviour of the SUI is of interest. Often, the specification of this interval, which we denote by I_o, is clearly apparent in the statement of the project goal. There are, however, many important circumstances where this does not occur simply because of the nature of the output data requirements. Nevertheless, it is essential that information about the observation interval be properly documented.

The starting point of this interval (its left boundary) almost always has an explicitly specified value. The endpoint (the right boundary) may likewise be explicitly specified, but it is not uncommon for the right boundary to only be implicitly specified. The case where a service facility (e.g. a grocery store) closes at a prescribed time (say 9:00 p.m.) provides an example of an explicitly specified right boundary. Similarly, a study of the morning peak-period traffic in an urban area may be required, by definition, to terminate at 10:00 a.m.

Consider, on the other hand, a study of a manufacturing facility that is intended to end when 5,000 widgets have been produced. Here the right-hand boundary of the observation interval is known only implicitly. Likewise, consider a study of the performance of a dragster. Here a simulation experiment ends when the dragster crosses the finish line of the racing track, but the value of time when this occurs is clearly not known when the experiment begins. In both these examples, the right boundary of the observation interval is implicitly determined by conditions defined on the variables of the model.

Another case of implicit determination occurs when the right-hand boundary is dependent on some integrity condition on the data that is being generated by the model's execution. The most typical such situation occurs when there is a need to wait for the dissipation of undesired transient effects. Data collection cannot begin until this occurs. As a result, what might be called the 'data collection interval' has an uncertain beginning. The situation is further complicated by the fact that the duration of the data collection interval (once it begins) is likewise uncertain because of the difficulty in predicting when sufficient data of suitable 'statistical quality' has been accumulated. Both these effects contribute to an uncertain right boundary for I_o. For example, consider a requirement for the steady-state average throughput for a communications network model following the upgrade of several key nodes with faster technology. The initial transient period needs to be excluded from the acquired data because of the distortion which the transient data would introduce. These are important issues that relate to the design of simulation experiments, and they are examined in detail in Chap. 7.

Nevertheless, a basic point here is simply that only portions of the data that is available over the course of the observation interval may have relevance to the achievement of the project goal. Consider, for example, a requirement for the final velocity achieved by a dragster when it crosses the finish line. The velocity values prior to the final value are not of any particular significance. The point illustrated here is that the observation interval and data collection interval are not necessarily the same. It is not uncommon for I_o to be substantially larger than the data collection interval.

Figure 2.1 illustrates some of the various possibilities relating to the observation interval that have been discussed above.

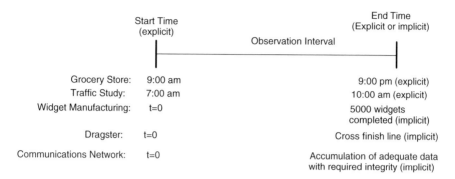

Fig. 2.1 The observation interval

2.2.2 Entities and Their Interactions

We have emphasized that a model is a representation of dynamic behaviour. This is a very general assertion and certainly provides no particular insight into the model building process itself. A useful beginning is an examination of some components that can be used as building blocks for the specification that we seek to develop. It is the dialogue about these components that begins the synthesis of a model.

Within the modelling and simulation context, dynamic behaviour is described in terms of the interactions (over time) among some collection of entities that populates the space that the SUI embraces. The feature of these interactions that is of particular interest is the fact that they produce change.

The entities in question typically fall into two broad types, one permanent (intrinsic to the SUI itself) and the other transient. With respect to the latter, it needs to be recognized that the SUI, like all systems, is a piece of a larger 'universe'; in other words, it functions within an environment. Certain aspects of this environment have an impact upon the behaviour that is of interest and these need to be identified. Some of these aspects map onto transient entities. Examples here are the ships that arrive at a maritime port within the context of a port model developed to evaluate strategies for improving the port's operating efficiency or alternately, the features of a roadway (curves, bumps) within the context of a model being used to evaluate high-speed handling and stability properties of automobiles. Note that any particular ship that arrives at the port usually exists as an integral part of the model only over some portion of the observation interval. When the service which it seeks has been provided, the ship leaves the realm of the model's implicit behaviour.

We note furthermore that the entities within any particular modelling context typically belong to categories; e.g. customers, messages, orders, machines, vehicles, manufactured widgets, shipments, predators, bacteria, pollutants, forces, etc. Interaction among entities can occur in many ways. Frequently, this interaction occurs because the entities compete for some set of limited resources (a type of entity), e.g. servers (within banks, gas stations, restaurants, call centres), transport

Table 2.1 Examples of entities

System under investigation (SUI)	Entity categories	Special types of entities	
		Resource	Queue
Gas station	Cars Trucks	Pumps Attendants	Queue of cars at each pump Queue of cars waiting for an attendant
Widget manufacturing	Parts Broken machines	Machines Repair technicians	List of component parts List of broken machines
Restaurant	Customers	Tables Servers Kitchen Cooks	Queue of customers waiting for tables Customers at tables waiting for service
Hospital emergency room	Patients	Doctors Nurses Examination rooms Ambulances	Waiting room queue Critical patient queue List of patients in examination rooms waiting for the doctor
Ecological system	Predator population Prey population		

services (cranes, trucks, tugboats) or health services (ambulances, operating rooms, doctors/nurses). This competition can give rise to a type of entity (called a queue) in which some entities are obliged to wait for their respective turn to access the resources (sometimes there are priorities that need to be accommodated). On the other hand, entities may exert influence upon other entities in a manner that alters things such as acceleration, direction of motion, energy loss, etc. Some examples of this range of possible entity categories and types are provided in Table 2.1.

We have indicated above that the representation of dynamic behaviour that we seek to develop begins with a description of the change-producing interactions among some set of entities within the SUI. The nature of these interactions is unrestricted, and this 'inclusiveness' is one of the outstanding features of the modelling and simulation approach to problem-solving. In fact, because of the complexity of the interactions that often need to be accommodated, alternate solution procedures (e.g. analytic) are usually unfeasible.

2.2.3 Data Requirements

Some entities are distinctive inasmuch as they give rise to data requirements. Although these naturally enter into the dialogue about the interactions that need to be identified in model formulation, this occurs only at a relatively abstract level.

This abstract view is adequate up to a point, but actual behaviour generation cannot take place until the data requirements are accommodated. In effect, the data serves to 'energize' the overall model specification.

Such data requirements can exist in a variety of forms. Consider, for example, a customer entity. Typically, the data requirement here is the characterization of the customer arrival rate or, equivalently, the time between successive arrivals. This commonly is a random phenomenon and consequently gives rise to the need to identify an appropriate probability distribution function. A similar example can be found in a manufacturing process where a machine entity is subject to failure. The characterization of such failure is typically in terms of rate of occurrence and repair duration which are again random phenomena and hence require appropriate specification in terms of probability distributions. Or, alternatively, consider the characterization of the flow through a 'pipe' entity in a chemical process. A data requirement here could be a single scalar value representing the coefficient of friction associated with the flow. Note that this scalar value could be dependent on other properties within the model such as temperature. As yet another example, consider the two-dimensional array of intercity flight times that would likely be necessary for a study of an airline's operational efficiency. In this case, this data object would probably be shared by all 'flight' entities. These examples demonstrate that data requirements can be associated with both permanent entities (e.g. the machines) and transient entities (e.g. the customers).

The detailed specifications for each such data requirement can be viewed as a *data model*. Each of these plays the role of a specialized sub-model that has localized relevance. Their elaboration can be separately undertaken and even temporarily deferred without compromising the main thread of model development. In this sense, their development can be associated with the software engineering notion of 'encapsulation'. Each data model is an accessory to the bigger picture of characterizing the relevant entity interactions that exist within the SUI.

The formulation of a data model can be a challenging task; furthermore, its correctness is essential to the quality of the results flowing from the modelling and simulation project. The task is of particular relevance in the context of DEDS models because of the multiplicity of random phenomena that need to be represented in such models. We explore this topic in Chap. 3.

To illustrate the various notions introduced above, let's return to the gas station example introduced in Chap. 1. The permanent entities include the attendants and the queue in front of each of the four pumps. There is only one transient entity category, namely, the vehicles that enter the gas station. Notice that the need for a customer entity is redundant because its role would be indistinguishable from the vehicle entity. Notice also that the pumps themselves are likely of no consequence. They would have to be included among the entities only if it was deemed that the possibility of their failure was sufficiently high to have relevance to the credibility of the model from the perspective of the project goal. Data models would have to be developed to deal with the characterization of the arrival rate of the vehicles and their separation into vehicle types and also the service times for each of the three phases of the service function.

The vehicles that enter (and subsequently leave) the gas station as discussed above provide an example of a distinctive and important feature of most models that emerge from the DEDS domain. Associated with almost all such models is at least one set of transient entities; e.g. the vehicles in the gas station model or the ships entering the port in an earlier example. We refer to any such collection of transient entities which flow through the model as an 'input entity stream'.

2.2.4 Constants and Parameters

The constants and parameters of a model have much in common. In particular, constants and parameters both serve simply as names for the values of features or properties within a model which remain invariant over the course of any particular experiment with the model, e.g. g could represent the force of gravity or *NCkOut* could represent the number of checkout counters in a supermarket. In the case of a constant, the assigned value remains invariant over all experiments associated with the modelling and simulation project. On the other hand, in the case of a parameter, the intent embedded in the project goal is to explore the effect of a specified set of values for the parameter upon behaviour, e.g. how the performance of an automobile is affected by the value of the parameter C_f that represents the friction coefficient associated with a tire rolling over a road surface.

Sometimes a parameter serves to characterize some 'size attribute' of the SUI, e.g. the number of berths at a seaport or the number of (identical) generators at a hydroelectric power-generating station. In many cases, such a size parameter might be associated with a facet of a design objective, and the determination of the most appropriate value for it may be one of the reasons for the modelling and simulation project. Consider, for example, a parameter denoted by L_C which represents the passenger load capacity of each elevator in a proposed elevator system for an office tower. The goal of the project could be to evaluate how each of the members of a particular set of possible values for L_C impacts a performance measure defined for the elevator system.

Often there exist alternate behavioural options to be considered within a model, e.g. using different triage methods within an emergency hospital. In such a circumstance, the alternate behaviour is associated with the notion of a 'policy option', where the range of alternate behaviour options is linked to the range of values for some parameter. A typical goal of a modelling and simulation project could be to evaluate a range of feasible policy options upon one or possibly several performance measures with the intent of choosing a 'best' option.

2.2.5 Time and Other Variables

As indicated earlier, the 'end point' of the modelling process is a computer program that embraces the specification of the dynamic behaviour that we seek to study, in other words, the simulation program. A prerequisite for the correct development of any computer program is a high degree of precision in the statement of the specifications that the program must meet. Generally speaking, this corresponds to raising the level of abstraction of the presentation of these specifications, in other words, elimination of unnecessary detail. A variety of means are available, but perhaps the most fundamental is the incorporation of carefully defined variables that enable the formulation of the specifications in a precise and unambiguous way.

Within the modelling and simulation context, the particular purpose of these variables is to facilitate the formulation of those facets of the SUI's behaviour that are relevant to the goal of the project. In fact, meaningful dialogue about most aspects of a modelling and simulation project is severely hampered without the introduction of variables. They provide the means for discussing, elaborating, clarifying and abstracting the dynamic behaviour that is of interest within the context of the goal of the project.

Variables provide an abstraction mechanism for features of the model whose values typically change as the model evolves over the course of the observation interval. Variables fall into a variety of categories, and some of these are examined in the discussion below. As we shall see, *time* itself is a very special variable that is common to all dynamic models.

2.2.5.1 Time

Because our interest is exclusively with dynamic systems, there is one variable that is common to all models that we consider, namely, time (which we generally denote with the symbol t). Apart from its pervasiveness, time is a special variable for two additional reasons. First of all, it is a 'primitive' variable in the sense that its value is never dependent upon any other variable. Second, and in direct contrast, the value of most other variables varies with time, i.e. they are functions of time.

It needs to be emphasized here that the variable, t, represents 'virtual time' as it evolves within the model. Except for certain special cases, this has no relation to 'wall clock' (i.e. real) time.

2.2.5.2 Time Variables

As indicated above, most of the variables within the context of our discussion of dynamic models are functions of time, i.e. they are time-dependent variables or simply *time variables*. If X is designated as a time variable, then this is usually made explicit by writing $X(t)$ rather than simply X. Within the context of our modelling and simulation discussions, a time variable is frequently regarded as representing a time trajectory. Standard mathematical convention associates with $X(t)$ a statement about the set of values of t for which there exists a defined value for X. This set is called the domain of X. In our case, the most comprehensive domain for any time variable is the observation interval, I_o, and some of the time variables that we discuss do have this domain.

However, because of the underlying nature of the computational process, a time variable $X(t)$, which is the outgrowth of behaviour generation, will have defined values at only a discrete and finite subset of I_o, i.e. the domain set has a finite size. The underlying implication here is that from the *perspective of observed behaviour*, the value of a time variable $X(t)$ will be associated with a finite sequence of ordered pairs, namely,

$$< (t_k, x_k) : k = 0, 1, 2, \ldots >$$

where the t_k's correspond to the domain set for $X(t)$, $x_k = X(t_k)$ and $t_{k+1} > t_k$. In effect, the sequence shown above can be viewed as an 'observation' of the time variable $X(t)$.

The reader is cautioned not to attempt an incorrect generalization of the above. Our claim is simply that time variables reflected in the observed behaviour of a dynamic model (as acquired from a simulation experiment) have a finite domain set or can be represented with a finite domain set. We do not claim that all time variables have a finite domain set (consider, for example, an input variable *growth* whose value is $e^{0.1t}$).

2.2.5.3 Input, State and Output Variables

There are three important types of time variable normally associated with any model. We examine each in turn.

Input Variables

Input variables are intended to serve as the means for characterizing the inputs to a SUI during the model building process. However, a meaningful discussion of input variables cannot be undertaken without a clear understanding of what exactly constitutes an input to a dynamic model. This is necessary as well for ensuring that references to input are properly interpreted in discussions about the model. In spite of the fundamental nature of the notion of input, the matter of explicitly defining it is largely overlooked in the modelling and simulation literature. Discussions essentially rely on the reader's intuitive interpretation.

The presentation in this book adopts a specific and straightforward interpretation of the input to a SUI. Our perspective originates from the acknowledgement that any SUI is surrounded by an environment which can influence the SUI's behaviour without, in turn, being influenced by that behaviour. Such influences which flow across the boundary that separates the SUI from its environment constitute inputs to the SUI. While the SUI's environment may be the predominant source of such unidirectional influences, we need to emphasize that influences that impact upon the behaviour of the model but are isolated from that behaviour can also originate within the SUI itself (this is explored further in Chap. 3).

To illustrate the nature of the confusion that can arise, consider a variable, $D(t)$, which is intended to represent the force of air friction upon the motion of a high-performance aircraft. The value of $D(t)$ depends on both the aircraft's altitude and its velocity. To be more specific, the value of $D(t)$ is typically taken to be

$$D(t) = \mu V^2(t)$$

where $\mu = \hat{C}_D \rho$. Here $\hat{C}_D$ is a constant that depends on the physical (aerodynamic) shape of the aircraft, and ρ is the local air density at the aircraft's current altitude. Local air density is a physical property of the aircraft's environment; in particular, it monotonically decreases with distance above sea level.

A question that could be reasonably asked at this point is the following: can $D(t)$ be regarded as an input variable within a model of the aircraft's flight? The conclusion which flows from our adopted definition as outlined above (input is an effect that impacts upon the behaviour of the model being developed but is not itself influenced by it) is that $D(t)$ does *not* represent an input to the model of the aircraft's behaviour. This is because $D(t)$ depends on state variables of the model, namely, velocity and altitude (the latter via the dependency of air density, ρ, on altitude). In other words, while it is true that $D(t)$ has an impact on the aircraft's flight, it is likewise true that the aircraft's changing altitude and velocity have an impact on the value of $D(t)$. (The constant $\hat{C}_D$ has no consequence here because, by convention, a predetermined constant value is never regarded as an input.)

The notion of an 'input entity stream' was introduced earlier in Sect. 2.2.3. Such a collection of transient entities of a particular category flowing into a DEDS model most certainly corresponds to an input and consequently needs to have a characterization in terms of an input variable. The two key features of any input entity stream are the time of the first arrival and the inter-arrival times or, more simply, the arrival times of the entities within the stream. The formulation of a suitable input variable that captures these features is explored in Chap. 3.

State Variables

The notion of state variables is generally considered to be a key aspect in the development of models for dynamic systems. The basis for this importance arises from several properties that are ascribed to this special set of variables. One of these is a 'constructive property'. This property simply means that the model's dynamic behaviour can be entirely defined in terms of the state variables together with (not surprisingly) the model's input variables and parameters. Another generally assumed property of the set of state variables is that they capture the effects of past behaviour insofar as it impacts upon future behaviour (both observed and unobserved; i.e. from the perspective of model output). We note also that the frequently used phrase: 'state of the model' is a reference to the value of its state variables at some time t that is either explicitly or implicitly specified.

Because of these significant properties that are generally associated with the model's state variables, it would be reasonable to assume that any discussion of this special class of variables would naturally begin with an examination of the means for their identification within the model. Unfortunately, however, there are very few rules for dealing with the task of state variable identification that have general applicability. Exceptions, however, do exist in highly structured situations and one such situation is where the model being studied is formulated in terms of ordinary differential equations. Because of the important insights provided by this special case, we begin with an examination of it.

The discussion that follows relates to a dynamic system model, M, specified by the set of first-order differential equations given in Eq. (2.1).

$$\dot{\mathbf{x}}(t) = \mathbf{f}(\mathbf{x}(t), \mathbf{u}(t), \mathbf{p}) \tag{2.1}$$

Here the vector[2] $\mathbf{x}(t)$ represents a candidate set of state variables while $\mathbf{u}(t)$ and $\mathbf{p}$ are the input variables and the parameters of the model, M, whose behaviour is to be observed over the observation interval $I_o = [T_A, T_B]$. Suppose T is a point in time between T_A and T_B and suppose the behaviour of the model has evolved up to $t = T$ and is stopped. If it is true that restarting the model with knowledge of *only* $\mathbf{x}(T)$, $\mathbf{u}(t)$ for $t \geq T$, and $\mathbf{p}$ would produce the same observed behaviour over $[T, T_B]$ as if the behaviour generating process (in this case, the solution of Eq. (2.1)) had not been stopped at $t = T$ then we shall refer to the variable collection $\mathbf{x}(t)$ as having *Property* $\sum$. If, furthermore, it is true that no member of $\mathbf{x}(t)$ can be removed without destroying *Property* $\sum$ then $\mathbf{x}(t)$ is said to have *Property* $\sum*$. This property has very special significance within the context of models of the form given in Eq. (2.1) because it provides the basis for identifying a set of state variables. More specifically we assert that the set of variables in the vector $\mathbf{x}(t)$ qualifies as a set of state variables for the model, M, if and only if it has Property $\sum*$.

The application of the above 'qualification criterion' in the general case of a dynamic model (more specifically, in the DEDS context) is generally not meaningful. Nevertheless, some useful guidelines can be extracted. For example, it is not unreasonable to assert that values must be specified for all state variables before a unique experiment with the model can be initiated. Viewed from an inverse perspective, if an experiment can be duplicated without the need to previously record the initial value of some variable, then that variable likely does not qualify as a state variable.

The formulation of straightforward 'rules' for state variable identification for models that have their origin within the DEDS domain is difficult for a number of reasons. Recall, for example, that in the DEDS domain, there typically exists, at any moment in time, a collection of 'transient entities' whose members contribute to its behaviour over a limited time interval; e.g. the entities associated with an input entity stream. Linked to these entities are attributes and some of these have an impact upon the manner in which the model's components interact. In other words the attribute values of these flowing entities influence the model's behaviour and therefore contribute to the constructive property that characterize the model's state variables. A possible interpretation here is that the state variable collection is not static but rather is time-varying.

---

[2]A collection of variables organized as a linear array is called a *vector variable*. We generally use bold font to indicate vector variables. The number of variables represented is called the dimension of the vector. Sometimes the actual size of the vector is not germane to the discussion and is omitted.

Output Variables

The output variables of a model reflect those features of the SUI's behaviour that originally motivated the modelling and simulation project; in other words, it is the project goal that effectively gives rise to the model's output variables. Because of this, it is not unreasonable to even regard a model as simply an outgrowth of its output variables. The output variables serve as conduits for transferring information about the model's behaviour that is relevant to achieving the project goal. Notice that there is an implication here that a simulation experiment that generates no output information serves no practical purpose!

Many output variables are time variables. This should not be surprising inasmuch as any dynamic model evolves over time, and consequently, it is reasonable to expect that the behaviours of interest are likewise functions of time. The output variable $y(t)$ could, for example, represent the velocity of a missile over the course of its trajectory from the moment of firing to the moment of impact with its target or $y(t)$ might represent the number of messages waiting to be processed at a particular node of a communications network viewed over the duration of the observation

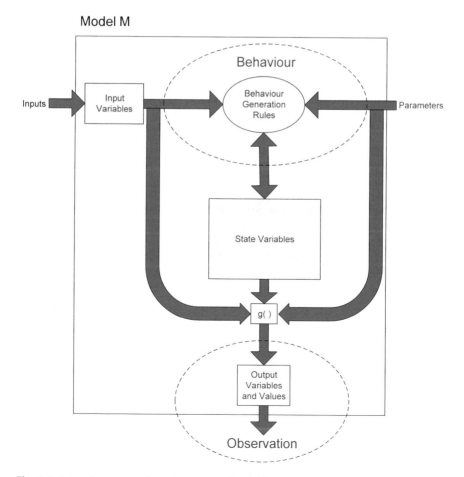

Fig. 2.2 Interaction among the various types of variables

interval. It is important to note that within highly structured model building contexts, output variables that are time variables can generally be expressed as functions of the model's state variables, sometimes together with the model's input
variables and parameters as shown in Eq. (2.2) below.

$$\mathbf{y}(t) = \mathbf{g}(\mathbf{x}(t), \mathbf{u}(t), \mathbf{p}) \qquad (2.2)$$

In the DEDS domain, the characterization of output variables is, however,
generally not as straightforward as Eq. (2.2) suggests. One reason is clearly the
difficulty in identifying state variables as previously noted. Another reason arises
because of the distinctive nature of DEDS output requirements. Consider, for
example, a requirement for the average waiting time of customers in a queue
waiting for service. Such an output requirement does not lend itself to the characterization suggested in Eq. (2.2).

We conclude this section by including Fig. 2.2 that illustrates how the parameters and the various types of variables interact to generate behaviour and how the
observation of behaviour is provided via the outputs of the model.

2.2.6 An Example: The Bouncing Ball

The example we consider below illustrates how the notions of parameters and
variables play a crucial role in enabling a clear and succinct formulation of a model
for a dynamic system.

A boy is standing on the ice surface of a frozen pond and (at $t = 0$) throws a ball
into the air with an initial velocity of V_0. When the ball leaves the boy's hand, it is a
distance of y_0 above the ice surface. The initial velocity vector makes an angle of θ_0
with the horizontal. The boy's objective is to have the ball bounce at least once and
then fall through a hole that has been cut in the ice. The hole is located at a distance
H from the point where the boy is standing. There is a wind blowing horizontally.
By way of simplification, we assume that the ball's trajectory lies within a plane
that passes through the boy's hand and the hole in the ice surface. Furthermore, the
direction of the wind is parallel to this plane. Consequently, the ball will never
strike the ice surface 'beside' the hole. If it does not enter the hole, then it can only
fall in front of it or beyond it. The general configuration is shown in Fig. 2.3.

Fig. 2.3 The bouncing ball

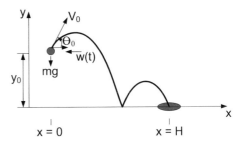

The goal of our modelling and simulation project is to evaluate values for the release angle, θ_0, in order to find a trajectory for the ball that satisfies the boy's objective.

A model for the underlying dynamic system can be formulated by a straight-forward application of Newton's second law (the familiar $F = ma$). We begin by introducing four state variables to characterize the ball's flight trajectory, namely:

$x_1(t)$ – the ball's horizontal position at time t
$x_2(t)$ – the ball's horizontal velocity at time t
$y_1(t)$ – the ball's vertical position at time t
$y_2(t)$ – the ball's vertical velocity at time t.

There are two forces acting on the ball. The first is gravity and the second is the force resulting from the wind. In order to proceed to the next level of refinement, two assumptions are in order:

- We assume that when the ball is released from the boy's hand, the velocity vector, V_0, lies in a vertical plane that passes through the boy's position and the location of the hole in the ice (this ensures that the ball is heading, at least initially, in the direction of the hole)
- We assume that the horizontal wind velocity is parallel to the plane specified above (this ensures that the wind will not alter the 'correct' direction of the ball's initial motion, i.e. towards the hole).

With these two assumptions, the ball's initial motion can be described by the following set of differential equations:

$$\frac{dx_1(t)}{dt} = x_2(t)$$
$$\frac{dx_2(t)}{dt} = -W(t)/m$$
$$\frac{dy_1(t)}{dt} = y_2(t) \qquad (2.3)$$
$$\frac{dy_2(t)}{dt} = -g$$

where $W(t)$ represents the force of the wind acting on the ball's horizontal motion and g represents the gravity force acting on the ball. Each of these four first-order differential equations needs to have a specified initial condition. These are $x_1(0) = 0$, $x_2(0) = V_0\cos(\theta_0)$, $y_1(0) = y_0$, $y_2(0) = V_0\sin(\theta_0)$, where y_0 is the height above the ice surface of the boy's hand when he releases the ball (The value assigned to $x_1(0)$ is arbitrary and zero is a convenient choice.).

In view of the boy's objective, it is reasonable to assume that the ball leaves the boy's hand with an upward trajectory (in other words, $\theta_0 > 0$). Sooner or later however, gravity will cause the ball to arc downwards and strike the ice surface (hence $y_1 = 0$). At this moment (let's denote it by T_C), the ball 'bounces', and this represents a significant discontinuity in the ball's trajectory. A number of additional assumptions must now be introduced to deal with the subsidiary modelling requirement that characterizes this bounce. These are as follows:

- We assume that the bounce takes place in a symmetric way in the sense that if the angle of the velocity vector (with respect to the horizontal) at the moment prior to the collision is θ_C, then immediately after the collision the velocity vector has an orientation of $-\theta_C$.
- Energy is lost during the collision, and we take this to be reflected in a reduction in the magnitude of the velocity vector (loss in kinetic energy). More specifically, if $|V_C|$ is the magnitude of the velocity vector immediately prior to the collision, then we assume that $\alpha |V_C|$ is the magnitude after the collision, where $0 < \alpha < 1$.

In effect, the two above assumptions provide a specification for the behaviour that characterizes the dynamics of the ball at the point of collision with the ice surface. More specifically, we have formulated a model of the bounce dynamics which is

$$\theta_C^+ = -\theta_C$$
$$\left|V_C^+\right| = \alpha|V_C|$$

Here θ_C^+ is the angle of the velocity vector at time T_C^+, which is the moment of time that is incrementally beyond the moment of contact, T_C. Similarly, $\left|V_C^+\right|$ is the magnitude of the velocity vector at time T_C^+.

Although the underlying equations of motion remain unchanged following the bounce, there is a requirement to initiate a new trajectory segment that reflects the changes that occur due to the collision. This new segment of the ball's trajectory begins with 'initial' conditions that incorporate the assumptions outlined above, namely,

$$
\begin{aligned}
x_1(T_C^+) &= x_1(T_C) \\
x_2(T_C^+) &= \alpha x_2(T_C) \\
y_1(T_C^+) &= 0 \\
y_2(T_C^+) &= -\alpha y_2(T_C)
\end{aligned}
\tag{2.4}
$$

Our model for the trajectory of the ball is given by Eqs. (2.3) and (2.4). In this model, $W(t)$ (the force exerted by the wind) represents an input variable, and α

could be regarded as a parameter for an alternate case where the project goal is to evaluate the relationship between α and the problem solution θ_0^*. The state variables are x_1, x_2, y_1 and y_2. This is clearly reflected in their essential role in re-initializing the ball's trajectory following each collision with the ice surface.

Before leaving this example, it is of some interest to revisit the stated goal of this modelling and simulation project. In particular, consider the fundamental issue of whether or not the underlying problem has a solution. It is important to recognize here that the existence of an initial release angle of the ball (θ_0) that will cause the ball to fall through the hole in the ice is not guaranteed. This is not a deficiency in the model but is simply a consequence of the underlying physics. The boy's throwing action gives the ball kinetic energy, which is dependent on the release velocity (V_0). This energy may simply be insufficient to accommodate the energy losses encountered by the ball over the course of its trajectory and the ball may not even be able to travel the distance H where the hole is located. Furthermore, there is the effect of the wind force that is restraining movement towards the hole.

2.3   The Modelling and Simulation Process

An outline of the essential steps involved in carrying out a modelling and simulation study is provided in this section. While the initial steps can be effectively presented using various notions that have been previously introduced, there are several aspects of the latter stages that require extensive elaboration. This is provided in the discussions that follow. An overview of the process is provided in Fig. 2.4.

We begin by noting that the overview of Fig. 2.4 does not include a preliminary phase during which solution alternatives for the problem are explored and a decision is made to adopt a modelling and simulation approach. The option of simultaneously carrying out other complementary approaches is entirely reasonable and is often prudent for some portions of the problem. Although this preliminary phase is not explicitly represented in Fig. 2.4, its existence and importance must nevertheless be recognized.

It should be emphasized that a modelling and simulation project of even modest size is often carried out by a team of professionals, where each member of the team typically contributes some special expertise. There is, therefore, a need for effective communication among team members. Some facets of the discussion in this book have their basis in this communication requirement.

2.3.1   The Problem Description

The first important task of the modelling and simulation team is the development of a document called the *problem description*. This key document evolves from the elaboration of the rudimentary information received from the client concerning the SUI and the problem of interest in whatever detail is deemed essential by the project

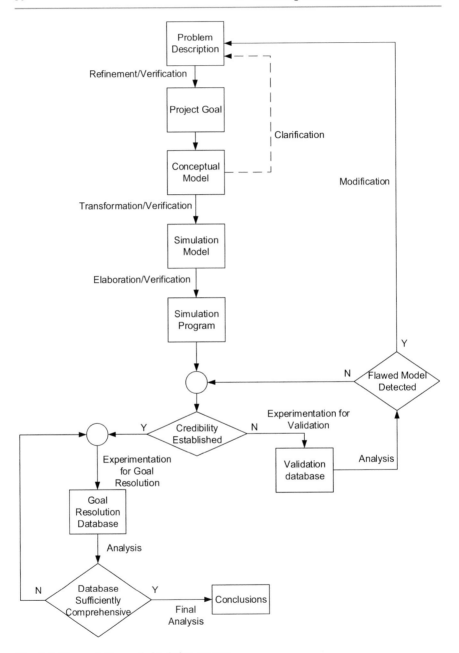

Fig. 2.4 The modelling and simulation process

team. Possible solutions to be evaluated might also be provided. The intent of the elaboration process is to ensure that the problem description document can serve as a meaningful foundation for undertaking a modelling and simulation study.

Balci [1] defines a 'problem formulation' phase as 'the process by which the initially communicated problem is translated into a formulated problem sufficiently well-defined to enable specific research action'. The problem formulation phase is meant to provide a clear and complete statement of the problem such that a modelling and simulation exercise (i.e. Balci's 'research action') can be undertaken. We identify the outcome of Balci's problem formulation phase with the problem description document.

Although we associate the notion of 'completeness' with the problem description document some qualification is required. There is, first of all, the understanding that the document is not static but rather is subject to refinement as the model building effort progresses. Nevertheless, its initial rendition has sufficient content to enable the initiation of the model development phase. The main focus is on behavioural features which are typically formulated in terms of the various objects that populate the space that the SUI embraces. Particular emphasis is on the interactions among these objects. The description is organized so that all SUI specific jargon is clarified.

Apart from its behavioural features, the SUI's structural features are also included because these provide the context for the interactions among the objects (e.g. the layout of the pumps at a gas station or the topology of the network of streets being serviced by a taxi company). Informal sketches are often the best means of representing these structural features (see, e.g. Fig. 2.3). These are an important part of the presentation because they provide a contextual elaboration that can help to clarify the nature of the interaction among the entities. Because of these contributions to the understanding, such sketches are a legitimate component of the problem description.

Note also that we adopt the perspective that various simplifications that may be deemed appropriate and assumptions that may be required (see Sect. 2.1) are not part of the problem description document but rather evolve as part of the subsequent model development phase. They must nevertheless be agreed upon by all project stakeholders and be carefully documented.

2.3.2   The Project Goal

Inasmuch as the project goal is to evaluate a set of possible solutions to the problem (see Sect. 1.1), the first step in the development of a modelling and simulation study is the formulation of such a set. Note that these solution possibilities are, almost always, expressed in terms of values specified for parameters. Strategies for achieving the goal include details of the experimentation process relating to the parameter variations together with the identification of the output variables that produce, during experimentation, the data required for analysis and evaluation. The project goal and the associated strategies have a fundamental impact upon model development and must, therefore, be regarded as prerequisites to that development process.

2.3.3 The Conceptual Model

The information provided by the problem description is, for the most part, relatively informal. Because of this informality, it is generally inadequate to support the high degree of precision that is required for creating a credible (simulation) model embedded within a computer program. A refinement phase must be carried out in order to add detail where necessary, incorporate formalisms wherever helpful and generally enhance the precision and completeness of the accumulated information. Enhanced precision is achieved by moving to a higher level of abstraction than that provided by the problem description. The reformulation of the information within the problem description in terms of parameters and variables is an initial step since these notions provide a fundamental means for removing ambiguity and enhancing precision. They provide the basis for the development of the simulation model that is required for the experimentation phase.

There are a variety of formalisms that can be effectively used in the refinement process. Included here are mathematical equations and relationships (e.g. algebraic and/or differential equations), symbolic/graphical formalisms (e.g. Petri nets, finite state machines), rule-based formalisms, structured pseudo code and combinations of these. The choice depends on suitability for providing clarification and/or precision.

The result of this refinement process is called the *conceptual model* for the modelling and simulation project. The conceptual model may, in reality, be a collection of partial models each capturing some specific aspect of the SUI's behaviour. The representations used in these various partial models need not be uniform.

The conceptual model is a consolidation of all relevant structural and behavioural features of the SUI in a format that is as concise and precise as possible. It provides the common focal point for discussion among the various participants in the modelling and simulation project. In addition, it serves as a bridge between the problem description and the simulation model that is essential for the experimentation activity (i.e. the simulation phase). As we point out below, the simulation model is a software product, and its development relies on considerable precision in the statement of specifications. One of the important purposes of the conceptual model is to provide the prerequisite guidance for the software development task.

It can frequently happen that the formulation of the conceptual model is interrupted because it becomes apparent that the information flowing from the problem description is inadequate. Missing or ambiguous information are the two most common origins of this difficulty. The situation can be corrected only by acquiring the necessary clarification and incorporating it into the problem description. This possible 'clarification loop' is indicated with a dashed lined in Fig. 2.4.

In Fig. 2.4, a verification activity is associated with the transition from the problem description to the conceptual model. Both verification and the related notion of validation are examined in detail in Sect. 2.4. As will become apparent from that discussion, verification is included as part of the transition under consideration because it involves a reformulation of the key elements of the model from one form to another, and the correctness of this transformation needs to be confirmed.

In the modelling and simulation literature, the phrase 'conceptual model' is frequently reduced simply to 'model'. Our usage of the word 'model' without a modifier will generally imply a composite notion that includes a conceptual model and its simulation model counterpart as well as the simulation program.

As a concluding observation in this discussion, it is worth pointing out that there is by no means a common understanding of the modelling and simulation literature of the nature and role of a conceptual model. The overview presented by Robinson [21] gives considerable insight into the various perspectives that prevail.

2.3.4 The Simulation Model

The essential requirement for the experimentation phase of a modelling and simulation project is the existence of an executable computer program that embodies the conceptual model. This is, in fact, a two-step process: the first step is to develop the simulation model from the conceptual model and the second step is the creation of the simulation program. The simulation model is a representation of the model that properly respects the syntax and semantic constraints of some programming language or a simulation program development environment. The result of this transformation is the *simulation model* for the project.

The formulation of a simulation model can be undertaken in a variety of programming environments. Included here are general-purpose programming languages such as C++ and Java. Increasingly more common, however, is the use of programming environments that has been specifically designed to support the special requirements of simulation studies. Many such environments have appeared in recent years; some examples are SIMSCRIPT II.5, ProModel, GPSS, SIMAN, ACSL, Modelica, Arena, CSIM and SIMPLE++. Such environments generally provide features to support the management of time, collection of data and presentation of required output information. In the case of projects in the DEDS domain, additional features for the generation of random variates, management of queues and the statistical analysis of data are also provided.

The simulation model is the penultimate (rather than the ultimate) stage of a development process that began with the identification of the problem that gave rise to the decision to formulate a modelling and simulation project. By and large, its 'content' is not sufficient to serve as the behaviour generating mechanism required for the simulation activity. A variety of auxiliary services generally have to be incorporated. The result of augmenting the simulation model with complementary program infrastructure that provides essential functional services yields the *simulation program* as described in the following section.

Nevertheless, it should be noted that in Fig. 2.4 the transition from the conceptual model to the simulation model is associated with two activities, namely, transformation and verification. As in the earlier transition from problem description to the conceptual model, verification is required here to confirm that the transformation has been correctly carried out.

2.3.5 The Simulation Program

As noted above, the simulation model typically falls short of having all the functionality needed to carry out all aspects of a simulation study. When the programming environment used in developing the simulation model is specifically designed for simulation studies (e.g. the environments listed earlier), these auxiliary functionality requirements are minimal. This is in contrast to the complementary program infrastructure that is required if a general-purpose programming environment is being used (e.g. C++ or Java).

The services in question fall into two categories, one relates to fundamental implementation issues while the other is very much dependent on the nature of the experiments that are associated with the realization of the project goal. Included within the first category are such basic tasks as initialization, assignment of the observation interval, management of stochastic features (when present), solution of equations (e.g. the differential equations of a continuous system model), data collection, etc. Convenient program functionality to deal with these various tasks is normally provided in software environments specifically designed to support the simulation activity. But this is certainly not the case in general-purpose programming environments where considerable additional effort is often required to provide these functional requirements. However, program libraries may be available to satisfy some of these requirements.

The second category of functional services can include such features as data presentation (e.g. visualization and animation), data analysis, database support, optimization procedures, etc. The extent to which any particular modelling and simulation project requires services from this second category can vary widely. Furthermore, modelling and simulation software environments provide these services only to varying degrees, and consequently, when they are needed, care must be taken in choosing an environment that is able to deliver the required services at an adequate level.

The manner in which the support services to augment the simulation model is invoked varies significantly among software environments. Almost always there is at least some set of parameters that need to be assigned values in order to choose from available options. Often some explicit programming steps are needed. Considerable care must be taken when developing the simulation program in order to maintain as clear a separation as possible of the program code for the simulation model from the code required to invoke the ancillary services. Blurring this, separation can be detrimental because the resulting simulation program may become difficult to verify, understand and/or maintain. It has, in fact, been frequently noted (e.g. Ören [17]) that an important quality attribute of a simulation software platform is the extent to which it facilitates a clear separation of the code for the simulation model from the infrastructure code required for the experimentation that is required for the achievement of the project goal.

Figure 2.4 indicates that a verification activity needs to be carried out in the transition from the simulation model to the simulation program. This need arises because this transition typically involves a variety of decisions relating to the execution of the simulation model, and the correctness of these decisions must be confirmed. Consider, for example, a simulation model that incorporates a set of ordinary differential equations. Most modelling and simulation programming environments offer a variety of solution methods for such equations, and each has particular strengths and possibly weaknesses as well. If the equations in question have distinctive properties, then there exists a possibility of an improper choice of the solution method. The verification process applied at this stage would uncover the existence of such a flaw.

Finally, it should be stressed that the simulation program is a software product, and as such, the process for its development shares many of the general features that are associated with the development of any software product.

2.3.6 The Operational Phases

Thus far, our outline of the modelling and simulation process has focused on the evolution of a series of interdependent representations of SUI. However, with the existence of the simulation program, the stage is set for two operational phases of the process that we now examine. The first of these is the validation phase, whose purpose is to establish the credibility of each the model realizations, from the perspective of the project goal. The notion of validation is examined in some detail in Sect. 2.4 which follows below, and hence, we defer our discussion of this phase.

The second phase, which can begin only after the model's credibility has been established, is the simulation phase which can be regarded as the experimentation phase. This activity is presented in Fig. 2.4 as the task of 'goal resolution'. This is achieved via a series of experiments with the simulation program during which an ever-increasing body of data is collected and analysed until it is apparent that a 'goal resolution database' is sufficiently complete and comprehensive to permit conclusions relating to the goal to be confidently formulated.

2.4   Verification and Validation

A simulation model is a software product and like any properly constructed artefact, its development must adhere to design specifications. Assuring that it does is a verification task. All software products have a well-defined purpose (e.g. manage a communications network or ensure that an optimal air/fuel mixture enters the combustion chamber of an internal combustion engine). In the case of a simulation model, the purpose is to provide an adequate emulation of the behavioural features of some SUI, where 'adequate' is assessed from the perspective of the project goal. Assuring that this is achieved is a validation task.

Verification and validation are concerned with ensuring the credibility of the conclusions that are reached as a consequence of the experiments carried out with the simulation program. They can be reasonably regarded as part of the general thrust of quality assurance (the topic of the following section). However, the central role of these notions within the modelling and simulation process, as presented in Fig. 2.4, suggests that they merit special treatment. The range of perspectives relating to the processes associated with these notions further justifies an examination that extends beyond their obvious contribution to quality assurance.

The terms verification and validation are used in a variety of disciplines, notably software engineering. By and large, the distinction in the meaning of these two notions is often poorly understood. In the software engineering context, however, a remarkably concise and revealing presentation of the essential difference can be expressed in terms of two closely related questions. These (originally formulated by Boehm [6]) are:

- *Verification* – are we building the product right?
- *Validation* – are we building the right product?

The product referred to here is the software product being developed. Reinterpretation within a modelling and simulation context is straightforward. The 'product' is the model and the notion of 'building the right product' corresponds to developing a model that has credibility from the perspective of the project goal. On the other hand, 'building the product right' corresponds to ensuring that the artefact that begins as a meaningful and correct problem description and then undergoes various transformations that culminate in a simulation program is never compromised during these various transformations.

Verification is concerned with ensuring that features that should (by design) be clearly apparent in each manifestation of the model are indeed present. Whether or not these features properly reflect required/expected model behaviour (always from the perspective of the project goal) is an issue that falls in the realm of validation. By way of illustration, consider a modelling and simulation project whose primary purpose is to provide support for the design of the various control systems that are to be incorporated into a new, highly automated, thermoplastics manufacturing plant. Certain thermodynamics principles have been identified as being best suited as the basis for formulating the model of one particular aspect of the chemical kinetics that is involved in the process. The approach will, in all likelihood, give rise to a conceptual model that incorporates a set of partial differential equations. The task of ensuring that these differential equations are correctly formulated on the basis of the principles involved and ensuring that they are correctly transformed into the format required by the simulation software to be used falls in the realm of verification. Confirmation that the selected principles are indeed an adequate means of representing the relevant behaviour of the chemical process is a validation task.

Consider an alternate but similar example where a modelling and simulation project is concerned with exploring alternatives for enhancing the operating efficiency of a large metropolitan hospital. The model is to be organized as a number of interacting components. One of these will focus on the operation of the elevator system which has frequently been observed to be a point of congestion. The behaviour of any elevator system is described by a relatively complex set of rules. Ensuring that these rules are correctly represented in each of several realizations (e.g. natural language in the initial statement, then an intermediate and more formal representation such as a flowchart and finally in the program code of the simulation model) is part of the verification activity. Confirmation of the correctness of the rules themselves is a validation task.

A simulation model always functions in some software environment, and assumptions about the integrity of the environment are often made without any particular basis. Confirmation of this integrity is the verification task. Consider, for example, the adequacy of the numerical software for the solution of the differential equations in a continuous time dynamic system model or the adequacy of the mechanism used for generating a random number stream required in a DEDS model. Confirmation that such essential tools are not only available but are sufficiently robust for the project requirements is part of the verification activity.

It has been rightly observed (e.g. Neelamkavil [15]) that 'complete validation' of a model is an objective that is beyond the realm of attainability; the best that can be hoped for is 'failure to invalidate'. A related idea is contained in one of a collection of verification and validation principles suggested by Balci [1], namely, that the outcome of the validation activity is not binary valued. Degrees of success must be recognized and accepted and the credibility of the conclusions derived from the experiments with the model treated accordingly. The practical reality for accepting less than total success in the validation endeavour originates in the significant overhead involved. Carrying out validation to a level that totally satisfies all concerned parties can be both expensive and time-consuming (and possibly impossible!). A point of diminishing returns is invariably reached, and compromises, together with the acceptance of the attendant risk, often become unavoidable.

Validation must necessarily begin at the earliest possible stage of the modelling and simulation project, namely, at the stage of problem definition. Here the task is simply to ensure that the statement of the problem is consistent with the problem that the project originator wants to have solved. This is of fundamental importance because, for the members of the project team that will carry out the project, the problem statement *is* the problem. The documented problem statement is the only reference available for guidance. All relevant facets must, therefore, be included and confirmation of this is a validation task.

The problem definition has many facets, and most have direct relevance to the validation task. One which is especially relevant is the statement of the project goal. This statement has a profound impact that ranges from the required level of granularity for the model to the nature of the output data that needs to be generated. Consider, for example, the model of an airliner. A model that has been validated within the context of a project that is evaluating a business plan for a commercial airline will most likely not qualify as an adequate model within the context of a project that seeks to determine the aircraft's aerodynamic characteristics. In effect then, one of the most fundamental guiding principles of any validation activity is that it must be guided by the goal of the study.

One (essentially naïve) perspective that might be adopted for validating a model is simply to ensure that its observed behaviour 'appears correct', as reflected by animation, graphical displays or simply the values acquired by some set of designated variables. The assessment here is clearly entirely qualitative rather than quantitative and hence is very subjective. A far more serious shortcoming of this approach is that it makes no reference to the goal of the modelling and simulation study. As noted above, this is a very serious flaw. The absence of this context carries the naïve (and most certainly incorrect) implication that the model has 'universal applicability'. However, it is rare indeed that a model of 'anything' is appropriate for all possible modelling and simulation projects to which it might be linked.

Nevertheless, the relatively superficial approach given above does have a recognized status in the validation 'toolkit' when it is refined by including the understanding that the observers are 'domain experts' and that their judgement is being given with full understanding of the model's purpose. With these qualifiers, the approach is referred to as *face validation*.

It is reasonable to assume that within the framework of the project goal, a collection of (more or less) quantifiable anticipated behaviours for the model can be identified. These will usually be expressed in terms of input–output relationships or more generally in terms of cause/effect relations. Consider, for example, a model developed to investigate the aerodynamic characteristics of an aircraft. The occurrence of an engine failure during a simulation experiment should (after a short time interval) lead to a decrease in the aircraft's altitude. If this causal event does not take place, then there is a basis for suspicion about the model's adequacy.

Or consider introducing the occurrence of disruptive storms in a harbour model. It is reasonable to expect that this would result in a decrease in the operating efficiency of the harbour as measured by the average number of ships per day passing through the loading/unloading facilities.

As a final example, consider doubling the arrival rate of tourists/convention attendees in an economic model of an urban area. This should result in an approximate doubling in the occupancy rate of the hotels in the area. Furthermore, an occupancy rate increase of more than a factor of two should be cause for some reflection about possible flaws in the model's specification.

The general approach outlined above is often called *behaviour validation*. An implicit assumption in the approach is that a verified simulation program is available for experimentation. The approach has been examined in some detail by Birta and Ozmizrak [5], who incorporate the notion of a validation knowledge base that holds the collection of behavioural features that need to be confirmed. The investigation includes a discussion of a procedure for formulating a set of experiments that efficiently 'covers' the tests that are implied in the knowledge base. An accessory question that does need resolution prior to the implementation of the process relates to the level of 'accuracy' that will be expected in achieving the designated responses. Behaviour validation has several noteworthy aspects; for example, the knowledge base can conveniently accommodate insights provided by a domain expert, and as well, it can accommodate data acquired from an observable system when such an option exists.

In fact, this latter feature is closely related to a notion called *replicative validation*, i.e. confirming that the simulation program is capable of reproducing all available instances of the SUI's input–output behaviour. This notion is clearly restricted to the case where the SUI actually exists and behavioural data has been collected. But even in such circumstances, there remain open questions, e.g. could there not be 'too much' data available, and if so, how can it be organized into meaningful (non-redundant) classes, and how is the significance of project goal accommodated?

Validation in the modelling and simulation context must also embrace the data modelling task. For example, suppose that once ordered, the arrival time for a replacement part for a machine in a manufacturing process is random. There are at least two possible choices here for representing this situation. One is simply to use the mean delay time (assumed to be known) and an alternative is to use successive samples drawn from a correctly specified probability distribution. Ensuring that a satisfactory choice is made (as evaluated in terms of project goal) can be regarded as a validation task.

Accreditation is a notion that is closely related to validation. Accreditation refers to the acceptance, by a designated 'accreditation authority', of some particular simulation model for use within the context of a particular modelling and simulation project. Several important issues are associated with this notion, e.g. what guidelines are followed in the designation of the accreditation authority and how is the decision-making procedure with respect to acceptance carried out. These are clearly matters of critical importance, but they are, for the most part, very situation dependent and for this reason, we regard the topic of accreditation as being beyond the scope of our discussions. Certification is an equivalent issue which is explored in some detail by Balci [2].

We conclude this section by observing that the importance of model credibility has been recognized at even legislative levels of government because of the substantial government funding that is often provided in support of large-scale modelling and simulation projects. In 1976, the American Government's General Accounting Office presented to the US Congress the first of three reports that explored serious concerns about the management, evaluation and credibility of government-sponsored simulation models (see [8, 9, 10]). For the most part, these concerns were related to verification and validation issues in the context of modelling and simulation projects carried out by, or on behalf of, the United States Department of Defense. This latter organization is possibly the world's largest modelling and simulation user community and a comprehensive presentation of its perspective about verification and validation can be found in [25]. An overview of some of the material contained in [25] together with a discussion of verification and validation issues in some specialized circumstances (e.g. hardware-in-the-loop, human-in-the-loop, distributed environments) can be found in Pace [18]. A comprehensive examination of validation in the modelling and simulation context is provided by Murray-Smith [14]. Validation is, in fact, a topic with many facets and a recent volume that examines it from an interdisciplinary perspective is provided by Beisbart and Saam [4].

2.5 Quality Assurance

Quality assurance within the framework of modelling and simulation is a reference to a broad array of activities and methodologies that share the common objective of ensuring that the goal of the simulation project are not only achieved but are achieved in a timely, efficient and cost-effective manner. An interesting overview of these can be found in Ören [17]. A significant thrust of the quality assurance effort necessarily deals with ensuring the credibility of the simulation model, where credibility here must always be interpreted from the perspective of the goal of the project. Nevertheless, there are a variety of other important issues that fall within the realm of quality assurance. We examine some of these in the discussion below.

2.5.1 Documentation

A modelling and simulation project of even modest complexity can require many months to complete and will likely be carried out by a team having several members. Information about the project (e.g. assumptions, data sources, credibility assessment) is typically distributed among several individuals. Personnel changes can occur, and in the absence of documentation, there is a possibility that important fragments of information may vanish. Likewise, the reasons for any particular decision made during the course of the project may be completely obvious when it is made but may not be so obvious several months later. Only with proper documentation can the emergence of this unsettling uncertainty be avoided. Comprehensive documentation also facilitates the process of 'coming-up-to-speed' for new members joining the team.

   Project documentation must not only be comprehensive but must also be current. Deferring updates that outline changes that have occurred is a prescription for rapid deterioration in the value of the documentation since prospective users will become wary of its accuracy and hence will avoid reliance on it. In the extreme, documentation that is deferred until the end of the project essentially belies the intent of the effort.

We note finally that there is a fundamental need to ensure that a high standard of quality is maintained in the preparation and organization of project documentation. While certainly important from the perspective of contributing to clarity, quality documentation has, in addition, the implicit and desirable consequence of enhancing confidence in the conclusions drawn from a simulation study.

2.5.2 Program Development Standards

Premature initiation of the computer programing phase must be avoided. Often there is an urge to begin the coding task before it is entirely clear what problem needs to be solved. This can result in a simulation program that is poorly organized because it is continually being 'retrofitted' to accommodate newly emerging requirements. Any computer program developed in this manner is highly prone to error.

2.5.3 Testing

Testing is the activity of carrying out focused experiments with the simulation program with a view towards uncovering specific properties. For the most part, testing is concerned with establishing credibility and consequently, considerable care needs to be taken in developing and documenting the test cases. Testing activity that is flawed or inadequate can have the unfortunate consequence of undermining confidence in the results flowing from the simulation project.

Testing can have a variety of objectives. For example, regression testing is undertaken when changes have taken place in the simulation program. In such circumstances, it is necessary to confirm not only that any anticipated behavioural properties of the simulation model do actually occur but also that improper 'side-effects' have not been introduced. This implies carrying out some suite of carefully designed experiments before and after the modifications.

Another testing perspective is concerned with trying to acquire some insight into the boundaries of the usefulness of the simulation program relative to the goal of the project. This can be undertaken using a process called stress testing whereby the model is subjected to extreme conditions. For example, in the context of a manufacturing process, the effect of extremely high machine failure rates could be explored, or alternatively, in a communication system context, the impact of data rates that cause severe congestion could be explored. The intent of such testing is to create circumstances that provide insight into limits of the model's plausibility in terms of an adequate representation of the SUI's behaviour.

2.5.4   Experiment Design

We use the phrase 'experiment design' to refer to a whole range of planning activities that focus on the manner in which the simulation program will be used to achieve the project goal. The success of the project is very much dependent on the care taken in this planning stage. Poor experiment design can seriously undermine the conclusions of the study and in the extreme case may even cast suspicion on their reliability.

Some examples of typical matters of concern are:

- What output needs to be generated?
- How will initialization and transient effects be handled?
- Are there particular operational scenarios that are especially well suited to providing the desired insight in situations where the relative merits of a number of specified system design alternatives need to be examined?
- Is there a useful role for special graphics and/or animation and if so, what should be displayed?

Frequently, the project goal includes an optimization requirement and the difficulty of this task is often underestimated. Care is required both in the formulation of the optimization problem itself and in the identification of an appropriate optimization procedure for its solution. Problem formulation corresponds to the specification of a scalar-valued criterion function, whose value needs to be either minimized or maximized. The required nature of this function is often clearly apparent from the goal of the project. Nevertheless, care must be taken to avoid attempting to embed separate aspects of the project goal into various criterion functions. The parameters available for adjustment in the search for an extreme value for the criterion function need to be identified. Frequently, there are constraints in the admissible values for these parameters, and such constraints must be clearly identified. Alternately, there may be prescribed requirements on certain features of the model's behaviour that have to be incorporated.

The identification of an appropriate optimization procedure for solving the problem then has to be carefully considered. Numerous techniques can be found in the optimization literature, and these are generally directly applicable within the modelling and simulation context when stochastic effects are not present. Their applicability is, however, seriously undermined when stochastic behaviour is embedded in the criterion function's evaluation as is the usual case in the DEDS context.

While true optimality may be unfeasible, sometimes a 'sub-optimal' solution can be a reasonable expectation. In such circumstances it is highly desirable to have available a means for estimating the likely deviation from optimality of the accepted solution. In practice, when stochastic effects are present, the search for optimality may simply have to be abandoned and replaced with the relatively straightforward task of selecting the best alternative from among a finite collection of options.

The challenge in this option is the identification of an appropriate collection of options to be investigated.

A collection of techniques called DOE (design of experiments) is somewhat aligned with the task of system optimization and, in particular, with the task of identifying a collection of options that are worthy candidates for investigation These concepts are not new but nevertheless continue to be discussed in the literature [12, 13] and, as well, in modelling and simulation conferences [22]. The objective of this work is to provide help in formulating meaningful experiments that reduce the overhead of identifying most promising alternatives from among the large number that is often available. For example, the analysis can assist in the identification of the parameters that are most likely to have a significant impact on the search for optimality.

An introduction to the issues involved in optimization studies within the modelling and simulation context is provided in Part 4 of this book.

2.5.5 Presentation/Interpretation of Results

Often the person/organization that has commissioned the modelling and simulation project remains remote from the model development stage (both conceptual and simulation). Unless explicitly requested, great detail about the model's features should not dominate any presentation made as the project unfolds. The focus should be on results obtained from the simulation experiments that relate directly to the goal of the project. This is not to suggest that additional information that appears pertinent should not be presented but its possibly tangential relevance should be clearly pointed out. The wide availability of increasingly more sophisticated computer graphics and animation tools can be creatively incorporated, but the visual effects they provide should serve only to complement, but not replace, comprehensive quantitative analysis.

2.6 The Dynamic Model Landscape

Models of dynamic systems can be characterized by a wide variety of features. For the most part, these are inherited from the underlying system (i.e. the SUI) that the model emulates. We examine some of these characterizing features in the following discussion.

2.6.1 Deterministic and Stochastic

The system context for a large class of modelling and simulation projects includes random elements. Models that emerge from such contexts are called stochastic models, which are very distinct from deterministic models which have no random aspects. Values taken from any particular experiment with a stochastic model must be regarded as observations of some collection of random variables. The need to deal with random aspects of stochastic models (and the underlying SUI) has a very substantial impact on essentially all facets of both the modelling and the simulation phases of a project. A whole range of related considerations must be carefully handled in order to ensure that correct conclusions are drawn from the study. A few of these are listed below.

• Only aggregated results are meaningful, hence many simulation runs need to be carried out in order to realize an experiment; in other words, an experiment is a collection of runs which yields one or more confidence intervals each of which includes a point estimate (see Chap. 7 for more details).
• The 'start-up' issue which may require ignoring data from the initial portion of a simulation experiment.
• The need to formulate a collection of data models that capture the various autonomous random phenomena that are embedded in the model.

2.6.2 Discrete and Continuous

In models for discrete event dynamic systems (i.e. DEDS models), state changes occur at particular points in time whose values are not known a priori. As a direct consequence, (simulated) time advances in discrete 'jumps' that have unequal length.

In contrast, with models that emerge from the domain of continuous time dynamic systems (i.e. CTDS models), state changes occur continuously (at least in principle) as time advances in a continuous fashion over the length of the observation interval. It must, however, be stressed that this is an idealized perspective and runs counter to the realities introduced by the computational process. It is simply unfeasible for any practical procedure to actually yield data at every value of time within the continuum of the observation interval. Thus, from the perspective of the observer, state changes do apparently occur with discrete 'jumps' as the solution unfolds over the observation interval.

Our presentation in this book may give the erroneous impression that models neatly separate into the two broad categories that we refer to as DEDS models and CTDS models. This is an oversimplification. There is, in fact, a third category of models that are usually called combined models where the name reflects the combination of elements from both the discrete and continuous domains. As an illustration, consider the parts in a manufacturing plant that move from one workstation to another on the way to assembly into a final product. At these workstations, queues form and the service function provided by the workstation may have random aspects (or may become inoperative at random points in time). Thus, the basic elements of a DEDS model are present. At some workstations, the operation may involve heating the part to a high temperature in a furnace. This heating operation and the control of it would best fall in the realm of a CTDS model. Hence, the overall model that is needed has components from the two basic domains.

Work on the development of modelling formalisms and tools for handling this third category of combined models has a long history. The interested reader wishing to explore this topic in more detail will find relevant discussions in Cellier [7], Ören [16] and Praehofer [19].

2.6.3 Linear and Non-linear

The linearity and non-linearity properties of systems are basic considerations in many areas of analysis, e.g. mathematics, system theory, etc. However, because the experiments that are inherent within the modelling and simulation context are always assumed to be carried out in a computer environment, the information that is 'delivered' evolves entirely from computations that are generally unencumbered by numerical complexity. Hence, the inherent simplifications (and resulting approximations) to the analysis process that are associated with linearization generally have little or no value in the modelling and simulation realm. This absence of any need to distinguish between linear and non-linear systems/models is most certainly one of the noteworthy features of the modelling and simulation approach to problem-solving.

2.7 Exercises and Projects

2.1. Technical papers in the modelling and simulation applications literature are sometimes lacking in the clarity with which they deal with such fundamentals as:

(a) The goal of the modelling and simulation study;
(b) Outline of the conceptual model;
(c) Identification of input and output variables;
(d) Model validation efforts;
(e) Evaluation of success in achieving a solution to the problem that motivated the study.

Choose two papers in some particular application area of modelling and simulation and compare the effectiveness with which the authors have addressed the items listed above. Some application areas that could be considered are:

(a) Network management and control,
(b) Ecological and environmental systems,
(c) Biomedicine and biomechanics,
(d) Power generation and distribution,
(e) Automated manufacturing,
(f) Robotics and autonomous systems,
(g) Transportation and traffic,
(h) New product development.

Technical papers and/or pointers to technical papers in these areas can be found at web sites such as www.scs.org and http://ieeexplore.ieee.org/xpl/conhome.jsp?punumber=1000674. Reference [4] in Chap. 1 will also be of value.

References

1. Balci O (1994) Validation, verification, and testing techniques throughout the life cycle of a simulation study. Ann Oper Res 53:121–173
2. Balci O (2001) A methodology for certification of modeling and simulation applications. ACM Trans Model Comput Simul 11:352–377
3. Barlow J (2009) Simplification: ethical implications for modelling and simulation. In: Proceedings of the 18th world IMACS/MODSIM congress, Cairns, Australia, ISBN: 978-0-9758400-7-8. http://www.mssanz.org.au/modsim09/F12/kragt.pdf, pp 432–438
4. Beisbart C, Saam NJ (eds) (2019) Computer simulation validation. Springer, Switzerland
5. Birta LG, Ozmizrak NF (1996) A knowledge-based approach for the validation of simulation models: the foundation. ACM Trans Model Comput Simul 6:67–98
6. Boehm BW (1979) Software engineering: R&D trends and defence needs. In: Wegner P (ed) Research directions in software technology. MIT Press, Cambridge, MA
7. Cellier FE (1986) Combined discrete/continuous system simulation—application, techniques and tools. In: Proceedings of the 1986 winter simulation conference, Washington, IEEE Computer Society Press
8. General Accounting Office (1976) Report to the Congress: ways to improve management of federally funded computerized models. Report LCD-75-111. U.S. General Accounting Office, Washington, DC
9. General Accounting Office (1979) Guidelines for model evaluation. Report PAD-79-17. U.S. General Accounting Office, Washington, DC
10. General Accounting Office (1987) DOD simulations: improved assessment procedures would increase the credibility of results. Report GAO/PEMD-88-3. U.S. General Accounting Office, Washington, DC
11. Innis G, Rexstad E (1983) Simulation model simplification techniques. Simulation 41(7):7–15
12. Law A (2015) Simulation model and analysis, 5th edn. McGraw-Hill Education, New York, NY
13. Montgomery DC (2017) Design and analysis of experiments, 9th edn. John Wiley & Sons Inc
14. Murray-Smith DJ (2015) Testing and validation of computer simulation models. Springer
15. Neelamkavil F (1987) Computer simulation and modeling. Wiley, Chichester
16. Ören TI (1971) GEST: general system theory implementor, a combined digital simulation language. PhD dissertation, University of Arizona, Tucson, AZ

17. Ören TI (1981) Concepts and criteria to access acceptability of simulation studies. Commun ACM 24:180–189
18. Pace DK (2003) Verification, validation and accreditation of simulation models. In: Obaidat MS, Papadimitriou GI (eds) Applied system simulation: methodologies and applications. Kluwer Academic Publishers, Boston
19. Praehofer H (1991) System theoretic formalisms for combined discrete continuous system simulation. Int J Gen Syst 19:219–240
20. Ramadge PJG, Wonham WM (1992) The control of discrete event systems. In: Ho Y-C (ed) Discrete event dynamic systems. IEEE Press, Piscataway, pp 48–64
21. Robinson S (2006) Issues in conceptual modelling for simulation: setting a research agenda. In: Proceedings of the 2006 Operations Research Society simulation workshop, March, Lexington, England
22. Sanchez SM, Wan H. Work smarter, not harder: a tutorial in designing and conducting simulation experiments. In: Proceedings of the 2015 winter simulation conference, Huntington Beach, California, Dec 6–9, 2015, pp 1795–1809
23. Shannon RE (1975) Systems simulation: the art and science. Prentice Hall, Englewood Cliffs
24. Shannon RE (1998) Introduction to the art and science of simulation. In: Medeiros DJ, Watson EF, Carson JS, Manivannan MS (eds) Proceedings of the 1998 winter simulation conference, Washington, IEEE Computer Society Press, pp 7–14
25. Verification, Validation, & Accreditation (VV&A): Recommended Practices Guide (RPG) (2011). https://vva.msco.mil/

Part II
DEDS Modelling and Simulation

In the second part of this book (Chaps. 3–7), we examine the modelling and simulation process within the discrete-event dynamic systems (DEDS) domain. The presentation is, for the most part, guided by the general process presented in Fig. 2.4.

An important behavioural feature of DEDS models is the central role played by random phenomena. The inter-arrival times between messages entering a communication network and the time to service customers at the checkout counter of a grocery store are examples of such phenomena.

Data modelling is an important subtask of the conceptual modelling phase in the DEDS domain. To a large extent, it is concerned with correctly representing stochastic features of the SUI (System Under Investigation). Generally, it involves determining appropriate probability distributions from collected data. This can be a demanding and time-consuming task. The project goal guides the identification of the data models that are required. The data modelling task is considerably facilitated when the SUI currently exists and is accessible because then data collection is possible. When the SUI does not yet exist (or indeed, may never actually 'exist'), data modelling becomes a very uncertain undertaking and essentially depends on insight and intuition.

Chapters 3 and 4 deal with issues relating to conceptual modelling in the DEDS domain. Chapter 3 provides an overview of some key aspects of random behaviour and variables that are associated with this type of behaviour. It discusses data modelling and also lays a foundation for the important notions of input and output. The focus of Chap. 4 is the presentation of a comprehensive framework for conceptual model development in the DEDS domain. This activity-based framework is called the ABCmod conceptual modelling framework and its purpose is to capture the structural and behavioural features of the SUI that have relevance to achievement of the project goal.

Chapter 5 introduces the traditional World Views used in creating discrete-event simulation models and shows how a conceptual model that has been formulated in the ABCmod activity-based framework described in Chap. 4 can be transformed into a simulation model based on any of these World Views.

Chapter 6 presents a novel world view called the Activity-Object world view, which enables a straightforward creation of a simulation model directly from an ABCmod conceptual model. While traditional world views require that activities be broken down into underlying events, the Activity-Object world view benefits from the contemporary object-oriented paradigm which allows the preservation of the activity constructs as objects. This preservation feature is important because it considerably simplifies the transformation step into a simulation model and, as well, renders the simulation model far more transparent since the structural and behavioural constructs of the conceptual model are fully visible within it.

The project goal has a direct impact on the way in which experimentation is carried out with the simulation program. A basic objective of experimentation is to produce values for the performance measures stipulated in the project goal. In Chap. 7, we examine how the experimentation activity has to be organized in order to acquire meaningful values for these performance measures.

Documentation is an essential part of any modelling and simulation project. This documentation can be organized in many different ways. We summarize below a particular approach that is formulated in terms of sections for a modelling and simulation project document that is based on the stages of the modelling and simulation process shown in Fig. 2.4. Each of the sections of this document is briefly summarized.

1. Problem Description Section: An overview of this section was presented in Sect. 2.3.1 and we now elaborate somewhat on its contents. In particular, we adopt the perspective that it has two subsections called 'the problem statement' and 'the SUI details'. The problem statement provides a concise statement of the problem to be solved while the SIU details section outlines the known relevant information about the SUI.

   Our view of the SUI details subsection is very much influenced by the science research stream of thought on problem structuring outlined by Woolley and Pidd [Problem structuring—a literature review. J Oper Res Soc (32):197–206, 1981]. It 'sees problem structuring in terms of gaining an understanding of the problem situation mainly by gathering quantitative data, going out and looking, etc. There is a consciousness of finding out what is *really* happening, of discovering the *real* problem. Whatever is discovered in the course of this initial ferreting about will determine which aspects of the situation are modelled'. Note that the key implication from a modelling and simulation perspective is that model development flows from the insights thus obtained.

2. Project Goal Section: This section provides a statement of the goal of project. The goal has a fundamental impact on model development and effectively serve as a prerequisite for that development (Sect. 2.3.2).

3. Conceptual Model Section: This section is primarily concerned with the presentation of a conceptual model for the project as presented in the problem description (Sect. 2.3.1). In Chap. 4, we adopt the view that the conceptual model has two principal functions, namely to provide an effective communication vehicle for the various stakeholders in the project and to provide a specification for the computer program (i.e. simulation model) that will be used to carry out the experiments. Accordingly, we present the conceptual model at two distinct levels: the first (the 'high level') captures the features of the model at a level of detail that is both adequate for communication and comprehensive. The second level (the 'detailed level') provides the detail needed for simulation model development.

4. Simulation Model Section: The simulation model is a computer program artefact, and like all computer programs it is essential that adequate documentation about its overall organization be provided so that those involved with its modification and maintenance have adequate information to carry out required tasks. The development of program code from an ABCmod conceptual model is explored in some detail in Chaps. 5 and 6.

5. Validation and Verification Section: In spite of significant care taken during the development efforts that yield a simulation model, that model may be flawed. The origins of these flaws are typically categorized into one of two broad areas: inadequate understanding of the SUI's behaviour and errors introduced during computer program development. A variety of tests and safeguards are typically invoked to uncover these deficiencies and it is normally expected that these various measures be appropriately documented (Balci, The Handbook of Simulation, Chap. 10, John Wiley & Sons, New York, NY, pp. 335–393, 1998).

6. Experimentation and Output Analysis Section: This section reports on the experimentation that is carried out with the simulation program and on the analysis of the results that are obtained. The extent to which these results solve the original problem that motivated the study is typically included as well. The discussions of Chap. 7 explore this topic.

DEDS Stochastic Behaviour and Modelling

3

3.1 The Stochastic Nature of DEDS Models

This section explores some important aspects of the random (stochastic) nature of DEDS and introduces several assumptions that are typically made about it within the model development process.

We begin by noting that sequences of random variables, ordered in time, are a fundamental and recurring feature within the DEDS domain. Our particular interest, here, includes both the creation of such sequences and capturing values of such sequences generated within the model. Often, these sequences fall within the realm of stochastic processes (an overview is provided in Sect. A2.8), and this is reflected in our discussion below.

Consider a simple view of the operation of a delicatessen counter, which has one server. Customers arrive at the counter, wait in a queue until the server is available and then select and purchase items at the counter. Two important random phenomena drive this SUI. The first is the arrival of customers and the second is the time it takes to select and purchase items at the counter, which is referred to as the service time. Each of these phenomena (arrivals and servicing) is, in fact, a sequence of continuous random variables that are linked to the stream of customers.

The arrival of customers can be viewed as the stochastic process[1] $X = (X_1, X_2, X_3, \ldots, X_n)$ where the indices $j = 1, 2, 3, \ldots, n$ are implicit references to customers which arrive at times t_j and t_0 gives the start of the observation interval. The first customer arrives at time $t_1 = t_0 + X_1$. Subsequent customer arrival times ($j > 1$) are $t_j = t_{j-1} + X_j$ (note that the X_j's correspond to interarrival times). Figure 3.1 shows, on the horizontal axis, a typical instance (an observation) of this discrete-time stochastic process. The sequence of random variables associated with the servicing times for the n customers can similarly be regarded as a discrete-time stochastic process.

---

[1]This is an example of the statistical notion of an arrival process.

© Springer Nature Switzerland AG 2019
L. G. Birta and G. Arbez, *Modelling and Simulation*, Simulation Foundations, Methods and Applications, https://doi.org/10.1007/978-3-030-18869-6_3

In the simple delicatessen model above, both of these random phenomena (i.e. arrivals and servicing) would typically be independent of any other phenomena or interactions in the system. Throughout our discussions, we refer to stochastic processes with this independence attribute as *autonomous stochastic processes*. A stochastic process that is not autonomous is said to be *dependent*.

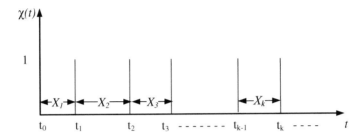

Fig. 3.1   Observation of an arrival stochastic process

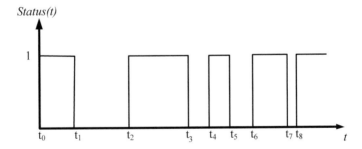

Fig. 3.2   Observation of an autonomous continuous-time stochastic process representing machine status

A continuous-time stochastic process is a collection of time functions (time trajectories) that are defined over the entire observation interval. In a DEDS model, this collection is a collection of piecewise constant time functions because changes in value occur only at discrete points in time. An observation of a continuous-time stochastic process is one particular member of this collection. To illustrate, consider the status of a machine in a manufacturing system which is subject to random failure due to parts deterioration. The time trajectory (*Status(t)*) shown in Fig. 3.2 can be regarded as an observation of the underlying continuous-time stochastic process. Here, we have chosen to use the value 1 to indicate that the machine is fully operational and the value 0 to indicate a failure status (the machine is either waiting for service or repair is underway).

In our delicatessen example, the autonomous arrivals and servicing stochastic processes give rise to *dependent* stochastic processes via the inherent behaviour properties of the model. Consider, for example, the waiting times of customers in the delicatessen queue and the length of the queue. This sequence of (random) waiting times is a discrete-time stochastic process; e.g. $W = (W_1, W_2, W_3, ..., W_n)$ where W_j is the waiting time of the jth customer. The length of the queue is likewise a dependent (but in this case continuous time) stochastic process. Figure 3.3 provides an example of an observation, $L(t)$, of this continuous-time stochastic process.

Dependent random phenomena can have fundamental importance in simulation studies in the DEDS domain. This is because some of their properties, typically called performance measures, often play a fundamental role in the resolution of the project goal. These properties are random variables because of the underlying dependence on other random variables. Often interest focuses on the change in the value of performance measures that result from some change in system operation or system structure. For example, the service time could be reduced by restricting customer choices to some collection of pre-packaged items. A change in the structure of the system would result if an additional person was hired to serve behind the counter and a two-queue service protocol was established. The goal of a modelling and simulation project could then be to evaluate the effect of such changes on customer waiting time.

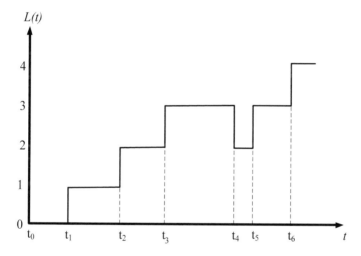

Fig. 3.3 An observation of a dependent continuous-time stochastic process representing queue length

The representation of autonomous random phenomena within a model often incorporates simplifying assumptions. For example, a common assumption is that customer inter-arrival times can be represented with a *homogeneous* stochastic process, that is, the sequence of random variables that constitute the stochastic process are independent and identically distributed (IID).

Unfortunately, there are many cases where such assumptions are simply not realistic. Consider our delicatessen counter example. There are 'busy periods' over the course of a business day, during which customer arrivals occur more frequently. This implies that the mean of the customer inter-arrival time distributions will be shorter during these busy periods. It is reasonable to assume that dependent stochastic processes such as waiting times will be affected and will also be non-stationary. When appropriate, this issue can be circumvented by redefining the observation interval so that the study is restricted, for example, to the busy period. Then, the validity of a homogeneous stochastic process assumption for the customer inter-arrivals time can be reasonably assured.

Even in cases where the autonomous stochastic processes within a DEDS model are stationary over the observation interval, dependent stochastic processes can still exhibit transient behaviour. These transient effects are a consequence of initial conditions, whose impact needs to dissipate before stationary behaviour evolves. Consider, for example, waiting times when the delicatessen of our example first opens in the morning. The first customer will experience no waiting time and receive service immediately. Subsequent customers will likewise experience short waiting times. As time progresses, more customers will enter the queue and waiting times could start to lengthen, but in any event, the exceptional circumstances immediately following the opening will disappear.

This behaviour will occur even when we assume that the customer inter-arrival times are represented by a non-homogeneous stochastic process. If the mean of the inter-arrival times changes to a new value at the start of the busy period, the waiting times will again go through a transient period prior to eventually reaching steady-state behaviour. Note that it can be reasonably assumed here that the waiting time stochastic process is positively correlated.

Many of the variables that are introduced during conceptual model development for a discrete-event dynamic system are associated with stochastic processes. These may be either autonomous or dependent. There is, however, an important distinction which relates to the continuous-time or discrete-time nature of the stochastic process in question. In the former case, an associated variable typically relates to a specific entity within the model, and the time trajectory of that variable captures the behaviour of some property of that entity over the course of the observation interval, e.g. the length of a queue entity. In effect, this time trajectory is an observation of a continuous-time stochastic process.

The interpretation is markedly different in the case, where the variable is associated with a discrete-time stochastic process. In this case, the variable is attached to each instance of a collection of like entities. Therefore, the 'value' of the variable relates specifically to a particular entity instance (such variables are often referred to as 'instance variables'). The consolidation of these values into a collective view yields an observation of a discrete-time stochastic process, e.g. the waiting times of the customers served at a delicatessen counter.

It must also be stressed that the variables outlined above are especially important because they provide the basis for the trajectory sequence and/or sample sequence output that is required for the achievement of the project goal. These issues are examined in Sect. 3.4.

A data model is required for each autonomous stochastic process that is identified in a conceptual model. Such data models provide the basis for generating the values that are associated with these processes. Such a process might, for example, represent the inter-arrival times of an input entity stream as introduced in Chap. 2.

Creating a data model consists of determining appropriate probability distributions for the constituent random variables of the stochastic process. Data models can be very complex. Such models could be required to represent non-stationary stochastic processes or even multivariate stochastic processes, where one stochastic process is correlated with another. However, our treatment of data modelling in this book will be restricted in scope. In particular, we limit our considerations to autonomous stochastic processes that are piecewise homogeneous (see Sect. A2.8).

The key feature of a homogeneous stochastic process is that the data modelling task reduces to the identification of a single underlying distribution function (because each of the constituent random variables has the same distribution). The situation is somewhat more demanding in the general piecewise homogeneous case, where the observation interval is divided into m segments inasmuch as there is a different distribution required for each of the m segments.

A useful overview of data modelling is provided by Biller and Nelson [2]. More comprehensive discussions can be found in a variety of references such as Banks et al. [1], Kleijnen [9], Law and Kelton [13] and Leemis and Park [14].

3.2   DEDS-Specific Variables

The central role of randomness and stochastic processes in the study of DEDS was emphasized in Sect. 3.1. Many of the fundamental concepts relating to these topics are reviewed in Annex 2. There are, however, additional variable-related notions that are pertinent to our presentation of the modelling and simulation activity within the DEDS domain. These will play an important role throughout our discussions in Part II of this book.

3.2.1   Random Variates and RVVs

The execution of a DEDS simulation model can be viewed, in somewhat idealized terms, as a mapping of prescribed random input behaviour into random output behaviour. The 'idealization' at play here arises from the observation that 'randomness' within the computing milieu is not, in fact, genuinely random. More specifically, the 'random' values that are required in the characterization of random input behaviour are, of necessity, samples taken from prescribed probability distributions *using algorithmic methods*. However, by their fundamental nature, these methods do not guarantee genuine randomness. The values thus generated are called *random variates*, and they are generally regarded as 'pseudorandom'. As might be correctly assumed, there has been an extensive study of the properties of these algorithms as well as their efficacy. An exploration of these topics is provided in Sect. 3.7.

The reader should, however, appreciate that the algorithmic basis for random variates does have an 'upside'. In many circumstances, there is a need to explore model behaviour under a variety of controlled conditions (e.g. comparing design alternatives). In these circumstances, it is critical to be able to 'duplicate' random input behaviour. This requirement is easily accommodated in the algorithmic milieu of pseudorandomness.

Random variates play a key role in simulation studies in the DEDS domain. In the interests of completeness, we provide the following definition: *a random variate is a sample from a prescribed probability distribution which has been generated using an algorithm-based procedure.* Furthermore, it is of considerable usefulness in the discussions which follow to introduce the notion of a *random variate variable* (RVV). We define an RVV to be a variable, whose values are random variates. It is important to note that any particular sequence of values acquired by an RVV can be interpreted as an observation of an autonomous stochastic process.

Suppose *delta* is an RVV. This implies that the values for *delta* are delivered via a named algorithmic procedure. To emphasize the special role of such a procedure, we attach a prefix 'RVP' to the assigned name. For example, the procedure associated with *delta* might be *RVP.GetDelta*(). We shall refer to such a procedure as an 'RVP'.

The simplest interpretation for any RVV is the case where the underlying random variates flow from a single probability distribution. While this interpretation is often adequate, it is unnecessarily restrictive. There are, in fact, circumstances where the underlying random variates for a particular RVV may flow from any one of several probability distributions in some well-defined manner. The RVP associated with the RVV then has the additional task of managing the distribution selection criterion.

The random behaviour inherent in DEDS models has its origins in randomness in the inputs that are associated with the particular SUI that is being modelled. While the randomness in input may be formulated in idealized terms, the reality of pseudorandomness as outlined above must be acknowledged. This, in particular, implies that RVVs become an important constituent in the specification of input-related notions that arise in the modelling process.

The formulation of appropriate data models is one of the key aspects of the modelling process in the DEDS domain. Data models provide the basis for emulating the random phenomena that stimulate the behaviour of the SUI. A variety of important facets of data modelling is explored in Sect. 3.6.

3.2.2 Entities and Their Attributes

The notion of developing a DEDS model in terms of entities which may either exist permanently within the model or alternatively have a transient existence within it was introduced in Chap. 2. When this perspective is adopted, the entities become the principal constituents in the model and the development of the rules of behaviour focuses on characterizing the manner in which the entities undergo changes over the course of the observation interval. Change is formulated in terms of the values of attributes that are associated with the entities. The identification of suitable attributes is very much dependent upon the specific nature of an entity and the requirements of the problem. In some cases, there may be a particular trail of data produced by these entities (stored in an entity attribute) that is relevant to the output requirements that are implicit in the project goal.

Three common attribute types for the transient entities can be identified. These are outlined below.

- Feature Reflectors: Transient entities may have properties or features that have direct relevance to the manner in which they are treated by the rules of behaviour. For example, a 'size' which may have one of the three values (SMALL, MEDIUM or LARGE) or a 'priority' which may have one of the two values (HIGH or LOW). The value assigned to an attribute that falls in the category of a feature reflector usually remains invariant over the course of the entity's existence within the scope of the model.
- Path Reflectors: A DEDS model often has an inherent network structure, and in such cases, it is sometimes necessary to maintain an explicit record of what nodes an entity has already visited or, alternately, what nodes remain to be visited. Attributes that have this function fall in the category of path reflectors. The values of these attributes naturally change as the entity flows through the model. The value of an attribute that is a path reflector plays a key role in initiating the transfer of an entity to its next destination.
- Elapsed-Time Reflectors: Output requirements arising from the project goal often need data that must be collected about the way entities from a particular category have progressed through the model. Frequently this requirement is for some type of elapsed-time measurement. The values assigned for such elapsed-time attributes often provide required output data. For example, it may be required to determine the average time spent waiting for service at a particular resource within the model. An attribute introduced in this context could function as a 'time stamp' storing the value of time, t, when the waiting period begins. A data value for a predefined data set would then be generated as the difference between the value of time when the waiting period ends and the time stamp value.

A number of attribute types for permanent entities can also be identified.

- Status Reflector: An attribute of this type is used to indicate the status of an entity; e.g. BUSY/IDLE/BROKEN.
- Entity Reflector: An attribute of this type provides a reference to another entity (normally a transient entity) which, for example, is being serviced by the permanent entity.
- Quantity Reflector: Sometimes an entity 'accommodates' a number of other entities. An attribute of this type is used to indicate the number of such entities. For example, such an attribute could indicate the number of entities in a queue, the number of available resources or the number passengers on a bus.
- Location Reflector: Consider, for example, an entity that represents a van that picks up clients at an airport to take them to any one of several car rental offices. An attribute for that entity which provides the location of the van would be of this type.

We conclude this discussion of entity attributes with two important observations. We note first that apart from certain special cases (e.g. feature reflectors outlined above), attributes have values that generally vary over the course of the observation interval; in other words, they are variables. Their changing values typically occur at random times and are an integral part of behaviour characterization; randomness is a consequence of the stochastic nature of model inputs. We note also that the existence of transient entities implies transient variables. In other words, the collection of variables that exist within a DEDS model as it evolves over (simulated) time is not fixed in either number or identity.

3.2.3   Discrete-Time Variables

A discrete-time variable is a time-dependent variable, whose value changes only at discrete points in time. Such variables play a fundamental role in the development of models for discrete-event dynamic systems. As will become apparent in the presentation that follows, such variables are also relevant to the characterization of both the input and the output of such models. In this regard, we note that both input and output are almost always closely related to stochastic processes as discussed earlier. In effect, these variables serve to capture observations of the stochastic processes with which they are associated. Examples of such variables are given in Figs. 3.1, 3.2 and 3.3.

We recognize two types of discrete-time variable which are called *sample discrete-time variables* (as in Fig. 3.4) and *piecewise constant discrete-time variables* (as in Fig. 3.5). We begin in Sect. 3.2.3.1 by presenting a simple scheme for representing the time evolution of both these types of discrete-time variable. This representation scheme is consistent with the requirements of the conceptual modelling process that is presented in Chap. 4.

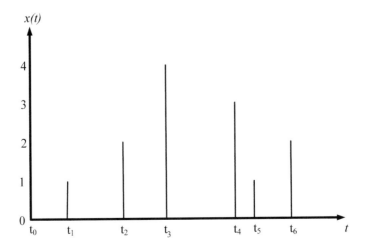

Fig. 3.4 A sample discrete-time variable

It was noted in Sect. 3.1 that a data model is required for each autonomous stochastic process that is identified within the realm of the SUI. This in effect implies a specification requirement for the discrete-time variable that yields observations of the stochastic process. The representation scheme that we outline below accommodates this specification requirement.

3.2.3.1   Representation

We consider first the sample discrete-time variable, $x(t)$, shown in Fig. 3.4. Note that values for $x(t)$ exist only at a finite set of points in time. This variable could, for example, represent the flow of orders for a particular product that is being marketed by an Internet-based distributing company. Orders arrive at the time points indicated and the number of units in the order is given by the value of x at those points in time.

The evolution of a sample discrete-time variable over time is associated with a finite sequence of discrete values of its argument, t, namely, $\langle t_r, t_{r+1}, t_{r+2}, \ldots \rangle$ where $t_{k+1} = t_k + \Delta_k$ with $\Delta_k > 0$, $k = r, r + 1, r + 2, \ldots$ and $r = 0$ or 1 (the first value in the sequence may or may not coincide with the left-hand boundary of the observation interval, namely, t_0). The key feature here is that the value of $x(t)$ is meaningful only for those values of t that belong to the time sequence. As a direct consequence, it follows that a convenient representation for $x(t)$ is the sequence of ordered pairs (when $r = 0$):

$$\mathrm{CS}[x] = \langle (t_k, x_k) : k = 0, 1, 2, \ldots \rangle$$

where $x_k = x(t_k)$. We call $\mathrm{CS}[x]$ the *characterizing sequence* for x.

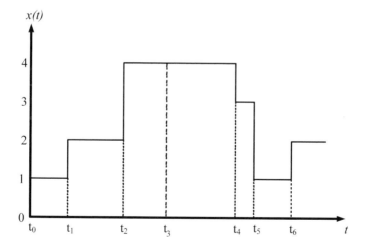

Fig. 3.5 A piecewise constant discrete-time variable

Note that there are two separate aspects of this representation; namely, the time sequence and the value sequence. In fact, it is convenient to separate the two underlying sequences in the following way:

$$\mathrm{CS_D}[x] = \langle t_k : k = 0, 1, 2, \ldots \rangle$$

$$\mathrm{CS_R}[x] = \langle x_k : k = 0, 1, 2, \ldots \rangle$$

which are called, respectively, the *domain sequence for x* and the *range sequence for x*.

A typical piecewise constant discrete-time variable is shown in Fig. 3.5. Here, $x(t)$ could represent the number of electricians, at time t, within the maintenance team of a large manufacturing plant that operates on a 24-hour basis but with varying levels of production (and hence varying requirements for electricians). The behaviour of the SUI likely depends directly on the value of $x(t)$ for all t; consequently, the representation of x as a piecewise constant time variable is an essential choice.

The key difference between the sample variable of Fig. 3.4 and the piecewise constant variable shown in Fig. 3.5 is simply that in the latter case, x has a defined value for all values[2] of t within the observation interval, while in the former case, x has a value only at the discrete points in time that are identified.

This difference between a sample discrete-time variable and a piecewise constant discrete-time variable as shown in Fig. 3.4 and Fig. 3.5, respectively, is certainly significant. However, from the perspective of a characterizing sequence representation, this difference is transparent. This allows for a uniform representation of discrete-time variables, and this simplifies many aspects of the conceptual model development as discussed in Chap. 4.

Nevertheless, there are differences in the manner in which sample discrete-time variables and piecewise constant discrete-time variables are handled in our conceptual modelling framework in Chap. 4. Consequently, we shall often refer to the characterizing sequence for a sample discrete-time variable, x, as a *sample sequence* (denoted by PHI[x]), and we shall refer to the characterizing sequence for a piecewise constant discrete-time variable, x, as a *trajectory sequence* (denoted by TRJ[x]).

3.2.3.2 Specification

As we have previously noted, there are important circumstances where a discrete-time variable from either of the two subclasses of interest needs to be specified. In view of the representation scheme we have chosen, this specification task corresponds to establishing the values within both the domain sequence and the

---

[2]Some care is nevertheless required in clarifying which of two possible values is to be taken at the discrete points in time where change occurs. Our convention will be that $x(t) = x_k$ for $t_k \leq t < t_{k+1}$ for $k = 0, 1, 2, \ldots$

range sequence for the variable, x, under consideration. This can be regarded as a data modelling task.

A variety of data model formats is possible. The wide range of alternatives falls into two broad categories: deterministic and stochastic. However, before we explore the data modelling task, we remind the reader that the conceptual model development process is inherently iterative and evolves in a manner that reflects progressively deeper insights into model requirements. Consequently at any particular point during this process, it may be that the required information to characterize either the domain sequence or the range sequence for some particular variable is not yet known. This should not be regarded as an inherent flaw but simply as an incomplete facet of model development.

We consider first some options for the data model associated with the domain sequence. Note that this effectively relates to the specification of the time increment, Δ_k, between successive time values in the sequence; i.e. $t_{k+1} = t_k + \Delta_k$. Perhaps the simplest case is when these values are explicitly provided in the form of a data list. In most other deterministic specifications, Δ_k can be expressed in terms of a known function, F, of the index k; i.e. $\Delta_k = F(k)$. The simplest case, however, is when $F(k) =$ constant; in other words, the values in the domain sequence are equally spaced.

The alternate possible specification format is stochastic. In this case, the values of Δ_k are established by selections from a specified probably distribution function, $\mathbf{P}$. In the simplest case, $\mathbf{P}$ is independent of k; in other words, the same probability distribution function applies for all k. A simple generalization occurs when the specification for Δ_k is based on a family of probability distribution functions. For example, if k is even, then the selection is from distribution $\mathbf{P}_E$, while if k is odd, the selection is from $\mathbf{P}_O$.

The data modelling options and formats for the range sequence are similar. The important difference is that the data models, in this case, deal with the range values directly; in other words, there typically is no counterpart to the generation of increments between successive values.

It is important to observe that the values within the domain sequence and the range sequence for a discrete-time variable will, in many cases, be treated as random variate variables. This will become apparent in the conceptual modelling discussions of Chap. 4.

3.3 Input

In Sect. 2.2.5.3, we outlined our perspective that 'input' is simply any facet of behaviour that impacts upon the SUI's behaviour but is not itself affected by it. This allows for a considerable range of interpretations and in this regard, we identify two broad categories of input that provide a natural dichotomy of the range of possibilities. These input categories are called exogenous and endogenous and each is outlined below:

(a) *Exogenous input*: Generally, it can be assumed that there exists a boundary that surrounds the SUI and separates it from its environment. Nevertheless, there are almost always aspects of the environment that extend across this boundary and influence the SUI's behaviour. This influence represents the *exogenous input* to the SUI. Examples of exogenous input are the occurrence of storms that disrupt the operation of a port or the arrival of customers at the counter of a fast-food outlet.

(b) *Endogenous input*: Often, there are facets of the SUI itself that have an impact upon its behaviour but are not themselves influenced by it. This 'embedded influence' represents the *endogenous input* to the SUI. Two examples of endogenous input are the varying number of part-time servers working at the counter of a fast-food outlet over the course of a day and the service times for the stream of customers that arrive at the counter of the outlet.

A fundamental initial step in the modelling process is the identification and characterization of the SUI's inputs. This necessarily takes the form of appropriately formulated input variables that capture the essential features of each identified input behaviour. Generally, such input variables are time dependent and their time dependency must be explicitly specified. Such variables are generally discrete-time variables, and consequently, they can be accommodated according to the conventions outlined earlier in Sect. 3.2.3. More specifically, their specification simply requires the specification of their respective characterizing sequences.

The preceding discussion is somewhat misleading inasmuch as it projects the impression that the identification of input variables is straightforward. This is not entirely the case and some guidelines are in order. These are given below:

(a) A constant is never an input variable (a constant is an identifier with which we associate a fixed value).

(b) A parameter is never an input variable (a parameter is an identifier whose assigned value is invariant over the course of the observation interval for any particular experiment with the simulation model but will change in other experiments in a manner consistent with the project goal).

(c) A sample discrete-time variable is an input variable if its range sequence values impact upon the behaviour of the model but are not themselves dependent on that behaviour.

(d) A piecewise constant discrete-time variable is an input variable if its range sequence values are not (by definition) identical and they impact upon the model but are not themselves dependent on that behaviour.[3]

---

[3]The intent here is to circumvent any possible misinterpretation with respect to guidelines (a) and (b).

3.3.1 Modelling Exogenous Input

We associate two types of input variable with the exogenous inputs to an SUI in order to accommodate significantly different requirements. These are described below:

(a) *Environmental input variables*: These variables are used in the model to characterize the environmental effects that have an impact on the SUI's behaviour. Earlier, we used the example of the occurrence of storms that disrupt the operation of a port to illustrate exogenous input to an SUI. In this case, one might introduce an environmental input variable named Storm, whose values alternate between TRUE and FALSE to reflect the existence or not of storm conditions. This variable would fall in the category of a piecewise constant discrete-time variable.

(b) *Entity stream input variables*: We've previously noted that transient entities that flow through the domain of the SUI are common elements in most modelling and simulation projects in the DEDS domain (e.g. customers at our fast-food outlet). We regard these as originating in the SUI's environment and generally returning to it; in other words, they flow across the boundary that separates the SUI from its environment and hence are regarded as exogenous input.

We use an entity stream input variable to represent the arrival aspect of this flow. This variable is a sample discrete-time variable. Its domain sequence corresponds to the points in time when entity arrivals (from a particular category) occur, and its range sequence provides the number of such arrivals at each arrival time point (typically 1). Note that the creation of the entity itself is necessarily a separate operation; the entity stream input variable simply serves to invoke this operation. Note also that the creation of each entity in such a stream requires, in particular, the initialization of its attributes.

3.3.2 Modelling Endogenous Input

Endogenous inputs are also discrete-time variables, which fall into two types that we call *independent input variables* and *semi-independent input variables*. In the case of an independent (endogenous) input variable, x, both the domain sequence, $CS_D[x]$, and the range sequence, $CS_R[x]$, are independent of model behaviour; hence they are established by predetermined data models. On the other hand, we refer to x as a semi-independent (endogenous) input variable when only its range sequence, $CS_R[x]$, is independent of model behaviour. This, in particular, implies that there is no explicit data model associated with the domain sequence; i.e. the time values in the domain sequence evolve implicitly in a manner dictated by the behaviour of the model.

To illustrate, consider the example introduced earlier (fast-food outlet) for which we might introduce the variable N to represent the number of part-time servers working at the counter. The implication here is that the predefined values of N vary over the course of a workday, changing at predefined discrete points in time. This piecewise constant discrete-time variable $N(t)$ falls in the category of an independent (endogenous) input variable.

Notice also that $N(t)$ might, in some circumstances, be included among a variety of policy options being considered for improving/optimizing the operation of the fast-food outlet. In effect, an independent (or semi-independent) input variable can sometimes be regarded as a generalization of a parameter in the sense that it too is available for modification within the context of an optimization task within the goal of the modelling and simulation study.

We indicated earlier that service time (e.g. of customers at the counter of a fast-food outlet) is also an endogenous input. A variable introduced to represent service time; e.g. $S(t)$, is a sample discrete-time variable. Note, however, that while the values acquired by S (its range sequence) flow from a prescribed data model, the corresponding values of t (its domain sequence) are dictated by model behaviour. Consequently, $S(t)$ is a semi-independent exogenous input variable.

3.4   Output

The fundamental outcome of a simulation run is provided by the output that is produced. Output is that set of generated data that has relevance to the achievement of the project goal. Output can take several forms, but its origins are almost exclusively linked to the changing values of entity attributes within the model. More specifically we regard any output variable to be a function of one or more attributes.

From a modelling perspective, output, like input, is identified using suitably defined variables. Because of the stochastic behaviour inherent in the study of DEDS, these output variables are random variables and this has significant implications on how meaningful data is acquired and interpreted.

From a simulation perspective, it is the values acquired by these variables that have relevance. However, the underlying interpretation here is not entirely straightforward. Consider, for example, a service depot that employs a technician who may be engaged in any one of four tasks (called T1, T2, T3 and T4). This resource entity (possibly called Technician) would likely have an attribute (possibly called *status*), whose value at any time t would be one of T1, T2, T3 or T4. Of interest might be the piecewise constant variable *specialStatus*(*t*) whose value at time t_k is 1 if *status* = T1 or *status* = T4 but 0 otherwise. The output variable *specialStatus*(*t*) has the characterizing sequence:

$$\text{CS}[specialStatus] = \langle (t_k, y_k) : k = 0, 1, 2, \ldots \rangle$$

where y_k is either 1 or 0. Recall (Sect. 3.2.3.1) that the piecewise constant discrete-time variable $specialStatus(t)$ can likewise be characterized as a trajectory sequence TRJ[$specialStatus$] where

$$\text{TRJ}[specialStatus] = \text{CS}[specialStatus]$$

As a second example, consider a queue entity that would typically be needed to temporarily hold entities that cannot immediately access a service provided by some resource. In such a case, it would be reasonable to define an attribute for this queue entity that serves to store the number of entities in the queue, possibly called n (referenced in ABCmod with Q.QueueName.n where QueueName is the name of this entity category; see Chap. 4). Suppose at time t_k, the number of entities in the queue changes to y_k due to either an insertion or a removal from the queue. If we assume that the value of the attribute n is of interest, then n becomes an output variable and is, in fact, a piecewise constant discrete-time variable whose characterizing sequence is

$$\text{CS}[\text{Q.QueueName.n}] = \langle (t_k, y_k) : k = 0, 1, 2, \ldots \rangle$$

with TRJ[Q.QueueName.n] = CS[Q.QueueName.n].

Consider now the messages that flow through a communications network. Each such message could be regarded as an instance of a (transient) entity category reasonably called Message. At each of the switching nodes through which a particular message passes, it generally encounters a delay and these delays would likely be accumulated in an attribute (possibly) called totDelay (referenced in ABCmod as iC.Message.totDelay; see Chap. 4) defined for the entity category Message. Following the final switching node visited by any particular message, its journey through the network ends and it exits the network. An increasing sequence of exit time values $\langle t_1, t_2, \ldots \rangle$ is, therefore, generated which is associated with a corresponding sequence of totDelay values: $\langle d_1, d_2, \ldots \rangle$. The attribute totDelay can be regarded as a sample discrete-time variable whose characterizing sequence is[4]:

$$\text{CS}[\text{C.Message.totDelay}] = \langle (t_1, d_1), (t_2, d_2), \ldots (t_N, d_N) \rangle$$

If totDelay is of interest, this characterizing sequence becomes an output and can be represented as a sample sequence PHI[C.Message.totDelay], that is,

$$\text{PHI}[\text{C.Message.totDelay}] = \text{CS}[\text{C.Message.totDelay}]$$

---

[4]The collection of all values of this output resulting from an experiment, namely, $\{d_1, d_2, \ldots, d_N\}$ (where N is the total number of messages passing through the network over the course of the observation interval), is an observation of a stochastic process.

Both trajectory sequences and sample sequences are collections of data values generated by a run of the simulation model. Often, it is not the collection itself that is of interest[5] but rather properties of the values in this collection, e.g. mean, maximum or minimum (we refer here to the second data element in the tuples in question). This gives rise to another category of output variable that we call a *derived scalar output variable* (DSOV) that is derived from data elements in a trajectory sequence or a sample sequence. For example, we could have $y_{av} = \text{AVG}(\text{TRJ}[y])$ or $y_{mx} = \text{MAX}(\text{PHI } [y])$, where AVG and MAX are operators that yield the average and maximum values, respectively, of the data within the referenced collection. Many such operators can be identified, e.g. MIN, NUMBER, etc. Notice that the values of DSOVs such as y_{av} and y_{mx} are established by a post-processing step at the end of a simulation run. Notice also that both y_{av} and y_{mx} are, in fact, random variables and any experiment with the model yields an observation of those random variables where an observation, in this case, is simply a value computed from data within a set.

The value assigned to a DSOV is acquired by processing a data set at the end of a simulation run. However, there are circumstances where it is far more straight-forward to carry out equivalent processing during the course of the run. In such circumstances, the processing yields a value for a scalar output variable at the end of the experiment without the need for a post-processing step. We call such a variable a *simple scalar output variable* (SSOV). This approach is especially useful when the required post-processing operation on a data set is highly specialized and not directly available. Consider, for example, a case where an output requirement is for the number of customers at a fast-food outlet, whose waiting time for service exceeded 5 minutes. This value could certainly be obtained via post-processing of the data within a sample sequence, but a more straightforward and explicit approach is simply to embed a counter at some appropriate point within the model.

In summary then, we deal with observations of output variables and these fall into one of four categories: trajectory sequences, sample sequences, DSOVs and SSOVs. Values are placed into trajectory sequences and sample sequences as they become available. An overview of the four categories is provided below:

(a) *Trajectory sequence*: A trajectory sequence captures the sequence of values acquired by an output variable, *y*, that is a piecewise constant discrete-time variable. The trajectory sequence for the output variable *y* is denoted by TRJ[*y*].
(b) *Sample sequence*: A sample sequence captures the sequence of values acquired by an output variable, *y*, that is a sample discrete-time variable. The sample sequence for the output variable *y* is denoted by PHI[*y*].
(c) *Derived scalar output variable* (*DSOV*): Normally, the entire collection of data accumulated in a trajectory sequence or a sample set is not of interest. Rather it is some scalar-valued property of the data that has relevance, e.g. average, maximum or minimum. The scalar value obtained by carrying out the

---

[5]The notable exception here is the case where a graphical presentation of the data is desired.

designated operation at the end of the simulation run is assigned to an output
variable called a derived scalar output variable (DSOV).

(d) *Simple scalar output variable* (*SSOV*): The role of an SSOV parallels that of a
DSOV inasmuch as both provide some specific measurable property of either
a trajectory sequence or a sample sequence. In the case of the SSOV, that
property is monitored during the course of a particular run and a final value is
available at the end of the run. This is in contrast to a DSOV whose value
flows from a post-processing step at the end of a run.

3.5   DEDS Modelling and Simulation Studies

The goal of a modelling and simulation project implicitly defines one of two possible
types of study. As will become apparent in later discussions (in particular, in Chap. 7),
these differences are easy to appreciate but have a significant impact on the nature of
the experimentation procedure that needs to be carried out. The procedure in one case
requires considerably more effort. The main differences arise from two mutually
interdependent features; one relates to specifications on the right-hand boundary of
the observation interval, and the other relates to possible constraints imposed on the
acquired data. We shall refer to the two alternatives as bounded horizon studies and
steady-state studies. They[6] are summarized below.

Bounded Horizon Study

- The right-hand boundary of the observation interval is specified in the problem
  statement either explicitly by a given value of time or implicitly by some combination of values acquired by the model's variables.
- There are no restrictions on the properties of the dependent stochastic processes
  that are of interest. Often transient behaviour dominates.

Steady-State Study

- The right-hand boundary of the observation interval is not provided in the
  problem statement. Its value emerges in an indirect fashion because the observation interval extends to a point where the time series of acquired data is long
  enough to reflect steady-state behaviour.
- Steady-state behaviour of the dependent stochastic processes that are of interest is
  essential. In other words, the focus is on behaviour patterns in the absence of
  transient effects.

---

[6]These two types of study are often referred to as 'terminating simulations' and 'nonterminating
simulations', respectively.

3.6 Data Modelling

Data models (expressed in terms of theoretical probability distributions) are provided for the various autonomous stochastic processes that appear in the DEDS examples presented in this book. This might give the erroneous impression that defining such data models is simple and straightforward. In reality, much time and effort are required for carrying out the data modelling task that gives rise to these data models. Furthermore, it must be strongly emphasized that improper data models can destroy the value of the results that flow from a simulation study. Data models play the role of input data to a computer program, and a long established principle in software engineering is that 'garbage in equals garbage out'!

When the SUI exists, it may be possible to obtain insights about its various autonomous stochastic processes by observing the existing system. Data can be gathered and analysed to obtain information necessary for the formulation of suitable data models. In other cases, when such data is not available (e.g. the SUI does not exist or the collection of data is impossible or too costly), data models may have to be constructed on the basis of the insight provided by domain specialists, i.e. 'educated guesses'.

In this section, we first consider the case when data can be obtained from the SUI, and we provide an overview of steps required to develop, from the collected data, the underlying distributions that serve as data models. Some general guidelines are given for formulating data models when no data exists.

Data models are integrated, in a modular fashion, into the conceptual modelling framework that is presented in Chap. 4. Data modelling is usually carried out in parallel with the conceptual modelling task. Both of these modelling exercises can give rise to refinements to the problem description and/or project goal.

3.6.1 Defining Data Models Using Collected Data

Our introduction to data modelling is restricted to modelling autonomous stochastic processes that are (piecewise) homogeneous. The objective is to formulate either an appropriate theoretical or empirical distribution derived from the collected data.

Collected data can be used to formulate a data model that is specified in terms of a cumulative distribution function that is called an empirical CDF. Section 3.6.4 discusses this approach. When the empirical distribution is continuous, an inherent disadvantage of the approach is that it will not generate values outside the limits of the observed values. It can nevertheless yield values other than those that have been observed. An alternate (and generally preferred) approach is to use statistical techniques to fit a theoretical distribution to the collected data. A theoretical distribution provides a number of advantages; e.g. it smooths out irregularities that could arise with the empirical alternative, and it can yield values outside the boundaries of the observed values. Theoretical distributions always have embedded parameters (see Sects. A2.3.6 and A2.4.4) which provide a simple means for adjusting the distribution to best fit the collected data.

Our first task is to determine if the collected data does indeed belong to a homogeneous stochastic process. This requires two tests: one to determine if the stochastic process is identically distributed and a second to determine if the constituent random variables are independent. A number of analysis techniques exist for testing for these properties and a few are presented in Sect. 3.6.2. The final task is to fit a theoretical distribution to the collected data. Software is available for analysing collected data and fitting theoretical distributions. Such software is available in standalone form, e.g. ExpertFit [12] and Stat::Fit [6]. This functionality is often integrated directly into simulation packages, e.g. Arena [8] or ProModel [6] (which includes Stat::Fit).

3.6.2 Does the Collected Data Belong to a Homogeneous Stochastic Process?

This section presents two techniques for evaluating the independence of collected data and a technique for evaluating stationarity.

3.6.2.1 Testing for Independence

Two graphical methods for evaluating independence are presented here: scatter plots and autocorrelation plots. In both cases, the objective is to investigate possible dependencies among the values in a time series obtained as an observation of a stochastic process (see Sect. A2.8). More specifically, we assume that our collected data is the time series $x = (x_1, x_2, x_3, \ldots, x_n)$, which is an observation of a stochastic process, $X = (X_1, X_2, \ldots, X_n)$.

A scatter plot is a display of the points $P_i = (x_i, x_{i+1})$, $i = 1, 2, \ldots, (n-1)$. If little or no dependence exits, the points should be scattered in a random fashion. If, on the other hand, a trend line becomes apparent, then dependence does exist. For positively correlated data, the line will have a positive slope, that is, both coordinates of the points P_i will be reasonably similar in value. If data is negatively correlated, the trend line will have a negative slope; i.e. a small value of x_i will be associated with a large value of x_{i+1} and vice versa.

We illustrate the method with two separate time series. The first consists of 300 data values generated from a gamma distribution (with $\alpha = 2$, $\lambda = 1/3$).[7] The second has 365 values representing the daily maximum temperatures in Ottawa, Ontario, Canada,[8] between 20 May 2005 and 20 May 2006.

Figure 3.6 shows the scatter plot for the first case. Clearly, there is no apparent trend and consequently independence can be assumed. The scatter plot shown in Fig. 3.7 for the temperature data shows a trend line with a positive slope. The implied positive correlation is to be expected inasmuch as there is a great likelihood that the temperature on successive days will be similar.

---

[7]Generated using the Microsoft® Office Excel 2003 Application.
[8]Source: http://ottawa.weatherstats.ca.

A scatter plot is a presentation of data values that are immediately adjacent, i.e. that have a lag of 1. An autocorrelation plot, on the other hand, is more comprehensive because it evaluates possible dependence for a range of lag values. An autocorrelation plot is a graph of the sample autocorrelation $\rho(k)$ for a range of lag values, k, where

$$\rho(k) = \frac{\sum_{i=1}^{n-k}(x_i - \bar{x}(n))(x_{i+k} - \bar{x}(n))}{(n-k)s^2(n)}$$

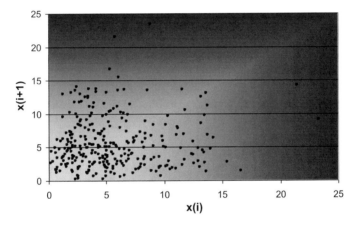

Fig. 3.6 Scatter plot showing uncorrelated data

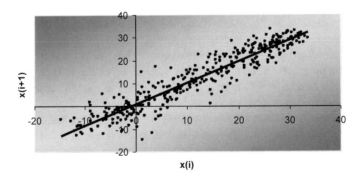

Fig. 3.7 Scatter plot showing correlated data

Here, $x(n)$ and $s^2(n)$ are estimates of the sample mean and sample variance, respectively, for the time series (some elaboration can be found in Sect. A2.6).

Figure 3.8 shows the autocorrelation plots for the time series obtained from the gamma distribution. The graph shows that the sample autocorrelation is low for all lag values, which reinforce the earlier conclusion that the data is independent. For the temperature time series, the autocorrelation plot in Fig. 3.9 shows high values (i.e. close to 1) for the sample autocorrelation for all lag values between 1 and 30, indicating a high level of correlation between temperatures over a month.

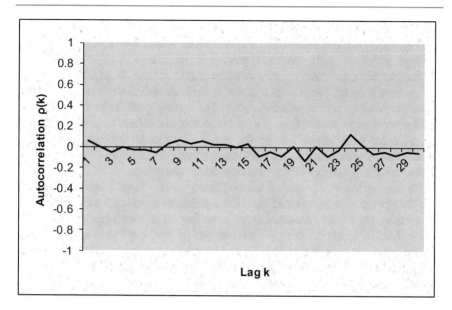

Fig. 3.8 Autocorrelation plot showing uncorrelated data

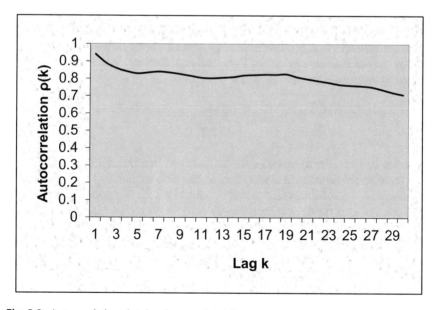

Fig. 3.9 Autocorrelation plot showing correlated data

3.6.2.2 Testing for Stationarity

We begin with the assumption that a collection of values $x = \{x_i : i = 1, 2, \ldots, m\}$ has been acquired for the stochastic process X. Each x_i is an $n(i)$-tuple of values obtained for the constituent random variables within X; i.e. $x_i = (x_{i,1}, x_{i,2}, \ldots, x_{i,n(i)})$. The process X could, for example, represent the inter-arrival times of customers at the delicatessen counter introduced earlier. The $n(i)$-tuple (or time series), x_i, could be interpreted as an observation of X on the ith day of a collection period extending over m consecutive days. Testing for stationarity can be a relatively elaborate process, but as a minimum it requires the assessment of the degree to which average values within the collected data remain invariant over time.

We outline below a graphical method that provides insight into the variation over time of the average value of the collected data, hence an approach for carrying out a fundamental test for stationarity. Assume that the collected data extends over a time interval of length T. The procedure begins by dividing T into a set of L time cells of length Δt. Recall that each data value (say $x_{i,j}$) is necessarily time indexed (either explicitly or implicitly) and consequently falls into one of the L time cells. The $n(i)$ values in the time series x_i can then be separated into disjoint subsets according to the time cell to which each value belongs. The average value of the data in each cell is computed for each of the m time series, and then a composite average over the m time series is determined and plotted on a time axis that is similarly divided into L cells. The resulting graph, therefore, displays the time behaviour of average values within the cells.

Within the context of our delicatessen example, the recorded data could represent observations of the customer inter-arrival times. For day i, we denote by $\bar{a}_{ci}$ the average inter-arrival time in cell c. Then, for each time cell c, compute an overall average inter-arrival time $\bar{a}_c$ using data from all m days; i.e.

$$\bar{a}_c = \frac{1}{m} \sum_{i=1}^{m} \bar{a}_{ci}$$

The value $\bar{a}_c$ provides an estimate for the mean of the underlying distribution of those inter-arrival time random variables, whose time value falls in the time cell c.

A plot of $\bar{a}_c$ versus time cell c provides a visual aid for evaluating how the mean of the distributions varies over time. Figure 3.10 shows such a plot computed from 3 days of observed inter-arrival times in the delicatessen between 9h00 to 18h00 (see Sect. 3.1). An interval Δt of 30 minutes was used. The plot clearly shows that the mean does vary since smaller averages occur around noon and at the end of the day, that is, during rush hour periods.

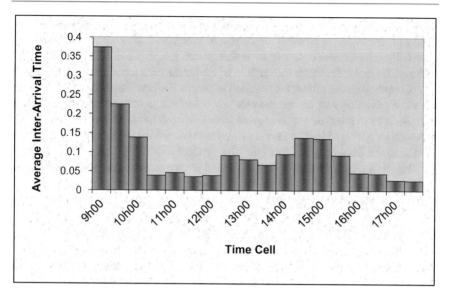

Fig. 3.10   Nonstationary inter-arrival times

3.6.3   Fitting a Distribution to Data

Fitting a theoretical distribution that matches time series data obtained from a homogeneous stochastic process is a trial-and-error procedure. The procedure usually begins with a histogram developed from the collection of n values belonging to some particular time series. If the objective is a continuous distribution, then the histogram provides a representation whose shape approximates the underlying probability density function. On the other hand, if the objective is a discrete distribution, then the histogram provides a representation whose shape approximates the underlying probability mass function. A graphical display of the associated cumulative distribution function can also be helpful for specifying empirical distributions.

The general shape of the histogram typically suggests possible theoretical distribution candidates. Parameters that are associated with theoretical distributions then need to be estimated. Goodness-of-fit tests are generally used to determine how well the parameterized distribution candidates fit the data. A selection is made based on the results of this analysis.

As an example, consider a time series obtained by observing the 'group sizes' that enter a restaurant over the course of a particular business day. The distribution of interest here is discrete, and the histogram shows the number of occurrences of each of the group sizes within the available time series data. The histogram shown in Fig. 3.11 illustrates a possible outcome. A group size of 4 clearly occurs most frequently. The associated cumulative distribution is also provided in Fig. 3.11, and it shows, for example, that just over 70% of the group sizes are equal to or less than 4.

An approximation process is required to handle the continuous case; i.e. the case where observed data values can assume any value in a prescribed interval of the real line. In this circumstance, a histogram is constructed by dividing the interval into subintervals called *bins*. The number of values that fall into each bin is counted and plotted as the frequency for that bin. The 'smoothness' of the graph that results is very much dependent on the bin size. If the bin size is too small, the resulting plot can be ragged. If the bin size is too large, the graph's value for inferring a candidate distribution can be compromised. Banks et al. [1] suggest choosing the number of bins to be $\sqrt{n}$, where n is the number of values in the time series observation. On the other hand, Stat::Fit [6] recommends $\sqrt[3]{2n}$ for the number of bins.

Figure 3.12 shows a histogram created using 100 data values ($n = 100$) generated from an exponential distribution using 22 bins. It illustrates how a ragged plot can occur when the bin size is too small. Figure 3.13 shows the histogram using 10 bins (value recommended by Banks et al.). Figure 3.14 shows the histogram using 6 bins (recommended value used in Stat::Fit).

Once the histogram has been created, the shape of the histogram is used to suggest one or more theoretical distributions as possible candidates for the data model. Values for the parameters of each of these candidates must be estimated, and a number of estimation methods developed for this purpose are available. In the discussion that follows, we briefly examine the category of estimators called maximum-likelihood estimators (further details can be found in Law [13]).

The sample mean $\bar{x}(n)$ and the sample variance $s^2(n)$ of the time series observation play a key role in the maximum-likelihood parameter estimation procedure.

Table 3.1 shows how maximum-likelihood estimators for several distributions are computed. Estimators for other distributions can be found in [1, 13].

Table 3.1 Maximum likelihood estimators

Distribution	Parameters	Estimators
Exponential	λ	$\hat{\lambda} = \frac{1}{\bar{x}(n)}$
Normal	μ, σ^2	$\hat{\mu} = \bar{x}(n)$ $\hat{\sigma}^2 = s^2(n)$
Gamma	α, λ	Compute $T = [\ln(\bar{x}(n)) - \frac{1}{n}\sum_{i=1}^{n} \ln(x_i)]^{-1}$ and find $\hat{\alpha}$ from Table 3.2 using linear interpolation. $\hat{\lambda} = \frac{\hat{\alpha}}{\bar{x}(n)}$

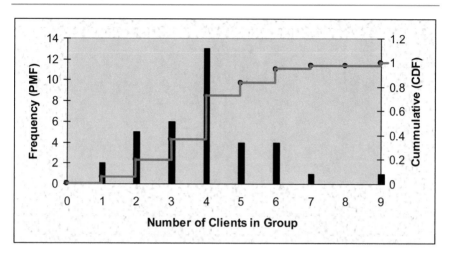

Fig. 3.11 Histogram for discrete valued data

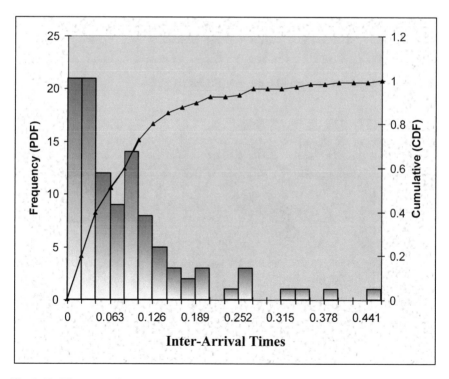

Fig. 3.12 Histogram of inter-arrival times with 22 bins

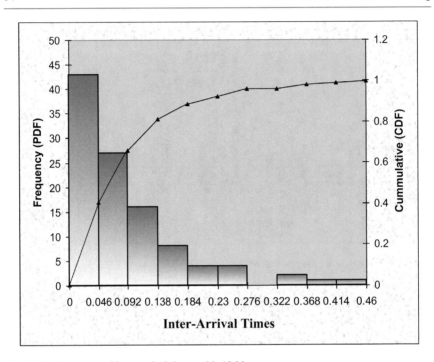

Fig. 3.13 Histogram of inter-arrival times with 10 bins

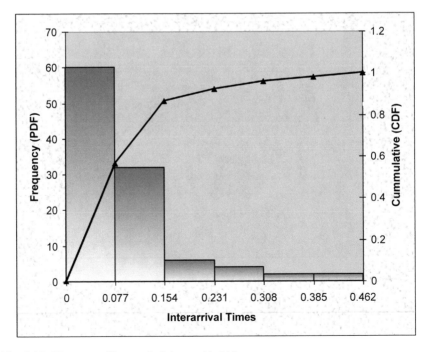

Fig. 3.14 Histogram of inter-arrival times with 6 bins

Table 3.2 Estimating the α parameter for the gamma distribution

1/T	$\hat{\alpha}$	1/T	$\hat{\alpha}$	1/T	$\hat{\alpha}$
0.02	0.0187	2.7	1.4940	10.3	5.3110
0.03	0.0275	2.8	1.5450	10.6	5.4610
0.04	0.0360	2.9	1.5960	10.9	5.6110
0.05	0.0442	3.0	1.6460	11.2	5.7610
0.06	0.0523	3.2	1.7480	11.5	5.9110
0.07	0.0602	3.4	1.8490	11.8	6.0610
0.08	0.0679	3.6	1.9500	12.1	6.2110
0.09	0.0756	3.8	2.0510	12.4	6.3620
0.1	0.0831	4.0	2.1510	12.7	6.5120
0.2	0.1532	4.2	2.2520	13.0	6.6620
0.3	0.2178	4.4	2.3530	13.3	6.8120
0.4	0.2790	4.6	2.4530	13.6	6.9620
0.5	0.3381	4.8	2.5540	13.9	7.1120
0.6	0.3955	5.0	2.6540	14.2	7.2620
0.7	0.4517	5.2	2.7550	14.5	7.4120
0.8	0.5070	5.4	2.8550	14.8	7.5620
0.9	0.5615	5.6	2.9560	15.1	7.7120
1.0	0.6155	5.8	3.0560	15.4	7.8620
1.1	0.6690	6.0	3.1560	15.7	8.0130
1.2	0.7220	6.2	3.2570	16.0	8.1630
1.3	0.7748	6.4	3.3570	16.3	8.3130
1.4	0.8272	6.6	3.4570	16.6	8.4630
1.5	0.8794	6.8	3.5580	16.9	8.6130
1.6	0.9314	7.0	3.6580	17.2	8.7630
1.7	0.9832	7.3	3.8080	17.5	8.9130
1.8	1.0340	7.6	3.9580	17.8	9.0630
1.9	1.0860	7.9	4.1090	18.1	9.2130
2.0	1.1370	8.2	4.2590	18.4	9.3630
2.1	1.1880	8.5	4.4090	18.7	9.5130
2.2	1.2400	8.8	4.5600	19.0	9.6630
2.3	1.2910	9.1	4.7100	19.3	9.8130
2.4	1.3420	9.4	4.8600	19.6	9.9630
2.5	1.3930	9.7	5.0100	20.0	10.1600
2.6	1.4440	10.0	5.1600		

Derived from table provided by Choi and Wette [3]

Once parameters for a candidate distribution have been estimated, a goodness-of-fit test needs to be used to determine how well the selected parameterized theoretical distribution fits the collected data. Various such tests are available and an overview can be found in Law [13] or Banks et al. [1]. Among the options is the chi-square test[9] which we summarize below.

Suppose D_c is the parameterized candidate distribution to be tested. The objective is to determine if there is a basis for rejecting D_c because it provides an inadequate match to the collected data. The first step in the test is to determine a value for an 'adequacy measure', A_m, that essentially compares the frequencies in the histogram formulated from the collected data to expected frequency values provided by D_c. The definition for A_m is

$$A_m = \sum_{i=1}^{k} \frac{(O_i - E_i)^2}{E_i} \tag{3.1}$$

where

- k is the number of class intervals which initially is equal to the number of bins in the data histogram. In other words, a class interval is initially associated with each bin.
- E_i is the expected frequency for the ith class interval based on D_c. It is defined as $E_i = np_i$, where p_i is the probability that a value falls into the interval, that is, $P[x_i < x < x_{i+1}]$ where x_i and x_{i+1} are the boundaries of the ith class interval and n is the number of values in the available time series data. The probability p_i can be computed using the cumulative density function $F(x)$ of D_c, i.e.

$$E_i = n(F(x_{i+1}) - F(x_i))$$

- When E_i is less than 5, the class interval is combined with an adjacent one (thus the new interval contains multiple adjacent bins), and the value of k is appropriately reduced. The E_i for the newly enlarged class interval is then re-evaluated. This step is repeated until all E_i values are greater than 5.
- The value O_i corresponds to the frequency observed in the histogram bin that corresponds to the ith class interval. For a class interval that contains more than one bin, the frequencies from the corresponding histogram bins are summed to provide the value for O_i.

Clearly, the intent is to have A_m small. The practical question, however, is whether A_m is sufficiently small. The decision is made by comparing the value of A_m with a 'critical value', χ^*, obtained from the chi-square distribution table as given in Table 3.3. The selection of χ^* depends on two parameters: the first is the degrees of freedom, $v = k - g - 1$, where g is the number of parameters embedded in D_c, and the second is α (the level of significance) for which 0.05 is a commonly used value.

---

[9]The test is shown for the continuous distribution. For discrete distributions, each value in the distribution corresponds to a class interval and $p_i = P(X = x_i)$.

If $A_m > \chi^*$, then A_m is not sufficiently small and D_c should be rejected as a distribution option.

The procedure outlined above for fitting a theoretical distribution to a time series observation is illustrated with the gamma-distributed data used in Sect. 3.6.2.1. Seventeen (the approximate square root of 300) bins were used to generate the histogram shown in Fig. 3.15. Two theoretical distributions are selected for consideration as candidates, namely, an exponential distribution and a gamma distribution.

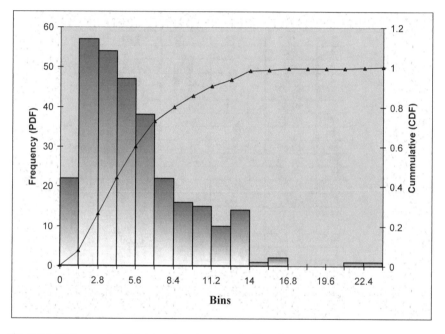

Fig. 3.15 Histogram for 300 data values (gamma distributed)

Values for the sample mean and sample variance of the data are $\bar{x}(300) = 5.58$ and $s^2(300) = 14.26$, respectively. The estimate of the exponential distribution's single parameter $\hat{\lambda}$ is equal to $1/5.58 = 0.179$. Table 3.4 shows the values for O_i derived from the data histogram and E_i derived from the exponential distribution candidate (with mean 5.58). Notice how the bins 12–17 are collapsed into 2 class intervals. Using the data in Table 3.2, it follows from Eq. (3.1) that $A_m = 37.30$. With $v = 11$ ($k = 13$ and $g = 1$) and $\alpha = 0.05$, the critical value (from Table 3.5) is $\chi^* = 19.68$. Because $A_m > \chi^*$, the exponential distribution candidate is rejected by the test.

Table 3.3 Chi-square distribution

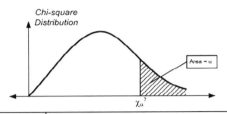

Degrees of freedom	Level of Significance (α)				
ν	0.005	0.01	0.025	0.05	0.1
1	7.88	6.63	5.02	3.84	2.71
2	10.60	9.21	7.38	5.99	4.61
3	12.84	11.34	9.35	7.81	6.25
4	14.86	13.28	11.14	9.49	7.78
5	16.75	15.09	12.83	11.07	9.24
6	18.55	16.81	14.45	12.59	10.64
7	20.28	18.48	16.01	14.07	12.02
8	21.95	20.09	17.53	15.51	13.36
9	23.59	21.67	19.02	16.92	14.68
10	25.19	23.21	20.48	18.31	15.99
11	26.76	24.72	21.92	19.68	17.28
12	28.30	26.22	23.34	21.03	18.55
13	29.82	27.69	24.74	22.36	19.81
14	31.32	29.14	26.12	23.68	21.06
15	32.80	30.58	27.49	25.00	22.31
16	34.27	32.00	28.85	26.30	23.54
17	35.72	33.41	30.19	27.59	24.77
18	37.16	34.81	31.53	28.87	25.99
19	38.58	36.19	32.85	30.14	27.20
20	40.00	37.57	34.17	31.41	28.41
21	41.40	38.93	35.48	32.67	29.62
22	42.80	40.29	36.78	33.92	30.81
23	44.18	41.64	38.08	35.17	32.01
24	45.56	42.98	39.36	36.42	33.20
25	46.93	44.31	40.65	37.65	34.38
26	48.29	45.64	41.92	38.89	35.56
27	49.64	46.96	43.19	40.11	36.74
28	50.99	48.28	44.46	41.34	37.92
29	52.34	49.59	45.72	42.56	39.09
30	53.67	50.89	46.98	43.77	40.26
40	66.77	63.69	59.34	55.76	51.81
50	79.49	76.15	71.42	67.50	63.17
60	91.95	88.38	83.30	79.08	74.40
70	104.21	100.43	95.02	90.53	85.53
80	116.32	112.33	106.63	101.88	96.58
90	128.30	124.12	118.14	113.15	107.57
100	140.17	135.81	129.56	124.34	118.50

Generated using Microsoft Excel function CHIINV

Table 3.4 Observed and expected frequency data for exponential distribution candidate

Class Interval				O_i		E_i		
1	0	-	1.4	22		66.56		
2	1.4	-	2.8	57		51.79		
3	2.8	-	4.2	54		40.30		
4	4.2	-	5.6	47		31.36		
5	5.6	-	7	38		24.40		
6	7	-	8.4	22		18.99		
7	8.4	-	9.8	16		14.78		
8	9.8	-	11.2	15		11.50		
9	11.2	-	12.6	10		8.95		
10	12.6	-	14	14		6.96		
11	14	-	15.4	1		5.42		
12	15.4	-	16.8	2	2	4.22	7.50	
	16.8	-	18.2	0		3.28		
13	18.2	-	19.6	0	1	2.55	7.29	
	19.6	-	21	0		1.99		
	21	-	22.4	1		1.55		
	22.4	-	23.8	1		1.20		

Table 3.5 Observed and expected frequency data for gamma candidate

Class Interval				O_i		E_i		
1	0	-	1.4	22		26.11		
2	1.4	-	2.8	57		52.17		
3	2.8	-	4.2	54		53.96		
4	4.2	-	5.6	47		46.17		
5	5.6	-	7	38		36.05		
6	7	-	8.4	22		26.66		
7	8.4	-	9.8	16		19.02		
8	9.8	-	11.2	15		13.23		
9	11.2	-	12.6	10		9.028		
10	12.6	-	14	14	15	6.07	10.10	
	14	-	15.4	1		4.034		
11	15.4	-	16.8	2		2.655		
	16.8	-	18.2	0		1.733		
	18.2	-	19.6	0	4	1.124	7.00	
	19.6	-	21	0		0.725		
	21	-	22.4	1		0.465		
	22.4	-	23.8	1		0.297		

Using the specifications given in Table 3.1, the estimates for the gamma distribution's two parameters are $\hat{\alpha} = 2.141$ and $\hat{\lambda} = 0.384$ (notice that these values are different from the parameters that were used in generating the data; namely, $\alpha = 2$ and $\lambda = 1/3$). Table 3.5 shows the values for O_i derived from the data histogram and E_i derived from the gamma distribution candidate. In this case, bins 10–17 are collapsed into 2 class intervals which become intervals 10 and 11. From Eq. (3.1), it follows that $A_m = 10.96$. Now $v = 8$ ($k = 11$, $g = 2$) and with $\alpha = 0.05$, the critical value $\chi^* = 15.51$. Because $A_m < \chi^*$, the gamma distribution candidate is not rejected by the chi-square test.

3.6.4 Empirical Distributions

When it is difficult to fit a theoretical distribution to the collected data, an empirical distribution can usually be formulated to serve as a data model. The procedure requires a cumulative distribution function (CDF) and this can be easily developed from the histogram of the collected data.[10] Observe, for example, that in Fig. 3.15 the CDF is already present and defined by a series of points. Values between these points can be obtained by interpolation (e.g. linear interpolation).

We use the Java Class *Empirical*[11] as a random variate procedure (RVP). It first creates an empirical distribution and then uses the resulting data model to generate samples. The main steps involved are as follows: an array of histogram frequencies is provided to an *Empirical* object. When instantiated, the object creates an internal representation of the CDF from these frequencies, and then the *Empirical* object generates random variates from the distribution using the CDF and the inverse transform method to be discussed in Sect. 3.7.2.

Figure 3.16 provides a Java method that instantiates an *Empirical* object referenced by *empDM*, extracts the CDF representation from *empDM* and then uses *empDM* to generate ten random numbers. The *empDM* object is instantiated with the Class constructor that has the following three arguments:

- The array *histogram* contains the frequency values from the data histogram of Fig. 3.15 with *histogram*[$i - 1$] equal to the frequency value of bin i with $i = 1, 2, ..., L$ (where $L = 17$).
- The argument *Empirical.LINEARINTERPOLATION* indicates that the distribution is continuous and linear interpolation is to be used to formulate the CDF. Otherwise, a discrete CDF is created and used.
- The third argument (*new MerseeneTwister*()) instantiates a uniform random number generator object.

---

[10]It is also possible to derive the CDF directly from the collected data.
[11]The *Empirical* Class is provided as part of the cern.colt Java package provided by CERN (European Organization for Nuclear Research). See Chap. 5 for some details on using this package.

```
class Emp
{
  public static void main(String[] args)
  {
    double randomValue;
    double[] histogram = {
              22, 57, 54, 47, 38, 22, 16,
              15, 10, 14, 1, 2, 0, 0, 0, 1, 1
    };
    double scaleFactor = 1.4; // Width of the histogram bin
    double xMax = 23.8;    // maximum data value

    // Create Empirical Object
    Empirical empDM = new Empirical(histogram,
                         Empirical.LINEAR_INTERPOLATION,
                         new MersenneTwister());

    // Get defining points on the CDF from empDM
    for(int i = 0 ; i <= pdf.length ; i++)
    {
      System.out.println(i + ", " + (i*scaleFactor) + ", " + empDM.cdf(i));
    }

    // Get empDM to generate 10 random numbers
    for(int i = 0 ; i < 10 ; i++)
    {
      randomValue = xMax*empDM.nextDouble();
      System.out.println(randomValue);
    }
  }
}
```

Fig. 3.16 Implementing an RVP using an empirical distribution

The object first creates a CDF, $F(y)$, by defining L points $(y_i, F(y_i))$, where

$$y_i = i \quad \text{with} \quad i = 1, 2, \dots \quad L = 17$$

$$F(y_i) = F(y_{i-1}) + \frac{\text{histogram}[i-1]}{K}$$

with

$y_0 = 0$, $F(y_0) = 0$ and K is the number of points (namely, $\sum_{k=1}^{L} \text{histogram}[k-1]$).

The method $cdf(i)$ returns the value $F(y_i)$. Table 3.6 shows the values returned by $cdf(i)$ for each i in the range 0–17. The returned values can be mapped onto the domain of the underlying CDF by multiplying each $y_i = i$ by the bin width which is 1.4. The value of $F(y)$ when y is not an integer in the 0–17 range is obtained by linear interpolation.

Table 3.6 The CDF used by the empirical object empDM

i	y	F(y)
0	0	0
1	1.4	0.073333
2	2.8	0.263333
3	4.2	0.443333
4	5.6	0.6
5	7	0.726667
6	8.4	0.8
7	9.8	0.853333
8	11.2	0.903333
9	12.6	0.936667
10	14	0.983333
11	15.4	0.986667
12	16.8	0.993333
13	18.2	0.993333
14	19.6	0.993333
15	21	0.993333
16	22.4	0.996667
17	23.8	1

The random numbers generated by an *Empirical* object vary between 0 and 1. Thus, random values returned by *empDM* must be multiplied by xMax = 23.8 to obtain values that fall into the domain of the CDF. The program shown in Fig. 3.16 produces the following values: 6.3445029, 9.3559343, 1.4540021, 8.0323057, 5.3995883, 6.0610477, 3.8339216, 6.385926, 1.9355035, 9.2322093, 1.5383085, 4.5596041, 3.2674369, 7.8097124, 4.0771284, 5.7625121, 4.9623389, 2.9529618, 3.1037785 and 10.877524. These correctly fall in the domain of the CDF; i.e. between 0 and 23.8.

3.6.5 Data Modelling with No Data

When data cannot be collected or is not available (e.g. the SUI does not exist), then 'educated guesses' provide the means of last resort for formulating data models for the autonomous stochastic processes that are required in the simulation model. These guesses can be based on research material and/or on information obtained from individuals who are particularly familiar with the SUI.

When only the minimum and maximum values for a random phenomenon can be realistically assumed then a reasonable distribution candidate is a uniform distribution (see Sect. A2.4.4.1). Because of this 'minimum knowledge' feature, the uniform distribution is sometimes referred to as the 'distribution of maximum ignorance'!

If the minimum, maximum and modal values of a distribution can be specified, then the triangular distribution provides a convenient choice (see Sect. A2.4.4.2). The beta distribution (see Sect. A2.4.4.8) can provide a variety of forms over the unit interval [0, 1] and can be easily shifted to accommodate other intervals.

Hundreds of distributions have been created to model different types of phenomena. The type of phenomena under consideration often suggests a particular group of candidate distributions that are especially relevant. A comprehensive discussion can be found in Banks et al. [1].

3.7 Simulating Random Behaviour

In this section, we provide a brief overview of some of the most common techniques for generating samples from a specified distribution, in other words, techniques for generating random variates. The presentation is not intended to be complete and comprehensive. Its purpose is primarily to provide some insights into the techniques that are widely implemented in simulation software environments and consequently are conveniently accessible. Nevertheless, some appreciation for the nature of the procedures being invoked can provide a basis for ensuring correct usage, understanding potential shortcomings and dealing with unanticipated results.

3.7.1 Random Number Generation

As will become apparent in the following subsection, the common methods for generating random variates depend on the availability of a stream of random values that are uniformly distributed on the unit interval. While a procedure for generating uniformly distributed random values has its own intrinsic importance, this secondary role considerably amplifies this importance. On first glance, the development of such a procedure might seem straightforward but this is far from being so. The subtle complexities that need to be addressed have given rise to a considerable body of research literature. The overview presented by L'Ecuyer [11] and the special journal issue [4] are recommended for readers wishing to explore the topic in more detail within a modelling and simulation context.

From a theoretical point of view, the basic requirement is that any value in the [0, 1] interval be equally likely and that there be no interdependence among the values that are generated (e.g. values to the right of the mean and to the left of the mean should not occur in batches, or values should not tend to have a pattern of successively diminishing or successively increasing). Furthermore, from a practical point of view, there are implicit requirements for:

- Computational efficiency because many thousands of variates may be needed for any particular simulation experiment.

- Reproducibility because it should be possible to replicate any particular random number stream in order to repeat experiments.
- Hardware independence; i.e. the procedure should not be intimately locked into the hardware architecture of any particular computer in order to ensure portability.

The reproducibility requirement might correctly suggest a fundamental contradiction to the reader. The important implication here is that we are obliged to abandon our original quest and be satisfied with the generation of *pseudo* random numbers, which provide reproducibility at the expense of genuine 'randomness'. We note furthermore that any implementation via an algorithmic process will intrinsically provide reproducibility. Thus, the challenge reduces to the search for a 'good' algorithm; i.e. one that yields a random number stream that has satisfactory statistical properties and also provides the efficiency and hardware independence that we seek.

The most widely used technique for generating streams of pseudorandom numbers is an approach called the *linear congruential* method. It is the remarkably simple iterative formula[12]:

$$K_i = (a\,K_{i-1} + c)\mathrm{mod}\,m\ \text{with}\ i = 1, 2, \ldots$$

where a and m are positive integers and c is a non-negative integer. The initial value K_0 (likewise a positive integer) is called the *seed*. It was first proposed by Lehmer [15] with $c = 0$ in which case the method is called the *multiplicative congruential* method. The case where $c \neq 0$ (which is called the *mixed congruential* method) was suggested by Rotenberg [17] and Coveyou [5]. As might be expected, the values chosen for the parameters a, c and m have a significant impact on the statistical quality of the sequence of numbers that are generated.

Several basic features of the formula should be noted:

- The integer values that are generated fall in the $[0,\ m-1]$ interval. They can be shifted into the $[0, 1)$ interval simply by dividing by m. In other words, the values $u_i = K_i/m$ fall in the range $[0, 1)$.
- Suppose that the qth value K_q in the sequence $K_1, K_2, \ldots, K_p \ldots, K_q$ is the first occurrence of equality to a previously generated value K_p; then the subsequence between K_p and K_q will be continually recycle as the sequence continues. Such an occurrence must happen sooner or later because there are at most m distinct values that can be generated. In other words, the process has a maximum *period* of m.
- Because at most m distinct values can be generated, there is an immediate divergence from the properties of a 'genuine' continuous random variable, U, that is uniformly distributed on the $[0, 1]$ interval. For example, suppose $m > 3$; then

---

[12]mod is the modulo operator; p mod q yields the remainder when p is divided by q where both p and q are positive integers.

the probability that U falls between $2.5/m$ and $2.6/m$ is $0.1/m$. However, there is zero probability that the linear congruential method will yield a value in this range.

These apparent shortcomings of the approach can, to a large extent, be overcome by suitable choices for the available parameters: a, c and m. Selection of a very large value for m has obvious advantages. In fact, m is typically chosen to be of the form 2^b, where b is the word length of the computer being used. That choice has the added advantage of simplifying the modulo calculation which can be carried out with a shift or mask operation.

As noted earlier, the longest possible period for the linear congruential method is m. An obvious question then is whether or not there exist parameter selections that will yield this limiting period. The following result due to Hull and Dobell [7] answers the question:

• The linear congruential method has a full period (a period of m) if and only if the following conditions hold (throughout, divides means **exactly** divides, i.e. zero remainder):

 – The only positive integer that divides both m and c is 1.
 – If q is a prime number (divisible only by itself and 1) that divides m, then q divides $(a - 1)$.
 – If 4 divides m, then 4 divides $(a - 1)$ (i.e. $a = 4k + 1$ for some positive integer k).

Notice that the first of these conditions precludes the existence of a full period multiplicative congruential method. If m is chosen to be a power of 2, e.g. $m = 2^b$ ($b > 0$), then a full period method results if and only if c is an odd integer and $a = 4k + 1$ for some positive integer k.

Although a full period multiplicative congruential generator is not possible (because the first condition listed above fails), the statistical properties of this method have generally proved to be superior to those of the mixed generators. Consequently, it is widely used. It can be shown (see e.g. Knuth [10]) that if m is chosen to be 2^b, then the maximum period is 2^{b-2}, and this period is achieved if K_0 is odd and $a = 8k + h$ where k is a non-negative integer and h is either 3 or 5.

A variety of tests have been formulated for assessing the acceptability of any random number stream, e.g. frequency test, runs test and poker test. These have been designed to detect a behaviour that is inconsistent with the statistical integrity of the stream. Some details of these tests can be found in Banks et al. [1] and Law [13].

3.7.2 Random Variate Generation

Our concern in this section is with generating samples from arbitrary, but specified, distributions. The technique we outline depends on the availability of random samples that are uniformly distributed. In practice, these random samples will originate from an algorithmic procedure which provides samples whose statistical properties only approximate those of a uniform distribution; e.g., the techniques outlined in the previous section. Consequently, we are obliged to acknowledge that the results generated will likewise fall short of being ideal.

One of the most common techniques for generating variates from a specified distribution is the *inverse transform method*. Its application depends on the availability of the CDF (cumulative distribution function), $F(x)$, of the distribution of interest. The method is equally applicable for both the case where the distribution is continuous or discrete. We consider first the continuous case.

Application of the method in the continuous case requires the assumption that $F(x)$ is strictly increasing; i.e. $F(x_1) < F(x_2)$ if and only if $x_1 < x_2$. A representative case is shown in Fig. 3.17. Because of this assumption, it follows that $P[X \leq x] = P[F(X) \leq F(x)]$. The procedure is illustrated in Fig. 3.17 and is as follows (F^{-1} denotes the inverse of the CDF):

- Generate a sample, u, from the uniform distribution on $[0, 1]$ (recall that the CDF for this distribution is $F_U(u) = P[U \leq u] = u$).
- Take $y = F^{-1}(u)$ to be the generated value (note that $F^{-1}(u)$ is defined because u falls in the range 0 and 1, which corresponds to the range of F).

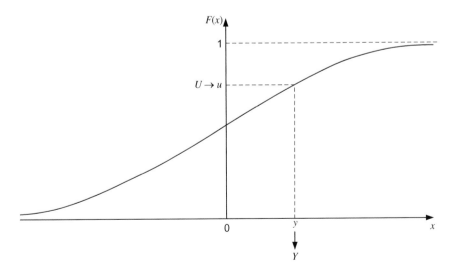

Fig. 3.17 Illustration of the inverse transform method for the case of a continuous distribution

The procedure, in effect, creates a random variable $Y = F^{-1}(U)$. To confirm that the procedure is doing what we hope it is doing, we need to demonstrate that $P[Y \leq y] = F(y)$, i.e. that the CDF for Y is the one of interest. This demonstration is straightforward:

$$
\begin{aligned}
P[Y \leq y] &= P[F^{-1}(U) \leq y] \\
&= P[F(F^{-1}(U)) \leq F(y)] \text{ (because } F \text{ is strictly increasing)} \\
&= P[U \leq F(y)] \\
&= F(y)
\end{aligned}
$$

When the CDF of interest can be explicitly 'inverted', the procedure becomes especially straightforward. Consider the case of the exponential distribution with mean $1/\lambda$; the CDF is $F(x) = 1 - e^{-\lambda x}$. We begin by setting $u = F(x) = 1 - e^{-\lambda x}$ and then obtaining an expression for x in terms of u. This can be readily achieved by taking the natural logarithm which yields

$$
x = -\ln(u - 1)/\lambda \tag{3.2}
$$

The implication here is that if we have a sequence of u's that are uniformly distributed random values on the interval $[0, 1]$, then the corresponding values, x, given by Eq. (3.2) will be a sequence of samples from the exponential distribution having mean $1/\lambda$.

The inverse transform method is equally applicable when the requirement is for random variates from a discrete distribution. Suppose that X is such a random variable whose range of values is $x_1, x_2, \ldots, x_n$. Recall that in this case the CDF, $F(x)$ is

$$
F(x) = P[X \leq x] = \sum_{x_i \leq x} p(x_i)
$$

where $p(x_i) = P[X = x_i]$. The procedure (which is illustrated in Fig. 3.18) is as follows:

- Generate a sample, u, from the uniform distribution on $[0, 1]$ (i.e. from the distribution whose CDF is $F_U(u) = P[U \leq u] = u$).
- Determine the smallest integer K such that $u \leq F(x_K)$ and take x_K to be the generated value.

Repetition of the procedure generates a stream of values for a random variable that we can represent by $\tilde{X}$. To verify the correctness of the procedure, we need to demonstrate that the values x_i that are generated satisfy the condition $P[\tilde{X} = x_i] = p(x_i)$, for $i = 1, 2, \ldots, n$. Observe first that $\tilde{X} = x_1$ if and only if $U \leq F(x_1)$.

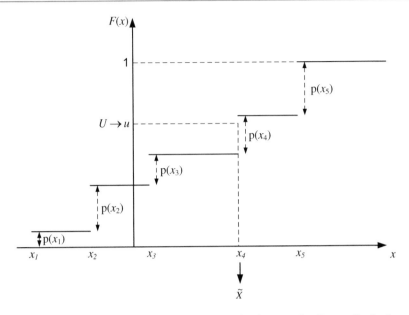

Fig. 3.18 Illustration of the inverse transform method for the case of a discrete distribution

Thus

$$P\big[\tilde{X} = x_1\big] = P[U \leq F(x_1)] = F(x_1) = p(x_1)$$

Assume now that $i > 1$. The procedure ensures that

$$\tilde{X} = x_i \text{ if and only if } F(x_{i-1}) < U \leq F(X_i)$$

Consequently

$$\begin{aligned}
P[\bar{X} = x_i] &= P[F(x_{i-1}) < U \leq F(x_i)] \\
&= F_U(F(x_i)) - F_U(F(x_{i-1})) \\
&= F(x_i) - F(x_{i-1}) \\
&= p(x_i)
\end{aligned}$$

which completes the demonstration. Note that the above relies on the fact that for any discrete CDF, $F_Y(y)$, it is true that $P[a < Y \leq b] = F_Y(b) - F_Y(a)$.

Because the CDF for the normal distribution cannot be written in closed form, the inverse transform method is not conveniently applicable for generating samples from that distribution. Fortunately, there are a number of alternative general methods available which can be used. In fact, there is one technique that is specifically tailored to the normal distribution. This is the *polar method* and a description can be found in Ross [16].

An alternate general technique is the *rejection–acceptance method*. Its implementation depends on the availability of the probability density function of the distribution of interest (hence, it can be used for generating samples from the normal distribution). It is equally applicable for both discrete and continuous distributions. In its simplest form, the method involves generating samples from a uniform distribution and discarding some samples in a manner which ensures that the ones that are retained have the desired distribution. The method shares some features with the Monte Carlo method for evaluating integrals (see Sect. 1.6). A more comprehensive treatment of the method can be found in Ross [16], where the presentation includes a more general approach for the underlying procedure than we give in the following discussion.

In the simplest form described here, the implementation depends on the assumption that the probability density function, $f(x)$, (or probability mass function) of interest is bounded on the left (by a) and on the right (by b). Consequently, if a long 'tail' exits, it needs to be truncated to create the values a and b. We assume also that the maximum value of $f(x)$ on the interval $[a, b]$ is c.

The procedure (which is illustrated in Fig. 3.19) is as follows:

- Generate two samples, u_1 and u_2, from the uniform distribution on $[0, 1]$ (i.e. from the distribution whose CDF is $F_U(u) = P[U \leq u] = u$).
- Compute $\tilde{x} = a + u_1(b - a)$.
- Compute $\tilde{y} = c u_2$.
- If $\tilde{y} \leq f(\tilde{x})$, accept $\tilde{x}$ as a valid sample; otherwise repeat the process with the generation of two new samples from U.

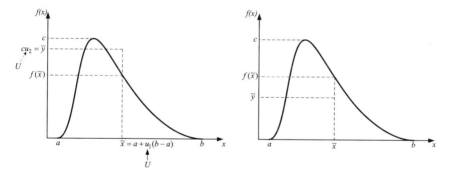

Fig. 3.19 Illustration of the rejection–acceptance method

While by no means a formal proof, the following observations provide some intuitive confirmation of the procedure's correctness. Note that the tuple $(\tilde{x}, \tilde{y})$ is a point in the abc rectangle (see Fig. 3.19). Because the values u_1 and u_2 are independent samples from the uniform distribution, each point in the rectangle is equally likely to occur. Suppose x_1 and x_2 are two distinct points in the $[a, b]$ interval with $f(x_1) > f(x_2)$ (see Fig. 3.20). Over a large number of repetitions of the procedure, $\tilde{x}$ will coincide with x_1 and x_2 the same number of times. However, the occurrence of x_1 is far more likely to be 'outputted' by the procedure than x_2. More specifically:

• Given that x_1 has occurred, the probability of it being outputted is

$$P[cU \le f(x_1)] = P[U \le f(x_1)/c] = f(x_1)/c$$

• Given that x_2 has occurred, the probability of it being outputted is

$$P[cU \le f(x_2)] = P[U \le f(x_2)/c] = f(x_2)/c$$

While the occurrence of x_1 and x_2 are equally likely, the relative proportion of x_1 outputs to x_2 outputs is proportional to $f(x_1)/f(x_2)$, which is consistent with an intuitive perspective of the distribution of interest.

Fig. 3.20 Illustration of relative output frequency

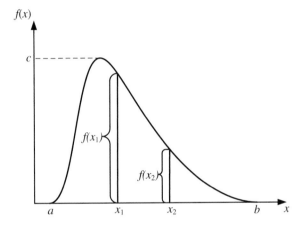

One shortcoming of the method is the uncertainty that any particular execution of the procedure will be successful (i.e. yield an acceptable sample). The probability of success is equal to the relative portion of the abc rectangle that is filled by the density function of interest. Since the area of the rectangle is $c(b-a)$ and the area of the density function is 1, the probability of success is given by the ratio $1/[c(b-a)]$. Two representative cases are shown in Fig. 3.21; the occurrence of rejections will, on average, be more frequent in the case of $f_2(x)$ than in the case of $f_1(x)$.

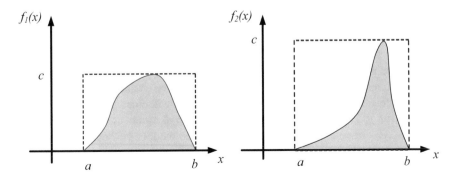

Fig. 3.21 Illustration of relationship between rejection probability and shape

References

1. Banks J, Carson JS II, Nelson BL, Nicol DM (2005) Discrete-event system simulation, 4th edn. Pearson Prentice Hall, Upper Saddle River
2. Biller B, Nelson BL (2002) Answers to the top ten input modeling questions. In: Proceedings of the 2002 winter simulation conference, published by ACM. San Diego, California
3. Choi SC, Wette R (1969) Maximum likelihood estimation of the parameters of the gamma distribution and their bias. Technometrics 11:683–690
4. Couture R, L'Ecuyer P (eds) (1998) Special issue on random variate generation. ACM Trans Model Simul 8(1)
5. Coveyou RR (1960) Serial correlation in the generation of pseudo-random numbers. J ACM 7:72–74
6. Harrell C, Ghosh BK, Bowden RO Jr (2004) Simulation using ProModel, 2nd edn. McGraw-Hill, New York
7. Hull TE, Dobell AR (1962) Random number generators. SIAM Rev 4:230–254
8. Kelton DW, Sadowski RP, Sturrock DT (2004) Simulation with Arena, 3rd edn. McGraw-Hill, New York
9. Kleijnen Jack PC (1987) Statistical tools for simulation practitioners. Marcel Dekker Inc, New York
10. Knuth DE (1998) The art of computer programming, vol 2: Seminumerical algorithms, 3rd edn. Addison-Wesley, Reading
11. L'Ecuyer P (1990) Random numbers for simulation. Commun ACM 33:85–97
12. Law Averill M, McComas MG (1997) Expertfit: total support for simulation input modeling. In: Proceedings of the 1997 winter simulation conference, published by ACM. Atlanta, GA, pp 668–673
13. Law AM (2015) Simulation modeling and analysis, 5th edn. McGraw-Hill, New York
14. Leemis LM, Park SK (2006) Discrete event simulation: a first course. Pearson Prentice Hall, Upper Saddle River
15. Lehmer DH (1949, 1951) Mathematical methods in large scale computing units. In: Proceedings of the second symposium on large-scale digital calculating machinery, 1949 and Ann Comput Lab (26):141–146 (1951). Harvard University Press, Cambridge, MA
16. Ross SM (1990) A course in simulation. Macmillan Publishing, New York
17. Rotenberg A (1960) A new pseudo-random number generator. J ACM 7:75–77

A Conceptual Modelling Framework for DEDS

4

4.1 Introduction

Conceptual modelling is concerned with developing a meaningful representation of the SUI. In spite of this rather fundamental purpose, neither the conceptual modelling process nor the artefact which it produces (namely, a conceptual model) has a universally accepted interpretation within the modelling and simulation literature. There are a variety of reasons for this. Perhaps, the most significant reason is the challenge presented by the diversity and the unbounded complexity that characterizes the DEDS domain. As well, there is the inherent ambiguity that is associated with the two keys words: 'meaningful' and 'representation'.

Nevertheless, the topic is correctly recognized as one of the significant importance, and consequently a wide range of perspectives do exist. A recent exploration of the conceptual modelling landscape can be found in [11]. In an overview article for that volume, Robinson [12] provides the following definition:

> The conceptual model is a non–software-specific description of the computer simulation model (that will be, is or has been developed), describing the objectives, inputs, outputs, content, assumptions, and simplifications of the model.

It is of some interest to observe that this definition includes a reference to 'objectives' which is a theme that has been stressed in our discussions in previous chapters. Note also that, as presented above, the conceptual model relates to the simulation model and not to the system/situation under investigation (SUI). As will become clear in the discussion to follow, our perspective is not entirely consistent with this view.

Closely related to the intent of a conceptual model is the format used to develop it. Again there are many options, and to a large extent their application is linked to the assumed purpose of the conceptual modelling task. Some formats are entirely informal, e.g. natural language or ad hoc graphical renditions, while others emerge

© Springer Nature Switzerland AG 2019
L. G. Birta and G. Arbez, *Modelling and Simulation*, Simulation Foundations,
Methods and Applications, https://doi.org/10.1007/978-3-030-18869-6_4

from existing formalisms, e.g. finite state machines or Petri nets [8, 16, 18]. In the case of natural language, inherent ambiguities usually undermine effectiveness. A difficulty with graphical renditions is the management of their complexity when the level of complexity within the SUI increases. In the case of finite state machines and Petri nets, a key issue that emerges relates to 'modelling power' which is often insufficient to accommodate the complexities of the SUI. In the case of Petri nets, various enhancements have been proposed to deal with this, e.g. targeted extensions [1], stochastic Petri nets [3], timed Petri nets [17] and coloured Petri nets [5].

The DEVS approach formulated by Zeigler [19, 20] provides a well-developed formal approach specifically oriented to the characterization of the behaviour of discrete-event dynamic systems. The approach is founded on concepts borrowed from systems theory and consequently adopts a perspective that is based on the notions of state, input and output and the transition functions that generate state changes and output values. The approach provides a foundation that allows for theoretical examination of properties which is most certainly a noteworthy distinction.

The perspective we adopt is that a conceptual model is a carefully constructed artefact that consolidates, in a consistent and coherent manner, those features of the SUI that are deemed to have relevance to the achievement of a project goal. As noted in Chap. 1, the project goal can generally be regarded as the evaluation of a set of possible solutions to a stated problem. Notice that a corollary to this assertion is the fact that there may be facets of the SUI that are purposely ignored because they are judged to be devoid of relevance (essentially the notion of simplification as discussed in Sect. 2.1).

Apart from the fundamental requirement to capture the essential behavioural features of the SUI, there are two important qualities that should be addressed in the construction of an appropriate conceptual model within the DEDS domain, namely,

(a) The conceptual model must be sufficiently transparent so that all stakeholders in the project can use it as a means for discussing those mechanisms within the SUI that have relevance to the characterization of its behaviour (as interpreted from the perspective of project goal).
(b) The conceptual model must be sufficiently comprehensive so that it can serve as a specification for the computer program that will provide the means for carrying out the simulation study.

With respect to (b), it should be noted that the incorporation of basic software development constructs (e.g. looping and decision-making) is of considerable value here. Note also that (b) necessarily implies that explicit consideration of output requirements for the study must be included.

Both of these requirements place constraints on the format which is used to formulate the conceptual model. Indeed these constraints are, to some extent, conflicting. The second constraint clearly implies a high level of clarity and precision. But such a representation may be regarded as tedious by a sizable community of the stakeholders in the project. Thus, care is needed in developing a framework that balances, but nevertheless accommodates, these requirements.

In this chapter, we present such a framework which we call the *ABCmod framework* (*A*ctivity-*B*ased *C*onceptual *mod*elling), and we refer to the conceptual model that emerges as an *ABCmod conceptual model*. The framework contributes to our basic intent in this book of providing a foundation for the successful completion of any modelling and simulation project in the DEDS domain.

The ABCmod framework is informal in nature and is driven by a desire for conceptual clarity. But at the same time, it has a high level of precision, generality and adaptability. To a significant extent, these features flow from the incorporation of software development concepts (looping, decision-making, data structures, etc.). The framework, as presented in the discussions which follow, can accommodate a wide range of project descriptions. However, it is not intended to be rigidly defined; it can be easily extended on an ad hoc basis when specialized needs arise.

The development of a conceptual model is not necessarily a unidirectional procedure that simply moves from one stage to another. Insights are often acquired which can reflect back to earlier phases of the project and indeed even to the statement of project description and project goal (see Fig. 2.4). This potential to refine the underlying purpose of the modelling and simulation study is an important feature that is inherent in the conceptual model-building process.

4.1.1   Exploring Structural and Behavioural Requirements

By way of setting the stage for the presentation of the ABCmod framework, we outline here some pertinent features of a particular discrete-event dynamic system, namely, the operation of a department store. With some elaboration, the outline could evolve into a modelling and simulation project but that is not of concern in this discussion.[1] The intent is simply to illustrate a variety of modelling requirements that are typical of the DEDS domain with a view towards identifying some desirable contents of a toolbox for conceptual model construction.

Customers arrive in the department store and generally intend to make purchases at various merchandise areas (departments) of the store. At each such area, a customer browses/shops and may or may not make one or more selections to purchase. If selections are made, they are paid for at one of several service desks located within the area. Following each visit to a merchandise area, customers either move to another department or, when their shopping task is completed, they leave the store.

In this fragment of the SUI's description, each customer corresponds to an entity which we might reasonably regard as a type of 'consumer' entity (note that such a

---

[1]A possible goal might be to determine how best to allocate a new part-time employee so that operational efficiency is maximized.

generalization is, in fact, an abstraction step). In a similar way, we could regard each of the service desks within the various merchandise areas as a resource entity (another abstraction) inasmuch as the consumer entities (the customers) need to access these resource entities in order to complete their purchase transactions (these transactions are, after all, the purpose of their visit to the department store). Because the service function at the resource entity (i.e. service desk) has a finite duration, there is a possibility that some consumer entities may not receive immediate attention upon arrival at the resource because it is 'busy' serving other customers. Hence, it is reasonable to associate a queue entity with each resource entity where consumer entities can wait for their turn to access the resource.

The merchandise areas where customers evaluate merchandise (with a view towards possibly making a purchase) are distinctive. On one hand, each can be regarded simply as a 'holding' area for a collection of consumer entities (i.e. the customers). With this perspective, a merchandise area can be represented as a particular type of aggregate which we call a group (an unordered collection of entities). On the other hand, a merchandise area is a prerequisite for an activity called 'shopping'. Hence, it has features of a resource. In effect, then, the merchandise areas within the department store have a dual role. This notion of duality is explored further in the discussions that follow.

It's of some interest to observe that the discussion above has identified two types of aggregate, namely, queues and groups. Entities within a group are not organized in a disciplined way as in the case of a queue, but rather they simply form an identifiable collection.

In effect, we have transformed various elements of the SUI (e.g. customers, merchandise areas, service desks) into generic elements that have broader, more general, applicability (consumer entities, queue entities, group entities, resource entities). This is an important abstraction step and lies at the core of our conceptual modelling process. These various generic entities that we have introduced are among the model-building artefacts that are explored in greater detail in the presentation of the ABCmod conceptual modelling framework that follows in the remainder of this chapter.

It needs to be appreciated, however, that the mapping process from elements in the SUI to generic elements is not always as straightforward as the preceding discussion might suggest. Consider, for example, a number of machines within a manufacturing environment that are subject to failure. A team of maintenance personnel is available to carry out repairs. While the machines are operating, they can certainly be viewed as resource entities in the manufacturing operation, but when they fail, they become consumer entities because they need the service function of the maintenance team. In other words, the machines can shift from one role to another. Such circumstances are not uncommon, and a means to deal with them in our modelling toolbox is essential.

The discussion above illustrates some of the *structural elements* that are needed for the development of a conceptual model for the department store. We note also that it is usually helpful to formulate a graphical representation of these structural features (a schematic view). Such a representation may also incorporate some aspects of behaviour, thereby providing preliminary insight into how the structural elements interact. Figure 4.1 shows such a representation for the case where we restrict our considerations to three separate merchandise areas in the department store.

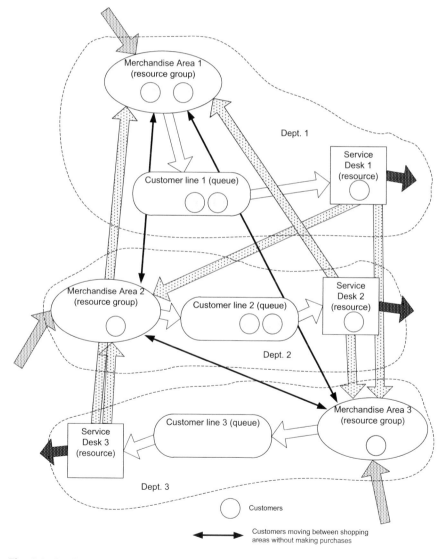

Fig. 4.1 A schematic view of the department store

In Fig. 4.1, the arrows indicate movement of the consumer entities. The dark shaded arrows indicate departure from the department store. The partially shaded arrows indicate movement from one merchandise area to another following a purchase, and the non-shaded arrows show movement within a particular department.

The discussion above has focused on some structural elements that provide a foundation for the modelling process. We explore this example further but now from the perspective of behaviour. A useful way to begin is to examine how various shoppers in the department store might interact with each other and the resources in the department store.

Figure 4.2 shows a possible interaction scenario for three shoppers all of whom enter shopping area 1 upon their arrival. The three shoppers (called A, B and C) arrive in the store at times A_0, B_0 and C_0, respectively, and leave the store at times A_5, B_7 and C_3, respectively.

There are a number of important observations that can be made about Fig. 4.2. Notice, in particular, that some type of transition occurs at each of the time points A_0 through A_5, B_0 through B_7 and C_0 through C_3. These transitions, in fact, represent changes that must be captured in the model-building process. Notice also that some of these time points are coincident, for example, $A_2 = B_2$, $A_3 = C_2$ and $A_5 = B_4$, suggesting that several different changes can occur at the same moment in time. It is also clear from Fig. 4.2 that there are intervals of time during which at least some of these three shoppers are engaged in the same activity, for example, between C_0 and B_1 all three customers are browsing in Area 1, and between C_1 and A_2 customers A and C are waiting in Queue 1.

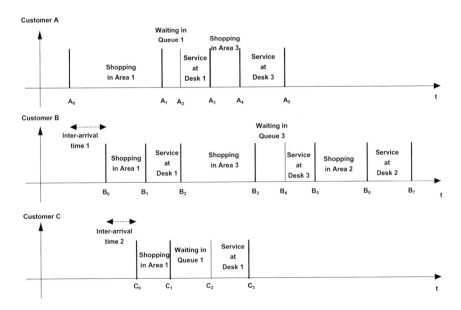

Fig. 4.2 Behaviour of three department store shoppers

We have previously noted that each of the service desks can be regarded as a resource and shoppers need to acquire ('seize') this resource in order to pay for items being purchased before moving on to another merchandise area. The payment activity at a service desk has several noteworthy features:

1. There is a precondition that must be TRUE before this service activity can begin (the server must be available and there must be a shopper seeking to carry out a payment transaction).
2. The service activity carries out a purposeful task and has duration, i.e. it extends over an interval of time.
3. Changes take place when the service function is completed (e.g. at time $A_3 = C_2$, the number of shoppers in merchandise Area 3 increases by one, and the number in the queue in front of service desk 1 decreases by one).

These features are directly reflected in one of the main ABCmod conceptual modelling artefacts used to characterize behaviour. This will become apparent in the discussion of Sect. 4.2.2.2.

From the perspective of the requirements of a modelling and simulation project, the description given above for the department store shoppers is incomplete in several respects. Many details need to be provided, for example, how is the set of merchandise areas that a particular customer visits selected? What is the order of the visitations? And how many servers are assigned to the service desks? Can a particular customer not make any purchase at some of the assigned merchandise areas and if so, then under what circumstances? The information for dealing with these questions is not provided in the descriptive fragment that is given here but would most certainly be necessary before a meaningful conceptual model could be formulated. Indeed, one of the important functions of the conceptual modelling process is to reveal the absence of such essential details.

Likewise, several data models need to be determined. Included here would be the characterization of customer arrival rates and service times at the service desks, allocation of the shopping areas to be visited by the arriving customers and the characterization of the duration of the browsing phase within each merchandise area. It is especially important to observe that these various data models will provide the basis for generating events that give rise to change. For example, the event associated with the end of a particular customer's browsing phase will generally (but not necessarily) result in that customer's relocation into the queue associated with the service desk of that service area.

The intent of the discussion in this section has been to explore some of the important facets of the modelling process within the DEDS domain, at least insofar as they are reflected in the department store example that was considered. This overview will, nevertheless, provide a foundation for the discussion in the remainder of this chapter.

4.1.2 Overview of the Constituents of the ABCmod Framework

The fundamental outcome of conceptual modelling is a conceptual model which provides a representation (i.e. abstraction) of the SUI's behaviour that is consistent with the requirements of the project goal. The SUI is generally a complex object, and an important asset in coping with complexity in any context is a collection of clearly defined concepts that map onto the various facets that are of interest. Our concern is with the modelling of behaviour; consequently, the need is for a framework that provides a consistent and coherent way of viewing and describing the mechanisms that give rise to behaviour.

We undertake that task by adopting the perspective that behaviour is the consequence of interaction among some collection of objects that populate the space of the SUI. A prerequisite for dealing with this interaction is a means for characterizing these objects. There are, therefore, two collections of modelling artefacts required for the conceptual modelling process. The first deals with the abstraction of the objects that are interacting within the SUI (in effect, the structure of the SUI), and the second focuses on the nature of these interactions (the behaviour of the SUI). These dual requirements are handled in the ABCmod conceptual modelling framework with two basic model-building artefacts called entity categories and behavioural artefacts.

The various entity categories within an ABCmod conceptual model are tailored to the specific nature of the SUI and the goal of the project. The entities within these categories are surrogates for the objects of interest within the SUI. The behavioural artefacts that provide the basis for modelling the SUI's behaviour fall into two categories called activities and actions. These behavioural artefacts can be viewed as a consolidation of SUI specific rules which govern the manipulation of entities over the course of the observation interval.

An activity is a fundamental behavioural artifact in the ABCmod framework which serves to encapsulate a specific 'unit' of behaviour that is judged to have relevance to the model-building task. In general, an activity characterizes a purposeful engagement among one or more entities which results in the completion of a specific task with a finite duration. The identification of these units of behaviour is the task of the conceptual model builder. An activity is instantiated as the consequence of an event that is encapsulated within its specification and this, in particular, implies that multiple instances of any activity can coexist.

In summary, there are three noteworthy features of an ABCmod activity:

- It represents an indivisible unit of behaviour that occurs within the SUI.
- It is associated with some purposeful task.
- It evolves over a non-zero (finite) interval of time.

Earlier, it was pointed out that any approach to formulating a DEDS model must necessarily be based on a coherent perspective of the constituents of a discrete-event dynamic system. A variety of such perspectives for simulation models have, in fact, emerged and these are referred to as 'world views'.[2] One of the 'views' that is typically mentioned is called *activity scanning* (which has evolved into the Three-Phase world view). The focus of the ABCmod conceptual modelling framework on activities could give the erroneous impression that this framework is an outgrowth of the Activity Scanning or the Three-Phase world view (in fact it is closer to the latter). While there are some similarities, there are nevertheless fundamental differences.

Perhaps, the most significant of these differences relates to the underlying intent. The ABCmod framework is concerned with conceptual modelling, while the Activity Scanning world view is a computer programming paradigm.[3] A key facet of this difference relates to the management of (simulated) time. A computer programming environment must carefully accommodate the challenges of time management, while in the ABCmod framework, time management issues are not addressed because they have no conceptual relevance. This, in particular, implies that there are no mechanisms to coordinate the attribute changes that are specified in the behavioural artefacts of any particular ABCmod conceptual model. Nevertheless, the specification of dynamic behaviour embedded in the ABCmod conceptual model can be readily integrated with a generic time management procedure to create a simulation program (executable program code). This topic is explored in detail in Chaps. 5 and 6.

4.2 Essential Features of an ABCmod Conceptual Model

4.2.1 Model Structure

4.2.1.1 Entities

An entity is a named *m*-tuple of attributes where we regard an attribute as a variable, i.e. a name (identifier) which has a value. The value of an attribute of an entity generally varies with time in accordance with the underlying rules of behaviour of the model. Nevertheless, it often does occur that an attribute's value remains unchanged after its initial assignment.

Each ABCmod entity belongs to a distinct named category. Sometimes there is more than one entity within a category and in such a circumstance each has the same set of attributes. Consider, for example, the collection of customers in the department store example. Each customer would be represented by an entity and each such entity would have the same collection of attributes. The model would capture this collection of customers within an entity category that could reasonably be named 'Customer'.

---

[2]These world views are examined in Chap. 5.
[3]A comprehensive presentation of the activity scanning from a programming perspective can be found in Kreutzer [6]. Examples of the utilization of this paradigm can be found in Martinez [7], Shi [15] and Gershwin [2].

It follows then that an important initial step in the development of an ABCmod conceptual model for any particular modelling and simulation project is the identification of an appropriate collection of entity categories, i.e. ones that accommodates the modelling requirements of the project. This collection, in effect, defines the structure of the conceptual model being constructed.

Each entity category has two properties which are called *role* and *scope*. The notion of *role* is intended simply to provide a suggestive (i.e. intuitive) link between the features of the SUI and the conceptual model-building environment provided by the ABCmod framework. The value assigned to *role* reflects the model builder's view of the entity category in question, and each entity in that category has that *role*. There are four basic alternatives (i.e. values for *role*) that align with a wide variety of circumstances, namely,

- Resource: When the entities (or entity) within a category provide(s) a service.
- Consumer: When the entities (or entity) within a category seek(s) one or more services (typically provided by a resource entity).
- Queue: When the entities (or entity) within a category serve(s) as the means for maintaining an ordered collection of other entities of the same entity category. The number of such entities that are accommodated at any point in time normally varies and often there is a maximum capacity; both these values are typically maintained in designated attributes of the entities (or entity) of the queue category in question.
- Group: When the entities (or entity) within a category serve(s) as the means for maintaining an unordered collection of other entities of the same category. As in the case of a queue entity, the number of such entities that are accommodated at any point in time normally varies, and often there is a maximum capacity; both these values are typically maintained in designated attributes of the entities (or entity) of the group category in question.

By way of an illustration, a frequently appropriate perspective is one where consumer entities flow from one resource entity to another resource entity accessing the services that are provided by them. At any particular point in time, however, access to a particular resource entity may not be possible because it is already engaged (busy) or is otherwise not available (e.g. out of service because of a temporary failure). Such circumstances can be accommodated by connecting the entity to an aggregate entity (queue or group) that is associated with the resource entity where they can wait until access to the resource entity becomes possible.

There is no reason to believe, however, the four alternatives listed above for *role* will necessarily encompass all possible circumstances. Note furthermore that it is often the case that an entity category's *role* may exhibit duality.

Consider, for example, a category intended to represent a set of machines which periodically break down and require repair (e.g. in a manufacturing process). These machines, because they provide a service, would reasonably be regarded as a collection of resources in the manufacturing process. However, they are subject to failure, and when they fail, they are likely placed on a repair list. In effect, they now become 'consumers' of the service function provided by a repair crew (another resource entity) which moves around the plant carrying out repairs according to the repair list. In an ABCmod conceptual model of this situation, the repair resource entity may simply cycle through a list of consumer entities that are in need of repair and provide the required function if feasible (a queue entity would be used to represent the repair list). Such situations where the value of *role* can vary over the course of the observation interval are not uncommon, and we view the *role* in such a case as having 'mutually exclusive duality'. In this example, the value of *role* assigned to the machine entities would be Resource Consumer.

It is interesting to observe in the above example that the queue entity in which the machines await service is essentially virtual. Furthermore, it is the service-providing resource entity (the repair crew) that is 'in motion', while the service-requesting consumer entities (the failed machines) are stationary.

There is likewise the possibility of *role* having 'concurrent duality'. This would occur, for example, in the case of a service counter at a fast-food outlet. The counter can be regarded as a group because it maintains a collection (unordered) of customers being served. But, at the same time, it can be regarded as a resource because customers must acquire a position at the counter (i.e. become a member of the group) as a prerequisite for the activity of 'getting served'. In this situation, we would assign the dual value of Resource Group to *role*.

In the discussions that follow, we shall frequently use phrases such as 'the resource entity called *X*' to imply that the *role* of the entity category to which *X* belongs is Resource. In a similar way, the phrase 'consumer entity' is a reference to an entity whose *role* is Consumer.

The *scope* of an entity category reflects upon the number and permanence of the entities within that category. In the case where exactly one entity exists in a category, that entity category is said to have *scope* = Unary. When an entity category contains a finite (and fixed) number, $N > 1$, then *scope* = Many[N]. If the number of entities in the category varies over time, then *scope* = Transient.

In the case where the *scope* of an entity category is either Unary or Many[N], the entities of the category remain within the realm of the model over the entire observation interval. The entities in this case are referred to as 'members' of the entity category. In the case where *scope* = Transient, the entities of the category typically have a transient existence (i.e. they cannot only be 'created' but they can likewise be 'eliminated'). These entities have no explicit identifier but can, nevertheless, be referenced (as discussed in the following section). They are referred to as 'instances' of the entity category. Naming conventions for entities are outlined Sect. 4.2.1.3.

4.2.1.2 Attributes

The identification of appropriate attributes for the entities within an entity category is governed to a large extent by the requirements that emerge in the process of characterizing dynamic behaviour. The collection of behavioural artefacts used in this characterization within the ABCmod framework reacts to and manipulates the values of the attributes of entities. It follows then that the selection of the attributes shared by the entities within an entity category is a fundamental concern. A discussion of some aspects of the selection of appropriate attributes was presented in Chap. 3 (Sect. 3.2.2). We explore this matter further in the discussion below.

The most common aggregate entity is a queue entity (i.e. an entity belonging to a category with *role* = Queue) where other entities are enqueued. From this observation, it is reasonable to suggest two particular attributes for any queue entity within the model, namely, *list* and *n*. Here list would serve to store the entities that are enqueued in a queue entity and *n* would be the number of entries in that list.

It needs to be stressed that the above selection of attributes for characterizing a queue entity is intended simply to be suggestive and is not necessarily adequate for all situations. In some cases, for example, it may be appropriate to include an attribute that permits referencing the specific resource entity associated with the queue entity or an attribute that provides the capacity of the queue entity.

The characterization of an entity category with *role* = Group is similar to that of a queue entity category but there is an important difference, that is, there is no intrinsic ordering discipline. As in the case of an entity category with *role* = Queue, the attributes for a group entity category could reasonably include *list* and *n*. In some situations, it may be useful to include an attribute that specifies the capacity of the group entity. This is very much context dependent and provides a further illustration of the need to tailor the attributes of entity categories to the specific requirements of a project.

4.2.1.3 Entity and Attribute Naming Conventions

Each entity category in an ABCmod conceptual model has a unique name. That name is an integral part of the identifier for individual entities that are members or instances of that category. The identifier for an entity has a format that reflects the properties of the entity category to which it belongs. In particular for the case where the entity category has *scope* = Unary, the unique entity in this category has the identifier *X.Name* where *Name* is the name of the category and *X* is one of R, C, Q, G (or some combination of these alternatives) depending on the value of *role*, e.g. $X = R$ if *role* = Resource, $X = C$ if *role* = Consumer and $X = RG$ if *role* = Resource Group.

When an entity category called *Name* has *scope* = Many[*N*], then *X.Name* designates the entire category. The *k*th member of the category is designated as *X.Name[k]*. We refer to *k* as the member identifier which can be either a numerical constant or a symbolic constant (e.g. names such as MILL, LATH, …, etc.). When *k* is numeric, its range is typically [1, *N*].[4] It is important to note that when an entity belonging to a category with *scope* = Many[*N*] is placed into a queue entity or into a group entity, it is the member identifier that is inserted into the queue or group.

---

[4]Alternate numerical identifiers for the members of a category with *scope* = Many[*N*] may be preferred, e.g. the range for *k* could be [0, *N* − 1] to accommodate common programming conventions for indexing arrays.

When *scope* = Transient, we use i*X.Name* simply as a reference to some particular instance of the entity category which is relevant to the context under consideration. It does not serve as a unique identifier; i*X.Name* will refer to different entity instances (of the same entity category) within different activity instances.

For example, consider the entity category with the name Tugboat that has *role* = Resource and *scope* = Many[2]. The identifiers for the two entities in this category are R.Tugboat[1] and R.Tugboat[2]. Alternately, if this entity category contained a single entity (*scope* = Unary), then the identifier R.Tugboat is used for that single entity. Continuing with this example, we shall frequently use phrases such as 'a Tugboat entity' to imply an entity that belongs to the entity category named Tugboat. Note, however, that this reference does not reveal the *scope* of the underlying entity category.

A means for referencing the attributes of entities is essential. Our convention in this regard is closely related to the convention described above for identifying entities. In particular, the convention endeavours to clearly reveal the name of the entity category to which the entity in question belongs (a specification for each entity category which lists and describes all attributes is included in the detailed ABCmod conceptual model (see Sect. 4.3.3.1)).

Attribute identifiers are constructed by appending the attribute name to the entity identifier or reference. Consider an entity category called *Name*, with *scope* = Transient. By our previously outlined convention, the identifier for an entity instance that belongs to this entity category has the generic form:

$$iX.Name$$

where X is the value of *role* and is one of R, C, Q, G (combinations of the form YZ, where each of Y and Z can be one of R, C, Q, G where $Y \neq Z$ are also possible). Suppose that *Attr* is an attribute of the entity category called *Name*. Then we use

$$iX.Name.Attr$$

to refer to that attribute of the entity referenced by i*X.Name*.

Alternately suppose we consider an entity category with *scope* = Many[N]. The generic identifier for the kth member of this entity category is

$$X.Name[k]$$

Again, if *Attr* is an attribute of the entities in this category, then we use

$$X.Name[k].Attr$$

to identify to that particular attribute of the kth member.

4.2.1.4 State Variables

References to the state of a model are an important and integral part of the discussions surrounding the model development process. Inasmuch as the model's
state at time t is simply the value of its state variables at time t, a prerequisite for
such discussions is a clear understanding of what constitutes a collection of state
variables for the model. As previously noted, however (Sect. 2.2.5.3), the identification of such a collection is not straightforward within the DEDS domain.

Nevertheless, it is helpful to observe that some subset of the attributes of the
entities, existing at any particular moment in time t, e.g. $A^*(t)$, can often be regarded
as the model's state variables. This follows from the observation that the information embedded in these attributes contributes to the constructive property (see
Sect. 2.2.5.3) that is a key characteristic of the model's state variables. Accordingly,
in the discussions that follow, a reference to the state of the model at time t can
generally be interpreted as the set of values of the attributes within $A^*(t)$.

4.2.2 Model Behaviour

We begin by noting that the characterization of behaviour within an ABCmod
conceptual model is, for the most part, aligned with changing attribute values.
These changes take place in concert with the traversal of the time variable, t, across
the observation interval and are a consequence of the occurrence of specific conditions. The identification of these conditions and the attribute value changes which
they cause are fundamental facets of behaviour characterization. Recall also that the
collection of entities that exist within the conceptual model can vary over the course
of the observation interval as transient entities enter and leave the model.

An important but implicit assumption relating to the variable t is the assumption
that within all sections of any ABCmod conceptual model, the units associated with
t are the same, for example, seconds, days, years and the like.

The characterization of behaviour in the ABCmod framework is carried out
using a collection of *behavioural artefacts*. Two principle types are *activities* and
actions. The notion of an 'event' is central to both these artefacts. We therefore
precede our discussion of these artefacts with an outline of our perspective of the
key notion of an event.

4.2.2.1 Events

The notion of an event is fundamental in any model-building discussion within the
DEDS domain. In spite of this importance, a universally accepted definition for this
notion has proven to be elusive. Our perspective in the ABCmod framework is that
an event is an instantaneous change in the status of the model. In effect, an event
has a time of occurrence (its *event time*) and an impact upon the status of the model.
This impact is outlined in the event's *status change specification (SCS)*. An SCS
generally includes, but is not restricted to, a change in the model's state. We explore
now some specification issues relating to event times of events.

Two types of event can be identified based on a difference in the manner in which event time is prescribed. The first is called a *scheduled event*. The essential feature of a scheduled event is that its occurrence happens at some known point in time. Typically, this time is specified relative to of the event time of another related event. To illustrate, consider a scheduled event B which will happen following an event A. If we let τ_A and τ_B denote the event times of events A and B, respectively, then $\tau_B = \tau_A + \Delta$ where $\Delta > 0$ is known either explicitly (e.g. $\Delta = 10$) or implicitly (e.g. Δ is a random variate variable[5]).

An alternate (and indirect) event time specification mechanism flows from the notion of a *precondition* which is a Boolean condition typically formulated in terms of one or more state and/or input variables. An event which has a precondition is called a *conditional event*. Such an event will occur if and only if its precondition evaluates to TRUE; in other words, only then will the event actually occur, thereby precipitating the change(s) specified in its associated SCS. The event time for a conditional event is the value of time when the event's precondition acquires a TRUE value.

Like any event, the SCS of a scheduled event (with event time τ) changes the state of the model. Such change can yield a TRUE value for the precondition of one or more conditional events, thereby enabling the occurrence of those events, i.e. enabling the changes embedded in their respective SCSs. These changes may, in turn, enable further conditional events; in other words, a cascading effect is clearly possible.[6] Note that the event times for all such these conditional events will be τ.

Scheduled and conditional events are certainly fundamental notions in the DEDS modelling context. There is, however, a variation that needs to be introduced, namely, the notion of a 'tentatively scheduled' event. Earlier we introduced the scheduled event B whose event time τ_B was given as $\tau_B = \tau_A + \Delta$ where τ_A denotes the event time of an event A and $\Delta > 0$. If we allow the *possibility* for an event C with event time $\tau_C = \tau_A + \Delta_1$ with $0 < \Delta_1 < \Delta$ to intercept (stop) the occurrence of event B, then event B becomes a *tentatively scheduled event*. We call the event C an *intercepting event*. Typically, intercepting events are conditional events. The notion of tentatively scheduled events has significance in our development of the activities which follows below.

4.2.2.2 Activities

The *activity* is the main modelling artefact within the ABCmod framework for characterizing change or, equivalently, behaviour. An ABCmod conceptual model typically incorporates a number of activities. Each activity has a name and serves to represent a unit of behaviour that has been identified as having relevance from the perspective of model formulation that is consistent with project goal. The notion of

---

[5]It is of some interest to note that scheduled events provide the basis for the time management of all DEDS simulation models. This is explored in some detail in Chaps. 5, 6 and 7.
[6]The reader should observe that, to be meaningful, a discussion of events within a DEDS modelling context cannot be exclusively concerned with conditional events, i.e. at least one scheduled event needs to be included.

'unit' here is intended to suggest 'minimality'; in other words, an activity should be viewed as being 'atomic' in the sense that it captures an aspect of the model's behaviour that is not amenable to subdivision (at least from the perspective taken by the model builder). An activity can also be regarded as an abstraction of some purposeful task that takes place within the SUI. Its achievement invariably requires at least one entity and usually involves interaction with other entities. The key consequence of both the initiation and the completion of this task generally takes the form of changes in the value of some of the attributes of the entities within the model.

It is important to note that each activity captures a particular type of task. Consequently, many 'instances' of a particular activity (i.e. task) could be simultaneously 'in progress'. For example, in the department store, it would be feasible, at any moment in time, for multiple customers to be in the process of being served at the same desk. In this case, several instances of the activity formulated to represent the payment task would be simultaneously underway.

As noted above, an activity serves to characterize a unit of behaviour that is associated with a purposeful task within the SUI. As might be expected, the purposeful tasks that need to be accommodated do have different requirements, and consequently we recognize four types of activity within the ABCmod framework. These all share a common structural organization that is formulated around two events.

The generic representation for an activity is shown in Fig. 4.3a. It consists of a starting event and a terminating event which are separated by a duration. As previously pointed out, an event has two fundamental attributes, namely, its event time (the time of its occurrence) and its status change specification (SCS) which is a specification of its impact upon the model's status. The feature that distinguishes the four activity types is the manner in which the activity's initiation takes place, or more specifically the manner in which the event time of the starting event (τ_S) is established. As will be apparent in the discussion to follow, this can take place either explicitly or implicitly. For convenience, we introduce the idea of an 'activity initiator' to serve as the generic mechanism that establishes the time of initiation of an activity instance; in effect, the event time of the starting event (τ_S) is shown in Fig. 4.3b.

Figure 4.3b makes explicit the two attributes of each of the events, namely, their respective event times and SCSs. The event time of the starting event of Fig. 4.3a (τ_S) is a consequence of the activity's Initiator. The terminating event is a scheduled event (see the description of a scheduled event in Sect. 4.2.2.1) and its event time (τ_T) is dependent upon the event time of the starting event and the value of a time duration Δ. Specifically, we have

$$\tau_T = \tau_S + \Delta$$

The representation given in Fig. 4.3b provides the basis for our presentation of the four types of activity within the ABCmod framework.

Fig. 4.3 Representation of the events that comprise an activity instance

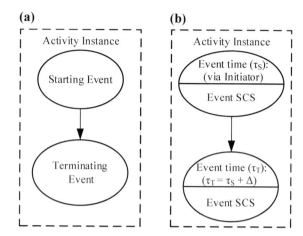

The conditional activity[7]: In this case, the Initiator is a precondition, i.e. a Boolean condition typically formulated in terms of one or more of the entity attributes/variables that exist within the model. It needs to be stressed, furthermore, that explicit reference to simulation time within a precondition must be made with care by always keeping in mind that the precondition is evaluated at discrete points in time. The event time of the starting event of any instance of a conditional activity is the value of time when its precondition evaluates to TRUE. Also, care must be taken to ensure that the SCS of the starting event eventually renders the precondition FALSE to prevent inappropriate re-initialization of the activity.

The scheduled activity: In this case, the Initiator is a time sequence.[8] Each value in the time sequence maps onto the event time of the starting event of an instance of the scheduled activity.

The sequel activity: Often there are circumstances where one identifiable unit of behaviour follows immediately upon completion of another and circumstances require a separation between these two units of behaviour. Consider, for example, the case where a resource used for the first unit of behaviour is not required for the second unit of behaviour. Figure 4.4 shows the relationship between the predecessor activity (*PreAct*) and the subsequent sequel activity (*SeqAct*). Note that the event time of the starting event of the sequel activity always coincides with the event time of the terminating event of the predecessor activity.

As an example, consider a port where a tugboat is required to move a tanker from the harbour entrance to a berth where a loading (or unloading) operation immediately begins. Here, the berthing and the loading operations both map onto activities, but the latter is distinctive because (by assumption) no additional

---

[7]Our notion of the conditional activity coincides with the perspective of an activity taken by Hills [4].

[8]A time sequence is a sequence of increasing positive real values. These values may either be explicitly given or be given implicitly via a data procedure (see Sect. 3.2.1). Sometimes, it is convenient to regard a time sequence as a vector.

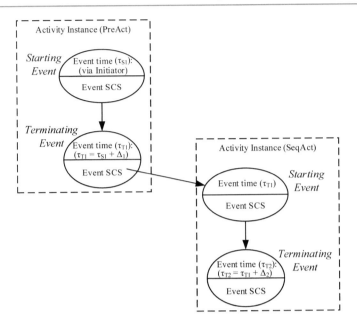

Fig. 4.4 Representation of the events that comprise a sequel activity instance

resource is required and hence it can be immediately initiated upon completion of the berthing operation. In an emergency healthcare context, the situation would likely arise when the activity associated with the emergency transport of a patient by ambulance to a hospital emergency room is immediately followed by the activity of triage care.

The scheduled sequel activity: This fourth ABCmod activity type is a generalization of the sequel activity outlined above. It provides for scheduling the sequel activity at some time point in 'the future' rather than having its initiation coincide with the event time of the terminating event of the initiating activity instance (which we assume to be τ_T). In effect, the scheduled sequel activity called *SchSeqAct* then functions as a scheduled activity whose time sequence is determined by adding a delay Δ^* (possibly random) to the terminating event time of each predecessor activity instance, i.e. $(\tau_T + \Delta^*)$. Note also that it is possible to schedule a sequel activity as part of an action instance (see discussion of the *action* in Sect. 4.2.2.3).

The common feature of each of the four activity types outlined above is that an activity instance incorporates exactly two events. It begins with an event (the starting event) and, after an interval of time (the duration), ends with a second event (the terminating event). There are circumstances, however, when this basic flow is inadequate for capturing the nature of the unit of behaviour that needs to be modelled. More specifically, the unfolding of this behaviour, in the context of a specific activity instance, may be <u>interrupted</u>.

A common source of interruption is a change in the value of an input variable. In response to such a change in value, one or more activity instances may need to alter the manner in which they complete the unit of behaviour that has already been initiated.

Pre-emption is a particular type of interruption. It would, for example, occur in a situation where the instantiation of an activity instance depends upon the availability of a limited resource that is required by several activities. A common method of resolution is to assign access priorities to competing activity instances. With this approach, an activity instance can interrupt some lower priority instance that is currently accessing the resource and take control of the resource. The interrupted activity instance is, in fact, suspended and its completion is not guaranteed. If completion does occur, then the duration of the suspended activity instance becomes distributed over at least two disjoint time intervals. In the extreme case, completion may never occur.

The requirement for incorporating a possible interruption can be embedded into the specification of any one of the four types of activities outlined earlier. This incorporation takes the form of one (or more) **conditional events** embedded into the specification. Such events are called *intercepting events.* . When such a requirement is introduced, the activity is referred to as an *extended activity*. Notice that the number of events within an extended activity increases from two to three or more.

Three possible intercepting events are shown in Fig. 4.5. In the first case, the intercepting event (a) might reflect the situation where the duration of the activity instance is simply extended and the terminal event still follows. In the second case (b), the implication is that the current activity instance is halted (in particular, the terminating event does not occur) and completion takes place when a second instance the activity is initiated (typically at some future time). The implication of the third intercepting event (c) shown in Fig. 4.5 is that the instance of the activity is halted without any expectation of resumption at some future time.

Restarting an extended activity after it has encountered an interruption requires a conditional activity. When the extended activity is a conditional activity, its precondition would necessarily include a suitable mechanism to enable the restarting. In the case of either a sequel activity or a scheduled sequel activity, a separate related conditional activity is required to restart and continue with the original activity instance, thereby completing the associated unit of behaviour.

4.2.2.3 Actions

The *action* is the second category of behaviour modelling artefact within the ABCmod framework. While activities serve to capture the various relevant tasks that are carried out within the SUI, an action simply provides the means for characterizing relevant events that are not embedded within activities. Because an action is simply an event, it unfolds at a single point in time and consequently the concept of 'instances' is not meaningful. However, multiple occurrences of the action are typical. Each action within an ABCmod conceptual model has a name.

There are three types of action called the *scheduled action*, the *scheduled sequel action* and the *conditional action*. Their differences parallel the differences between the scheduled activity, scheduled sequel activity and the conditional activity as outlined earlier.

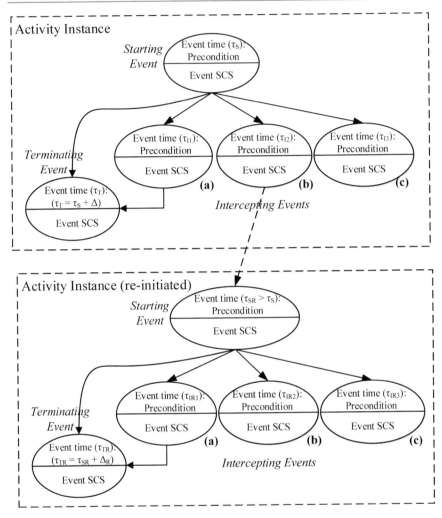

Fig. 4.5 A representation of three possible intercepting events within an extended activity

The *scheduled action* consists of a particular event that is scheduled at one or more points in time. There is a time sequence associated with the scheduled action which defines the points in time at which the scheduled action occurs. As will become apparent in the discussion of Sect. 4.3.3.3, the scheduled action has special relevance in providing the means for handling the notion of input within the ABCmod framework.

The *scheduled sequel action* allows the scheduling of an action at some time point in 'the future' when the event time of the action does not coincide with the event time of the terminating event of the initiating activity instance or the event of the initiating action.

The *conditional action* corresponds to a conditional event. The conditional action is frequently used to accommodate a circumstance where a specific status change must be inhibited until particular circumstances exist within the model. In effect, the need for a delay of uncertain length is thus introduced. In this circumstance, the conditional action serves as a 'sentinel' that awaits the development of the condition(s) that permit the change to occur.

4.3 Conceptual Modelling in the ABCmod Framework

We have previously noted that the conceptual modelling task is concerned with the characterization of SUI behaviour at a level of detail that is appropriate for achieving the goal of the modelling and simulation project. As discussed in Chap. 2, defining the project goal is the first step in the modelling activity.

In effect then, the underlying conceptual modelling task is concerned with capturing and organizing relevant detail consistent with the project goal. The discussions in the preceding sections have outlined a model-building framework that provides the building blocks for organizing this detail in a focused and reasonably intuitive fashion. Nevertheless, except for the simplest of cases, the artefact that emerges (i.e. the ABCmod conceptual model) can still become relatively complex.

Dealing with complexity in any context is considerably facilitated when it can be addressed at more than one level of detail. This view has, in fact, been adopted in the ABCmod framework inasmuch as conceptual model development evolves as a two-stage process. We refer to the initial stage as 'high level' given that its purpose is restricted to providing a preliminary perspective that is unencumbered by excessive detail. This version of the model is primarily intended for discussion among all project stakeholders. The necessary detail is introduced at a second stage of development which is called the 'detailed level'. This version of the model is a detailed specification primarily intended for the simulation modelling team which has the responsibility of transforming the conceptual model into a simulation program.

4.3.1 Project Goal: Parameters, Experimentation and Output

Typically, experimentation is concerned with exploring changes in behaviour that result from various predetermined values assigned to defined parameters. Parameters can emerge in either of the two contexts. The first context relates to the SUI/model itself and the second to the SUI/model input. In the first case, a parameter can alter either a structural aspect of the model or its rules of behaviour. As an example of a change in structure, consider the case of a port where the number of berths is changed to determine the impact on turnaround time of tankers using the berths for loading. An example of changing the rules of behaviour is the addition of a phone service at a delicatessen counter in which case one or more employees have the additional responsibility of answering the phone as well as serving customers at the counter.

Sometimes changes to the rules of behaviour relate to collections of coordinated modifications to the rules. These collections are frequently called 'policy options' and the evaluation of their impact is embedded in requirements stated either explicitly or implicitly in the project goal. In such a circumstance, they are managed by a parameter whose values essentially index the available options.

The second context where parameters can appear relates to the input to the SUI, more specifically to facets of the input variables that have been chosen to characterize input. It was indicated previously (see Sect. 3.3) that we regard service time (e.g. of customers at the counter of a fast-food outlet) as an endogenous input. An endogenous input variable introduced to represent service time is typically a random variate variable (RVV) whose values are random variates. It may be, for example, that the mean value of the underlying distribution for this RVV is a parameter.

Output is an essential property of a simulation run. It is that set of generated data that has relevance to the achievement of the project goal. This data is captured via one or more output variables embedded in the conceptual model. A recommended aspect of the project goal discussion is a list of the output variables that have relevance to achieving this goal. The manner in which each of these output variables acquires its value is part of the detailed conceptual model specification.

Experimentation may require additional outputs which also need to be identified. These additional outputs serve to support the chosen experimentation strategy for solving the problem. It is therefore important to consider the experimentation strategy before starting the development of the conceptual model. When developing an experimental strategy, the observation interval must be well defined. Also, it is clearly essential to ensure that the units associated with the time variable, t, are documented.

The example projects provided in Annex 1 illustrate a variety of project goals.

4.3.2 High-Level ABCmod Conceptual Model

4.3.2.1 Simplifications and Assumptions

The importance of properly documenting simplifications and assumptions was stressed in Chap. 2 (Sect. 2.1). Our convention in the ABCmod conceptual modelling framework is to provide this documentation as part of the high-level conceptual model.

Simplifications are difficult to classify or even to characterize because they are very much dependent upon the features of the SUI and the goal of the project. It is worth noting here Shannon's suggestion [14] that modelling is an art form that has its roots in the issue of simplification. An interesting exploration of several DEDS model simplification methods can be found in Robinson [11]. One of these is the notion of 'aggregation'. Consider, for example, the service positions at the counter of a fast-food outlet. Each position is, in effect, a resource because a customer must first occupy such a position before service can begin. But these positions are (at least in most cases) identical, and distinguishing them serves no useful purpose and needlessly complicates the model. In the context of the ABCmod framework, such a collection of 'positions' would be consolidated into an entity category with a dual *role* = Resource Group, thereby achieving a degree of simplification. It's also of some interest to note that this 'simplification' can also be regarded as either abstraction or 'granularity reduction'.

Another simplification technique is referred to by Robinson as 'excluding components and details'. This is illustrated in the example we present in Sect. A1.3. In that example, a 'harbour master' is introduced in the problem description. This individual directs the movement of tankers entering and leaving the port being considered. Our model development for the problem, in fact, ignores this individual since it is only the policies which he/she applies that have relevance.

The notion of 'rule set reduction' is also identified by Robinson. His illustrative example relates to the complexity of the 'rules' that could be formulated to govern how a customer selects a checkout queue in a supermarket. A range of options is possible (shortest queue, assessment of the movement rates of the queues, evaluation of the quantity of merchandise being purchased by those already in the queue, etc.). As well there is the matter of 'queue jumping'. The key question is simply what level of complexity is warranted by the specific goal of the project. A question that unfortunately is not always readily answered.

We note also that the characterisations adopted for many of the stochastic distributions within a conceptual model are often assumptions. This is especially common when the SUI does not yet exist.

4.3.2.2   Structural View

The first step in the development of a conceptual model is generally the identification of the entity categories that are required for its structural representation. A structural diagram, a high-level visual presentation of entities within these categories, is of considerable help in carrying out this first stage in the conceptual modelling process. A collection of icons, summarized in Table 4.1, is provided for this purpose.

The icons in Table 4.1 reflect the situation where *scope* ≠ Transient. For the case when *scope* does equal Transient, icons representing instances of that entity category are used and distributed within the diagram in a way that illustrates their participation in the model. A legend is used to clarify the entity references. Shading or other embellishments (that preserve the geometric shape) are often used to improve the diagram.

To illustrate, we consider the department store example introduced earlier in Sect. 4.1.1. Recall the SUI includes customers who shop in merchandise areas (three are available), possibly making selections, paying for them at service desks (one is available in each department) and then (possibly) moving on to another merchandise area. A structural diagram representation of the entities appropriate for the department store example is given in Fig. 4.6. Note that a collection of three shopping areas is shown, each of which has a service desk and a line in front of this desk. The Customer entities belong to a category with *scope* = Transient and representative instances are shown by shaded circles.

Practical constraints naturally arise when the number of departments in the department store example becomes large. However, since all departments have the same structure, it suffices to simply present one 'generic' department as shown in

Table 4.1 Summary of structural diagram icons

Icon	Description	Icon	Description
C.Name	A consumer entity	RG.Name	Entity with both resource and group *roles* (octagon)
R.Name	A resource entity	CQ.Name	Entity with both consumer and queue *roles* (union of circle and rounded rectangle)
G.Name	A group entity	CG.Name	Entity with both consumer and group *roles* (union of circle and ellipse)
Q.Name	A queue entity (rounded rectangle)	RCG.Name	Entity with resource, consumer and group *roles* (union of rounded rectangle and the ellipse)
RC.Name	Entity with both resource and consumer *roles* (rounded square)	RCQ.Name	Entity with resource, consumer and queue *roles* (union of rounded square and rounded rectangle)
RQ.Name	Entity with both resource and queue *roles* (elongated hexagon)		

Fig. 4.7. In this figure, ID is a member identifier whose value ranges between 1 and N where N is the number of departments that need to be accommodated.

The structural diagram is a collection of labelled icons as presented in Table 4.1 and represents the main aspect of the structural view. A brief outline of each of the entity categories that appears in the structural diagram augments the view. Placement of the icons in a manner that reflects the SUI layout is often helpful.

4.3.2.3   Behavioural View

Generally, the entity (or entities) of a any particular entity category become(s) involved in a sequence of activity instances. This involvement is an important behavioural feature and can take place in a variety of ways depending on underlying circumstances. It is helpful to have a high-level presentation of this essential

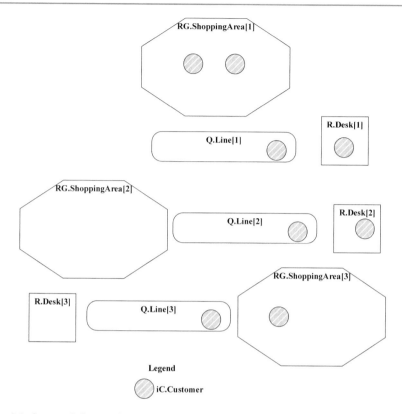

Fig. 4.6 Structural diagram for the department store example

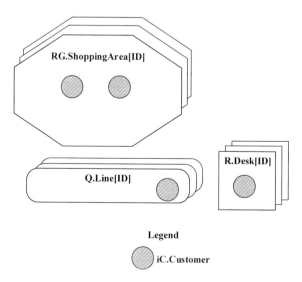

Fig. 4.7 Generic structural diagram for the department store example

feature. This presentation is called a life cycle diagram for the entity category. The collection of life cycle diagrams provides the behavioural diagram for the project.

Typically, there is a life cycle diagram for every entity category but there are exceptions. For example, when the entity (or entities) of an entity category participates only in various instances of *one* particular activity, then a life cycle diagram is redundant. Furthermore, life cycle diagrams are generally not relevant for the aggregate entity categories (*role* = Queue and *role* = Group).

Generally an entity participates, at different times, in a number of activity instances of which some may be instances of the same activity; this will be apparent in its life cycle diagram. The implication here is that an arbitrary instance/member of a particular entity category can participate in instances of several different activities. Note also that multiple instances/members of the same entity category can be involved in separate instances of the same activity (see the Conveyor Project in Sect. A1.6).

In general, a life cycle diagram has many segments where each segment shows the transition from one activity instance to another. Each activity in a life cycle diagram is shown as a labelled rectangle. The intent of the diagram is to show the possible sequences of activity instances in which the entity can become involved.

There are several possible circumstances whereby an entity instance can transit from an instance of activity A to an instance of activity B. Often, this transition encounters a delay. This delay can occur because of inhibiting preconditions and we indicate this possible delay with a circle which we call a *wait point*. The wait point simply shows that the entity's transition to a subsequent activity instance is not necessarily immediate. Furthermore, it is possible for several arrows to emanate from the wait point which shows that the entity may transit to any one of several activity instances. The transition will ultimately be to the one whose precondition first becomes TRUE.[9] Figure 4.8a illustrates this situation. The presence of a wait point thus implies that each possible subsequent activity following the wait point is a conditional activity. If the subsequent activity instance is a scheduled sequel activity, then an S is included within the wait point to indicate this situation. This is shown in Fig. 4.8b.

Often an entity can transit directly from activity instance **A** to activity instance **B**. There are two distinct possibilities as shown in Fig. 4.9. In Fig. 4.9a, the implication is that the transition is direct because **B** is a sequel activity. In Fig. 4.9b, the divided circle is intended to imply that the transition is direct even though **B** has a precondition. The situation that is being represented is the case where it is known that the precondition (of **B**) has a TRUE value at this point in this particular life cycle diagram because of circumstances established in the preceding activity (namely, **A**). This 'null wait point' is necessary to acknowledge that B is a conditional activity and thus maintain consistency with some life cycle diagram in which the transition to **B** is subject to its precondition.

---

[9]Note that it is implicitly assumed that race conditions will not occur in a properly formulated ABCmod conceptual model.

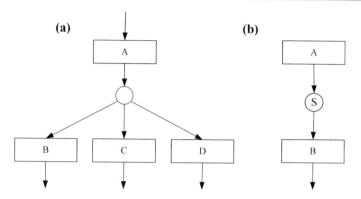

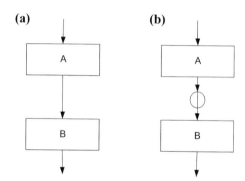

Fig. 4.8 Entity transition possibilities through a wait point

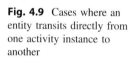

Fig. 4.9 Cases where an entity transits directly from one activity instance to another

An entity that transits to an activity instance may cause that activity instance to precipitate a pre-emption of another activity instance in which an entity of lower priority is being serviced by a resource required by both entities. Figure 4.10 illustrates this case. Note that the double arrow from the wait point indicates the possibility of the subsequent activity **B** pre-empting another activity instance (not shown in the diagram).

Interruption of an activity instance may result in one or more entities no longer being engaged in that activity instance. There are, however, several possibilities for re-engagement. Figure 4.11a shows the case where the entity becomes engaged in a new activity instance after the interruption. The dashed arrow leaving activity instance **A** indicates the path taken when interruption occurs, while the arrow exiting horizontally is taken when normal termination of **A** occurs. Figure 4.11b illustrates the case where the displaced entity encounters a delay and ultimately transits back to a new instance of the original activity (especially typical with pre-emption).

Because the characterization of behaviour within the ABCmod framework also relies on the action artefact, such artefacts will often appear in life cycle diagrams.

Fig. 4.10 Transition that
may precipitate pre-emption

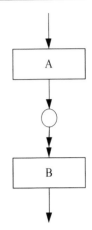

Fig. 4.11 Indication of the
possibility of interruption

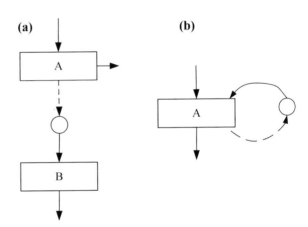

The action instance is represented in a life cycle diagram by a labelled rectangle
with a rounded corner. Three separate possibilities are shown in Fig. 4.12.

In Fig. 4.12a, the instance of the scheduled action labelled **A** is associated with an
input entity stream. This implies that the life cycle diagram of Fig. 4.12a relates to an
entity category with *scope* = Transient. Because a time sequence determines the event
times of a scheduled action, it is not meaningful for an instance of such an action to be
embedded within a life cycle diagram; in other words, the action instance labelled B in
Fig. 4.12b cannot be the instance of a scheduled action, i.e. the action instance labelled
B must be conditional because of the presence of the wait point that precedes it. Finally,
when a scheduled sequel action instance (B) follows activity instance (A), an S is
embedded in the wait point between A and B as shown in Fig. 4.12c.

Figure 4.13 illustrates a life cycle diagram for the Customer entity in the department store example. The Customer entity category has *scope* = Transient which
implies that there is an entity stream exogenous input which in turn implies the need
for an action (called Arrival in Fig. 4.13). Upon arrival, the Customer engages in
the shopping activity in some department and may or may not make a purchase. If a

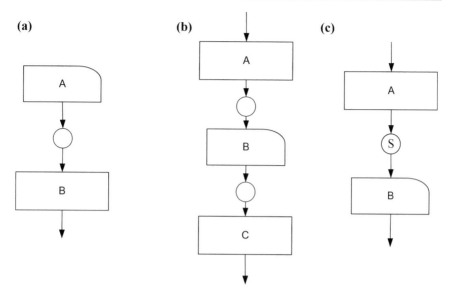

(a)   **(b)**   **(c)**

Fig. 4.12 Scheduled action (**a**), conditional action (**b**) and scheduled sequel action (**c**)

Fig. 4.13 Life cycle diagram
for the customer entity in
department store example

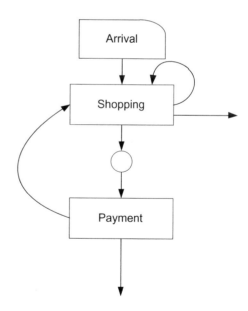

purchase is made, then the payment activity takes place followed (possibly) by
further shopping in another department or exit from the store. Shopping in any
particular department may not result in a purchase (hence, no payment activity is
required). In this circumstance, the Customer may either leave the store or move on
to another department to continue shopping.

The life cycle diagrams of Fig. 4.8 through to Fig. 4.13 have several superficial similarities with the activity cycle diagram that can be found in the modelling and simulation literature (e.g. Kreutzer [6] and Pidd [9, 10]). The fact that both diagrams are constructed from rectangles and circles provides the basis for the apparent similarity. There are, however, significant inherent differences as outlined below.

It is true that the rectangle in both an activity cycle diagram and a life cycle diagram represents an activity. In the case of the activity cycle diagram, this activity is a conditional activity (in our terminology), and the precondition is formulated strictly in terms of entity membership in one or more specified queues. Furthermore, the SCS of the terminating event in this activity is likewise restricted to the relocation of entities into appropriate queues. As is apparent from our earlier discussion, such restrictions are absent in the case of conditional activities in the ABCmod context and hence in the context of those represented in a life cycle diagram.

The circle in an activity cycle diagram has a very specific meaning, namely, it represents a queue. The circle in an ABCmod life cycle diagram is a 'wait point' which is a conceptual construct that indicates a delay (which is often, but not always, the result of having to wait in a queue).

Finally, we note that activity cycle diagrams for individual entity categories are typically interleaved, as appropriate, to convey an integrated view of entity inter-actions. The representation of such an integrative view is not explicit with ABCmod life cycle diagrams but is, nevertheless, apparent from the activity/action names that appear in the life cycle diagrams. For example, in the behavioural diagram of the Port project presented in Sect. A1.3, the presence of the Berthing and Deberthing activities in both the Tanker and Tug entity life cycle diagrams indicate an inter-action between the entities and thus an interleaving of their respective life-cycles.

4.3.2.4   Input

Recall from the discussion in Sect. 3.3 that our perspective of model development within the DEDS domain recognizes two broad classifications of input, namely, exogenous and endogenous. Input variables are associated with each of these classifications. In the case of exogenous input, these variables are called either environmental variables or entity stream variables, depending on the underlying role that they play. In the case of endogenous input, the associated variables are called either independent variables or semi-independent variables (a typical example of a semi-independent variable is when it provides the duration of an activity). The required input variables are typically inferred from features of the behavioural constructs (e.g. time sequences and durations) as formulated in the high-level behavioural view (Sect. 4.3.2.3).

The various types of input variable (environmental, entity stream and (semi-) independent) have distinctive properties and these impact upon their representation and likewise upon implementation requirements within a simulation program. An essential task in the conceptual modelling process therefore is to clarify these properties. In the discussion below, we outline this clarification process based on the modelling concepts and artefacts within the ABCmod framework.

We regard all input variables as discrete-time variables (see Sect. 3.2.3). This, in particular, means that they have a domain sequence and a range sequence. In the high-level representation, our concern is with identifying the data procedures

Table 4.2 Template for input variables

Exogenous Input (Environmental)			
Variable Name	**Description**	**Domain Sequence**	**Range Sequence**
uEnv	*Brief description of the role of this environmental input variable*	*x.DuEnv(), the data procedure that provides the values in the domain sequence* $CS_D [uEnv]$	*x.RuEnv(), the data procedure that provides the values in the range sequence* $CS_R [uEnv]$
Exogenous Input (Entity Stream)			
Variable Name	**Description**	**Domain Sequence**	**Range Sequence**
uES	*Name of the entity category associated with this entity stream variable*	*x.DuES(), the data procedure that provides the values in the domain sequence* $CS_D[uES]$	*Either a constant value or x.RuES(), the data procedure that provides the values in the range sequence* $CS_R [uES]$
Endogenous Input (Independent)			
Variable Name	**Description**	**Domain Sequence**	**Range Sequence**
uInd	*Brief description of the role of this independent input variable*	*x.DuInd(), the data procedure that provides the values in the domain sequence* $CS_D[uInd]$	*Either a constant value or x.RuInd(), the data procedure that provides the values in the range sequence* $CS_R [uInd]$
Endogenous Input (Semi-Independent)			
Variable Name	**Description**	**Value(s)**	
uInd	*Brief description of the role of this semi-independent input variable*	*Either a constant value or x.RuInd(), the data procedure that provides the values in the range sequence* $CS_R[uInd]$	

required for the generation of the values within the domain sequence and the range sequence for each of the input variables that have been identified.

There are two alternatives for the data procedures in question. If, for example, the values in a domain sequence have a stochastic origin, then the required data procedure is a *random variate procedure* (RVP, see Sect. 3.2.1). Otherwise (i.e. when the values have a deterministic basis), a *deterministic value procedure* (DVP) is required (the need for a procedure can clearly be circumvented when the value in question is a given constant). Corresponding options apply for the range sequence.

The pertinent details for all input variables are presented in the high-level view in a tabular format. A particular naming convention is recommended in order to clarify interrelationships. The associated template is provided in Table 4.2.[10]

The creation of this table begins with listing the input variables together with their descriptions. The balance of the table is updated with references to data procedures that emerge from a data modelling phase. The detailed specifications of these procedures (namely, the data models underlying each of the deterministic value and random variate procedures) are provided in appropriate templates in the detailed conceptual model.

4.3.3 Detailed ABCmod Conceptual Model

The distinction between structural and behavioural components of an ABCmod conceptual model is retained in the detailed level presentation. The discussion which follows is accordingly separated. In Sect. 4.3.3.1, we first outline the two structural components of an ABCmod conceptual model. The collection of behavioural components is outlined in Sect. 4.3.3.3 but is preceded in Sect. 4.3.3.2 with an outline of some behaviour modelling artefacts which provide the building blocks for the behavioural components.

The material presented in Sects. 4.3.3.1 and 4.3.3.3 essentially provides a template for the development of any conceptual model within the ABCmod framework.

4.3.3.1 Structural Components

Constants and Parameters
The constants and parameters embedded within an ABCmod conceptual model are collected together and presented in a single tabular format. The template is given in Table 4.3. Constants and parameters are not restricted to scalar values. Data structures, such as data sequences (arrays), can also be readily accommodated.

---

[10]In the procedure identifiers, '*x*' is a placeholder for either *DVP* or *RVP*, which indicates either a deterministic value procedure or a random variate procedure.

Table 4.3 Template for constants and parameters

Constants		
Name	**Description**	**Value**
cName	*Brief description of the significance of cName*	*Value assigned to cName*
Parameters		
Name	**Description**	**Values**
pName	*Brief description of the significance of pName*	*The collection of values of pName that are of interest*

Entity Structures

An *entity structure* provides the specification for each entity category. More specifically, it lists the names of the attributes defined for the entity category and provides a description for each attribute. Each entity structure has a label. The general format for this label is **Type**: *Name* where *Name* is the name of the entity category and **Type** = {*role*} {*scope*} as appropriate to the underlying entity category. For example, the entity structure label 'Resource Many [2]: Tugboat' relates to an entity category called 'Tugboat' which consists of a collection of two entities. Alternately, the entity structure label 'Resource Unary: Tugboat' relates to an entity category called Tugboat consisting of a single entity.

Inasmuch as an entity structure is a specification, it can be regarded as the template for 'deriving' entities. In other words, with this perspective, there is an implicit requirement that two 'tugboat entities' be derived from the entity structure having the label 'Resource Many [2]: Tugboat' and one from the entity structure having the label 'Resource Unary: Tugboat'. Each derived entity acquires the attributes listed for the entity structure. Entities belonging to categories with *scope* = Unary or *scope* = Many[N] are implicitly derived at the start of the observation interval, while those belonging to categories with *scope* = Transient must be explicitly derived within behavioural specifications (i.e. SCSs) using the standard procedure SP.Derive() (see Sect. 4.3.3.2).

A tabular format is used for each entity structure in an ABCmod conceptual model. The template for this specification is given in Table 4.4.

Table 4.4 Template for an entity structure

Type: Name	
A brief description of the entity category called Name.	
Attributes	Description
AttributeName1	*Description of the attribute called AttributeName1.*
AttributeName2	*Description of the attribute called AttributeName2.*
.	.
.	.
AtributeNamen.	*Description of the attribute called AttributeNamen*

where **Type** is: {*role*} {*scope*} as outlined earlier.

4.3.3.2 Behaviour Modelling Artefacts

The Activity Construct

The key role played in the ABCmod framework by the notion of an activity was outlined earlier in Sect. 4.2.2.2. We show in Table 4.5 the template for presenting an activity construct.

The *Initiation Basis Keyword* in the template of Table 4.5 reflects on the type of the activity construct that is being formulated (conditional, scheduled, sequel or scheduled sequel). The *basis specification* provides appropriate details. Elaboration of the features of these relationships is given in Table 4.6.

As shown in Table 4.7, the extended activity construct includes the specification of the intercepting event within the duration component of the activity construct. Clearly, the intercepting event is a conditional event since its specification includes a precondition.

An extended activity instance that does become interrupted terminates via appropriate specifications in the SCS of the intercepting event (see Table 4.7). In addition, all entities involved in the activity instance must be disposed of by the SCS of the intercepting event. Note also that the starting event of the extended activity construct must be able to deal with both instantiating a new instance of that activity and, as well, an instance that must complete the task already started within a previous instance that was interrupted.

It is important to appreciate that nothing in the formulation of the activity construct, in any of its specific forms as outlined above, precludes the possibility of the simultaneous (concurrent) existence of multiple instances of that construct. Also, note also that the possibility of an 'empty' SCS of the terminating event is allowed, but the reader is cautioned here to distinguish between the essential existence of a terminating event and the possibility of an empty SCS.

Table 4.5 Template for the activity construct

Activity: *Name*	
A description of the activity called Name.	
Initiation Basis Keyword	*Basis Specification.*
Event SCS	*SCS of the starting event*
Duration	*The duration of an activity instance (often an endogenous input variable whose value is acquired from an data procedure; see section 4.3.2.4)*
Event SCS	*SCS of the terminating event*

Table 4.6 Initiation basis keywords and their corresponding specifications (activity construct)

Initiation basis keyword	Activity type	Basis specification
Precondition	Conditional	Boolean expression that specifies the condition under which the starting event occurs
TimeSequence	Scheduled	Either an explicit or implicit specification of a sequence of times (e.g. via an input data procedure, see Sect. 4.3.2.4) that provide the event times for the activity's starting event
Causal	Sequel	Parameter list for information exchange
Causal	Scheduled Sequel	Parameter list for information exchange

Table 4.7 Template for the extended activity construct

Activity: *Name*	
A description of the extended activity called Name.	
Initiation Basis Keyword	*Basis Specification.*
Event SCS	*SCS of the starting event*
Duration	*The duration of an activity instance (typically an endogenous input variable whose value is acquired from a data procedure; see section 4.3.2.4)).*
Interruption Precondition	*Boolean expression that specifies the condition under which an intercepting event occurs*
Event SCS	*SCS of the intercepting event*
Event SCS	*SCS of the terminating event*

The Action Construct

The notion of an action (as outlined in Sect. 4.2.2.3) is likewise a key artefact for behaviour modelling in the ABCmod framework. The template for an action construct is shown in Table 4.8.

Table 4.8 Template for the action construct

Action: *Name*	
A description of the action called Name.	
Initiation Basis Keyword	*Basis Specification*
Event SCS	*SCS of the event that is embedded in the action construct*

The initiation basis keyword in the template of Table 4.8 can either be *Time-Sequence* or *Precondition* which, as in the case of the activity constructs, is determined by the type of action construct (scheduled or conditional) that is being formulated. The basis specification is, in turn, similarly dependent on the type of action construct. The pertinent features of these relationships are the same as those summarized in Table 4.6 for the activity constructs.

Procedures

Standard Procedures

A variety of 'standard' operations reoccur in the formulation of the SCSs within the various behaviour constructs that emerge during the development of any ABCmod conceptual model. We assume the existence of procedures to carry out these operations, and these are freely used wherever required. They are briefly outlined below:

- Derive(*EntityStructName*); usage is *Ident* ← Derive(*EntityStructName*): Derives an entity referenced by identifier *Ident* from the entity structure called *EntityStructName*.
- InsertGrp(*GroupName, Item*): Inserts *Item* into the *list* attribute of the group entity called *GroupName* and increments the n attribute of the group entity.
- InsertQue(*QueueName, Item*): Inserts *Item* into the *list* attribute of queue entity called *QueueName* according to the declared queuing protocol associated with *QueueName* and increments the attribute n of the queue entity.
- Leave(*Ident*): It frequently occurs that a specific entity's existence within the model comes to an end. This procedure explicitly indicates such an occurrence and its argument is the identifier of the entity in question. The procedure is typically invoked from within the SCS of the terminating event of an activity instance.
- Put(*PHY[y], val*): Places the value (*t, val*) into the sample sequence *PHY[y]* where t is the current time.
- Put(*TRJ[y], val*): Places the value (*t, val*) into the trajectory sequence *TRJ[y]* where t is the current time.

- RemoveGrp(*GroupName, Ident*): Removes an item from the *list* attribute of the group entity called *GroupName* and decrements the attribute *n* of the group entity. *Ident* is the identifier for the item to be removed from the group entity.
- *Ident* ← RemoveQue(*QueueName*): Removes the item which is at the head of the *list* attribute of the queue entity called *QueueName* and decrements the attribute *n* of the queue entity. *Ident* is the identifier for the returned item.
- StartSequel(*SeqAct, Param1, Param2, ...*): Initiates the sequel activity called *SeqAct* from within the terminating event of an antecedent activity. Whatever information exchange that is necessary is provided by *Param1, Param2*, etc.
- SchedSequel(*SchSeqAct, Delta, Param1, Param2, ...*): Initiates the scheduled sequel activity or the scheduled sequel action called *SchSeqAct* from within the terminating event of an antecedent activity after a delay *Delta*; whatever information exchange that is necessary is provided by *Param1, Param2*, etc.
- Terminate: An instance of an extended activity that undergoes an intervention must necessarily terminate. This is made explicit by ending the SCS of each intervention subsegment with a reference to the Terminate procedure.

User-Defined Procedures

Situations frequently arise where specialized operations need to be carried out within behaviour constructs—operations that are beyond the scope of the standard procedures listed above. Procedures to carry out these operations can be freely defined and referenced wherever necessary to facilitate the conceptual modelling task. They are called user-defined procedures (UDP), and they are summarized in a table whose template is given in Table 4.9.

Table 4.9 Template for summarizing user-defined procedures

User Defined Procedures	
Name	**Description**
ProcedureName(parameterlist)	*Specification of the user defined procedure (UDP) called ProcedureName.*

There are occasions when a user-defined procedure is referenced by only one of the behavioural constructs in the conceptual model. In such circumstances, it is convenient to simply append the specification of the UDP to the table that specifies the behaviour construct in question. The table designation that is used in this case is 'Embedded User-Defined Procedure'.

Note that by convention, we augment the names of standard procedures and user-defined procedures that are referenced in the body of behaviour constructs with the prefix 'SP.' or 'UDP.', respectively.

Input Data Procedures

Inputs that are discrete-time variables have data requirements. This data delivery requirement is encapsulated in one of the two types of input data procedure. If the data delivery has a deterministic basis, then a deterministic value procedure (DVP) is identified. If, on the other hand, the data delivery has a stochastic basis, then a random variate procedure (RVP) is identified (see Sect. 3.2.1). These procedures simply serve as 'wrappers' for the data specification. The rationale here is simply to facilitate modification of the actual source of the data if the need arises. The collection of such input data procedures that are required within an ABCmod conceptual model is summarized in templates given in Tables 4.10 and 4.11.

Table 4.10 Template for summarizing RVPs

Random Variate Procedures		
Name	**Description**	**Stochastic Data Model**
ProcedureName(parameterlist)	*An outline of the purpose of this procedure*	*Details of the mechanism that is invoked in order to generate the data values delivered by this procedure. Typically involves sampling values from one or more probability distributions.*

Table 4.11 Template for summarizing DVPs

Deterministic Value Procedures		
Name	**Description**	**Deterministic Data Model**
ProcedureName(parameterlist)	*An outline of the purpose of this procedure*	*Details of the mechanism that is invoked in order to generate the data values delivered by this procedure.*

Note that by convention, we augment the names of a deterministic value procedure and random variate procedure referenced in the body of behaviour constructs with either the prefix 'RVP.' or 'DVP.', as appropriate.

Output Constructs

The output specification at the detailed level in an ABCmod conceptual model elaborates on the output variables identified in the project goal. The specification has a tabular format and is organized along the lines of the various types of output outlined in Sect. 3.4. The template for this table is shown in Table 4.12.

The recording mechanisms for the values of output variables are of considerable relevance during the transition from an ABCmod conceptual model to a simulation model. In this regard, the following should be noted:

1. SSOV: The specification for acquiring the required value for an SSOV always appears in one of the ABCmod behaviour constructs.
2. TRJ[y]: A check for a value change in y must be made following each event. When a change has occurred, its new value must be duly recorded using the Put() standard procedure.
3. PHI[y]: Values for the sample variable y are explicitly deposited using the Put() standard procedure.
4. DSOV: Values for DSOVs are acquired during post-processing; this takes place following each run.

Table 4.12 Template for summarizing output

Output			
Simple Scalar Output Variables (SSOV's)			
Name	Description		
Y	Description of the simple scalar output variable Y		
Trajectory Sequences			
Name	Description		
TRJ[y]	Description of the piecewise constant discrete-time time variable y(t)		
Sample Sequences			
Name	Description		
PHI[y]	Description of the sample discrete-time variable y		
Derived Scalar Output Variables (DSOV's)			
Name	Description	Data Set Name	Operation
Y	Description of the value assigned to Y	The name of the data set from which the value of Y is derived.	The operation that is carried out on the underlying data set to yield the value assigned to Y

4.3.3.3 Behavioural Components

Initialization

There is almost always a need to assign initial values to some attributes of entities that exist within the model at the initial time, t_0. This is carried out within the SCS of the starting event of a scheduled action construct called Initialise which has a single value, t_0, in its TimeSequence. The template for this action construct is given in Table 4.13.

Table 4.13 The action template for initialization

Action: Initialise	
This scheduled action assigns initial values to some particular collection of attributes of the entities that exist at the initial time, t_0.	
TimeSequence	$<t_0>$
Event SCS	*Assignment of appropriate initial values to each of some particular collection of entity attributes.*

Output Construct

The specific outputs that are required are summarized using the template given in Table 4.12.

User-Defined Procedures

User-defined procedures that are required in the conceptual model are presented using the template given earlier in Table 4.9.

Input Constructs

At the detailed level of conceptual model development in the ABCmod framework, the handling of input focuses on the development of the action constructs (or possibly scheduled activity constructs) that are appropriate to the requirements of the various input variables that have been identified in the tabular summary developed in the high-level view as given in earlier in Table 4.2. RVPs and DVPs that are required in the development of these action constructs are presented using the templates given earlier in Table 4.10 and Table 4.11, respectively.

There are close relationships between the features of the action construct and the nature of the input variable under consideration. Recall, for example, that the time sequence for the action construct associated with a discrete-time variable evolves from the domain sequence of that variable.

We show in Table 4.14, Table 4.15 and Table 4.16 the templates for the action constructs that are associated with environmental input variables, entity stream input variables and independent input variables, respectively (input variables are specified using Table 4.2). The TimeSequence specification for each of these variables is provided by a data procedure that encapsulates the domain sequence for these variables (as before, x is a placeholder for either the prefix RVP or DVP). The event SCS typically makes reference to the range sequence for these variables.

Table 4.14 Action construct for the environmental input variable called uEnv

Action: Act_uEnv	
This scheduled action construct updates the value of the input variable uEnv.	
TimeSequence	*x.DuEnv*
Event SCS	*Assignment of a value to uEnv based on x.RuEnv*

Table 4.15 Action construct for the entity stream input variable called uES

Action: Act_uES	
This scheduled action creates one or more instances of the entity structure associated with the entity stream input variable uES.	
TimeSequence	*x.DuES*
Event SCS	*Assignment of values to the attributes of one or more new instances of the entity structure corresponding to this entity stream variable.* *The number of entity instances is provided by x.RuES.*

Table 4.16 Action construct for the independent input variable called uInd

Action: Act_uInd	
This scheduled action construct updates the value of the independent input variable uInd.	
TimeSequence	*x.DuInd*
Event SCS	*Assignment of a value to uInd based on x.RuInd*

Behavioural Constructs

The final components in the ABCmod conceptual model presentation are the specifications for the required collection of behaviour constructs (activities and actions) that are as yet unspecified. These are presented using the templates given earlier in Sect. 4.3.3.2.

4.4 Examples of ABCmod Conceptual Modelling: A Preview

In this section, we provide a brief introduction to several examples that illustrate various aspects of the use of ABCmod conceptual modelling features to carry out conceptual model development. A comprehensive presentation for each of the ABCmod conceptual modelling projects outlined below is provided in Annex 1.

We consider first a straightforward queuing system formulated around a fast-food outlet in a shopping mall (Kojo's Kitchen: Version 1). The project goal is to evaluate the extent to which an additional employee during two busy periods of the day will reduce the percentage of customers who wait longer than five minutes for service. The ABCmod conceptual model illustrates the use of the following ABCmod features.

- An exogenous input (for the arriving customers) and, as well, an independent endogenous input and two semi-independent endogenous inputs:
- SSOV outputs are used for providing the required percentage of customers who wait longer than 5 minutes.
- Entity categories make use of all basic ABCmod *roles* (two *roles* are combined) as well as *scope* = Unary and *scope* = Transient.
- Two scheduled actions are used, one for customer arrivals and the other for carrying out staff changes.
- A conditional activity is used to handle the customer-serving requirement.

A second version of Kojo's Kitchen is presented in which a more comprehensive solution strategy is examined. In that study, the effectiveness of each of three different employee schedule options is compared with the existing base case. One schedule adds one employee during the busy times as in the option examined in Version 1. The two other schedules likewise increase the number of employees during the busy time periods, but in addition also reduce the number of employees during the non-busy periods. The total number of person-hours that employees work is considered for each of these options as well as for the base case. This information contributes a valuable perspective on the assessment of the results of the study. The comparison relative to the base case provides the basis for a comprehensive analysis of experimentation results as presented in Sect. 7.4.2

Three versions of a Port Project are investigated. Each is concerned with the operation of a port where oil tankers are loaded. Tankers (of three different sizes) arrive in the harbour and must be towed by a tugboat from the harbour to the berths where they are loaded with oil. Upon completion of the loading operation, the tankers once again must wait for the tug to tow them back to the harbour entrance. The determination of a way to reduce the waiting times of tankers is the goal of the project. In this regard, all three versions of the project explore the extent to which the addition of a fourth berth to the three that now exist will reduce the tanker waiting times.

The initial version of this example illustrates the following ABCmod features not previously illustrated:

- Several constants are specified in the Constants table. These constants are used as durations in various activities.
- A sequel activity (called Loading) is used in the conceptual model. This activity is instantiated from the terminating SCS of the activity instance that carries out the berthing operation (called Berthing). A reference to the tanker entity instance that is being berthed is necessarily passed to the Loading activity instance by the Berthing activity.
- A trajectory sequence and a sample sequence are utilized to provide output data. DSOVs derived from the collected data provide the required outputs to assess the achievement of the project goal, namely, the average number of occupied berths and the average tanker waiting time.

- Two conditional activities in the port problem involve only a single entity, namely the tug that moves tankers from the harbour to the berths and then back to the harbour. This demonstrates that it is possible for an activity to involve only a single entity.
- The port project demonstrates that an activity instance can also reference a transient entity (iC.Tanker). In contrast, for the Kojo Kitchen examples, entities with *scope* = Transient are only referenced by another entity during their lifetime.

The travel protocol of the tug is altered somewhat in Version 2 of the Port Project. Specifically, circumstances are introduced which require the tug to return the harbour part way through its journey from the harbour to the berths. This introduces the need to embed an interruption into the Berthing activity of Version 1. In effect, the instance of the berthing activity is terminated and an instance of a (sequel) activity called ReturnToHarbour is initiated.

The occurrence of storms is introduced in Version 3 of the Port Project. This in turn introduces the need for interruptions in all activities, with the exception of the Loading activity. As a result one of the activities incorporates two interruptions (one previously added in Version 2 and the second due to the storm). Also demonstrated in Version 3 of the Port Project is the use of the scheduled activity in modelling an exogenous input (namely, the storm).

The sixth example, the Conveyor Project, provides an example of an entity category with *scope* = Many[N]. The context is a manufacturing process that involves three machines that process two types of components. The first machine (M1) processes both types of components as they arrive in the system. When processing at M1 is completed, each type of component is sent to one of two specific machines on a designated conveyor. However, if the designated conveyor is full, machine M1 must stop and is considered 'down'. The goal of this project is to determine the smallest capacity for the two conveyors connecting M1 to downstream machines so that the 'downtime' of the machine M1 does not exceed a prescribed limit over the course of the observation interval. The development of the conceptual model recognizes the symmetries that exist among the three machines and among the three conveyors by defining entity categories with *scope* = Many [N] for both the machines and the conveyors. This makes possible a significant simplification of the conceptual model that would otherwise be required. Such simplification is achieved by ensuring that the member identifiers for the Machine category and the Conveyor category are assigned in a manner that reflects, in a meaningful way, their physical association which is typically presented in the structural diagram. In this circumstance, a user-defined procedure must be used to select member identifiers to which an activity or action instance applies.

The final conceptual modelling example that is presented in Annex 1 relates to an optimization problem which again has a manufacturing process context (ACME Manufacturing). This project is outlined in detail in Chap. 12. The formulation of the conceptual model serves to further illustrate the usefulness of entity categories with *scope* = Many[N].

4.5 Exercises and Projects

4.1   An alternate approach for dealing with the excessive wait times of concern in the port project (Version 1) is to explore the impact of adding a second tugboat rather than a fourth berth. Formulate an ABCmod conceptual model that is based on this alternative. Note that this will require a specification of the manner in which the tugboats will cooperatively operate. In other words, an appropriate modification of the *Tug Operation* section under *SUI Details* will be required.

4.2   The dining philosophers' problem is a classic vehicle for illustrating the occurrence of deadlock in an operating system and, as well, for exploring strategies to avoid its occurrence. The concern in this problem is to explore the dining philosophers' problem from the perspective of a modelling and simulation project.

We imagine five philosophers seated around a circular table. Between each pair of philosophers, there is a single fork, and in the middle of the table is a large bowl of spaghetti. These philosophers have a very focused existence which consists of a continuous cycle of thinking and eating. There is, however, a basic prerequisite for the eating phase, namely, a philosopher must be in possession of both the fork on his[11] right and the fork on his left in order to access and eat the spaghetti at the centre of the table. Inasmuch as the philosophers are an orderly group, they have a protocol for acquiring the forks. When any particular philosopher finishes his thinking phase, he must first acquire the fork on his right and only then can he seek to acquire the fork on his left (whose acquisition enables the initiation of the eating phase). When the eating phase is complete, the philosopher replaces both forks and begins his thinking phase.

The situation outlined above can, however, lead to *deadlock*. This is a situation where no philosopher is eating and no philosopher is thinking, but rather they are all holding their 'right fork' and waiting to eat. Under these circumstances, the left fork will never become available for any of them and hence none will ever eat!

Suppose that the eating time, ET, for each of the philosophers is an exponentially distributed random variable with the same mean of μ_E minutes. Likewise, suppose that the thinking time, TT, for each of the philosophers is an exponentially distributed random variable with the same mean of μ_T minutes. It has been conjectured that there is an 'interesting' relationship between the ratio (μ_E/μ_T) and the time it takes for deadlock to occur (we denote this time interval by T_{dead}). The goal in part(a) is to carry out a modelling and simulation study to determine if there is indeed a noteworthy relationship that can be identified. The assessment is to be based on two graphs. The first is a graph of T_{dead} versus

---

[11]While the presentation suggests a group of male philosophers, this should not be taken literally since the group is, in fact, gender balanced.

(μ_E/μ_T) with (μ_E/μ_T) in the range 1–10, and the second is a graph of T_{dead} versus (μ_E/μ_T) with (μ_E/μ_T) in the range 0.1–1.

Formulate an ABCmod conceptual model for the project as outlined above. By way of initialization, assume that the five philosophers enter the SUI in a sequential manner at times: $t = 0$ (the left-hand boundary of the observation interval), $t = 0.1 \; \mu_E$, $t = 0.2 \; \mu_E$, $t = 0.3 \; \mu_E$ and $t = 0.4 \; \mu_E$. Upon entry, each philosopher begins a thinking phase.

4.3 The repeated occurrence of deadlock has greatly annoyed the dining philosophers described in Problem 4.2. After due consideration, they agreed to alter their fork acquisition protocol in one small (but significant) way. Instead of having to first acquire the fork on his right, the *fifth* philosopher will henceforth be required to first acquire the fork on his *left*, and only then can he seek to acquire the fork on his right (It can be readily demonstrated that with this altered rule, deadlock will indeed be avoided.). In this modified context, the goal of the modelling and simulation project is to develop a graphical presentation of the average waiting time to eat as a function of (μ_E/μ_T) where waiting time is measured from the moment a philosopher stops thinking to the moment when he begins eating.

Formulate an ABCmod conceptual model for these modified circumstances of the dining philosophers. Use the same initialization procedure that was outlined in Problem 4.2.

4.4 A lock system in a waterway provides the means for diverting boat traffic around a section of turbulent water. One (or more) lock is placed in a man-made parallel water channel, and each functions like an elevator, raising or lowering boats from one water level to another. In this way, boat traffic is able to bypass the non-navigable portion of a river. A representation of a typical lock's operation is given in Fig. 4.14a.

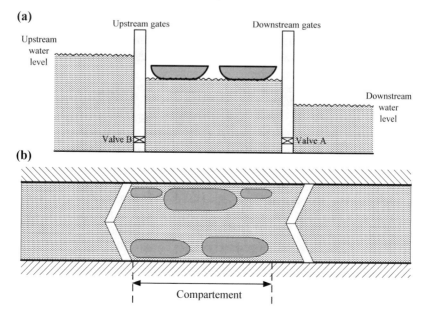

Fig. 4.14 (a) Lock operation, (b) Boat configuration in lock

The lock's operation can be divided into two very similar cycles which we refer to as the 'up-cycle' and the 'down-cycle'. The up-cycle begins with the upstream gates closed, valve A is opened until the water within the compartment reaches the downstream level and then the downstream gates open. Boats waiting at the downstream end to move upstream enter the compartment. When the compartment is suitably filled with boats, the downstream gates close, valve B opens (valve A is closed) and water fills the compartment to raise the boats to the upstream level. The upstream gates then open and the boats exit and continue on their journey. The stages of the down-cycle are the reverse of those described for the up-cycle.

The number of boats that can be accommodated within the lock compartment during either phase is naturally restricted by the physical size of the compartment. This is basically dependent on the length of the lock because boats must be moored along the edges of the compartment for safety reasons. Hence, the linear length of the compartment is a major constraining parameter. This constraint cannot be directly translated into a specific number of boats because boats have varying individual lengths. Furthermore, there is a requirement for a one meter separation between boats and between the boats adjacent to the moving gates and the gates themselves. A typical configuration of boats in the lock during an up phase is shown in Fig. 4.14b.

Both the up-cycle and the down-cycle have the same three phases which we call the loading phase, the transport phase and the exit phase. The loading phase is the phase when boats from the waiting queue position themselves within the lock compartment. The duration of this phase is dependent on the number of boats that enter the compartment. A reasonable approximation for this loading duration is $(d_1 + nd_2)$ where $d_1 = 4$ min, $d_2 = 2$ min and n is the number of entering boats. The boats selected to enter the compartment are taken sequentially from the waiting queue in the order of their arrival until the 'next' boat cannot fit any available space. Then, the next boat in the queue that *can fit* into the available space is selected, and this continues until no further boats can be accommodated.

The transport phase includes the closing of the open gate of the lock compartment and the filling/emptying of the water from the compartment by appropriate opening/closing of valves A or B. The duration of this phase is 7 min. The exit phase, during which the boats leave the compartment, has a duration of 5 min. Thus the total cycle time is $(d_1 + n d_2) + 7 + 5 = (16 + n d_2)$ min.

The management of the lock system (the particular lock in question is one of a series of locks on the waterway) has been receiving complaints about long delays during the peak traffic period of the day which extends from 11:00 a.m. to 5:00 p.m. (the lock system operates from 8:00 a.m. to 8:00 p.m.). Fortunately, traffic rapidly diminishes after 5:00 p.m., and the queues that normally exist at the end of the busy period generally empty by the 8:00 p.m. closing time.

The usable length of the lock compartment is 40 m. One option that is being considered by management to resolve the excessive delay problem is to increase the compartment length to 50 m. There are significant costs involved in such a reconfiguration, and there is uncertainty about what impact it would have on the delays experienced by the various boat categories. A modelling and simulation study has been proposed as a means for acquiring insight into the effectiveness of the plan. Furthermore, an increase in boat traffic is anticipated over the short term, and it has been suggested that the proposed compartment extension would also be able to accommodate at least a 15% increase. This possibility is also to be investigated by the study.

The boats travelling in this waterway fall into three types which we refer as 1, 2 and 3. These reflect a size attribute (i.e. small, medium and large, respectively). The actual length of arriving boats in category k is uniformly distributed in the range $[L_k - \Delta_k, L_k + \Delta_k]$ metres. During the high-traffic portion of the day, the inter-arrival time for boats in category k is exponentially distributed with mean μ_k (minutes). The values of the various constants are given in Table 4.17.

Table 4.17 Size and arrival attributes of the three boat types

Size	μ_k (minutes)	L_k (meters)	Δ_k (meters)
Small (k = 1)	5	6	1
Medium (k = 2)	15	9	1.5
Large (k = 3)	45	12	1.5

(a) Formulate a set of performance measures that would likely be of value for assessing the effectiveness of the proposed lock extension within the context of a modelling and simulation study.

(b) Develop an ABCmod conceptual model that captures the various relevant aspects of the problem.

4.5 HappyComputing Inc. is a personal computer service, sales and rental shop. Customers who arrive at the shop fall into one of four categories depending on the nature of the 'work' which results from their visit. These are labelled as follows:

- C1: This customer wishes to purchase or rent a PC.
- C2: This customer is returning a rental PC.
- C3: This is a customer who has brought in a PC that requires service of a relatively minor nature (e.g. upgrade of hard drive or installation of additional memory). The customer typically waits for the service to be completed or possibly returns later in the day to pick up the machine.
- C4: This is a customer whose PC has a problem that needs to be diagnosed before repair can be undertaken. In this case, the customer leaves the PC in the shop with the understanding that he/she will be telephoned when the problem has been corrected.

The shop has three employees: one is a salesperson and the other two are technicians. One technician (the senior technician) has extensive training and considerable experience. The other (the junior technician) has limited training and skills. The salesperson is the initial point of contact for all arriving customers. The needs of both type C1 and type C2 customers are handled exclusively by the salesperson.

The shop is open on Monday through Saturday inclusive from 9:00 a.m. to 6:00 p.m. The salesperson (or a substitute) is always present. The senior technician has a day off on Mondays, while the junior technician's day off is Thursday. Each employee has a 1-hour lunch break.

Customers of the type C3 category are handled by the junior technician on a first-in, first-out (FIFO) basis. However, in about 20% of the cases, the junior technician is obliged to consult with the senior technician in order to deal with some aspect of the servicing requirement. This draws the senior technician away from his normal work activity which is the servicing of the PCs that are brought to the shop by category C4 customers. Note that on Thursdays the senior technician takes responsibility for the C3 work on a priority basis (i.e. he or she always interrupts his C4 task to accommodate the C3 customer).

It is the policy of the shop to carry out a comprehensive examination of all rental PCs when they are returned and to carry out any necessary refurbishing before they placed back into the rental pool. The refurbishing includes a range of possible tasks that usually involve hardware and software upgrades. This refurbishing activity is the responsibility of the junior technician, but it is carried out only when there are no PCs from category C3 customers who require service.

The inter-arrival times of each customer category over the course of a business day are uniformly distributed; however, the parameters of the distributions vary according to customer type. The boundaries of the various uniform distributions are summarized in Table 4.18.

Each arriving customer, in effect, generates service requirement which requires time to complete. The service time requirement for each customer category is a random variable. The assumed distribution for each of the categories is given in Table 4.19 (together with associated parameter values).

The owner of the shop (who, in fact, is the senior technician) wants to decrease the turnaround time for the PCs brought in by C4 customers because that is the part of the business that he is especially interested in 'growing'. He has, furthermore, noticed that the current workload of the junior technician often leaves him with significant amounts of idle time. He is therefore considering asking the junior technician to take a number of courses in order to upgrade his technical skills. This will have two benefits. First, it will enable the junior technician to deal with the service requirements of C3 customers without having to request assistance, and second, it will enable the (upgraded) junior technician to assist with the servicing of the PCs brought to the shop by the C4 customers when he is not otherwise occupied with his current responsibilities. The owner anticipates that the net impact will be a reduction of about 25% in the turnaround time for the C4 category of service.

The goal of this modelling and simulation project is to determine if the owner's expectation is correct.

Table 4.18 Inter-arrival times for customer types

Customer type	Min (minutes)	Max (minutes)
C1	70	130
C2	110	170
C3	180	260
C4	120	210

Table 4.19 Service time requirements for each customer type

Customer type	Distribution of service time requirement	Distribution parameters[a] (minutes)
C1	Normal	$\mu = 25$, $\sigma^2 = 10$
C2	Uniform	min = 25, max = 35
C3	Triangular	a = 30, b = 75, c = 45
C4	Triangular	a = 45, b = 175, c = 140

[a]See Sect. A1.4.4 of Annex A

(a) The problem statement as given above omits several details that need to be provided before the project can be realistically undertaken. Identify these and suggest meaningful clarifications.

(b) What is a realistic performance measure for this study? Do you regard this as a bounded horizon study or a steady-state study?

(c) Develop an ABCmod conceptual model.

4.6 Reneging occurs when a customer in a queue (or anything else that is enqueued) has waited too long for service and abandons the queue. The length of the wait time that triggers the reneging event may be fixed or may be a random variable. There are various ways in which reneging can be handled within the ABCmod framework, and the purpose of this exercise is to formulate at least one such approach.

In Sect. A1. and Sect. A1.2, we outline a simple modelling and simulation project formulated around a fast-food outlet called Kojo's Kitchen. An ABCmod conceptual model that evolves from the project description is also presented. The SUI, as presented, does not include customer reneging. Your task is to introduce this feature and appropriately modify/extend the given ABCmod conceptual model so that reneging is incorporated.

Suppose we use the variable *renegeTime* to represent the length of time that a customer will wait in the queue before becoming disgruntled and leaving.

Fig. 4.15 Distribution
for the random variable
renege time

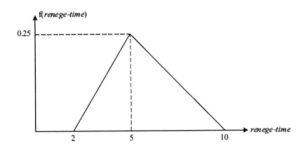

For definiteness, assume that *renegeTime* is a random variable and that it has a triangular distribution with parameters as shown in Fig. 4.15.

Note finally that a necessary part of any reneging specification is the clarification of what happens to a 'customer' that reneges. Assume that in the case of Kojo's Kitchen, such a customer simply disappears from the SUI. Consider using a scheduled sequel action.

References

1. Birta LG, Zhang M (1998) SimNet a conceptual modeling tool for simulation. In: Proceedings of the summer computer simulation conference, 1998
2. Gershwin SB (1991) Hierarchical flow control: a framework for scheduling and planning discrete events in manufacturing. In: Yu-Chi H (ed) Discrete event dynamic systems. IEEE Press, Piscataway
3. Haas PJ (2002) Stochastic Petri nets: modelling, stability, simulation. Springer, New York
4. Hills PR (1973) An introduction to simulation using SIMULA. Publication No. S55. Norwegian Computing Center, Oslo
5. Jensen K (1997) Coloured Petri nets, basic concepts, analysis methods and practical use (3 volumes). Springer, London
6. Kreutzer W (1986) System simulation: programming styles and languages. Addison-Wesley, Sydney/Wokingham
7. Martinez JC (2001) EZStrobe: general-purpose simulation system based on activity cycle diagrams. In: Proceedings of the 2001 winter simulation conference, IEEE Press, Arlington, VA, pp 1556–1564
8. Peterson JT (1981) Petri net theory and the modeling of systems. Prentice-Hall, Englewood Cliffs
9. Pidd M (2004) Computer simulation in management science, 5th edn. Wiley, Chichester
10. Pidd M (2007) Dynamic modeling in management science. In: Fishwick PA (ed) Handbook of dynamic system modeling. Chapman & Hall/CRC, Boca Raton
11. Robinson S (2004) Simulation: the practice of model development and use. Wiley, Chichester
12. Robinson S (2011) Conceptual modeling for simulation: definition and requirements. In: Robinson S, Brooks R, Kotiadis K, van der Zee D-J (eds) Conceptual modelling for discrete-event simulation. CRC Press, Boca Raton, pp 3–30
13. Robinson S, Brooks R, Kotiadis K, van der Zee D-J (eds) (2011) Conceptual modelling for discrete-event simulation. CRC Press, Boca Raton
14. Shannon RE (1975) Systems simulation: the art and science. Prentice Hall, Englewood Cliffs

15. Shi JJ (2000) Object-oriented technology for enhancing activity-based modelling functionality. In: Jones JA, Barton RR, Kang K, Fishwick PA (eds) Proceedings of the 2000 winter simulation conference, Orlando, Florida, 10–13 Dec 2000, pp 1938–1944
16. Tadao M (1989) Petri nets: properties, analysis and applications. Proc IEEE 77:541–580
17. Wang J (1998) Timed Petri nets, theory and application. Kluwer, Boston
18. Wang J (2007) Petri nets for dynamic event-driven system modeling. In: Fishwick PA (ed) Handbook of dynamic system modeling. Chapman & Hall, Boca Raton, pp 24(1)–24(17)
19. Zeigler BP (1976) Theory of modeling and simulation. Wiley, New York
20. Zeigler BP, Praehofer H, Kim TG (2000) Theory of modeling and simulation: integrating discrete event and continuous complex dynamic systems. Academic, San Diego

DEDS Simulation Model Development

5

5.1 Constructing a Simulation Model

The simulation model associated with a modelling and simulation project is a computer program that captures the behavioural and structural details of the SUI as specified by an ABCmod conceptual model. There are two important features of this computer program. The first is simply the fact that, like any computer program, its development must respect the rules of the programming language/environment chosen as the vehicle for implementation. The second feature, however, is distinctive to the modelling and simulation paradigm. It relates to the perspective the program writer takes with respect to the manner in which the model dynamics are 'packaged'; in particular, the management of time is a major concern.

For example, a reasonable choice might appear to be simply the incrementing of time in small steps coupled with an implementation of the behavioural constructs formulated in the ABCmod conceptual model, as outlined in Chap. 4. This, however, is not a practical choice because it does not lend itself to an efficient time advance mechanism which is a key constituent in the execution of any simulation model. With this approach at each time step, it must be determined if any event specified in the model's behavioural constructs occurs. This approach is exceedingly inefficient.

At its most fundamental level, the behaviour generated by a DEDS simulation model evolves as a consequence of discrete events at points in time which change the value of some of the model's variables. As discussed in Sect. 4.2.2.1, there are two main types of events: conditional events and scheduled events.[1] The execution of a simulation model must include the means for managing these events, i.e. ensuring their correct sequential execution, in other words the correct sequence of changes to the model's status as stipulated in the event specifications.

---

[1]Tentatively scheduled events and intercepting events are not considered here since traditional world views do not typically deal explicitly with interruptions.

© Springer Nature Switzerland AG 2019
L. G. Birta and G. Arbez, *Modelling and Simulation*, Simulation Foundations,
Methods and Applications, https://doi.org/10.1007/978-3-030-18869-6_5

Four word views (i.e. perspectives that can be adopted by the model builder in the simulation model formulation process) are outlined in Sect. 5.2. These are the Activity Scanning world view, the Event Scheduling world view, the Three-Phase world view, and the Process-Oriented world view. Each has its own approach for organizing these event specifications for execution. Much has been written on these world views, e.g. Banks et al. [1] and Pidd [8]. An overview is presented here in terms of the concepts presented in Chaps. 3 and 4, that is, in terms of scheduled and conditional events together with activities.

As we have emphasized in Chap. 4, an important purpose for a conceptual model for any particular SUI is to provide a specification for a simulation model. This specification necessarily identifies the events that underlie the behaviour that needs to be generated by the simulation model and these must be accommodated by whatever world view is chosen to characterize the simulation model. An implication here is that a conceptual model formulated within the ABCmod framework will be sufficient for the development of a simulation model based on any of the world views. This is indeed the case. Sections 5.3 and 5.4 outline how to transform an ABCmod conceptual model into a Three-Phase simulation model and into a Process-Oriented simulation model, respectively.

5.2 The Traditional World Views

The implementation of behavioural rules and time advance management in the various world views are based on list processing. One important list is the list of scheduled events frequently called the future event list (FEL). The list entries are commonly referred to as event notices. A notice contains the two components of a scheduled event: its event time and a reference to the event SCS (implemented by an 'event routine' in the program). Such a list is ordered according to event time. Notices are selected sequentially from the top of the list and when appropriate, the system clock which is a variable that contains current (simulated) time, is advanced to the event time found in the notice. This approach clearly offers significant efficiency relative to a rudimentary approach where time is simply incremented in small steps.

Any number of other lists can be used such as one that contains the references to the model's conditional events. After each change in the state of the model, preconditions in this list are tested and when appropriate, conditional events occur.

For example, GPSS (General-Purpose Simulation System) is basically a list processor that moves transactions from chain to chain[2] (GPSS *Transactions* are entities). In the case of the Process-Oriented world of GPSS, the transactions contain 'system' attributes that indicate both the next event within its process and that event's event time in the case of scheduled events (details about the Process-Oriented world view are provided in Sect. 5.2.4). Thus, transactions can be

---

[2]GPSS supports the Process-Oriented world view and uses the term *chain* instead of list.

processed like event notices as described above. Among the many GPSS chains
(lists) are:

- The Future Event Chain (FEC): a list that plays the same role as the FEL as
  described above.
- The Current Event Chain (CEC): a list of transactions that reference a conditional
  event.

The discussion that follows provides more details about each of the traditional
world views with a particular focus on time advance.

5.2.1   The Activity Scanning World View

As with the ABCmod framework, the Activity Scanning world view also models a SUI
behaviour as a set of activities (using the activity cycle diagram described at the end of
Sect. 4.3.2.3). In the implementation of an Activity Scanning world view model, these
activities are separated into events. Time advance consists of incrementing current time
in small steps as described in Sect. 5.1 and illustrated in Fig. 5.1. The execution and
time advance algorithm used in the traditional Activity Scanning world view is often
referred to as the two-phase approach. The two phases of the algorithm are:

- Phase A: Increment simulated time with a small time step (The size of this step is
  a fundamental implementation issue!).
- Phase B: Evaluate the preconditions of each event and process (via the execution
  of an event routine) the SCS of those whose precondition has become TRUE.
  Note that even a scheduled event is treated as a conditional event where '*current
  time* $\geq$ *event time*' is the precondition.

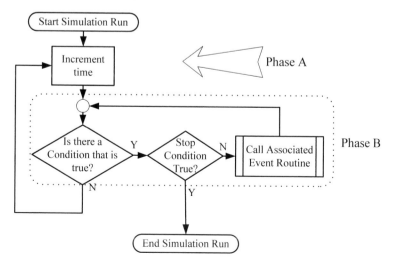

Fig. 5.1 Execution and time advance algorithm for the Activity Scanning world view

The Activity Scanning world view does not use a FEL to advance time. This is a distinctive feature of this world view. The inefficiency of the time advance mechanism that is used represents the fundamental shortcoming of this world view.

5.2.2 The Event Scheduling World View

In the Event Scheduling world view a DEDS simulation model is formulated in terms of a set of future events.[3] A future event encompasses a scheduled event together with all conditional events that could be affected by the occurrence of the scheduled event. A future event begins with a scheduled event (e.g. the terminating event within an ABCmod activity is a scheduled event) whose SCS might enable the precondition of one or more conditional events which may, in turn, cause further status changes. These changes may enable more conditional events, and this cascade may continue. This can lead to additional future events being scheduled as will become apparent in the discussion that follows. Thus, a future event is a collection of tasks that includes model variable changes and the possible scheduling of other future events. These actions all occur at the same value of (simulated) time. The simulation model's behaviour is formulated in terms of these future events, and a simulation run therefore unfolds as discontinuous jumps in time over the course of the observation interval.

An Event Scheduling world view requires specification at an event level rather than an activity level. The modeller must define the set of future events in order to capture the model's behaviour; any notion of an activity or some other higher level modelling construct is not present in this world view. This can make the task of model building more challenging than an activity-based approach [7].

Time advance in the Event Scheduling world view is achieved using a future event list (FEL) as illustrated in Fig. 5.2. The procedure is much more efficient than the approach that prevails in the Activity Scanning world view outlined earlier.

The execution and time advance algorithm contains a loop that processes the event notices on the FEL until a stop event notice or stop condition is encountered. The processing of an event notice has two steps: (1) the clock is advanced to the value of the event time, and (2) the referenced event routine is called to change the model's status and possibly add new notices to the FEL. The notice at the head of the FEL is the next one to be processed and is called the *imminent event notice*.

Stopping the simulation run can be achieved using either of two methods. The first is to run the simulation until a predetermined time. In this case, a stop event notice is placed on the FEL; its event time is the prescribed termination time for the simulation run. In the second method, a specified stop condition on the model's status is used to define termination of the simulation run.

---

[3]The term 'future event' is introduced here to distinguish it from the scheduled event and the conditional event that we have previously defined (Sect. 4.2.2.1). The word 'event' is traditionally used in discussing Event Scheduling models.

Fig. 5.2 Execution and time advance algorithm for the Event Scheduling world view

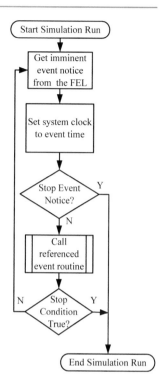

5.2.3 The Three-Phase World View

The Three-Phase world view is a variation of the Activity Scanning world view outlined earlier. The activity-based perspective is retained, but the rudimentary time advance algorithm is replaced with a significantly more efficient approach.

The main difference from the Activity Scanning world view is that it maintains a distinction between scheduled events and conditional events. Furthermore, it uses a future event list to advance time as in the Event Scheduling world view. As in the case of the Activity Scanning world view, activities are broken up into constituent events. The terminology in this world view includes 'B events' (for bound) which are scheduled events and 'C events' (for conditional) which are conditional events. This world view has much in common with our Activity Object World view (see Chap. 6). The three phases of the execution and time advance algorithm are illustrated in Fig. 5.3 and are summarized below:

- Phase A: This is the time advance phase. Remove the B event notice (i.e. imminent event notice) from the FEL and adjust the current time to its event time.
- Phase B: Processing the B event consists of executing its referenced event routine.

- Phase C: Processing the C events consists of testing preconditions of conditional events and executing the associated event routines of those whose precondition evaluates to TRUE. The loop in this phase is repeated until all preconditions are FALSE or until a Stop condition is encountered.

A model based on the original Activity Scanning world view can easily be cast into the Three-Phase world view by separating the events into B events (i.e. the scheduled events) and C events (i.e. the conditional events). Notice that if the C events are incorporated into appropriate B events to create a set of future events, then an Event Scheduling world view emerges.

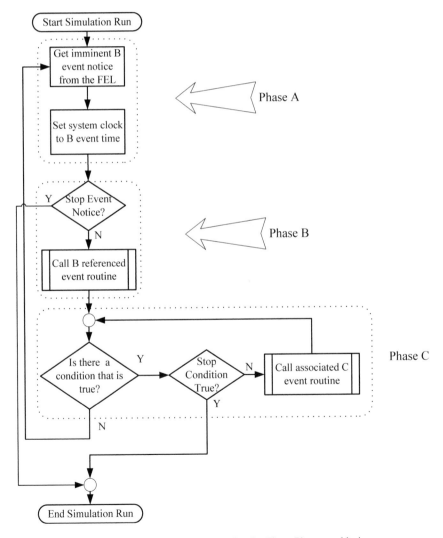

Fig. 5.3 Execution and time advance algorithm for the Three-Phase world view

5.2.4 The Process-Oriented World View

While the world views presented so far use approaches that break activities down into constituent parts, the Process-Oriented approach interconnects activities into units that correspond to entity life cycles very similar to those presented in Chap. 4. However, the life cycle presented in the Process-Oriented world view is typically a sequence of events that emerge from the activities in which an entity engages. The 'process' in this world view corresponds to such a life cycle. There are many of these and they interact as simulated time unfolds.

A commonly occurring example is a process that captures the entry of a consumer entity instance into the model, its participation in a number of activities and finally its departure from the model. Processes can also be defined for resources that participate in one activity after another (possibly in a circular manner) without ever leaving the model. Consider Fig. 4.13 that presents the life cycle of the shoppers moving from activity to activity in the department store model. A Process-Oriented simulation program would 'execute' a separate life cycle for each instance of a shopper entity.

GPSS (General-Purpose Simulation System) [9] is one of the oldest simulation environments (it turned 50 in 2011 [10]), but its basic Process-Oriented principles are still relevant, and versions of GPSS continue to be available [6]. An overview of GPSS is presented in Annex 3. Section 5.1 presented the GPSS notion of the transaction (entity) which is moved from chain (list) to chain in the simulation model; this movement is governed by a sequence of program statements called Blocks. Blocks are associated with functions that can act upon the transaction that 'enters' the Block, including the scheduling of the transaction on the FEC (Future Event Chain), or act upon other transactions located elsewhere in the model (e.g. moving another transaction from a retry chain to the CEC (Current Event Chain)). A transaction contains special system parameters (i.e. attributes) used for 'moving' transactions along its process. The most notable are the BDT (block departure time), the Current Block and the Next Block. Whenever a transaction is placed on the FEC, an event time is inserted into the BDT. The Current Block and Next Block contain, respectively, block numbers (statement numbers) indicating the block where the transaction is currently located and the next block the transaction will attempt to enter. These three parameters play a pivotal role in moving a transaction along its life cycle.

SIMSCRIPT [4] is a simulation environment that offers both the capability of developing models according to the Event Scheduling world view or the Process-Oriented world view. SIMSCRIPT has evolved into a general-purpose computer programming language, and its most recent version SIMSCRIPT III (Release 5.0) supports object-oriented programming. SIMSCRIPT documentation uses the following terminology (Law and Larmey [4]):

- Process: A time-ordered sequence of interrelated events separated by passages of time (either predetermined or indefinite), which describes the entire experience of an "entity" as it flows through a system.
- Process entity: An entity for which a process has been defined.
- Process routine: A segment of code, a subroutine, that provides the description of a process. This routine with multiple entry points plays the same role as the sequence of GPSS blocks described earlier. It contains statements that can act on a process entity or other entities in the model. SIMSCRIPT schedules processes, each of which involves the execution of a process routine for a specific process entity.
- Process notice: An entry added to a list that is used for scheduling a process in a manner that parallels the GPSS scheduling of a transaction on the FEC (recall that a transaction is associated with a process). Thus, scheduling a process in SIMSCRIPT or a transaction entity in GPSS are equivalent concepts.

Although a different terminology is used, both GPSS and SIMSCRIPT use the same approach in advancing time and updating the simulation model according to event SCSs. The following presents the execution and time advance algorithm in general terms for both these simulation environments.

We begin with the assumption that there is a 'process record' (PR) associated with each entity, e_k, that exists within the simulation model at (simulated) time t. The process record for e_k [denoted by $PR(e_k)$] has two essential components, namely, the life cycle of the entity e_k or, more specifically, a representation of the sequence of events that comprise its life cycle (i.e. its process) and a pointer to the current location of the entity in this life cycle, namely, either an event initiation basis or an event SCS (we refer to this pointer as the 'location pointer').

The events which punctuate the path of the entity's process cause changes to the model status, and it is these changes encapsulate the model's behaviour. The specifications for these events are embedded within each process record.

$PR(e_k)$ has either an 'unconditional status' or a 'conditional status'. The former indicates that the location pointer refers the initiation basis of a scheduled event, while the latter indicates that the location pointer refers the initiation basis of a conditional event (or to the members of a group of possible alternatives).

The task of the execution and time advance algorithm for a Process-Oriented simulation model is to correctly manage the progress of the entities based on the information in their respective process record. We consider two lists that are central to the algorithm's operation; these are the Scheduled List (SL) and the Pending List (PL). The SL is a list of notices, while the PL is a list of the process records that have a conditional status. A notice consists of a process record with an unconditional status and the event time of the referenced event. In effect, the SL is a time-ordered list used to advance time in a similar fashion to the FEL in other world views. The PL is a list of process records for which the location pointer points to conditional events (either its initiation basis or its SCS; however there is one exception, namely, it points to the SCS of a scheduled event after the process record is moved from the SL to the PL as described in step 3 below). These conditional

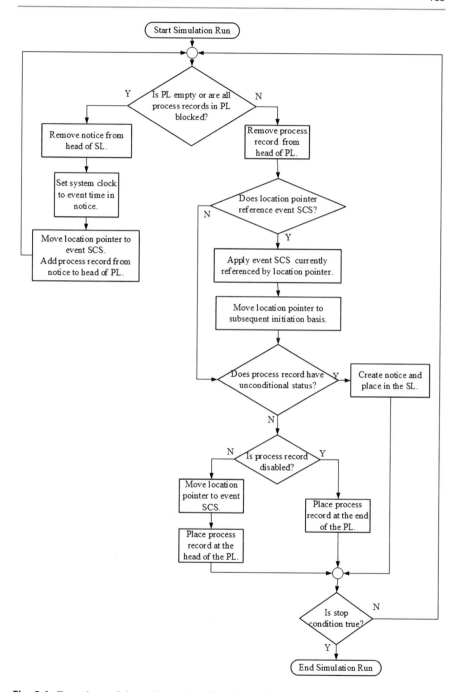

Fig. 5.4 Execution and time advance algorithm for the Process-Oriented world view

events may or may not be blocking progress of some entity because the model status is such that the associated precondition for these events is not enabled (i.e. does not have a TRUE value). When the progress of an entity in a process record is blocked, then we say the process record itself is blocked.

We begin with the assumption that there exist notices in the SL, and all process records on the PL are blocked. The steps of the execution and time advance algorithm (illustrated in Fig. 5.4) are as follows:

1. Remove the notice at the head of the SL.
2. Set current time to the event time of that notice.
3. Move the location pointer in the process record referenced in the notice to the event SCS of the scheduled event and place the process record at the head of the PL.
4. If the PL is empty or all process records on the PL are blocked, then go to step 1.
5. Remove the process record at the head of the PL.
6. If the location pointer is referencing an event SCS, then apply the event SCS to the model and move the location pointer to the initiation basis of the subsequent event.
7. If the process record is in an unconditional status, create a notice with the process record and an event time using the referenced initiation basis; place the notice on the SL and go to step 9.
8. Here the process record is in a conditional status, that is, the location pointer is referencing the precondition of a conditional event. If the precondition is TRUE (i.e. process record is not blocked), then move the location pointer to the event's SCS and place the process record at the head of the PL. Otherwise, move the blocked process record to the end of the PL.
9. If the stop condition is TRUE (i.e. the end of the observation interval has been reached), end the simulation run; otherwise, go to step 4.

Transforming a simulation model formulated in a non-Process-Oriented world view into a Process-Oriented simulation model is not always straightforward. For example, an event SCS in the originating ABCmod conceptual model is often separated into different processes of the Process-Oriented simulation model. This is illustrated in Sect. 5.4.

5.3 Transforming an ABCmod Conceptual Model into a Three-Phase Simulation Model

Figure 5.5 provides an overview of how the various components in an ABCmod conceptual model relate to those of a Three-Phase simulation model. The structural components of an ABCmod conceptual model consist of the entity structures, constants, parameters, input variables and output variables/sequences; these become data structures in the simulation model. Components representing behaviour are

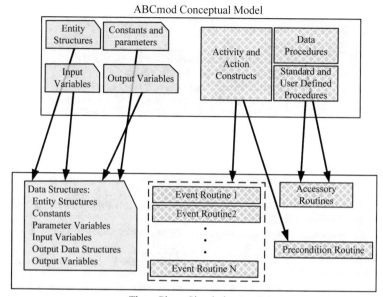

Fig. 5.5 Relating ABCmod components to Three-Phase simulation model components

transformed into 'routines'. We shall use Java [3, 5, 8] as our target environment, and hence the routines correspond to Java methods.

Translating the structural components is relatively straightforward. Typically, there is a direct mapping from the conceptual model components to appropriate data structures. In the Java programming environment, entity categories may be represented as classes defined by the simulation modeller (typical for the case of resource and consumer entity categories) or as one of the standard Java classes (typical for the case of queue and group entity categories). Table 5.1 lists a set of possible Java classes that can be used and extended to represent the various ABCmod entity structures. It is by no means complete, and the reader is encouraged to review available Java classes that could be appropriate for representing entities.

Translating the behavioural components, consisting of the actions, activities, data procedures (random variate procedures, deterministic value procedures) and user-defined procedures, is not as straightforward. Dealing with data procedures and user-defined procedures is relatively easy since they are typically coded as routines (e.g. Java methods). Often libraries are available that can provide routines for supporting required simulation functionality such as random variate generation and list processing. In the case of Java, the Java package cern.colt[4] provides a number

---

[4]CERN (European Organization for Nuclear Research) makes available a number of Java packages as Open Source Libraries for High Performance Scientific and Technical Computing in Java [2]. It provides a number of classes that implement stochastic data models. Version 1.2.0 was used during the writing of this book.

Table 5.1 Representing ABCmod entities with Java objects

Role of ABCmod entity structure	Java standard class
Queue	*ArrayList* (implementation of a resizable array) *LinkedList* (a linked list implementation) *ArrayBlockingQueue* (bounded FIFO queue) *ConcurrentLinkQueue* (thread-safe FIFO queue using linked nodes) *DelayQueue* (unbounded queue of delayed elements – when an elements' delay has expired, it is moved to the head of the queue in FIFO fashion) *LinkedBlockingQueue* (bounded queue implemented with linked nodes) *ProrityBlockingQueue* (similar to *PriorityQueue*, but offers also blocking retrieval functions when queue is empty) *PriorityQueue* (unbounded queue that orders elements according to priority) *AbstractQueue* (skeletal queue implementation that can be extended to create user-specific queue)
Group	*AbstractSet* (skeletal set implementation that can be extended to user-specific set) *HashSet* This class is an extension of the *AbstractSet* and supports the set methods using a hash table *LinkedHashSet* Similar to the *HashSet* class but stores objects in the set using a doubly linked list. This provides an order to the added objects, that is, it is possible to remove the objects in the same order they were stored

of classes for implementing random variates. The challenge in transforming behaviour constructs from an ABCmod conceptual model to a simulation model is with capturing the behaviour construct specifications in a set of event routines.

Each activity and action in an ABCmod conceptual model is separated into its constituent events, and these give rise to the Three-Phase C events (conditional events) and the Three-Phase B events (scheduled events). Fortunately, there exists a pattern in this reorganization of the constituent events as shown in Fig. 5.6. Notice, in particular, that an event routine is defined for the SCS of each scheduled action, for the SCS of the starting event of each scheduled activity and for the SCS of the terminating event of each activity that exist in the conceptual model. Likewise, the SCSs of scheduled sequel actions, the starting events of sequel activities and the starting events of scheduled sequel activities are transformed into event routines (not shown in Fig. 5.6). Thus, each B event in the Three-Phase simulation model is an implementation of the SCS of a scheduled event from within the activities and actions of the ABCmod conceptual model. The SCS of the events of conditional actions and the SCS of starting events of conditional activities are translated to C events which are collected into a precondition routine. Not shown in Fig. 5.6 are the intercepting events for extended activities that are also translated to C events.

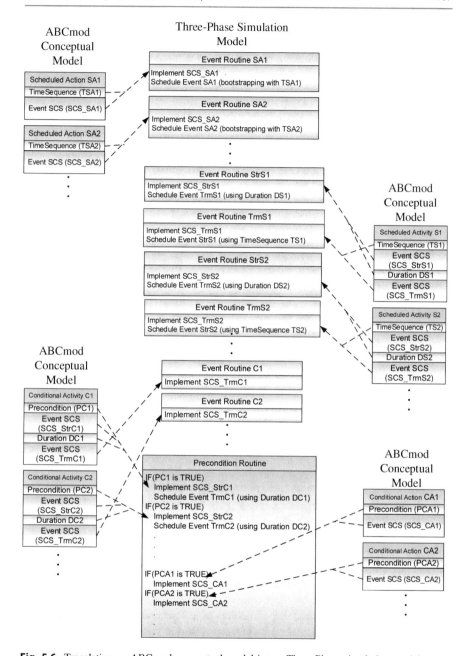

Fig. 5.6 Translating an ABCmod conceptual model into a Three-Phase simulation model

The B event routine associated with a scheduled action implements the action's SCS and schedules subsequent B events using the action's time sequence specification which is transformed into a routine that returns the next event time for the B event. This self-scheduling is referred to as *bootstrapping*.

The B event routine associated with the starting event of a scheduled activity implements the SCS of that activity's starting event and schedules the activity's terminating event using the activity's duration specification. The duration is added to the current time to produce the event time of the terminating event. The B event associated with the terminating event of any activity implements the event SCS of the terminating event. In the case of the scheduled activity, the B event associated with the terminating event also schedules the B event associated with the starting event using the time sequence in a similar fashion to bootstrapping outlined earlier for the B event routine associated with the scheduled action.

Figure 5.6 shows a precondition routine in which all C events are implemented. Calling the precondition routine represents the third phase of the Three-Phase execution and time advance algorithm presented in Sect. 5.2.3. This routine is called after the occurrence of each B event.

It should be noted, however, that this is not the most efficient approach because checking all preconditions of each C event routine may not be necessary. Typically, any particular B event routine can affect only a limited number of preconditions. Distributing the C events (i.e. the preconditions and SCSs) directly into the appropriate B event routines would likely increase efficiency. Note however that such an approach gives rise to an Event Scheduling simulation model.

The precondition routine has the task of implementing the conditional actions by carrying out the SCS of any action whose precondition acquires a TRUE value. It is also responsible for starting conditional activities. In other words, testing preconditions and, when the corresponding precondition acquires a TRUE value, carrying out starting event SCSs. It also handles each activity instance's duration by scheduling its terminating event, i.e. the corresponding B event.

Finally, the intercepting events (not shown) are also implemented in the precondition routine. When the intercepting precondition has a TRUE value, its event SCS is implemented and this typically includes the termination of the interrupted activity. This is accomplished by removing from the FEL the B event associated with that activity instance's terminating event.

The steps for creating a Three-Phase simulation model from an ABCmod conceptual model can be summarized as follows:

- Step 1 – Define appropriate data structures to accommodate the entity structures, constants, parameters and input variables within the conceptual model.
- Step 2 – Formulate appropriate implementations for the data procedures and user-defined procedures specified in the conceptual model. Data models such as random variate generators are typically implemented using available libraries; for example, in Java the COLT [2] classes *Exponential* and *Uniform* provide random variates according to the exponential and uniform distributions, respectively.

User-defined procedures are implemented as routines (i.e. as methods in the case of Java).

- Step 3 – Define a B event (an event routine) that implements the SCS of (1) each scheduled action, (2) each scheduled sequel action, (3) the starting event of each scheduled activity, (4) the starting event of each scheduled sequel activity and (5) the terminating event of each activity within the ABCmod conceptual model. In the case of scheduled actions and scheduled activities, the B event routine also implements the bootstrapping requirements.
- Step 4 – Implement a precondition routine which will, when corresponding preconditions acquire a TRUE value, (1) execute conditional actions, (2) initiate conditional activities and (3) invoke interruptions.
- Step 5 – Develop appropriate program code to generate the required output data. For generating sample sequence output, the SCSs in the ABCmod conceptual model explicitly indicate the output requirements [i.e. SM.Put()]. However, for trajectory sequence output, the requirement is often implicit. Program code must be developed to output the time/value pairs corresponding to changes in the variable associated with each trajectory sequence. For example, after the execution of a scheduled event SCS and all triggered conditional event SCSs, update all piece-wise constant output variables whose value has changed.

We illustrate the application of these steps in the context of the ABCmod conceptual model formulated for Kojo's Kitchen in Sect. A1.1. The Java implementation consists of a number of classes, each containing the implementation of different components of the ABCmod conceptual model. The principle simulation model class, *KojoKitchen*, instantiates all required objects. The *KojoKichen* class is an extension of the *ThreePhase* class which provides all required methods to execute the simulation model (e.g. scheduling events and running the simulation). In other words, instantiating a *KojoKitchen* class creates an object that is able to execute the simulation model.

Starting at Step 1, the ABCmod entity structures are implemented using a number of Java classes. The instances of the Customer entity structure (i.e. iC. Customer) are represented in Java as *Customer* objects (see the *Customer* class in Fig. 5.8) with two attributes called '*uType*' (set to a value of the enumerated type *Type*) and '*startWaitTime*' (contains the *double* value corresponding to the time that the customer enters the customer line). *Customer* objects are instantiated during the execution of the B events associated with the Arrivals scheduled action.

The 'RG.Counter' entity structure is implemented as a *Counter* object referenced by the variable *rgCounter* declared in the *KojoKitchen* class (see Fig. 5.7). In Fig. 5.8, the *HashSet* object referenced by *counter* provides the methods to add a *Customer* object to the group (*add*), remove a *Customer* object from the group (*remove*) and get the number of *Customer* objects in the group (*size*). The three methods *getN*, *spInsertGroup* and *spRemoveGroup* are used, respectively, to obtain the value of '*n*', to implement the standard procedure SP.InsertGrp and to

```
// The Simulation model Class
class KojoKitchen extends ThreePhase
{
    /* Group/Queue Entities Categories*/
    Counter rgCounter = new Counter();
    ArrayList<Customer> qCustLine =
                    new ArrayList<Customer>();
    /* SSOVs */
    public int ssovNumServed = 0;
    public int ssovNumLongWait = 0;
    public double ssovPropLongWait= 0;
    // Random Variate Procedures
    RVP rvp;
    // Model Behaviour
    // Constructor
    public KojoKitchen(double t0time, double tftime,
                    int addE, Seeds sd)
    { … }

    // B Events
    BEvents bevents = new Bevents(this);
    public void executeEvent(int evName, Object obj)
    {
        bevents.processEvent(evName, obj);
        preConditions();    // preconditions
    }

    // Check for starting event of Serving
    void preConditions()  { … }

    // termination explicit
    public boolean implicitStopCondition( )
    { return(false); }
}
```

Fig. 5.7 The KojoKitchen class

implement the standard procedure SP.RemoveGroup. Finally, the queue entity structure 'Q.CustLine' is implemented as an *ArrayList* object referenced by the variable *qCustLine* in the *KojoKitchen* class.

Step 1 is completed by encoding the constants, parameters and input variables that are specified in the ABCmod conceptual model. There are no constants in this example, and the parameter, RG.Counter.addEmp, is declared as the instance variable *addEmp* in the *Counter* class. The endogenous input variable RG.Counter.

```
public class Customer
{
        enum Type {W, U}; // W - sandwich, U - sushi
        Type uType;
        double startWaitTime; // Time start waiting
}

public class Counter
{
        public HashSet<Customer> counter =
                            new HashSet<Customer>(3);
        int uNumEmp; // Number of employees
        int addEmp; // Parameter

        public int getN() {return counter.size();}
        public void spInsertGroup(Customer icCustomer)
        { counter.add(icCustomer); }
        public boolean spRemoveGroup(Customer icCustomer)
        { return(counter.remove(icCustomer)); }
}
```

Fig. 5.8 Customer class and Counter class

uNumEmp defined in the conceptual model is declared as the instance variable *uNumEmp* in the *Counter* class.

For Step 2, the random variate procedures are implemented in the class *RVP* as three Java methods as shown in Fig. 5.9. The reference variable *rvp* in the *Kojo-Kitchen* class is used to access these methods. Four objects serve to implement the data models, one for inter-arrival times (an *Exponential* object), one random number generator (*MersenneTwister* object) for determining the type of an arriving customer and two others for service times (two *Uniform* objects referenced by the variables *wSrvTm* and *uSrvTm*). Three methods *duC()*, *customerType()* and *uSrvTm()* implement, respectively, the random variate procedures RVP.DuC(), RVP.CustomerType() and RVP.uSrvTm() that appear in the conceptual model. The class constructor *RVP* contains the necessary code to set up the various data model objects as well as the required initialization. Step 2 is now complete.

In Step 3 B events and corresponding event routines are implemented. Table 5.2 provides the names of the three B events for the Kojo's Kitchen model and the names of the corresponding event routines implemented as Java methods. The *processEvent* method, implemented with a simple *switch* statement, is shown in Fig. 5.10. Table 5.2 shows the association between B event names and integer identifiers. The *processEvent* is part of the class *BEvents* that contains the code implementing the B events. This method is called by the *executeEvent* method in the *KojoKitchen* class. The *executeEvent* method is called as part of the *runSimulation* method that implements the Three-Phase execution and time advance algorithm. The B event integer identifiers are stored in the event notice objects located in a FEL.

```java
public class RVP
{
    KojoKitchen m; // reference to the model
    /**  Implementation of Random Variate Procedures **/
    public RVP(KojoKitchen mod, Seeds sd)
    {
        m = mod;
        // Set up distribution functions
        interArrDist = new Exponential(1.0/Mean1,
                            new MersenneTwister(sd.arr));
        typeRandGen = new MersenneTwister(sd.type);
        wSrvTm = new Uniform(STWMIN,STWMAX,sd.wstm);
        uSrvTm = new Uniform(STUMIN,STUMAX,sd.ustm);
    }
    // Random Variate Procedure: RVP.DuC()
    // Means for sandwich customers inter-arrival times
    private final double Mean1 = 10;
    ...

    /* Data Models for implementing time sequence */
    private Exponential interArrDist; // Data model
    // Method
    public double duC()
    { ... }

    // Random Variate Procedure: RVP.CustomerType()
    private final double PROPW = 0.65;
    MersenneTwister typeRandGen;
    // Method
    public Customer.Type customerType()
    { ... }

    // Random Variate Procedure: RVP.uSrvTm()
    // Min and Max for service times
    private final double STWMIN = 3;
    ...
    private Uniform wSrvTm;  // Data model
    private Uniform uSrvTm;  // Data model
    // Method
    public double uSrvTm(Customer.Type type)
    { ... }
}
```

Fig. 5.9 Implementation of random variate procedures in the RVP class

Table 5.2 Simulation Model B events for the Kojo's Kitchen project

B event name	Identifier	Java method	ABCmod behavioural constructs
Arrivals	1	addCust	Arrivals action
StaffChange	2	staffChange	StaffChange action
EndServing	3	finishServing	Serving activity

```java
public void processEvent(int eventNum, Object obj)
{
    switch(eventNum)
    {
      case Arrivals: addCust(); break;
      case EndServing: finishServing((Customer) obj) ;
                       break;
      case StaffChange: staffChange() ; break;
      default:
              System.out.println("Bad event identifier"
                                 + eventNum);
             break;
    }
}
```

Fig. 5.10 Implementation of the *processEvent* method for Kojo's Kitchen

Figure 5.11 shows the three Java methods that implement the three B event routines; the methods are also part of the *BEvents* class. In the *addCust* method, an iC.Customer entity is instantiated as a *Customer* object and the '*type*' and '*start-WaitTime*' attributes are initialized appropriately (see Fig. 5.11). The *Customer* object is then added to the customer line using the *qCustLine.add* method. Bootstrapping is used in *addCust* to create a stream of arriving customers using the *duC()* method which gives the next arrival time. Scheduling the next arrival is accomplished with the *addEventNotice* method (provided within the *ThreePhase* class). The *Arrivals* integer identifier is passed to *addEventNotice* to be stored in the B event notice and it is passed back to the *processEvent* method (as shown in Fig. 5.10) when it reaches the head of the FEL. Note that if the *duC()* method returns a −1, the Arrivals B Event is no longer scheduled; this corresponds to the situation where simulated time has advanced beyond the closing time.

Figure 5.11 also shows the *staffChange* method which implements the scheduled action StaffChange as the B event *StaffChange*. The input variable *uNumEmp* (instance variable of the *rgCounter* object) is updated according to the current time as specified in the event SCS of the StaffChange action of the conceptual model. Again, bootstrapping is used to schedule the StaffChange action in order to update the value of the input variable. Note that in this case, however, a time sequence is implemented as an explicit array of times (*stChTmSeq*), and the instance variable *sctIx* is used to traverse that array. When the value −1 is reached, the action is no longer scheduled.

```
// Arrivals event
private void addCust()
{
    // Arrival Action Event SCS
    Customer icCustomer = new Customer();
    icCustomer.uType = m.rvp.customerType();
    icCustomer.startWaitTime = m.clock;
    m.qCustLine.add(icCustomer);
    // Schedule next arrival
    double nxtArrival = m.rvp.duC();
    if(nxtArrival != -1)
        m.addEventNotice(Arrivals,nxtArrival);
}
// Implementation of timeSequence
private double[] stChTmSeq = {0,90,210,420,540,-1};
private int sctIx = 0;
private void changeNumEmp()
{
    if(m.clock == staffChTmSeq [0])
                            m.rgcounter.uNumEmp = 2;
    else if(m.clock == staffChTmSeq [1])
        m.rgcounter.uNumEmp += m.rgcounter.addEmp;
    else if(m.clock == staffChTmSeq [2])
        m.rgcounter.uNumEmp -= m.rgcounter.addEmp;
    else if(m.clock == staffChTmSeq [3])
        m.rgcounter.uNumEmp += m.rgcounter.addEmp;
    else if(m.clock == staffChTmSeq [4])
        m.rgcounter.uNumEmp -= m.rgcounter.addEmp;
    else System.out.println("Invalid time: " + m.clock);
    sctIx++;
    if(stChTmSeq[sctIx] != -1.0)
    {
      sctIx++;
      m.addEventNotice(StaffChange, stChTmSeq[sctIx]);
    }
}
// Serving Terminating Event
private void finishServing(Object customer)
{
    // Serving Terminating Event SCS
    if(m.rgCounter.remove(customer)==false)
      System.out.println("Error: Couner empty");
}
```

Fig. 5.11 Encoding B event routines

Finally, Fig. 5.11 provides the details of the *finishServing* method (the event routine associated with the *EndServing* B event). This method illustrates how a reference to an object in the *EventNotice* is used. As shown in Fig. 5.12, when an event notice for the *finishServing* event is created, a reference to the customer, *iCCustomer* is added to the *rgCounter* object, and included in the notice. When *processEvent* (see Fig. 5.10) is called this reference is passed until it reaches the *finishServing* method that will use the reference to remove the appropriate *Customer* object from *rgCounter* object. This completes Step 3.

Step 4 provides the precondition routine which tests the preconditions of starting events. Because of the simplicity of Kojo's Kitchen project, only one precondition is tested in this routine, namely the precondition for the Serving activity. The *preConditions* method, shown in Fig. 5.12, is called by the *executeEvent* method in the *KojoKitchen* class after the *processEvent* is called (to process the imminent scheduled event). The *getN()* and the *size()* methods are available, respectively, in the *rgCounter* and *qCustLine* objects to determine the number of Customer objects

```
// The starting event SCS of the Serving activity
 void preConditions()
 {
     Customer icCustomer;
     double serviceTime;
     if(rgCounter.getN() < rgCounter.uNumEmp &&
        qCustLine.size() != 0) // Space at counter
     {
        icCustomer = qCustLine.remove(0);
        rgCounter.spInsertGroup(icCustomer);
        if( (clock-icCustomer.startWaitTime) > 5)
           ssovNumLongWait++;
        ssovNumServed++;
        ssovPropLongWait =
              (double)ssovNumLongWait/ssovNumServed;
        switch(iCCustomer.uType)
        {
          case 'W': serviceTime = clock+rvp.wSrvTm.nextDouble();
                    break;
          case 'U': serviceTime = clock+rvp.uSrvTm.nextDouble();
                    break;
          default: serviceTime = 1.0;
             System.out.println("Invalid customer type found " +
                          iCCustomer.uType + " ignored\n");
             break;
        }
        addEventNotice(bevents.EndServing,
                    serviceTime,
                    (Object)iCCustomer);
     }
 }
```

Fig. 5.12 Implementation of *preConditions()*

currently within each of these objects. The *remove(0)* method is used to remove the *Customer* instance from the *qCustLine* object. The starting event includes specifications to update the three SSOV output variables (which is part of Step 5 described below). The *addEventNotice* method schedules the Serving activity's terminating event, that is, it schedules the associated B event with its event time constructed from the activity's duration (either the *rvp.wSrvTm()* method or the *rvp.uSrvTm()* method according to the customer type given by *iCCustomer.uType*).

Step 5 deals with the output of sample sequence and trajectory sequence data. However, there is no requirement for neither a sample sequence nor a trajectory sequence in the Kojo's Kitchen project. Note in Fig. 5.12, that the requirement for recording output data is included in the starting event SCS of the Serving activity (implemented in the *preConditions* method). Here the values of the three SSOV variables *ssovNumLongWait* (number of customers that waited longer than 5 minutes), *ssovNumServed* (number of customers served) and *numPropLongWait* (proportion of customers that waited longer than 5 minutes) are updated.

5.4   Transforming an ABCmod Conceptual Model into a Process-Oriented Simulation Model

5.4.1   Process-Oriented Simulation Models

The process view for a simulation model begins with a collection of process specifications. Each of these can be formulated as an interconnection of some of the activities within an ABCmod. In one of the most common circumstances, the interconnections in a process specification are organized to reflect the life cycle of one of the transient consumer entities that have been identified. The implication here is that there generally is a process specification for each entity with *role* = Consumer and *scope* = Transient. An ABCmod conceptual model provides the entity life cycles in the high-level behavioural view. These provide the basis for organizing the activities in an ABCmod conceptual model into a collection of process specifications. In our examples below, the life cycles provided in the examples of Annex 1 are used.

It important to bear in mind that each consumer entity instance will "live" its own life cycle. This implies that each will have its own process instance and these can interact often by sharing (usually competing for) the various resources in the model.

An entity's flow from one activity to another within a process depends on the status of the 'downstream' activity's precondition. If it is FALSE, then flow is halted and a delay occurs. However a change from FALSE to TRUE as a consequence of a change in the model's status can occur. But what changes the model's status? Status changes result from the occurrence of events. When some ongoing activity terminates within an ABCmod activity instance the SCS of its terminating event usually changes some of the model's variables and this change could enable many pending activities within the various process instances. This is an example of how processes interact with each other. This situation is illustrated by the shopper entity that reaches the point in its process where it must wait in a queue. Its movement to the head of the queue and subsequently acquiring the service desk resource are consequences of other shopper entities progressing through their respective processes.

As an illustration of how processes can be defined for entities with *role* = Resource, consider the Port project discussed in Sect. A1.4. The activities MoveToHarbour and MoveToBerths involve only the tug resource entity. Because they do not involve the tankers, these activities are not part of the tanker process. They are, however, present in the tug process. Furthermore, the tanker process instances will interact with the tug process instance. For example, the arrival of the tanker in the port changes the model's status and this could result in a TRUE value for the precondition of the MoveToHarbour activity, thereby initiating it. Similarly, when the tug completes the Deberthing activity (which always involves a tanker), a Berthing activity could be initiated to move a waiting tanker in the harbour to the berths. Because activities such as Berthing and Deberthing involve both the tug and a tanker they become part of both the tanker process and tug process. This illustrates how interaction between processes can occur.

These process specifications are not linked to any particular simulation language or environment. The next section shows how the process specifications are used in formulating a simulation model in GPSS.

5.4.2 Overview of GPSS

We provide a brief introduction to the GPSS simulation environment. More extensive background material is provided in the GPSS primer given in Annex 3. Readers who are not familiar with GPSS should take the time to review Annex 3.

GPSS provides a Process-Oriented environment for developing simulation models. These models are formulated in terms of processes for GPSS *Transactions*. A Transaction is composed of a collection of parameters (these correspond to attributes in an ABCmod conceptual model). Each Transaction includes several additional standard parameters that support GPSS processing, e.g. a time parameter for scheduling and two references to GPSS Blocks (one that references the current Block in which the Transaction resides and the other that references the next Block the Transaction wishes to enter). GPSS manages Transactions on a number of lists using list-processing techniques. Two lists are especially important in GPSS, namely, the Future Event Chain (FEC) and the Current Event Chain (CEC). The FEC contains a list of Transactions ordered according to a GPSS standard time parameter. To advance time, the GPSS Scheduler will move the Transaction at the head of the FEC to the CEC and updates the simulation clock to the Transaction's standard time parameter (see Annex 3 for details). Transactions are scheduled by placing them on the FEC with a future time value stored in its time parameter. The Scheduler processes the Transactions on the CEC by invoking the functions of the Block referenced in the Transaction Next Block parameter. When the CEC becomes empty, the Scheduler returns to the FEC to fetch another Transaction.

GPSS Blocks are associated with specific functions, and they provide the basic processing elements for executing a simulation model. Conceptually, Transactions trigger these functions as they traverse the Blocks. For example, when a Transaction enters an ADVANCE Block, it will be delayed for some defined time before exiting the Block. The ADVANCE Block in fact schedules the entering Transaction on the FEC. This Block provides a natural mechanism for implementing the duration of an ABCmod activity.

The GPSS Block functions operate on structural entities which are internal data structures. It is important to be aware of such structural entities when creating a GPSS simulation model. For example, when a Transaction traverses the ENTER Block, the Block inserts the Transaction in a Storage structural entity. Table 5.3 shows possible mapping from ABCmod consumer entities and resource entities to the most common GPSS structural entities. Annex 3 provides a complete list of GPSS structural entities.

The mapping in Table 5.3 is not perfect. For example, consider the case where a resource entity is represented by a Transaction and it is necessary to attach a consumer entity (also represented by a Transaction) to the resource entity. In the ABCmod framework, the consumer entity can be referenced by an attribute of the resource entity. In GPSS, Transaction parameters are simple numeric values and consequently it is not possible to assign the Transaction representing the consumer entity to a parameter in the 'resource Transaction'. An alternate mechanism to accommodate this 'attachment' needs to be identified (a possible approach is outlined in the following section).

Development of a GPSS structural diagram as presented in Annex 3 is a recommended first step in creating a simulation model in GPSS. GPSS Blocks operate on the structural entities that appear in this diagram. Such diagrams can be inferred from the structural view in an ABCmod high-level conceptual model.

Elements of behaviour in a GPSS simulation model are expressed in terms of sequences of GPSS Blocks, each of which is called a GPSS Block segment (or simply a segment). A process in GPSS is formulated in terms of one or more such segments and has a graphical representation (see Fig. 5.13). In many cases, a process has a single segment. Segments start with a GENERATE Block through which Transactions enter the simulation model (this provides the means for handling the input entity stream specification of the ABCmod framework) and end with the TERMINATE Block when Transactions leave the simulation model. An

Table 5.3 Mapping ABCmod entities to GPSS structural entities	Role of an ABCmod entity	GPSS structural entity options
	Consumer	Transaction
	Resource	Facility or transaction
	Group	Storage or transaction group
	Queue	User chain

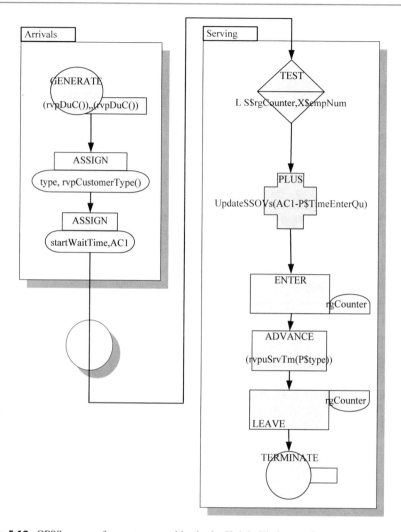

Fig. 5.13 GPSS process for customer entities in the Kojo's Kitchen project

example is shown in Fig. 5.13. The collection of segments that completely implements a process diagram can be regarded as GPSS process.

A GPSS simulation model is a collection of GPSS processes. Each Block within the graphical view of a GPSS process corresponds to a GPSS statement, thereby facilitating the construction of the corresponding GPSS program code. As an illustration, Fig. 5.14 shows the GPSS program fragment that corresponds to the Customer life cycle in ABCmod conceptual model for Kojo's Kitchen that is presented in Sect. A1.1.

```
*******************************************************
*   Customer Process
*******************************************************

***Arrival Activity
Customer GENERATE (rvpDuC()),,(rvpDuC()) ; Bootstrap. Block
        ; Scheduled Action SCS
        ASSIGN type,(rvpCustomerType())  ; Set type
        ASSIGN startWaitTime,AC1   ; Mark with current time
***Serving Activity
        ; Precondition
        TEST L S$rgCounter,X$uNumEmp  ; Precondition
        ; Starting Event SCS
        PLUS updateSSOVs(AC1-P$WaitTime) ; Updates SSOVs
        ENTER rgCounter    ; Enters the resource group
        ; Duration
        ADVANCE (rvpuSrvTm(P$type))
        ; Terminating Event SCS
        LEAVE rgCounter              ; Leaves the counter
        TERMINATE                    ; Leave the model
```

Fig. 5.14 GPSS code for the GPSS process shown in Fig. 5.13

There are many situations where a TERMINATE Block is not part of a process because the entity never leaves the simulation model. This is illustrated in the tug process for the Port project that is presented in the following section.

5.4.3 Developing a GPSS Simulation Model from an ABCmod Conceptual Model

In this section, we illustrate the procedure for transforming an ABCmod conceptual model into a Process-Oriented simulation model in GPSS. The ABCmod structural view is used to develop the GPSS structural diagram, and life cycles in the ABCmod behavioural view become GPSS processes. There are two basic steps:

(a) The entity structures specified in the ABCmod conceptual model are mapped to GPSS structural entities (the result is a GPSS Structure Diagram)
(b) Each ABCmod life cycle diagram (including the ABCmod behaviour constructs that are referenced in them) is transformed into a GPSS process. When an activity is present in more than one life cycle diagram, its content is distributed over the processes corresponding to those life cycles.

Blocks are used to implement the preconditions, SCSs and durations that appear in the ABCmod activities. The GPSS simulation modeller must carefully consider how the ABCmod constructs impact upon the GPSS structural entities and then implement event SCSs using GPSS Blocks. When an activity appears in multiple process diagrams, its various elements need to be separated and translated into Blocks within different segments. These segments appear in the GPSS process that are the counterparts to the several processes where the activity is located. These various notions are illustrated in the discussion that follows which is based on the Port project (Version 1) introduced in Sect. 4.4 and presented in Sect. A1.3.

Figure 5.15 shows a GPSS Structure Diagram for the Port project. It provides a GPSS-oriented representation of the ABCmod entities: R.Tug, Q.BerthingList, RG.Berths and Q.DeberthingList and, as well, representative tanker Transactions (called T1, T2, T3, …). Note that a User Chain is used to reference a tanker Transaction (i.e. Tanker entity instance) when it is in tow during a Berthing or Deberthing activity.

We consider first the ABCmod scheduled action TankerArrival that characterizes the arrivals of the Tanker entity instances. It can be directly translated into the first segment of the GPSS process as shown in Fig. 5.16. This segment generates arrivals of tanker Transactions within the GPSS simulation model. The section begins with the GENERATE Block that provides the necessary function to create an input stream of tanker Transactions (i.e. Tanker entity instances). The SCS that initializes the three tanker attributes is transformed into three ASSIGN Blocks. The procedure SP.InsertQue becomes an LINK Block which adds the tanker Transaction to the QBerthingList User Chain (see Fig. 5.15). Notice how the RVP.TankerSize data model is implemented using the GPSS Function Entity with a similar name (FN$rvpUTankerSize) and is used as an argument in the first ASSIGN Block. The rectangle backdrop in Fig. 5.16 corresponds to the scheduled action rectangle from the tanker life cycle diagram given in Sect. A1.3. The main rectangle is divided into two parts that correspond to the standard components of a scheduled action, namely, the precondition (PR) and the action's event SCS (E).

In subsequent figures that illustrate the GPSS processes, a backdrop of rectangles and circles is added to help illustrate Step (b) (as in Fig. 5.16). The rectangles and circles are organized to reflect process diagrams where a rectangle corresponds to an activity. These rectangles are broken down into the components of the various ABCmod constructs. The following labels are used to identify these components:

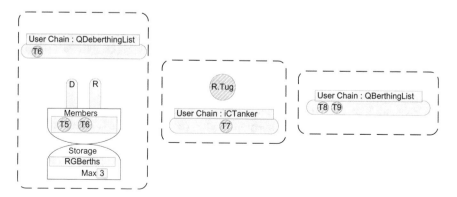

Fig. 5.15 Components of the GPSS structure diagram for the Port project

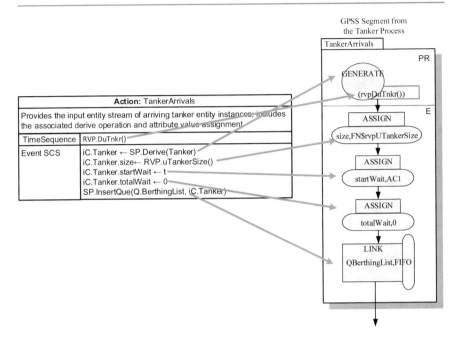

Fig. 5.16 Translating the TankerArrival action sequence

- PR—**Pr**econdition
- E—**E**vent (for Action Sequences)
- SE—**S**tarting **E**vent
- DU—**DU**ration
- TE—**T**erminating **E**vent
- IPR—**I**nterrupt **PR**econdition
- IE—**I**nterrupt **E**vent.

We consider now the case where a particular activity appears in more than one process diagram. The Deberthing activity is such a case because it is found in both the tanker and tug process diagrams (see the ABCmod behavioural diagrams presented in Sects. A1.3). Figure 5.17 shows how this ABCmod activity is transformed into GPSS Blocks located within the GPSS tanker and tug processes. The following comments elaborate on this transformation:

(a) Precondition: The tanker Transactions are placed in the QDeberthingList User Chain waiting for the availability of the tug (see Fig. 5.15). Consequently, the tug Transaction is responsible for initiating the Deberthing activity, i.e. implementing the precondition using a TEST Block in the tug process. The BV$DeberthingCnd is the GPSS Variable entity that represents the expression '(CH$QDeberthingList 'NE' 0)'. Note that checking the tug status

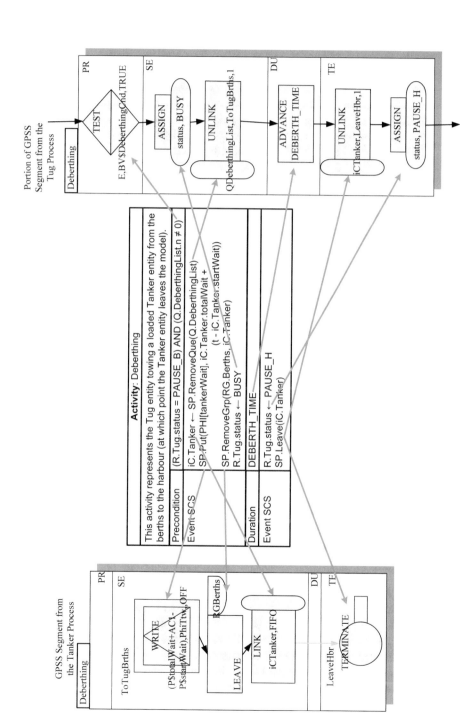

Fig. 5.17 Translating the Deberthing activity

parameter is not required in the expression since the tug Transaction tries to enter the TEST Block only when its status parameter is set to PAUSE_B.

(b) Starting event SCS: One part of the SCS in the starting event applies to the tug, while another part applies to the tanker. The part that applies to the tug is implemented in the tug process, e.g. setting the status parameter to BUSY, and the other part that applies to the tanker is implemented in the tanker process, e.g. manipulation of the tanker attributes (i.e. Transaction parameters). The WRITE Block saves specified values in a data stream labelled PhiTtw that represents the output sample sequence, called PHI[tankerWait]. SP.RemoveGrp (RG.Berths, iC.Tanker) is implemented by having the tanker Transaction traverse a LEAVE Block that refers to the GPSS RGBerths Storage entity.

The SP.RemoveQue in the starting event SCS is more complex because it carries out two tasks. The first removes the tanker from Q.DeberthingList, and the second references the tanker with iC.Tanker. This corresponds to removing a tanker Transaction from the QDeberthingList User Chain and placing the Transaction in the iCTanker User Chain (see Fig. 5.15). Two different Blocks are required: the UNLINK Block, traversed by the tug Transaction, carries out the first action, and a LINK Block, traversed by the tanker Transaction, places the tug Transaction in the iCTanker User Chain.

(c) Duration: Since the tanker Transaction is in the iCTanker User Chain, the ADVANCE Block, required for carrying out the duration, is placed in the tug process.

(d) Terminating event: The SCS instruction in the terminating event that updates the status attribute becomes an ASSIGN Block in the tug process. Recall that SP.Leave(iC.Tanker) specifies that the tanker leaves the ABCmod conceptual model. The UNLINK Block in the tug process is used to remove the tanker Transaction from the iCTanker User Chain and send it to the TERMINATE Block (label LeaveHbr) which then removes it from the GPSS simulation model.

The complete GPSS process for the tanker derived from the tanker life cycle diagram of the ABCmod conceptual model for Port Version 1 (together with associated ABCmod constructs) is given in Fig. 5.18. Similarly, the GPSS process for the tug is given in Fig. 5.19 (based on the tug life cycle diagram for the ABCmod conceptual model of Port Version 1. The transformation is based on the principles described in the previous paragraphs. By way of additional clarification, we note the following:

(a) From the behavioural view specified in the conceptual model, it's apparent that the Loading activity is specific to the tanker process and the MoveToHarbour and MoveToBerths activities are specific to the tug process. These activities (and the TankerArrivals scheduled action as well) consequently each is implemented in a single GPSS process.

(b) The Berthing activity, like the Deberthing activity examined in detail earlier, is distributed into both the GPSS tanker and GPSS tug processes.

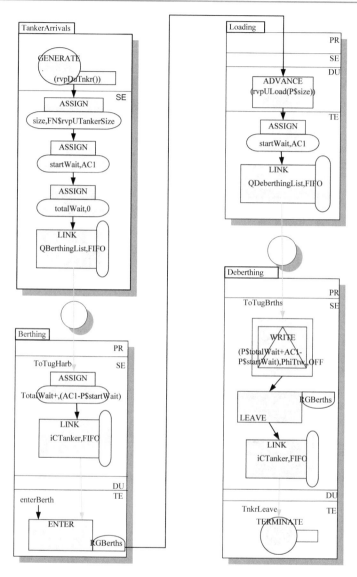

Fig. 5.18 GPSS TankerProcess

(c) Movement of a Transaction from a LINK Block to a subsequent Block requires a number of actions (as shown by the grey arrows Fig. 5.18). A LINK Block moves the tanker Transaction into one of the User Chains shown in Fig. 5.15. The Transaction is moved out of a User Chain by another Transaction (in this case the tug Transaction) when it traverses an UNLINK block. The tanker transaction is then 'sent' (see Annex 3 for details) to some GPSS Block. The grey arrows in Fig. 5.18 represent both the action of the LINK Block that

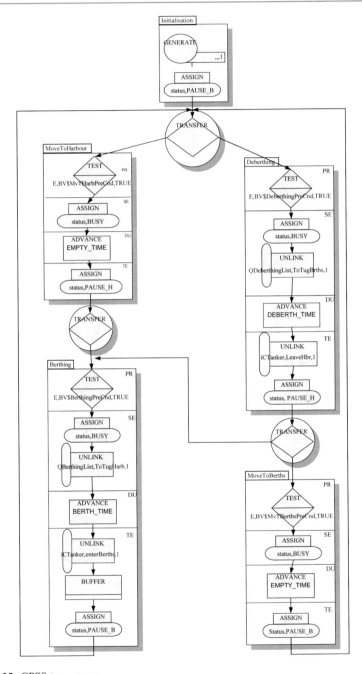

Fig. 5.19 GPSS tug process

moves the tanker Transaction into the referenced User Chain and the action of the UNLINK Block traversed by the tug Transaction to move the tanker Transaction out of the referenced User Chain to the referenced GPSS Block. For example, the grey arrow leading from the TankerArrival scheduled action to the Berthing activity represents the following actions:

– Upon entering the LINK Block, the tanker Transaction is moved into the QBerthingList User Chain.
– When the tug Transaction enters the appropriate UNLINK Block, the tanker Transaction is moved from the QBerthingList User Chain to the ASSIGN Block labelled: ToTugHrb (see Fig. 5.19).

(d) The Loading activity is implicitly started since the tanker Transaction moves automatically from the ENTER Block to the ADVANCE Block that implements the duration of the Loading activity. No explicit action is required (this is the commonly used GPSS counterpart for a sequel activity).

(e) The RVP.uLoad specified in the ABCmod conceptual model is implemented using a GPSS Plus Procedure with a similar name, i.e., rvpULoad.

(f) The GPSS tug process has only one GPSS Block segment in which the tug Transaction circulates without ever leaving the simulation model. Testing preconditions of the ABCmod activities is implemented using TEST Blocks as shown in Fig. 5.19. Each precondition is implemented with a Boolean Variable Entity that is referenced by a GPSS TEST Block. The definitions for these Variable Entities (see Annex 3) are as follows (as previously noted, testing the status parameter is not required):

– MvTHarbPreCnd BVARIABLE ((CH$QBerthingList 'NE' 0) 'AND' (S$RGBerths 'L' MaxBerth) 'AND' (CH$QDeberthingList 'E' 0))
– BerthPreCnd BVARIABLE (CH$QBerthingList 'G' 0)
– DeberthPreCnd BVARIABLE (CH$QDeberthingList 'NE' 0)
– MvTBerthsPreCnd BVARIABLE (CH$QBerthingList 'E' 0) 'AND' (R$RGBerths 'G' 0)

(g) Note the use of the BUFFER Block in the terminating event area of the Berthing segment of the tug process. When the tug Transaction enters this Block, it allows the tanker Transaction to be processed first in order to ensure that the tanker Transaction traverses the ENTER Block (and hence becomes member of the RGBerths Storage entity) before the tug Transaction moves on. This is important since the tug Transaction eventually tries to enter the TEST Block that implements the MoveToHarbour activity's precondition. This precondition includes evaluating the number of tankers in the RGBerths Storage entity.

A number of Blocks in GPSS are especially relevant for implementing interruptions that may occur during an ABCmod extended activity instance. In particular, PREEMPT and RETURN Blocks can be used to implement pre-emption (a particular form of interruption), while the DISPLACE Block can be used to implement a general interruption. Both the PREEMPT and DISPLACE Blocks remove Transactions that are on the FEC as a result of having entered an ADVANCE Block, i.e. Transactions whose duration is currently 'in progress'. Additional details about the operation of these Blocks can be found in Annex 3 or in the GPSS references that are provided there.

Version 2 of the Port project introduces the possibility of an interruption. In this case the tug, while carrying out its MoveToBerths activity, can (under certain conditions) become obliged to return to the Harbour to service a tanker that has arrived. Because of this interruption possibility, the ABCmod extended activity construct is required and Fig. 5.20 shows how it is translated into two GPSS segments that make up the GPSS tug process. An additional segment is required for a special Transaction, called the interruption Transaction, to monitor the interruption condition because the tug Transaction cannot monitor itself when scheduled on the FEC during the extended activity's duration. The intercepting event and its precondition are implemented within this additional segment.

The MoveToBerths intercepting precondition is implemented with the TEST Block that the interruption Transaction traverses only when the interruption condition has a TRUE value.[5] This can occur when the tug is involved in the Move-ToBerths activity (has entered the corresponding ADVANCE Block and is scheduled on the FEC).

The DISPLACE Block implements the actions of the SCSs contained in the MoveToBerths intercepting event (i.e. *SP.Start(ReturnToHabour)* and *SP.Terminate*). The DISPLACE Block uses the *X$TugId* argument to identify the tug Transaction that is to be displaced from the FEC and the *ReturnToHarbour* argument as the label of the destination Block for the displaced Transaction. Thus, the Block sends the tug Transaction to the first Block of the realization of the ReturnToHabour activity and by doing so terminates the MoveToBerths activity. The BUFFER Block that follows the DISPLACE Block allows the tug Transaction to move so that the interruption condition becomes FALSE (the tug status is changed) before the interruption Transaction can test the state of the model. Figure 5.21 shows how the above changes fit into the overall GPSS implementation of the tug process.

---

[5]A PLUS procedure, IsMvToBerthInt, is called to evaluate the status of the model. Using this procedure provides a clearer means of expressing the precondition than the use of a Boolean Variable entity. All SNAs needed for testing are passed as arguments to the Procedure. Note that it is necessary to represent the tug attributes as SaveValue entities instead of Transaction parameters to support the testing of these attribute values. GPSS places the interrupt Transaction on the Retry chains of the Tug_Status SaveValue entity, Tug_StartTime SaveValue entity, HarbourQu User Chain entity and DeberthQu User Chain entity. Whenever any of these entities change, GPSS moves all transactions from the corresponding retry chain to the CEC, including the interrupt Transaction which will try again to traverse the TEST Block, i.e. re-evaluate the TEST.

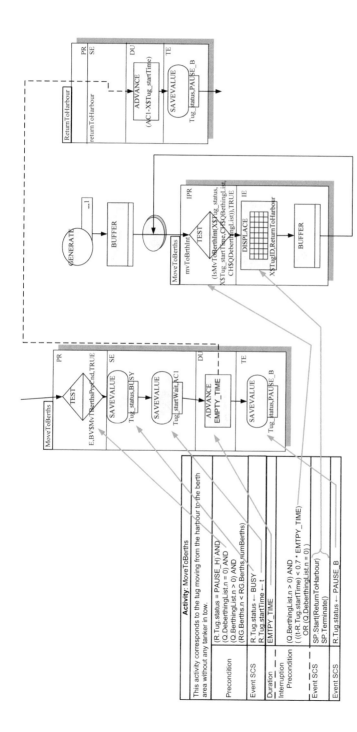

Fig. 5.20 Translating the MoveToBerths extended activity

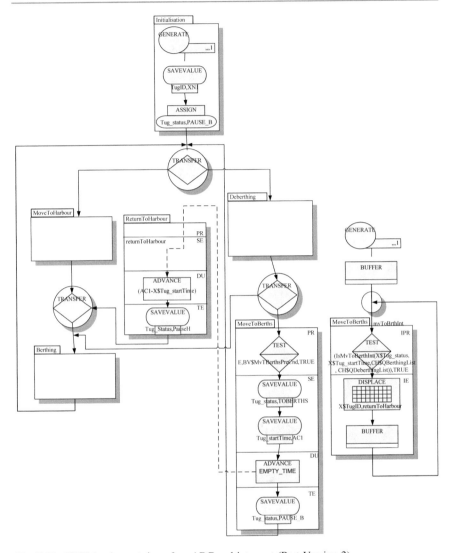

Fig. 5.21 GPSS implementation of an ABCmod interrupt (Port Version 2)

Version 3 of the Port project introduces the occurrence of storms which represent inputs to the SUI. The ABCmod conceptual model that is formulated in Sect. A1.5 handles the situation by the use of interrupts in the various activities that are affected by storms. Implementation of these various interrupts in the GPSS simulation model can be accomplished using the same approach that is outlined above. Details are left as an exercise for the reader.

5.5 Exercises and Projects

5.1 Develop a simulation program based on the Three-Phase world view (and/or Process-Oriented world view) for the ABCmod conceptual model formulated in Problem 4.1 of Chap. 4.

5.2 Develop a simulation program based on the Three-Phase world view (and/or Process-Oriented world view) for the ABCmod conceptual model formulated in Problem 4.2 of Chap. 4.

5.3 Develop a simulation program based on the Three-Phase world view for the ABCmod conceptual model formulated in Problem 4.4 of Chap. 4.

5.4 Assume that the development of the conceptual model of Problem 4.4 and the development of the simulation program of Problem 5.3 have been carried out by two teams where team A has the primary responsibility for the ABCmod conceptual model and team B has the primary responsibility for the simulation program. Carry out a verification exercise by

(a) Giving Team B the task of reviewing the conceptual model before developing the simulation program based on the Three-Phase world view,

(b) Giving Team A the task of reviewing the simulation program once it has been completed.

5.5 Develop a simulation program based on the Process-Oriented world view for the ABCmod conceptual model formulated in Problem 4.4 of Chap. 4.

5.6 Repeat the verification exercise of Problem 5.4 in the context of Problem 5.5.

5.7 Develop a simulation program based on the Three-Phase world view (and/or Process-Oriented world view) for the ABCmod conceptual model formulated in Problem 4.5 of Chap. 4.

5.8 Develop a simulation program based on the Three-Phase world view (and/or Process-Oriented world view) for the modified ABCmod conceptual model formulated in Problem 4.6 of Chap. 4.

References

1. Banks J, Carson JS II, Nelson B, Nicol DM (2005) Discrete-event system simulation, 4th edn. Prentice-Hall, Upper Saddle River
2. CERN, The Colt Project (2004) Version 1.2.0. http://dsd.lbl.gov/~hoschek/colt/
3. Horatmann C (2003) Computing concepts with Java essentials, 3rd edn. Wiley, New York
4. Law AM, Larmey CS (2002) An introduction to simulation using Simscript II.5. CACI Products Company, San Diego
5. Liang YD (2017) Introduction to Java programming: comprehensive version, 11th edn. Prentice-Hall, Upper Saddle River
6. Minuteman Software, GPSS World (2010) http://www.minutemansoftware.com/simulation.htm

7. Oracle, Java 2 Platform Standard Edition 12.0. http://www.oracle.com/technetwork/java/javase/documentation/index.html
8. Pidd M (2004) Computer simulation in management science, 5th edn. Wiley, Chichester
9. Schriber TJ (1991) An introduction to simulation using GPSS/H. Wiley, New York
10. Ståhl I, Henriksen JO, Born RG, Herper H (2011) GPSS 50 years old, but still young. In: Proceedings of the 2011 winter simulation conference, pp 3952–3962

The Activity-Object World View for DEDS

6

6.1 Building Upon the Object-Oriented Programming Paradigm

In Chap. 5, the important matter of implementing an ABCmod conceptual model as a simulation model was explored. Implementation procedures within the context of the various traditional world views are outlined. The feasibility of carrying out these implementations should not be surprising. The aspect that is not entirely satisfying, however, is the need to extract and separately manage the underlying events within the activities that exist within the conceptual model. This unfortunately undermines the semantic clarity of the activity-based approach upon which the ABCmod framework is based.

The Activity-Object world view is a natural evolution of the ABCmod conceptual modelling framework made especially convenient by the underlying concepts of the object-oriented programming paradigm.[1] The distinctive feature of the Activity-Object world view is that it preserves the two of the most fundamental modelling artefacts of the ABCmod framework, namely, entities that are instances of entity structures and activities that are instances of activity constructs. This not only retains the integrity of the ABCmod building blocks but is also significant from the perspective of the verification process inasmuch as the implementation of a conceptual model (formulated in the ABCmod framework) as a simulation model (based on the Activity-Object world view) is straightforward and transparent.

In this chapter, we outline the concepts upon which the Activity-Object world view is based. Our particular interest is with its implementation in the Java programming environment [5, 6]. Support to facilitate the implementation of ABCmod conceptual models in this environment has been developed, and we refer to this enhanced environment as ABSmod/J (a Java package). Details of ABSmod/J are provided in this chapter together with examples of its application.

---

[1]Initial work on the Activity-Object world view was reported in [1].

© Springer Nature Switzerland AG 2019
L. G. Birta and G. Arbez, *Modelling and Simulation*, Simulation Foundations, Methods and Applications, https://doi.org/10.1007/978-3-030-18869-6_6

We begin below with a brief summary of the key features and associated terminology of the object-oriented programming paradigm. The reader unfamiliar with this paradigm will better appreciate the presentation in this chapter by first acquiring a reasonable understanding of the paradigm by consulting the extensive available literature, e.g. [12].

Object: A *thing* in the *software* system that may represent a real-world object. It has a *state* represented by its attributes (data) and it has *behaviour* that relates to the way it acts (undergoes changes in state) and reacts to message passing. An object has an *identity* that allows it to be distinguished from other objects. Several references (sometimes called pointers) can be used to refer to an object with a specific identity, and thus an object can be accessed by many reference variables. The above concepts paraphrase the presentation in [12].

Class: A construct that provides the means to create objects. Objects are instances of classes. A class represents a complex data type consisting of a set of attributes (data structures) as well as methods which are subprograms used to manipulate the object's data structures.

Inheritance: A class can be defined as a subclass of another. The subclass inherits all defined attributes and methods defined in its parent class (superclass). It is common to define classes that have some methods for which only the method header appears. These are called abstract classes and the methods with only method headers are called abstract methods. Objects cannot be instantiated from abstract classes. A subclass that extends an abstract class must override the abstract methods thereby providing a class that can be instantiated.

Polymorphism: Given that a class can be defined as a subclass of another, instantiating an object means that it can be viewed an as object of any of its superclasses. This has implications on the attributes/methods that can be accessed in the subclasses (i.e. could be hidden).

6.2   Implementing an ABCmod Conceptual Model Within the Activity-Object World View

6.2.1   Overview

In the object-oriented environment, ABCmod entity structures become classes, and the objects instantiated from these classes during execution of the simulation model[2] correspond to the entities of the conceptual model.

The Activity-Object world view sees the instances of ABCmod behaviour constructs (both activity and action) as objects in the object-oriented programming paradigm. Consequently, the constructs are implemented as classes, and

---

[2]Recall that the execution of a simulation model can only be accomplished from within a simulation program. Consequently whenever the text within this book suggests that the stimulation model is executing, that execution is taking place as an integral part of the execution of a simulation program.

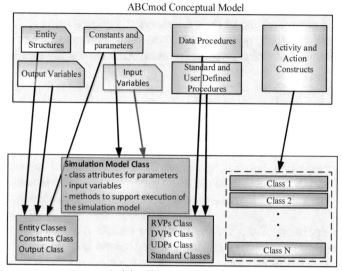

Fig. 6.1 Implementing an ABCmod conceptual model as an activity-object simulation model

activity/action instances then become instances of these classes (i.e. objects). Such instances are called activity (or action) objects or, in general, behaviour objects.

The especially noteworthy feature is that ABCmod behaviour constructs are implemented in their entirety in the simulation model, i.e. their structure is not dismantled. ABCmod preconditions, time sequences, event SCSs and durations are implemented as methods in the activity and action classes. This enables a straightforward and seamless transformation of an ABCmod conceptual model into a simulation model within the Activity-Object world view.

Figure 6.1 provides an overview of how the various components of an ABCmod conceptual model relate to those of a simulation model formulated from the perspective of the Activity-Object world view. The general steps are as follows:

1. Create a simulation model class. The object that results from the instantiation of this class manages the simulation and provides the mechanisms for experimentation.
2. Create the equivalent of the ABCmod conceptual model structural components. For example, represent entity structures as entity classes, constants as attributes in a constants class and parameters as attributes in the simulation model class.
3. Implement appropriate data procedures (RVP and DVP) and user-defined procedures as methods; these could be collected in the classes *RVPs*, *DVPs* and *UDPs*, respectively, as shown in Fig. 6.1. The ABCmod standard procedures are implemented either as methods found in standard Java classes (e.g. the *add* method from the *ArrayList* class that provides the functionality for SP.InsertQue) or methods that are provided in the ABSmod/J package (e.g. the *unscheduled-Behaviour* method that provides the functionality for SP.Terminate).

4. Define a class for each activity and action construct specified within the ABCmod conceptual model (Fig. 6.1 shows *N* such classes).
5. Implement a method, typically part of the simulation model class, for testing preconditions in the simulation model class; more specifically to:

   (a) Test preconditions of conditional activities and actions.
   (b) Test intercepting preconditions of extended activities.

6. Develop the *Output* class which manages output. It will contain output variables as attributes. In addition, it will contain methods to record output data of trajectory and sample sequences as well as methods for generating DSOV values from these sequences. The simulation model class provides methods that manipulate the output object in order to acquire the desired output after a simulation run.

The manipulation of the various behaviour objects relies on inheritance and polymorphism. For example, scheduled action objects and the various types of activity objects are all maintained indirectly on the same list, the *scheduled behaviour list* (discussed later). These are viewed and referenced as *Behaviour* objects. The *Behaviour* class is the superclass to all classes used in representing behaviour. Classes, tailored to the problem at hand, are created by the simulation modeller by extending subclasses of the *Behaviour* class which, when instantiated, can be referenced as *Behaviour* objects.

6.2.2   Execution and Time Advance Algorithm

Behaviour is achieved when an execution and time advance algorithm advances simulated time and applies event SCSs in a logically consistent manner, i.e. in a manner that correctly sequences the specifications encapsulated in the event SCSs of the ABCmod conceptual model under consideration. This implies creating and scheduling behaviour objects. Traditional world views use a list for scheduling events; this happens in the event scheduling world view and even in the case of activity-oriented approaches (e.g. the two-phase and three-phase approaches) [2]. The Activity-Object world view on the other hand uses a list for scheduling behaviour objects, not events. As is the case with event notices, the 'scheduled behaviour notices' (SB notices) are ordered on the scheduled behaviour list (SBL) according to a time stamp attribute stored in the SB notice. As expected, the time stamp gives the time at which one of the referenced behaviour object's events will occur. In the case of conditional and sequel activities, the terminating event is scheduled, while in the case of scheduled and scheduled sequel activities, both the starting event and the terminating event are scheduled; scheduling alternates between the starting and terminating events. Scheduled actions and scheduled sequel actions can also be added to the SBL, in this case it is the action's event that is scheduled.

The execution and time advance algorithm which is shown in Fig. 6.2 is similar to that of the three-phase algorithm (see Fig. 5.3). In phase A, a scheduled behaviour notice is removed from the top of the SBL. Its time attribute is used to update the simulation time. In phase B, event routines are executed, thereby executing event SCSs found in the corresponding ABCmod behaviour constructs. In the case of a

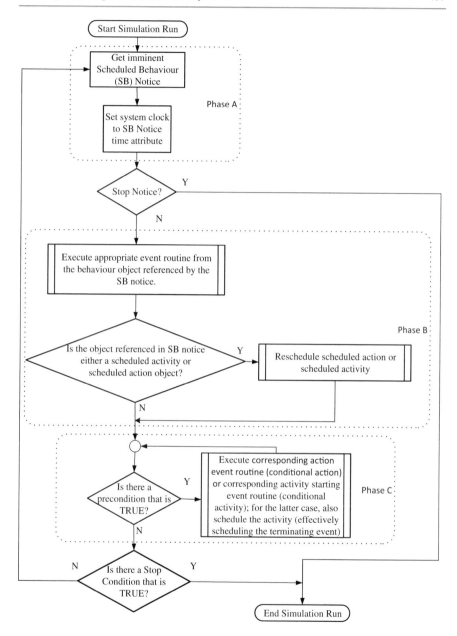

Fig. 6.2 Execution and time advance algorithm for the Activity-Object world view

scheduled action object, its event routine (event SCS) is executed. In the case of the scheduled activity or the scheduled sequel activity, either the starting event routine or terminating event routine is executed (starting event SCS or terminating event SCS). In fact, for each scheduled activity, event scheduling alternates between the starting event and the terminating event. In the case of the scheduled sequel activity, the activity must be explicitly scheduled using the standard procedure SP.SchedSequel() (see Sect. 4.3.3.2). In the case of conditional or sequel activity, the terminating event routine is executed (terminating event SCS). The Activity-Object world view does, in fact, schedule events, however what is significant is that the events remain packaged within the behaviour objects. This is the main difference from the three-phase world view.

In the second part of Phase B, if the behaviour object referenced by the SB notice is either a scheduled action or a scheduled activity, it is rescheduled (the time sequence and duration specifications provided in the ABCmod conceptual model is used to determine the scheduled time). In Phase C, all preconditions from conditional activities and conditional actions are evaluated. When a precondition is evaluated to TRUE, the corresponding behaviour object is instantiated, and in the case of conditional activities, the activity object is scheduled (effectively scheduling its terminating event). Intercepting preconditions (not shown in Fig. 6.2) are also evaluated and appropriate processing undertaken.

6.3   ABSmod/J

Event scheduling simulation programs are frequently created using general programming languages. Often, libraries are available for supporting this development. They provide functions for random number generators, random variate generators, list processing, etc. The Java package ABSmod/J includes list processing, time management and other functions to allow the creation of a simulation model. It is augmented with random variate functions from the Colt [3] Java package. ABSmod/J is a comprehensive environment that facilitates the transformation of an ABCmod conceptual model into a simulation program based on the Activity-Object world view.

6.3.1   The *AOSimulationModel* Class

In order to create the simulation model class (see Fig. 6.1), the ABSmod/J *AOSimulationModel* abstract class is extended (see the UML class diagram in Fig. 6.3). The class contains the variables, abstract methods and methods for the execution and time advance algorithm presented in Fig. 6.2. The class is intended for use with other classes (e.g. *SBNotice* and *OutputSequence*) used to define SB notices, activity objects and the collection of output (*Output* class). Abstract methods must be defined when extending this class.

This abstract class is somewhat misnamed. It is not, in fact, a class that corresponds to the simulation model for a given conceptual model. It simply provides the means to create a class (referred to as the simulation model class) that encompasses the core of the simulation model. Adding other classes (such as the entity and activity classes) as shown in Fig. 6.1 results in the simulation model. For experimentation, the simulation model class is instantiated to give an object that can execute the simulation model. The simulation program results from the addition of the ABSmod/J package, the Colt Java package and any experimentation code to these classes.

AOSimulationModel
-sbl : PriorityQueue
-esbl : LinkedList
-clock : double
-time0 : double
-timef : double
-StopBehaviour : Behaviour = null
#initAOSimulModel(in startTime : double)
#initAOSimulModel(in startTime : double, in endTime : double)
+setTimef(in endTime : double)
#runSimulation()
#scheduleAction(in action : ScheduledAction)
#scheduleActivity(in activity : Activity)
#reschedule(in beh : Behaviour)
#unscheduleBehaviour(in beh : Behaviour)
#getClock() : double
#testPreconditions(in beh : Behaviour)
#addSBNotice(in act : Behaviour, in tm : double)
#showSBL()
#showESBL()
#getCopySBL() : PriorityQueue
#getCopyESBL() : LinkedList
#eventOccured()
#implicitStopCondition() : boolean

Fig. 6.3 The AOSimulationModel abstract class

Below we briefly outline the components (attributes and methods) of the *AOSimulationModel* abstract class shown in Fig. 6.3:

(a) The *sbl* variable references a *PriorityQueue* object that serves as the scheduled behaviour list. The *PriorityQueue* class provides the functionality to rank SB notices in a time-ordered list (see *SBNotice* class below).

(b) The *esbl* variable references a *LinkedList* object to maintain a list of extended activity objects (when required). This list facilitates the evaluation of the intercepting preconditions of extended activity objects that have been scheduled on the *sbl*.

(c) The *clock* variable maintains the current value of simulated time.

(d) The variables *time0* and *timef* correspond to the boundaries of the observation interval. The value of the *timef* variable is set when the simulation run ends.

(e) The constant *StopBehaviour* (*null* reference) serves as the reference value in an SB notice to schedule termination of the simulation at a predefined time (explicit right-hand boundary of the observation interval).

(f) The method *initAOSimulModel* initializes the *sbl* and *esbl* variables, the simulation *clock* and *time0* variable, and optionally sets up a *StopBehaviour* SB notice. Both forms of the method *initAOSimulModel* initialize *clock* and *time0* to its *startTime* argument. The second form provides *endTime* that is used to create an SB notice on the *sbl* with the reference value *StopBehaviour* and *timeAttr* set to *endTime*.

(g) The method *setTimef* provides the means to change the right-hand boundary of the observation interval. Its effect is to add an SB notice containing the value of the *StopBehaviour* constant. This method can be used to implement warm-up periods or to increase the run length as described in Chap. 7.

(h) The simulation model execution is centred on the *runSimulation* method that implements the execution and time advance algorithm (see Fig. 6.2).

(i) The methods *scheduleAction* and *scheduleActivity* are used to schedule action objects and activity objects, respectively. The *reschedule* method is used to reschedule activities when necessary; the method is typically called at the start of the *testPreconditions* method.

(j) The *unscheduleBehaviour* method supports the implementation of interruptions. Recall from Chap. 4 that when an interruption of an extended activity occurs that activity instance is typically terminated. Terminating an extended activity object requires removal of the reference to the SB notice referencing the activity object (see Sect. 6.3.5 for details) from the *sbl* list and *esbl* list.

(k) The *getClock* method provides the means for accessing the value of simulated time as stored in *clock*. The *clock* variable is protected from direct access and this method provides the means for obtaining its value.

(l) The *testPreconditions* is an abstract method. Because it is specific to the model being implemented, it is necessarily developed by the modeller. It is called by the *runSimulation* method to test all preconditions of conditional activities and conditional actions as well as the intercepting preconditions of extended activities.

(m) The *showSBL* and *showESBL* methods display the contents of the *sbl* and *esbl* lists. When used in the *eventOccured* method, it is possible to trace the change in these lists after the occurrence of each completion of Phase C (Fig. 6.2), that is, at the end of each iteration of the execution and time advance algorithm.

(n) The *getCopySBL* and *getCopyESBL* methods provide a copy of the *sbl* and *esbl* lists, respectively. This allows for the possibility of customizing the display of the *sbl* and *esbl* lists to focus on behaviour notices of special interest.

(o) The *eventOccured* is an abstract method and is developed by the simulation modeller. It is executed at the end of each iteration of the execution and time advance algorithm. The method can be used to update the trajectory sequences. Used in conjunction with the *showSBL, showESBL, getSBLCopy* and *getESBLCopy* methods, the *eventOccured* method is useful for tracing the execution of a simulation model.

(p) The *implicitStopCondition* is a method which always returns *false*. It can, however, be overridden by the simulation modeller in order to implement an implicit right-hand boundary for the observation interval. This method accesses model variables that have relevance to the condition(s) that determine simulation run termination. For example, the method could return *true* when the 100th widget is completed in a manufacturing model.

Fig. 6.4 The SBNotice class

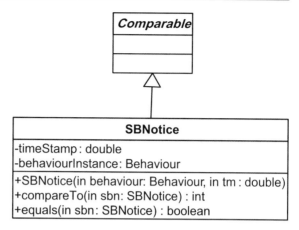

The *SBNotice* class, shown in Fig. 6.4, enables the instantiation of scheduled behaviour notices. *SBNotice* objects are maintained on the *sbl* list. Two variables are defined in the class: *behaviourInstance* (reference to a behaviour object) and *timeStamp* (scheduled time). The *SBNotice* class implements the *Comparable* interface which is a requirement for use with the *PriorityQueue* class (see the description of the *sbl* attribute above). The method *compareTo* is used by the *PriorityQueue* object to order the SB notices according to their *timeStamp* attribute.

The *runSimulation* method is an implementation of the execution and time advance algorithm shown in Fig. 6.5. The key aspects of this method are summarized below:

(a) The *sbl.poll()* method call removes the SB notice at the head of the scheduled behaviour list (namely, the *SBNotice* object referenced by the variable *nxtBeh*).

(b) The *clock* variable is assigned the value of the event notice attribute *nextBeh.timeStamp*. This implements the underlying time advance of the algorithm.

(c) If the SB notice is a Stop Notice, the *runSimulation* method breaks out of its processing loop (while loop). Recall that calling the *initAOSimModel(double startTime, double endTime)* places a Stop Notice on the *sbl* list. The Stop Notice is an *SBNotice* object with its attribute *behaviourInstance* set to the constant *StopBehaviour* and its *timeStamp* attribute set to a specific time when the simulation run is to terminate. This provides the mechanism for dealing with an explicit right-hand boundary for the observation interval. Notice also that before breaking out of the loop, the *timef* variable is updated with the current time (value of *clock*). Thus, at the end of a simulation run, the values contained in the variables *time0* and *timef* correspond to the endpoints of the observation interval.

(d) If the SB notice is not a Stop Notice, then the type of the behaviour object referenced by *nxtBeh.behaviourInstance* must be determined (is it a reference to an action object or an activity object and in the latter case, which subtype of activity (scheduled, conditional, sequel, extended)). This information is required in order to determine which event routine to execute. Java provides Class objects

```
public void runSimulation()
{
  while (true) // set up loop
  {
      /*-----------------Phase A ----------------------*/
      // 1) Get next SBNotice and update clock
      SBNotice nxtBeh = sbl.poll();
      clock = nxtBeh.timeStamp; // update the clock
      /*-----------------Phase B ----------------------*/
      // 2) If the stop event - terminate the simulation
      if (nxtBeh.behaviourInstance == StopBehaviour)
      {
          timef = clock;
          break;
      }
      // 3) Check for the ScheduledActivity
       if (ScheduledActivity.class.isInstance(nxtBeh.behaviourInstance))
      { /* Execute either startingEvent()  or terminatingEvent() method */ }
      // 4) Check for Scheduled Sequel Activity
      else if(ScheduledSequelActivity.class.isInstance(nxtBeh.behaviourInstance))
      { /* Execute either startingEvent()  or terminatingEvent() method */ }
      // 5) Check for other activities (ConditionalActivity and SequelActivity)
      else if(Activity.class.isInstance(nxtBeh.behaviourInstance))
      { /* Execute terminatingEvent() method */ }
      // 6) Check for ScheduledAction or ScheduledSequelAction
      else if (ScheduledAction.class.isInstance(nxtBeh.behaviourInstance) ||
          ScheduledSequelAction.class.isInstance(nxtBeh.behaviourInstance))
      { /* Execute actionEvent() method */  }
      else { /* Print error message */ }
      /*---------------------Phase C ----------------------*/
      // 7) Call testPreconditions() method, must also reschedule activities and actions
      testPreconditions(nxtBeh.behaviourInstance);
      // 8) Call eventOccured method
      eventOccured();
      // 9) Check if implicitStopCondition true and stop.
      if (implicitStopCondition())
      {
          timef = clock;
          break;
      }
  }
}
```

Fig. 6.5 Java code for the runSimulation method

as the means for achieving this. For example, *ScheduledActivity.class*[3] is a reference to such a class object. The method *isInstance* is available in this class object and in this example returns *true* when the object that is referenced by the method's parameter is a member of the *ScheduledActivity* class. The scheduled event routine (namely the corresponding method in the Behaviour object) that is executed is determined as follows:

- *ScheduledActivity*: The *ScheduledActivity* objects contain an attribute *eventSched* that is set to either *StartingEvent* or *TerminatingEvent*. It is the value of this attribute that determines which event routine (either the *startingEvent()* or *terminatingEvent()* method) has been scheduled and hence which event routine is to be executed.
- *ScheduledSequelActivity*: The *ScheduledSequelActivity* is treated the same way as the *ScheduledActivity*, that is, the *eventSched* attribute determines which event method is executed.
- *ConditionalActivity* or *SequelActivity*: The terminating event routine (*terminatingEvent()* method) is executed.
- *ScheduledAction or ScheduledSequelAction*: The event routine (*actionEvent()* method) associated with the *ScheduledAction* or *ScheduledSequelAction* object is executed.

(e) The *testPreconditions* method is called to evaluate the preconditions in the ABCmod conceptual model. This method is also responsible for rescheduling scheduled actions and scheduled activities (bootstrapping). The simulation modeller thus retains control on how preconditions are evaluated and of bootstrapping. A method *reschedule* is available to automatically reschedule actions and activities.

(f) Finally, the *implicitStopCondition* method is called to test for an implicit stop condition. When this method returns *true*, *runSimulation* breaks out of its processing loop. As in step (c), the *timef* variable is assigned the current time (namely the value of *clock*).

The *AOSimulationModel* class is extended to form the simulation model class shown in Fig. 6.1. For example, to implement the Kojo's Kitchen simulation model, a class called *KojoKitchen* could be introduced to extend the *AOSimulationModel* class; it therefore inherits all the functionality of the *AOSimulationModel* class described above. To complete a simulation model class, the simulation modeller will (as appropriate):

- Declare the conceptual model parameters as instances variables.
- Declare instance variables that reference the entity objects (see below) representing entity structures with *scope* = Unary and *scope* = Many[*N*].
- Declare instance variables that represent required input variables.

---

[3]The *class* attribute is made available by the Java virtual machine for all Java classes used in a program. Details can be found in the Java API documentation that describes *Class* [8].

- Declare instance variables that reference the *RVPs*, *DVPs*, *UDPs* and *Output* objects.
- Develop appropriate methods for supporting output.
- Develop a constructor to initialize the simulation model object.
- Develop the *testPreconditions* method for testing the model preconditions.
- Develop the *eventOccured* method to update trajectory sequences (with a call to a method in the Output object) and debugging logic.
- Develop the *implicitStopCondition* method if required to accommodate an implicit right-hand side of the observation interval.

6.3.2   Equivalents to ABCmod Structural Components

The next step in developing the simulation model is to represent the structural components specified in the ABCmod conceptual model.

Entity structures can be implemented with classes provided by the Java standard development kit [7]. New classes can also be created by the simulation modeller to meet the requirements of a specific ABCmod conceptual model. In the sequel, we shall refer to an object representing an entity as an entity object and a Java class representing an entity structure as an entity class. Table 6.1 lists a set of possible Java classes that can be used and extended to represent the various ABCmod entity structures with *role* = Queue and *role* = Group. Such Java classes provide the required functionality to manage groups and queues, i.e. lists of objects. It is by no means complete, and the reader is encouraged to review the available Java classes that might be appropriate for representing the entity structures in the conceptual model under consideration.

The constants in the ABCmod conceptual model are collected as static variables in the *Constants* class.[4] Conceptual model parameters are implemented as instance variables in the simulation model class.

6.3.3   Equivalents to ABCmod Procedures

The ABCmod conceptual model may include random variate procedures, deterministic value procedures and user-defined procedures. These procedures are implemented as methods in the classes called *RVPs*, *DVPs* and *UDPs*, respectively.[5] These classes are instantiated when the simulation model class is instantiated. Each of these objects contains a reference to the simulation model object. This reference is required to access data such as the simulation clock and entity object attributes.

---

[4]This class need not be instantiated. The class variables can be referenced using the class name as in *Constants.constant1*.

[5]It is also practical to associate RVP's, DVP's and UDP's with actions and activities when they are used only by the associated action/activity (see the Conveyor Project in Annex 1). When implementing such procedures, they are implemented directly within the action/activity class itself.

Table 6.1 Representing ABCmod entity structures with Java objects

Role of ABCmod entity structure	Java standard class
Queue	*ArrayList* (implementation of a resizable array)
	LinkedList (a linked list implementation)
	ArrayBlockingQueue (bounded FIFO queue)
	ConcurrentLinkQueue (Thread-safe FIFO queue using linked nodes)
	DelayQueue (unbounded queue of delayed elements – when an element's delay has expired, it is moved to the head of the queue in FIFO fashion)
	LinkedBlockingQueue (bounded queue implemented with linked nodes)
	ProrityBlockingQueue (similar to PriorityQueue, but offers also blocking retrieval functions when queue is empty)
	PriorityQueue (unbounded queue that orders elements according to priority)
	AbstractQueue (skeletal queue implementation that can be extended to create user-specific queue)
Group	*AbstractSet* (skeletal set implementation that can be extended to create user-specific set)
	HashSet – this class is an extension of the AbstractSet and supports the set methods using a hash table.
	LinkedHashSet – similar to the HashSet class but stores objects in the set using a doubly linked list. This provides an order to the added objects, that is, it is possible to remove the objects in the same order they were stored

6.3.4 Behaviour Objects

The hierarchy of classes shown in Fig. 6.6 has been designed to facilitate the implementation of ABCmod conceptual model behaviour constructs in ABSmod/J. This hierarchy of classes supports the manipulation of the behaviour objects for execution of event routines and time advance as presented in the description of the *runSimulation* method. The transformation of any particular type of behaviour construct in an ABCmod conceptual model into an ABSmod/J equivalent consists of defining a class which extends the corresponding class shown in Fig. 6.6.

The *Behaviour* class is the superclass for all objects used to implement the ABCmod conceptual model behaviour constructs. Thus, all *Behaviour* objects (i.e. objects derived from any subclass of the *Behaviour* class) can be referenced and placed on a single list such as the *sbl*[6] list.

As shown in Fig. 6.6, two direct subclasses are defined for the *Behaviour* class, namely, the *Action* class and the *Activity* class. The *Action* class includes the

---

[6]The *sbl* variable in the AOSimulation object is of type *PriorityList* <*SBNotice*>; and the *SBNotice* can reference any object which is a subclass of *Behaviour*.

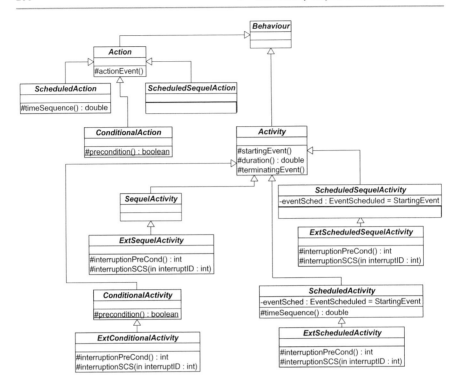

Fig. 6.6 ABSmod/J behaviour class hierarchy

actionEvent abstract method which is inherited by the subclasses *ConditionalAction,*
ScheduledAction and ScheduledSequelAction. The *ConditionalAction* and *Sched-*
uledAction classes include abstract methods that are specific to them, namely, the
precondition method and the *timeSequence* method, respectively. Similarly, subclasses of the *Activity* class include the *ScheduledActivity,* the *ConditionalActivity,* the
SequelActivity and *ScheduledSequelActivity* classes which inherit from the *Activity*
class the abstract methods *startingEvent, duration* and *terminatingEvent.* Available
classes for implementing extended activities are also shown in Fig. 6.6.

For each behaviour construct specified in the ABCmod conceptual model, the
simulation modeller extends one of the classes *ScheduledAction, ConditionalAction,*
ScheduledSequelAction, ConditionalActivity, SequelActivity, ScheduledActivity,
ScheduledSequelActivity, ExtConditionalActivity, ExtSequelActivity, ExtScheduled
Activity or *ExtScheduledSequelActivity* as appropriate. As an example, consider an
extended conditional activity called MoveToBerths in an ABCmod conceptual
model. Its implementation would correspond to a class (likely called *MoveToBerths*)
which extends the class *ExtConditionalActivity.* The modeller must then implement
the following methods: *precondition, startingEvent, duration, interuptionPreCond,*
interruptionSCS and *terminatingEvent* to reflect the corresponding specifications for
MoveToBerths provided in the conceptual model.

Notice that the *precondition* methods in the *ConditionalAction, Condi-*
tionalActivity and *ExtConditionalActivity* classes are static methods (shown

underlined) and thus can be executed without the existence of objects. This is an essential requirement for alignment with the ABCmod framework where conditional actions and activities are instantiated only when the respective precondition is *true*.

6.3.5 Bootstrapping and Evaluating Preconditions

The *testPreconditions* method (in the simulation model class) is responsible for the bootstrapping functionality to reschedule scheduled actions and activities as well as to test all preconditions within the conditional action and activity classes and the interrupt preconditions in the extended activity objects. The simulation modeller has complete control over the approach used to implement bootstrapping and evaluate preconditions. For simple bootstrapping, the method *reschedule(behObj)* is available for default bootstrapping (namely rescheduling the behaviour object referenced by *behObj*). In other words, it calls the *timeSequence()* method either to get the time for rescheduling the scheduled action or to get the time to reschedule the starting event of the scheduled activity. In the case of rescheduling the terminating event of the scheduled activity, its *duration* method is used.

Typically it is necessary to repetitively retest preconditions until all are *false*. Recall that each time a *precondition* method of a conditional activity is evaluated to *true*, an activity object is instantiated from the corresponding activity class, its *startingEvent* method is executed and the activity object is scheduled. The change made to the model's status by the *startingEvent* method may cause preconditions that were previously *false*, to become *true*, thus making necessary the re-evaluation of preconditions.

A different mechanism is used for testing interrupt preconditions. Activity objects corresponding to extended activities are referenced by the extended scheduled behaviour list (*esbl* list) as well as the *sbl* list. To evaluate the intercepting preconditions, the *testPreconditions* method must include code to scan through the *esbl* list and execute the *interruptionPreCond* methods. The *interruptionPreCond* method returns an integer value which is 0 when no intercepting precondition is *true* and a positive value when an intercepting precondition is true. The integer identifies which intercepting event has occurred (recall that it is possible to specify one or more intercepting events in an extended activity). The integer value is passed to the *interruptionSCS* method so that the appropriate status change (as specified in the SCS of the corresponding intercepting event within the ABCmod conceptual model) can be made.

6.3.6 Output

The class *Output* is provides a convenient means for centralizing all variables and methods related to output for the simulation model. SSOVs are represented using instance variables in this class. For implementing trajectory and sample sequences, objects are instantiated from the class *OutputSequence* and referenced by the *Output*

Table 6.2 DSOV surrogate in the OutputSequence class

DSOV surrogate	computeTrjDSOVs	computePhiDSOVs
sum	$y_K(t_f - t_K) + \sum_{i=1}^{K} y_{i-1}(t_i - t_{i-1})$	$\sum_{i=0}^{K} y_i$
sumSquares	$y_K^2(t_f - t_K) + \sum_{i=1}^{K} y_{i-1}^2(t_i - t_{i-1})$	$\sum_{i=0}^{K} y_i^2$
mean	$\dfrac{sum}{t_f - t_0}$	$\dfrac{sum}{K}$
max (maxTime)	Maximum value in the trajectory sequence (together with time it was recorded)	Maximum value in the sample sequence (together with time it was recorded)
min (minTime)	Minimum value in the trajectory sequence (with time it was recorded)	Minimum value in the sample sequence (with time it was recorded)
meanSquares	$\dfrac{sumsquares}{t_f - t_0}$	$\dfrac{sumsquares}{K}$
variance	$meanSquares - mean^2$	$meanSquares - mean^2$
stdDev	$\sqrt{variance}$	$\sqrt{variance}$

In this table, the pair (t_i, y_i) is a recorded time/value pair, K is the total number of recorded pair values (i.e. equals *number*) and t_0, t_f are passed to the *computeTrjDSOVs* method.

object (i.e. declarations of reference variables to *OutputSequence* objects are included in the *Output* class). Another useful method found in the *Output* class updates all trajectory sequences; this method is called from the *eventOccured* method so that trajectory sequences are correctly updated prior to each advance of simulated time.

The *OutputSequence* class provides the means both for collecting output data during the simulation run (with the method *put*[7]) and for computing all the scalar values listed in Table 6.2 (using methods *computeTrjDSOVs* or *computePhiDSOVs*). For example, the class offers access to the instance variable *number* that contains the number of members in the output data set being processed. Table 6.2 shows the instance variables that contain DSOV values after *computeTrjDSOVs* or *computePhiDSOVs* is executed. The values of the instance variables are accessed using Java getter methods.

The data values of the sequence are recorded in a file with the name provided in the constructor's argument (*OutputSequence(File-Name)*). The class also provides a number of other methods to facilitate collection and analysis of the output data namely:

- *clearSet*: This method removes all data currently recorded. This method can be used to accommodate warm-up periods (see Chap. 7) or to re-initialize the simulation program.

---

[7]Note that the data collection process for both trajectory sequences and sample sequences uses the same method; both time value and data value are stored for both types of output.

- *rewindSet*: This method provides the means to point to the beginning of the recorded data (i.e. beginning of the file that contains the data). A call to this method is required before making calls to the *get* method.
- *get*: This method makes possible the sequential reading of all the recorded output data so that operators other than those provided in Table 6.2 can be applied to generate project-specific DSOV values.

6.3.7   Summary

Figure 6.7 provides a snapshot overview of a generic ABSmod/J simulation model during its execution and shows typical relationships among the various objects. Objects are represented by rectangles with titles (see legend) where each title provides the Java class from which the object was instantiated. Some of the titles (shown in italic) are specific to the simulation model being developed, e.g. '*Entity 1*', '*Entity 2*' and '*Activity 2*'. (Notice that multiple objects can be instantiated from the same class, e.g. two '*Entity 1*' objects are shown.) Other classes (in regular font) such as *SBNotice*, *AOSimulationModel* and *PriorityQueue* are either Java provided classes or are part of the ABSmod/J package. The *SimModel* object shown in Fig. 6.7 is instantiated from the class '*SimModel*' which extends *AOSimulationModel*.

The simulation model object normally contains references to entity objects. These references can be stored in reference variables to implement entity structures with *scope* = Unary or in arrays to implement entity structures with *scope* = Many[N]. In both these cases, the entity objects (e.g. '*Entity 1*' and '*Entity 2*' objects shown in Fig. 6.7) exist during the lifetime of the simulation model object and are instantiated when the simulation model object is itself instantiated.

Objects that are referenced at different points in time by other entity objects or activity objects have transient existence. The '*Entity 2*' object (which could have *role* = Queue) is referencing two '*Entity 3*' objects (an indication that the '*Entity 3*' instances have a transient existence). Another '*Entity 3*' object is referenced by an '*Activity 2*' object, thereby associating this entity instance with the activity instance.

The *SimModel* class incorporates equivalents to a number of the ABCmod constructs, such as constants, parameters, data procedures (that encapsulate random number generators and random variates for supporting data models) and user-defined procedures. Grouping constants and procedures into separate classes simplifies the *SimModel* class. ABCmod parameters become attributes in the *SimModel* class.[8]

Many of the ABCmod constructs are implemented by a simulation modeller while other components of the simulation model class are inherited from the *AOSimulationModel* Java class, e.g. the methods *runSimuation*, *scheduleActivity* and *scheduleAction*. Note that the *testPreconditions* method is also part of the *SimModel* class.

---

[8]Parameters that are entity attributes will be located in the corresponding entity classes.

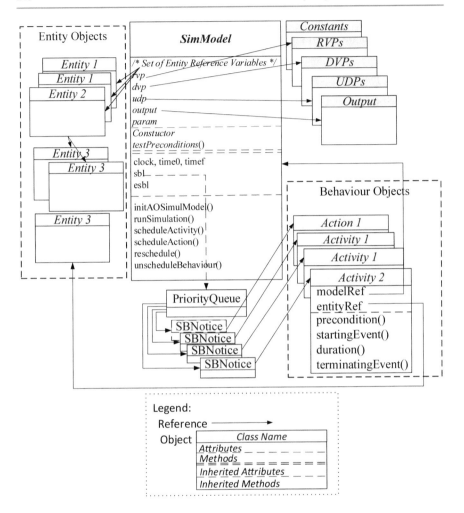

Fig. 6.7 Overview of ABSmod/J Activity-Object world view

Figure 6.7 shows how the scheduled behaviour list is supported in the ABSmod/J package. A *PriorityQueue* object is referenced by the *sbl* variable; this object maintains an ordered list of *SBNotice* objects each referencing some behaviour object. In Fig. 6.7, three '*Activity 1*' objects and one '*Activity 2*' object have been scheduled on the *sbl* list. The extended scheduled behaviour list is supported in the same fashion, but a *LinkedList* object is used instead of a *PriorityQueue* object because the *esbl* list is not time ordered.

Activity objects contain a variety of methods: *precondition, startingEvent, duration, terminatingEvent*, etc. (depending of the type of activity, see Fig. 6.6). In addition to these methods, the '*Activity 2*' contains two variables (*modelRef* and *entityRef*). The reference variable *modelRef* provides the reference to the *SimModel* object, which is needed in order to access the status of the simulation model. Making this variable static, i.e. a class variable, allows it to be initialized by the *SimModel* constructor so that it can be used by the static *precondition()* methods in conditional activity classes as well as by all instance methods in all instantiated objects. The second reference variable *entityRef* is an instance variable and is used to reference and access a transient entity (*Entity 3*) that is involved in the activity.

6.4 Examples of Activity-Object Simulation Models

6.4.1 Kojo's Kitchen

In this section we develop an Activity-Object simulation model for the ABCmod conceptual model for the Kojo's Kitchen project given in Sect. A1.1 using the ABSmod/J package. The first step is to create a simulation model class which we call *KojoKitchen* as shown in Fig. 6.8. Note first the manner in which the conceptual model parameter, the input variable as well as the group and queue entities are represented. The parameter and input variable are mapped onto the instance variables *addEmp* and *uEmpNum,* respectively (in the *rgCounter* object). Two reference variables are added to the *KojoKitchen* class to reference the objects that represent the RG.Counter and Q.CustLine entity structures (these both have *scope* = Unary in the ABCmod conceptual model).

Figure 6.8 also shows two other reference variables *rvp* and *output* that reference, respectively, an *RVPs* object and an *Output* object.[9] The contents of these classes are described in the discussion to follow. A getter method is included in the output section in the *KojoKitchen* class to provide the proportion of customers that waited more than 5 minutes, namely the value of the SSOV called *propLongWait* in the conceptual model.

The constructor method *KojoKitchen* initializes the simulation model. More specifically, it:

- Initializes parameter *rgCounter.addEmp* with the value of its method parameter *addE*.
- Calls the method *initialiseClasses* which will initialize the static variable *model* to the value of *this*,[10] in all classes to reference the *KojoKitchen* object. Most objects such as the *RVPs* object and the behaviour objects need to reference the *KojoKitchen* object to access the status of the model, such as the simulation clock or attributes of entity objects.
- Instantiates the *RVPs* class to initialize the *rvp* reference variable.[11]

---

[9]Note that *output* is initialized as part of its declaration, and thus the *Output* object is created upon the instantiation of the *KojoKitchen* class.

[10]The keyword *this* is used with most of the constructors shown. It is a reference variable available in all Java objects; its value is a reference to the object itself.

[11]The *RVPs* class is instantiated in its constructor since a set of seeds are required to initialize the various random variate generators and these are passed to the constructor. The *Seeds* object provides all required seeds.

```java
public class KojoKitchen extends AOSimulationModel
{
  // Parameter
  // See Counter Class RG.Counter.addEmp;
    // addEmp:  = 0 for Base Case ; = 1 for Alternate Case
  /*------------Entity Structures------------------*/
  // Group and Queue Entities
  protected Counter rgCounter = new Counter(); // Counter group
  protected ArrayList<Customer> qCustLine =
            new ArrayList<Customer>; // Line to counter
  // Input Variable
  protected int numEmp;
  // References to RVPs and DVPs objects
  protected RVPs rvp;
  // Output
  protected Output output = new Output();  // Output object
  public double getPropLongWait()
  { return output.propLongWait; }
  // Constructor - Initialises the model
  public KojoKitchen(double t0time, double tftime,
                     int addE, Seeds sd)
  {
    rgCounter.addEmp = addE; // Initialise parameter
    initialiseClasses(sd);
    rvp = new RVPs(this,sd); // Create RVPs object with seeds
    // Initialise the simulation model
    initAOSimulModel(t0time, tftime+60);
    // record the closing time - see implicitStopCondition
    closingTime = tftime;
    // Schedule the first arrivals and employee scheduling
    Initialise init = new Initialise();
    scheduleAction(init);  // Should always be first
    Arrivals arrival = new Arrivals();
    scheduleAction(arrival);  // customer
    StaffChange staffChangeAction = new StaffChange();
    scheduleAction(staffChangeAction); // change in employees
  }
  void initialiseClasses(Seeds sd)
  { // Sets all static references, model, to this object }
  // Testing preconditions
  protected void testPreconditions(Behaviour behObj)
  { /* ... */ }
  // Implicit stop condition
  private double closingTime;  // closing time of the Deli
  protected boolean implicitStopCondition( )
  {
    boolean retVal = false;
    if(getClock() > closingTime && rgCounter.getN() == 0)
      retVal = true;
    return(retVal);
  }
  // For debugging
  protected void eventOccured()
  { /* ... */ }
}
```

Fig. 6.8 *KojoKitchen* simulation model class

- Calls the *initAOSimulModel* method to initialize the inherited *AOSimula-tionModel* attributes such as the *sbl* list.
- Saves the closing time *tftime* in the instance variable *closingTime* for use by the *implicitStopCondition* method.
- Creates and schedules all *ScheduledAction* objects which include the *Initialise*, *Arrivals* and *StaffChange* objects (note that the *Initialise* object must be scheduled first).

An implicit stop condition is used in the Kojo's Kitchen simulation model. A simulation run is terminated when Kojo's Kitchen has closed (21h00) and all customers who joined the the queue before 21h00 have been served. The method *implicitStopCondition*, shown in Fig. 6.8, contains the code to meet these specifications. The *KojoKitchen* class includes the methods *testPreconditions* and *eventOccured*. Each of these methods is discussed later in this section. The completion of the *KojoKitchen* class should be deferred because many of the variables and the constituents of required methods have not yet been identified. For example, the *testPreconditions* method should be developed only after all action and activity classes have been developed. Similarly, the scheduling of scheduled actions are added to the *KojoKitchen* constructor only after the scheduled action classes have been developed.

The next step in developing the simulation model is to incorporate the conceptual model's structural components. No constants are defined for this conceptual model and thus the *Constants* class is not required. The parameter (*rgCounter.addEmp*) has already been included in the *Counter* class shown below. This leaves the development of the classes for representing the ABCmod entity structures.

The reference variable *rgCounter* references a *Counter* object (instantiated from the Counter class shown below) to represent the RG.Counter entity instance and the reference variable *qCustLine* references an *ArrayList* object to represent the Q.CustLine entity instance.

```
class Counter
{
    // Attributes
    protected int uNumEmp;  // Number of employees
    protected int addEmp;   // Number of employees to add
    // For implementing the group, use a HashSet object.
    protected HashSet<Customer> group = new HashSet<Customer>();
    // Required methods to implement standard procedures
    protected void spInsertGrp(Customer icCustomer)
        { group.add(icCustomer); }
    protected boolean spRemoveGrp(Customer icCustomer)
        { return(group.remove(icCustomer)); }
    protected int getN() { return group.size(); }
}
```

The Java standard classes *HashSet*[12] and *ArrayList* were selected from Table 6.1 to implement these entity instances. The *HashSet* class provides the methods to add an object to the set (*add*), remove an object from a set (*remove*) and get the number of elements in the set (*size*). The *ArrayList* object provides similar methods.

The *Customer* class, shown below, is a class developed for the Kojo's Kitchen simulation model.

```
class Customer
{
   double startTime; //Time when a customer enters the line
   enum Type {W, U};
   Type uType; //Type of customer
}
```

The entity attributes are represented with the *startTime* and *uType* instance variables. Note that an enumerated type (*Type*) is used to restrict the *uType* variable to one of its two admissible values of W (sandwich customer) or U (sushi customer). The *Customer* objects are created by the *Arrival* object and subsequently referenced.

The next step in the simulation model development is to deal with the set of procedures specified in the Kojo's Kitchen conceptual model. Three RVP procedures are specified in the ABCmod conceptual model. These are transformed to three methods shown in Fig. 6.9, namely, *duC*, *uCustomerType* and *uSrvTm*. A number of statistical data models are required to support these methods. The Colt package provides classes for this purpose. In our specific case, we instantiate an *Exponential* object (exponential distribution data model), a *MersenneTwister* object (a random number generator) and two *Uniform* objects (uniform distribution data models). These objects are instantiated in the class constructor and referenced by the instance variables *interArrDist*, *typeRandGen*, *wSrvTm* and *uSrvTm*.

As a specific illustration, the RVP.DuC() procedure is implemented with the *duC* method to produce the time sequence for the *Arrivals* class. It is supported with the *Exponential* object referenced by *interArrDist* to obtain a random inter-arrival time that is added to the current time to obtain the time of the next arrival. Thus, the method requires access to the simulation clock. The *AOSimulationModel* method *getClock* inherited by the *KojoKitchen* object provides the value of the clock. Thus, the *duC* method uses the reference variable *model* to obtain the clock value with the call *model.getClock()*. Note that the code for the *duC* method directly reflects the specifications found in the conceptual model for RVP.DuC().

The development of the behaviour classes that provide the behaviour objects is the next step in the development of Kojo's Kitchen simulation model. These classes are shown in Figs. 6.10, 6.11, 6.12 and 6.13. Note that these classes all contain the *model* reference variable which references the *KojoKitchen* object (as in the case of the *RVPs* class).

---

[12]The type parameter <*Customer*> present in the declarations determines the type of objects that can be added to the *HashSet* and *ArrayList* objects.

```
class RVPs
{
  static KojoKitchen model; // for accessing the clock
  // Constructor
  public RVPs(Seeds sd)
  {
    // Set up distribution functions
    interArrDist = new Exponential(1.0/MEAN1,
                                  new MersenneTwister(sd.arr));
    typeRandGen = new MersenneTwister(sd.type);
    wSrvTm = new Uniform(STWMIN,STWMAX,sd.wstm);
    uSrvTm = new Uniform(STUMIN,STUMAX,sd.ustm);
  }
  // RVP.DuC - Customer arrival times
  protected final static double MEAN1 = 10.0;
  protected final static double MEAN2 = 1.75;
  protected final static double MEAN3 = 9.0;
  protected final static double MEAN4 = 2.25;
  protected final static double MEAN5 = 6.0;
  protected Exponential interArrDist;  // Data Model
  protected double duC()
  {
    double nxtArrival;
    double mean;
    if(model.getClock() < 90) mean = MEAN1;
    else if (model.getClock() < 210) mean = MEAN2;
    else if (model.getClock() < 420) mean = MEAN3;
    else if (model.getClock() < 540) mean = MEAN4;
    else mean = MEAN5;
    nxtArrival = model.getClock() +
                        interArrDist.nextDouble(1.0/mean);
    if(nxtArrival > model.closingTime)
      nxtArrival = -1.0;  // Ends time sequence
    return(nxtArrival);
  }

  // RVP.uCustomerType - type of customer
  private final double PROPW = 0.65;
  private final double PROPU = 0.35;
  MersenneTwister typeRandGen;
  public Customer.Type uCustomerType()
  {
    double randNum = typeRandGen.nextDouble();
    Customer.Type type;
    if(randNum < PROPW) type = Customer.Type.W;
    else type = Customer.Type.U;
    return(type);
  }
  // RVP.uSrvTm - service time
  private final double STWMIN = 3;
  private final double STWMAX = 5;
  private final double STUMIN = 5;
  private final double STUMAX = 8;
  // Data Models
  private Uniform wSrvTm; // Sandwich cust. service time
  private Uniform uSrvTm; // Sushi customer service time
  // Method
  public double uSrvTm(Customer.Type type)
  {
    double srvTm = 0;
    if(type == Customer.Type.W)
      srvTm = wSrvTm.nextDouble();
    else if(type == Customer.Type.U)
      srvTm = uSrvTm.nextDouble();
    else
      System.out.println("rvpuSrvTm - invalid type"+ type);
    return(srvTm);
  }
}
```

Fig. 6.9 The RVPs class for the Kojo's Kitchen simulation model

```
class Initialise extends ScheduledAction
{
  static KojoKitchen model;
  // Time Sequence
  double [] ts = { 0.0, -1.0 }; // -1.0 ends scheduling
  int tsix = 0;  // set index to first entry.
  public double timeSequence()
  {
    return ts[tsix++];
  }
  // Event SCS
  public void actionEvent()
  {
    // System Initialisation
    model.rgCounter.numEmp = 2; // Input variable
    model.rgCounter.group.clear();  // empties the group
    model.qCustLine.clear();   // empties the line
    model.output.numServed = 0;
    model.output.numLongWait = 0;
  }
}
```

Fig. 6.10 Initialise class for the Kojo's Kitchen simulation model

Three scheduled actions (Initialise, Staffchange and Arrivals) are specified in the ABCmod conceptual model of Sect. A1.1. In the corresponding classes that are developed (Figs. 6.10, 6.11 and 6.12), the *timeSequence* method is used to implement the time sequences of the three scheduled actions and the *actionEvent* method is used to implement their respective event SCSs, as specified in the conceptual model. Each time sequence is implemented as an array in the *Initialise* and *StaffChange* classes. Note that the last value in both the arrays *ts* and *staffChangeTImeSeq,* is −1.0 which the *runSimulation* method interprets as the end of the time sequence. Instance variables (e.g. *tsix* in the *Initialise* class) are used to traverse the elements of these arrays. The time sequence in the case of the *Arrivals* class is implemented using the *duC* method found in the *RVPs* class. When closing time has been reached, this method returns −1.0.

The following, in particular, should be noted:

(a) In Fig. 6.10, the method *clear* is used to empty the counter resource group and the customer line, which corresponds to setting the attributes RG.Counter.n and Q.CustLine.n to zero in the ABCmod conceptual model.

(b) In Fig. 6.12, the ABCmod standard procedure SP.Derive is implemented using the *new* operator in Java which creates the *Customer* object. Note also that the RVP.uCustomerType() is mapped onto the method call *model.rvp.uCustomerType()* in the simulation model and the SP.InsertQue(Q.CustLine, iC.Customer) is mapped onto the method call *model.qCustLine.add(icCustomer).*

```
class StaffChange extends ScheduledAction
{
  static KojoKitchen model;  // reference to model object
  // Time Sequence
  private double[] staffChangeTimeSeq = {0,90,210,420,540,-1};
  private int sctIx = 0;
  public double timeSequence()
  {
    double nxtTime = staffChangeTimeSeq[sctIx];
    sctIx++;
    return(nxtTime);
  }
  // Event SCS
  protected void actionEvent()
  {
    if(model.getClock() == staffChangeTimeSeq[0])
      model.numEmp = 2;
    else if(model.getClock() == staffChangeTimeSeq[1])
      model.numEmp += model.addEmp;
    else if(model.getClock() == staffChangeTimeSeq[2])
      model.numEmp -= model.addEmp;
    else if(model.getClock() == staffChangeTimeSeq[3])
      model.numEmp += model.addEmp;
    else if(model.getClock() == staffChangeTimeSeq[4])
      model.numEmp -= model.addEmp;
    else
      System.out.println("Invalid time to schedule employees:"
                             +model.getClock());
  }
}
```

Fig. 6.11 StaffChange class for the Kojo's Kitchen simulation model

```
class Arrivals extends ScheduledAction
{
    static KojoKitchen model; // reference to model object
    // Time sequence
    public double timeSequence()
    {
        return model.rvp.duC();
    }
    // Event SCS
    public void actionEvent()
    {
        // WArrival Action Sequence SCS
        Customer icCustomer = new Customer();
        icCustomer.uType = model.rvp.uCustomerType();
        icCustomer.startTime = model.getClock();
        model.qCustLine.add(icCustomer);
    }
}
```

Fig. 6.12 Arrivals class for the Kojo's Kitchen simulation model

```
class Serving extends ConditionalActivity
{
  static KojoKitchen model; // for referencing the model
  private Customer iCCustomer; // Ref. to customer entity

  // Precondition
  protected static boolean precondition()
  {
    boolean returnValue = false;
    if((model.rgCounter.getN() < model.rgCounter.uNumEmp) &&
       (model.qCustLine.size() != 0)) returnValue = true;
    return(returnValue);
  }
  // Starting Event SCS
  public void startingEvent()
  {
    Output output = model.output; // ref. to output object
    iCCustomer = model.qCustLine.remove(0);
    model.rgCounter.spInsertGrp(iCCustomer);
    if((model.getClock()-iCCustomer.startTime) > 5)
       output.numLongWait++;
    output.numServed++;
    output.propLongWait =
            (double) output.numLongWait/output.numServed;
  }
  // Duration
  protected double duration()
  {
    return (model.rvp.uSrvTm(iCCustomer.uType));
  }
  // Terminating Event SCS
  protected void terminatingEvent()
  {
    if(model.rgCounter.spRemoveGrp(iCCustomer) == false)
      System.out.println("Error: Customer not at counter");
  }
}
```

Fig. 6.13 Serving class for the Kojo's Kitchen simulation model

There is only one activity in the Kojo's Kitchen ABCmod conceptual model, namely, the conditional activity called Serving. Figure 6.13 shows the *Serving* class that implements this activity. The four methods *precondition, startingEvent, duration* and *terminatingEvent* can be readily associated with the components of the corresponding ABCmod conditional activity construct. Recall the requirement for the *precondition* method to be a static method. The *Serving* class provides examples of how the various *ArrayList* methods are the surrogates for the ABCmod standard procedures and entity attributes, and how the methods in the *Counter* class play the same role, e.g.:

- In the *precondition* method, the *rgCounter.getN()* method call and the *qCustLine.size()* method call are used to obtain the value of the number being served at the counter (i.e. RG.Counter.n) and the value of the number in the customer line (i.e. Q.CustLine.n), respectively.
- In the *startingEvent* method, the *qCustLine.remove(0)* method call and *rgCounter.spInsertGrp(iCCustomer)* method call correspond to the execution of SP.RemoveQue(Q.CustLine) and SP.InsertGroup(RG.Counter, iC.Customer), respectively.
- In the *terminatingEvent* method, the rgCounter.*spRemoveGrp(iCCustomer)* method call corresponds to the execution of SP.RemoveGroup(RG.Counter, iC.Customer).

Notice also the use of the *iCCustomer* reference variable. This variable associates a specific *Customer* object with a specific *Serving* object which parallels an equivalent concept in the ABCmod conceptual model. In the conceptual model for Kojo's Kitchen, there may be up to three Serving activities (i.e. *Serving* objects) each associated with different customer entities (i.e. each referencing a different *Customer* object).

The next step in the development of the ABSmod/J simulation model for Kojo's Kitchen is the completion of the *testPreconditions* method. This is particularly straightforward because there is only one conditional activity (Serving) and hence only one precondition. The implementation is shown in Fig. 6.14, where it should be noted that the *testPreconditions* method first calls the *reschedule* method to reschedule either of the two scheduled actions specified in the model (i.e. *behObj* is referencing an *Arrivals* object or a *StaffChange* object). The method then simply calls the *precondition* method in the *Serving* class, and if it returns *true*, an instance of the *Serving* class is created, the *startingEvent()* method of the newly created *Serving* object is called, and finally the *Serving* object is scheduled using the *scheduledActivity* method.

```
/* Testing preconditions */
protected void testPreconditions(Behaviour behObj)
{
  reschedule(behObj);  // For bootstrapping
  if(Serving.precondition() == true)
  {
    Serving act = new Serving();
    act.startingEvent();
    scheduleActivity(act);
  }
}
```

Fig. 6.14 The testPreconditions method in the KojoKitchen class

Output requirements in the Kojo's Kitchen project involve only SSOV's, and consequently the ABSmod/J implementation is especially straightforward. The instance variables corresponding to the specified SSOV's are declared and initialised in the *Output* class as shown in Fig. 6.15.

The last method we consider is the *eventOccured* method shown in Fig. 6.16. This method has no counterpart in the Kojo's Kitchen ABCmod conceptual model because its purpose is to provide a means for tracing the execution of the simulation model. In Fig. 6.16, it can be noted that the status information selected here has four components, namely, the current value of simulated time, the length of the customer queue, the number of customers being served and the number of employees serving at the counter. The method also displays the content of the *sbl* list via a call to the *showSBL()* method, thereby, displaying those activities that are currently scheduled.

A typical trace output is shown in Fig. 6.17. The presentation is a sequence of information 'segments'. The first line in each segment provides the specified status information (beginning with the current value of simulated time, i.e. clock). Multiple lines follow (beginning with the label SBL) and these correspond to the contents of the *sbl* list at the time given by 'clock'.

```
class Output
{
  /* SSOVs */
  protected int numServed = 0;
  protected int numLongWait = 0;
  protected double propLongWait= 0;
}
```

Fig. 6.15 Output class

```
/* For debugging */
protected void eventOccured()
{
  System.out.println("Clock: "+getClock()+
    ", Q.CustLine.n: "+qCustLine.size()+
    ", RG.Counter.n: "+rgCounter.getN()+
    ", input numEmp: "+numEmp);
  showSBL();
}
```

Fig. 6.16 The eventOccured method in the KojoKitchen class

```
Clock: 0.0, Q.CustLine.n: 0, RG.Counter.n: 0, input numEmp: 2
------------SBL----------
TimeStamp:2.359 Activity/Action: kkSimModel.Arrivals
TimeStamp:90.0 Activity/Action: kkSimModel.StaffChange
----------------------
Clock: 2.359, Q.CustLine.n: 0, RG.Counter.n: 1, input numEmp: 2
------------SBL----------
TimeStamp:6.534 Activity/Action: kkSimModel.Arrivals
TimeStamp:7.317 Activity/Action: kkSimModel.Serving
TimeStamp:90.0 Activity/Action: kkSimModel.StaffChange
----------------------
Clock: 6.534, Q.CustLine.n: 0, RG.Counter.n: 2, input numEmp: 2
------------SBL----------
TimeStamp:6.563 Activity/Action: kkSimModel.Arrivals
TimeStamp:7.317 Activity/Action: kkSimModel.Serving
TimeStamp:9.790 Activity/Action: kkSimModel.Serving
TimeStamp:90.0 Activity/Action: kkSimModel.StaffChange
```

Fig. 6.17 Tracing an execution of the KojoKitchen simulation model

6.4.2 Port Version 2: Selected Features

Because of the simplicity of the Kojo's Kitchen project in Sect. 6.4.1, many important features of ABSmod/J were not illustrated in the preceding example. In this section, we use the example of Sect. A1.4, Port Version 2, to illustrate the following features: handling constants, starting a sequel activity, handling an extended activity, handling a collection of preconditions, handling trajectory and sample sequences.

We begin by showing below the *Constants* class developed for the Port Version 2 simulation model:

```
class Constants
{
    /* Constants */
    final static double BERTH_TIME = 2;
    final static double DEBERTH_TIME = 1;
    final static double EMPTY_TIME = 0.25;
}
```

The constants from the conceptual model are apparent. The constants are declared as static variables and thus are accessed using the class name. More specifically, the expression *Constants.BERTH_TIME* accesses the *BERTH_TIME* constant from within any method in the simulation model. The following *duration* method (taken from the *Berthing* class) illustrates such an access to *BERTH_TIME:*

```
public double duration()
{
    return Constants.BERTH_TIME;
}
```

The Port Version 2 ABCmod conceptual model includes two sequel activities: Loading and ReturnToHarbour. Figure 6.18 shows the *Loading* class used to implement the Loading sequel activity. Note that there is no *precondition* or

```
class Loading extends SequelActivity
{
    static PortVer2 model;
    // Tanker that is loading
    Tanker iCTanker;
    // Constructor
    public Loading(Tanker tnkr)
    {
        iCTanker = tnkr;
    }
    // Starting event SCS
    public void startingEvent()
    {
        /* Empty event */
    }
    // Duration
    public double duration()
    {
        return model.rvp.uLoad(iCTanker.size);
    }
    // Terminating Event
    public void terminatingEvent()
    {
        iCTanker.startWait = model.getClock();
        model.qDeberthList.add(iCTanker);

    }
}
```

Fig. 6.18 Loading class (extends the SequelActivity class)

timeSequence method in this class because it is an implementation of a sequel activity. The causal initiation basis (see Table 4.6) in the Loading activity identifies one parameter—a reference to the Tanker entity that is involved in the Loading activity. The parameter list is translated into the constructor parameter list in the *Loading* class; thus, a reference to a *Tanker* object is passed to the class constructor.

The Loading activity is started within the SCS of the terminating event of the Berthing activity. The following shows the *terminatingEvent* method included in the *Berthing* class; it illustrates how the Loading sequel activity instance is started, that is, how the standard procedure SP.Start(Loading, iC.Tanker) is transformed into an ABSmod/J counterpart.

```
public void terminatingEvent()
{
    //Berthing Activity Terminating Event SCS
    model.rgBerths.add(iCTanker);
    //Start Sequel Loading Activity
    Loading loadAct = new Loading(iCTanker);
    model.startSequelActivity(loadAct);
    //R.Tug.Status<-PAUSEB
    model.rTug.status = Tug.StatusVals.PAUSEB;
}
```

First, the *Loading* class is instantiated to produce a *Loading* object whose reference is passed to the *startSequelActivity* method of the *PortV2* simulation model class. Note that the *iCTanker* argument in the call to the *Loading* constructor passes the reference to the *Tanker* object that is associated with the *Berthing* object, and thus the *Tanker* object becomes associated with the newly created *Loading* object. The *startSequelActivity* method shown below (implemented in the *PortVer2* simulation model class – not shown) executes the *startingEvent* method of the sequel activity object and then schedules the sequel activity with a call to the *scheduleActivity* method. This method can be used to 'start' any sequel activity once the sequel activity object has been created.

```
protected void startSequelActivity(SequelActivity seqAct)
{
    seqAct.startingEvent();
    scheduleActivity(seqAct);
}
```

We now consider the extended activities within the Port Version 2 conceptual model and the interrupted processing that is required. Figure 6.19 shows the *MoveToBerths* class that represents the extended conditional activity MoveToBerths specified in the conceptual model. The methods *interruptionPreCond* and *interruptionSCS* implement the intercepting precondition and the SCS of the intercepting event, respectively. Both these methods can handle more than one intercepting event specification. The *interruptionPreCond* method returns an integer value that identifies which of the possible intercepting preconditions has a *true* value (there is only one in this example). This integer value is passed to the *interruptionSCS* method and is used to determine which intercepting event SCS is to be executed. If all intercepting preconditions have *false* values, the integer value 0 is returned by *interruptionPreCond*.

The intercepting preconditions are tested in the *scanInterruptionPreconditions* method shown in Fig. 6.20. All behaviour objects referenced from the *esbl* list are scanned. If the behaviour object has a *MoveToBerths* subclass, then its *interruptionPreCond* is called. If a nonzero value is returned then the following steps are taken:

- The object's *interruptionSCS* method is called.
- The object is unscheduled (removed from the *sbl* list and the *esbl* list[13]).

The *scanInterruptionPreconditions* method is called when the *testPreconditions* method is executed.

---

[13]Notice that the index *i* and the size of the *esbl* list *num* are both decremented because the *esbl* list is decremented by the *unscheduleBehaviour* method.

```
public class MoveToBerths extends ExtConditionalActivity
{
  static PortVer2 model; // For referencing the model
  // Precondition
  public static boolean precondition()
  { ... }
  // Starting Event SCS
  public void startingEvent()
  { ... }
  // Duration
  public double duration()
  { ... }
  // Testing intercepting preconditions
  public int interruptionPreCond()
  {
    int retVal = 0;
    if( model.qBerthList.size() > 0 &&
        (
          (model.getClock()-model.rTug.startTime)<
                              0.7*Constants.EMPTY_TIME ||
          model.qDeberthList.size() == 0
        )
      ) retVal = 1;
    return(retVal);
  }
  // All SCS events that cause an interruption
  public void interruptionSCS(int id)
  {
    // Only one valid id possible - so no check required
    // Start the ReturnToHarbour Activity
    ReturnToHarbour rtHbr = new ReturnToHarbour(model);
    model.startSequelActivity(rtHbr);
  }
  // Terminating Event SCS
  public void terminatingEvent()
  { ... }
}
```

Fig. 6.19 ExtendendActivity class: MoveToBerths

```
// Scan interruptions in extended activities
private boolean scanInterruptPreconditions()
{
  int num = esbl.size();
  int interceptingNum;
  SBNotice nt;
  Behaviour obj;
  boolean statusChanged = false;
  for(int i = 0; i < num ; i++)
  {
    nt = esbl.get(i);
    obj = (esbl.get(i)).behaviourInstance;
    if(MoveToBerths.class.isInstance(obj))
    {
      MoveToBerths mvTberths =
              (MoveToBerths) nt.behaviourInstance;
      interceptingNum= mvTberths.interruptionPreCond();
      if(interceptingNum!= 0)
      {
        mvTberths.interruptionSCS(interceptingNum);
        unscheduleBehaviour(nt);
        statusChanged = true;
        i--; num--;
      }
    }
    else System.out.println(
            "Unrecognized behaviour object on ESBL: "
            +obj.getClass().getName());
  }
  return(statusChanged);
}
```

Fig. 6.20 Testing intercepting preconditions

The approach used in testing the preconditions for the Port Version 2 conceptual model is general. It should be stressed however that other more refined approaches can be formulated by the simulation modeller to accommodate knowledge specific to the problem being considered. Preconditions and intercepting preconditions are re-evaluated as long as one of them evaluates to *true*. The evaluation of preconditions and intercepting preconditions takes place in two separate methods: *scanPreconditions* (see Fig. 6.21) and *scanInterruptionPreconditions* (discussed earlier and shown in Fig. 6.20). These methods scan the preconditions and intercepting preconditions, respectively, and if any evaluates to *true* appropriate steps are taken. Furthermore these methods return a *true* value if at least one of the preconditions or intercepting preconditions evaluates to *true*.

As shown in Fig. 6.21 the *testPreconditions* method is a simple loop that keeps looping until the *scanPreconditions* method returns a value *false*.

```
public void testPreconditions(Behaviour behObj)
{
  while(scanPreconditions() == true) /* repeat */;
}
// Checks preconditions and instantiates activities
// for each true precondition found.
// Returns true if an activity was instantiated
// and false otherwise.
private boolean scanPreconditions()
{
  boolean statusChanged = false;
  if(Berthing.precondition(this) == true)
  {
    Berthing act = new Berthing(this);
    act.startingEvent();
    scheduleActivity(act);
    statusChanged = true;
  }
  if(Deberthing.precondition(this) == true)
  {
    Deberthing act = new Deberthing(this);
    act.startingEvent();
    scheduleActivity(act);
    statusChanged = true;
  }
  if(MoveToBerths.precondition(this) == true)
  {
    MoveToBerths act = new MoveToBerths(this);
    act.startingEvent();
    scheduleActivity(act);
    statusChanged = true;
  }
  if(MoveToHarbour.precondition(this) == true)
  {
    MoveToHarbour act = new MoveToHarbour(this);
    act.startingEvent();
    scheduleActivity(act);
    statusChanged = true;
  }
  // Do not change the status if already true
  if(statusChanged) scanInterruptPreconditions();
  else statusChanged = scanInterruptPreconditions();
  return(statusChanged);
}
```

Fig. 6.21 PortVer2 testPreconditions method

The *scanPrecondition* method calls the *scanInterruptionPreconditions* method. If a value of *false* is returned by *scanPrecondition* the implication is that all preconditions and intercepting preconditions evaluate to *false*.

The trajectory sequence and a sample sequence specified in the three Port conceptual models are implemented as *OutputSequence* objects; this makes possible the application of the averaging operator to obtain values for the two DSOVs of interest (avgOccBerths and avgWaitTime). Figure 6.22 shows the *Output* class developed for the Port simulation model. Notice that two *OutputSequence* objects are created and referenced by the variables *trjNumOccBerths* and *phiTankerWait*. Notice also that the method *updateSequences* adds a value to the trajectory sequence only when a change in RG.Berths.n occurs (namely, the value returned by the method *model.rgBerths.size()* is different from the value in the variable *lastGBerthsN*).

```
class Output
{
    PortVer2 model; // reference to the PorVer2 model
    // Output Variables
    OutputSequence trjNumOccBerths;    // TRJ for size of BerthGrp
    int lastGBerthsN; // value of RG.Berths.n last stored
    OutputSequence phiTankerWait;  // PHI tanker wait time
    // Constructor
    public Output(PortVer2 model)
    {
        this.model = model;
        // Setup TRJ set
        trjNumOccBerths = new OutputSequence("gBerthsN");
        // First point in TRJ - R.BerthGroup is empty at t=0
        lastGBerthsN = 0;
        trjNumOccBerths.put(0.0,0.0);
        phiTankerWait = new OutputSequence("TankerTotalWait");
    }
    // Update the trajectory sequences
    protected void updateSequences()
    {
        // Update the berthGrpN Trajectory set
        int n = model.rgBerths.size();
        if(lastGBerthsN != n)
        {
            trjNumOccBerths.put(model.getClock(), (double) n);
            lastGBerthsN = n;
        }
    }
}
```

Fig. 6.22 Output class

The *PortVer2* simulation model class contains the following output section that uses the *Output* object:

```
//Output Section
Output output = new Output(this);
public void computeDSOVs()
{
   output.trjNumOccBerths.computeTrjDSOVs(time0,timef);
   output.phiTankerWait.computePhiDSOVs();
}
public double getAvgOccBerths()
{return output.trjNumOccBerths.getMean();}
public double getAvgWaitTime()
{return output.phiTankerWait.getMean();}
public void clearOutput()
{output.phiTankerWait.clearSequence();}
public void clearAllOutput()
{
   output.phiTankerWait.clearSequence();
   output.trjNumOccBerths.clearSequence();
}
```

A number of public methods are defined to support the experimentation phase:

- *computeDSOVs*: This method is called after a simulation run to invoke calculation of DSOV values by the *OuputSequence* objects;
- *getAvgOccBerths*: This method returns the value of the DSOV avcOccBerths;
- *getAvgWaitTime*: This method returns the value of the DSOV avgWaitTime;
- *clearOutput*: This method clears the recorded output in the*OutputSequence* object referenced by *phiTankerWait* (for removing data after a warm-up time—see Chap. 7);
- *clearAllOutput*: This method clears the recorded output in both*OutputSequence* objects. This is required for determining the warm up time (see Chap. 7).

6.4.3 Advantages of Using Entity Categories with *scope* = Many[*N*]

The Conveyor project introduced in Sect. 4.4 and presented in Sect. A1.6 illustrates the advantages of using of entity categories with *scope* = Many[*N*] as well as embedded UDPs and RVPs. This section illustrates how such ABCmod conceptual model features are implemented with ABSmod/J.

We begin with the translation of the CompProcessing activity to a class in the ABSmod/J simulation model. This activity construct from Annex 1 is reproduced in Fig. 6.23 and its implementation as the *CompProcessing* class is shown in Fig. 6.24.

The method *MachineReadyForProcessing* (the implementation of the equivalent ABCmod UDP) provides the means to specify a search of the model's structure for a set of machines/queues that meets the precondition to initiate the CompProcessing activity. The UDP gives the identifier of such a pair if it exists and returns the value NONE otherwise. It is used in the precondition to determine if an instance of the activity occurs and in the starting event SCS to obtain the member identifier of the

Activity: CompProcessing	
Processing components at any of the machines. In the case of machines M2 and M3, the component leaves the system after processing. In the case of machine M1, the component is left in the machine for removal by the conditional action.	
Precondition	UDP.MachineReadyForProcessing() ≠ NONE
Event	id←UDP. MachineReadyForProcessing() R.Machines[id].busy ← TRUE R.Machines[id].component ← Q.RemoveQue(Q.Conveyors[id])
Duration	RVP.uProcTime(id, R.Machines[id].component.uType)

Event	R.Machines[id].busy← FALSE
	IF(id ≠ M1) THEN *(Conditional action will remove*
	                    component from machine M1)
	        R.Machines[id].component ← NOCOMP
	        SP.Leave(R.Machines[id].component)
	ENDIF

Embedded User-Defined Procedures	
Name	**Description**
MachineReadyForProcessing()	Returns the member identifier, *id*, of a member of the R.Machines/Q.Conveyor categories under the following conditions: 1) Machine is not busy (R.Machines[id].busy = FALSE) 2) No component in the machine (R.Machines[id].component = NOCOMP) 3) Conveyor leading to machine is not empty (Q.Conveyors[id].n ≠ 0) If no machine is ready for component processing, NONE is returned.

Embedded Random Variate Procedures		
Name	**Description**	**Data Model**

uProcTime(machineId, type)	Provides the processing times for machines M1, M2 and M3; *machineId* has one of the id values M1, M2, or M3. The processing time for M1 is dependent on the component type given by *type whose value is either A or B*	Exponential(MEAN) where the value of MEAN depends on the procedure parameters as follows:
		
		machineId type MEAN
		  M1 A 2.1
		  M1 B 4.2
		  M2 *x* 9.4
		  M3 *x* 10.5
		Where *x* means that the value of *type* is disregarded.

Fig. 6.23 The ABCmod CompProcessing activity from the Conveyor project

```java
class CompProcessing extends Activity
{
  static Manufacturing model;
  int id; // identifiers for R.Machines and Q.Conveyors
  public static boolean precondition() {
    boolean retVal = false;
    if(machineReadyForProcessing() != Constants.NONE)
      retVal = true;
    return(retVal);
  }
  public void startingEvent() {
    id = machineReadyForProcessing();
    this.name = "M" + (id + 1);  // Adds name to object
    model.rMachines[id].busy = true;
    model.rMachines[id].component =
                      model.qConveyors[id].spRemoveQue();
  }
  public double duration() {
    return uProcTime(id, model.rMachines[id].component.uType);
  }
  public void terminatingEvent() {
    model.rMachines[id].busy = false;
    if(id != Constants.M1)
      model.rMachines[id].component = Machines.NO_COMP;
  }
  //---------------------Embedded UDP---------------------
  static protected int machineReadyForProcessing() {
    int machineId = Constants.NONE;
    for(int id = Constants.M1;
            id <= Constants.M3 && machineId == Constants.NONE;
            id++)
      if(!model.rMachines[id].busy &&
         model.rMachines[id].component == Machines.NO_COMP &&
         model.qConveyors[id].getN() != 0) machineId = id;
    return(machineId);
  }
//---------------------Embedded RVP---------------------
  // Embedded RVP
  static Exponential procM1A, procM1B;
  static Exponential procM2, procM3;
  // Initialise the RVP
  static void initRvp(Seeds sd) { /*Initialise*/ }
  static public double uProcTime(int machineId,
                                 Component.CompType type)
  { /* Computes and returns processing time */ }
}
```

Fig. 6.24 The *CompProcessing* Java class

entities from the R.Machine/Q.Conveyor categories that will be involved in the activity. Notice that the 'id' variable serves as a 'parameter' of the activity and hence the activity is said to be 'parameterized'.

Observe that the inclusion of UDPs in an action/activity class as opposed to the UDPs class poses no additional requirements. On the contrary, including an RVP in an action/activity class requires an additional method to initialize the random variate generator. In Fig. 6.24, the method *initRvp* is included to seed the *Exponential* random variate generator objects referenced by *procM1A*, *procM1B*, *procM2* and *procM3*. This method is called when the *Manufacturing* model class constructor is executed.

6.5 Closing Comments

Readers with significant proficiency and experience with object-oriented programming may have observed some deviation from strict adherence to object-oriented programming principles in the perspective adopted by the authors in developing ABSmod/J.

Experienced object-oriented (and Java) software developers may find it counter-intuitive to use entity classes (and objects) as simple data structures without any methods to modify the attributes declared in the class; indeed, the methods to change the attributes (i.e. the status of the entity objects) are part of the activity classes. This approach does not respect the encapsulation (or 'information hiding') principle where a set of public methods provide an API (applications programming interface) for a class and attributes are kept private within the class.

Adhering strictly to the encapsulation principle leads to a Process-Oriented world view. However, as Pidd [9, 11] notes, an object-oriented paradigm for the development of a simulation model should not restrict consideration to a process-oriented world view. In fact, Hills [4] demonstrates how Simula, a programming language that parented modern objected-oriented approaches, provides the means to develop a simulation model based on the three-phase (activity-based) world view. Pidd [10] also developed simulation models based on the three-phase world view in Java.

As previously mentioned, our ABSmod/J environment for simulation model implementation does not strictly respect some of the object-oriented practices (in particular, encapsulation[14]). However, the authors do acknowledge the value of modular programming practices (which traditionally employs encapsulation). Thus, the simulation model examples shown in this section have been developed as packages (the package *kkSimModel* for the Kojo's Kitchen simulation model and the package *portV2* for the Port Version 2 simulation model). Encapsulation is applied at

---

[14]It would be possible to apply encapsulation by using getter and setter methods. But this would only mask the issue since the rules (i.e. methods) for modifying the attributes of entities reside in the *Behaviour* classes.

the package level, that is, the simulation model class provides the API for experimentation with the simulation model (other objects instantiated from other classes in the simulation model package can only be accessed from the simulation model object and thus are protected from direct access from outside the package). Experimentation code is limited only to instantiating the simulation model class and to calling certain methods in this object for initializing the simulation model (program), executing simulation runs and obtaining desired output upon their completion.

6.6   Exercises and Projects

6.1   Develop a simulation program based on the Activity-object world view for the ABCmod conceptual model formulated in Problem 4.1 of Chap. 4.

6.2   Develop a simulation program based on the Activity-Object world view for the ABCmod conceptual model formulated in Problem 4.2 of Chap. 4.

6.3   Develop a simulation program based on the Activity-Object world view for the ABCmod conceptual model formulated in Problem 4.4 of Chap. 4.

6.4   Assume that the development of the conceptual model of Problem 4.4 and the development of the simulation program of Problem 6.3 have been carried out by two teams where team A has the primary responsibility for the ABCmod conceptual model and team B has the primary responsibility for the simulation program. Carry out a verification exercise by

(a)   Giving Team B the task of reviewing the conceptual model before developing the simulation program based on the Activity-Object world view.

(b)   Giving Team A the task of reviewing the simulation program once it has been completed.

6.5   Develop a simulation program based on the Activity-Object world view for the ABCmod conceptual model formulated in Problem 4.5 of Chap. 4.

6.6   Develop a simulation program based on the Activity-Object world view for modified ABCmod conceptual model formulated in Problem 4.6 of Chap. 4.

References

1. Arbez G, Birta, LG (2010) An activity-object world view for ABCmod conceptual models. In: Proceedings of the summer computer simulation conference, Ottawa, Ontario, July 2010
2. Banks J, Carson JS II, Nelson B, Nicol DM (2005) Discrete-event system simulation, 4th edn. Prentice-Hall, Upper Saddle River
3. CERN, The Colt Project (2004) Version 1.2.0. http://dsd.lbl.gov/~hoschek/colt/

4. Hills PR (1973) An introduction to simulation using SIMULA. Publication No. S55. Norwegian Computing Center, Oslo
5. Horatmann C (2003) Computing concepts with Java essentials, 3rd edn. Wiley, New York
6. Liang YD (2017) Introduction to Java programming: fundamentals first, 11th edn. Prentice-Hall, Upper Saddle River
7. Oracle, Java Development Kit Standard Edition 12.0. http://www.oracle.com/technetwork/java/javase/overview/index.html
8. Oracle, Java "Package java.lang" API documentation, Java SE 12.0 https://docs.oracle.com/en/java/javase/12/docs/api/java.base/java/lang/package-summary.html
9. Pidd M (2004) Computer simulation in management science, 5th edn. Wiley, Chichester
10. Pidd M, Cassel RA (1998) Three phase simulation in Java. In: Proceedings of the 1998 winter simulation conference, Washington DC, USA
11. Pidd M (1995) Object-orientation, discrete simulation and the three-phase approach. J Oper Res Soc 46(3):362–374
12. Stevens P, Pooley RJ, Pooley R (2006) Using UML: software engineering with objects and components. Addison-Wesley, New York

Experimentation and Output Analysis

7

7.1 Overview of the Issue

In this chapter we explore the experimentation and the output analysis phases of a modelling and simulation project which are both central to the success of the project. In other words, we examine the process of correctly formulating and carrying out goal-directed experiments with the simulation program and then extracting meaningful information from the data acquired via its output variables. The underlying complexity here arises from the uncertainty that is superimposed on all variables in any DEDS model by the random nature of input variables and by the random behaviour of 'internal' processes (e.g. message service time at the nodes of a communication network or failure characteristics of machines in a manufacturing plant). As we have previously noted, these random phenomena represent one of the essential differences between models arising from within the DEDS context and those that are typical within the realm of continuous-time dynamic systems.

A simulation program provides an observation window onto a variety of random phenomena that unfold as a result of the model's execution. These can be linked to random variables that have been identified as output variables and consequently are of special interest from the perspective of the project goal.

The notion of output variables was explored in the discussions of Chaps. 2, 3 and 4 where it was stressed that any model necessarily has one or more such variables associated with it. This follows simply because they serve as the conduits for the information that is essential for the achievement of the project's goal. In these earlier discussions, two categories of output variable were identified, namely, discrete-time variables (either piecewise constant or sample) that exist as time-indexed collections of values collected from some particular execution of the model and scalar output variables (SOVs) which can be either DSOVs (derived scalar output variables) or SSOVs (simple scalar output variables).

© Springer Nature Switzerland AG 2019
L. G. Birta and G. Arbez, *Modelling and Simulation*, Simulation Foundations,
Methods and Applications, https://doi.org/10.1007/978-3-030-18869-6_7

As we have previously noted, the collection of values referenced by a discrete-time output variable is rarely of interest. Instead, what is of interest is typically some property of this accumulated data, e.g. minimum, maximum, average and number (a count of the number of values in the collection). Such a value is computed and assigned to a designated scalar variable which is, in fact, a DSOV. SSOVs are likewise scalar output variables but their values, in the interest of practicality, are established 'on the fly' rather than via a post-processing step. Our interest throughout this chapter is primarily with SOVs.

Several examples of SOVs that might arise in an ABCmod conceptual model are listed below. The list demonstrates the most fundamental feature of any such variable, namely, that it always has a 'definition', i.e. a meaning in terms of the behaviours that are represented within the conceptual model. While this may appear obvious, it is a feature that must be unambiguously documented in the statement of the project goal.

- An output variable Y_A, which represents the average time spent waiting for tugboat service by the tankers that pass through an ocean port model.
- An output variable Y_B, which represents the maximum number of messages in the input buffer of a particular node, P, of a communications network, over a 24-hour period.
- An output variable Y_C, which represents the portion of time that all the attendants in a full-service gas station are busy, over the course of a business day.
- An output variable Y_D, which represents the proportion of customers that waited for more than 5 minutes for service at Kojo's Kitchen in a food court.

Some details for Y_A, Y_B and Y_C which are DSOVs are given in Table 7.1 in terms of the notions in our ABCmod framework. In particular, the table shows how a value might be established for these DSOVs by carrying out an operation on some underlying output sequence of data values. These data values are obtained from the execution of its respective simulation program (i.e. a 'simulation run' or simply a 'run'). The output variable Y_D would typically be treated as an SSOV and details of how its value would be established can be found in Sect. A1.1.

As previously noted, every DSOV is a random variable. This implies that the value of a DSOV acquired from a single simulation run is simply a sample of a random variable. It falls far short of providing useful information from the perspective of achieving the project goal. The information that is needed typically relates to the values of the properties of the distribution of the DSOV, and meaningful estimates of such properties can only be formulated from the results obtained from multiple runs as shown in Fig. 7.1, organized to yield independent observations as outlined in the following section.[1] Note that in this figure, the outputs shown (i.e. y_1, y_2, ..., y_n) could equally be sample values of an SSOV.

---

[1] Strictly speaking, this is not entirely correct. In the context of a steady-state study, there does exist an approach called the method of batch means where all required data is generated from a single "long" simulation run. A brief discussion can be found in Sect. 7.3.2.

Table 7.1 Elaboration of representative output variables

Output variable	SUI context	Relationship to model attributes	Output sequence (trajectory or sample sequence)	Operator on output sequence
Y_A	Ocean port	The value of the tanker's attribute iC.Tanker.totalWait is deposited into the sample sequence when the tug begins deberthing that tanker (see example 2 of Sect. 4.4.2).	*PHI*[iC.Tanker.totalWait]	AVG
Y_B	Communication network	Attribute Q.Pnode.n of the queue entity structure Pnode representing the input buffer of the particular node of interest, is a piecewise constant discrete-time variable and its trajectory sequence is recorded.	*TRJ*[Q.Pnode.n]	MAX
Y_C	Full-service gas station	We assume the existence of an attribute RG.Attendants.nBusy which represents the number of busy attendants and the parameter RG.Attendants.numAvail which gives the number of available attendants. The piecewise constant discrete-time output variable attendAllBusy is given the value 1 when all attendants are busy and 0 otherwise. When RG.Attendants.nBusy = RG.Attendants.numAvail, then all attendants (associated with the resource group entity) are busy.	*TRJ*[attendAllBusy]	AVG

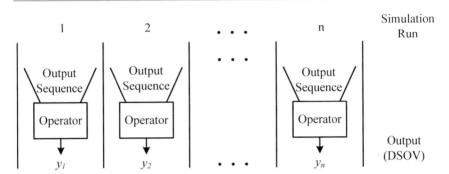

Fig. 7.1 Generation of data from multiple simulation runs

A frequent misunderstanding occurs when the project goal requires a mean value estimate that appears to coincide directly with some SOV (for definiteness, let's call it Y) that is defined as an average. Consider, for example, the port project (as discussed in Sect. A1.3) where the mean waiting time of tankers is required. A particular simulation run will yield a sample sequence whose members are the waiting times of the tankers that passed through the port during that run. The average of the values in this sample sequence (which we denote by $\hat{y}$) would represent an observation of the output variable, Y. One might be tempted here to use $\hat{y}$ as an estimate of the mean value that we seek (namely, the mean waiting time of tankers that pass through the port). However, there generally is a correlation among the values because, for example, a long wait by some particular tanker will likely result in long waits by succeeding tankers. This circumstance precludes the use of the basic methods from statistics which assume data values that are independent, e.g., for the determination of a confidence interval that we discuss below. It is for this reason that suitable simulation runs (or other equivalent approaches) are required which will generate a collection of independent observations of Y from which the desired mean value estimate together with a confidence interval can be formulated. This is achieved by proper management of the seeds (e.g. use of uncorrelated seeds) used in the random variate generation procedures.

For the most part, our interest in experimentation focuses on the determination of mean values for designated output variables within the simulation model. It needs to be recognized however that the determination of an exact value for these is rarely feasible. An experiment composed of a collection of simulation runs with the simulation program can, at best, deliver the data from which an estimate of the mean (called a *point estimate*) can be formulated together with an assessment of the quality of the estimate (i.e. a *confidence interval*). Guidance for determining how an experiment needs to be carried out and how the acquired data needs to be handled in order to obtain credible estimates is provided by some of the fundamental results from probability theory. An overview of these can be found in the latter sections of Annex 2. The topic is, however, explored in the discussions below.

7.2   Bounded Horizon Studies

We now consider the basic problem of correctly evaluating the data acquired by output variables in the context of a bounded horizon study. As outlined below, the general approach is to first determine an estimate of the mean from a collection of n simulation runs (a "point estimate", whose validity has some credible basis) and then formulate an interval that, with a prescribed degree confidence, contains the point estimate.

The considerations that follow rely heavily on the results presented in Annex 2, in particular, the results in Sects. A2.5, A2.6 and A2.7.

7.2.1   Point Estimates

Suppose Y is one of possibly many output variables of the simulation model **M** and we seek an estimate of the mean of the distribution of Y, namely, an estimate of $\mu = E[Y]$. The fundamental result from probability theory upon which we rely is the Strong Law of Large Numbers and the Central Limit Theorem (see Sect. A2.5). The interpretation in our context is that $\overline{Y}(n)$ approaches μ as n becomes large where

$$\overline{Y}(n) = \frac{1}{n}\sum_{k=1}^{n} Y_k$$

and we regard the Y_k's as surrogate random variables for Y that are associated with a sequence of n runs called *replications* (which we regard as an experiment) with **M** (the variable Y_k is associated with replication k). All Y_k's have the same distribution as Y because they reflect the same process (namely, the simulation model **M**). Furthermore, because they are linked to a sequence of replications, we can assume that the Y_k's are independent and normally distributed (Central Limit Theorem).

Correct replications are a key requirement in formulating the estimate that we seek. The implication here is that there is appropriate management of the seeds used to initialize the various random number generators from replication to replication to create a meaningful set of independent and identically distributed observations (initial conditions, however, must remain invariant except when their values required to be random).

On the basis of the above, a point estimate of μ can be obtained in the following way:

1. Choose a suitable value for n, the number of replications (in principle, n needs to be large, but a value in the order of 30 is generally considered satisfactory).
2. Collect the n observed values $y_1, y_2, \ldots, y_n$ for the random variables $Y_k, k = 1, 2, \ldots, n$, that result from n replications of the simulation model, **M**.

3. Compute

$$\bar{y}(n) = \frac{1}{n}\sum_{k=1}^{n} y_k$$

The numerical value that results for $\bar{y}(n)$ is regarded as the point estimate for $\mu = E[Y]$ that we seek.

7.2.2 Interval Estimation

We now expand our task by undertaking to find a suitable value for the number of replications n, which will ensure a particular 'quality' for the estimate $\bar{y}(n)$. We know from Sect. A2.7 that an interval (called the *confidence interval*) can be established within which μ falls with a prescribed level of confidence. This interval has the form $[\bar{y}(n) - \zeta(n), \bar{y}(n) + \zeta(n)]$ and $\zeta(n) = \left(t_{n-1,a} s(n)/\sqrt{n}\right)$ and $\zeta(n)$ is called the *confidence interval half-length* whose value is clearly dependent on the student's t-distribution value $t_{n-1,a}$ (see Table A2.4).

The quality criterion we introduce is the requirement that with confidence $100C\%$ $(0 < C < 1)$, $|\bar{y}(n) - \mu| < \zeta^*$. In other words, we want to ensure that (with a pre-scribed level of confidence) the interval half-length $\zeta(n)$ is less than a specific value denoted by ζ^*. A possible choice for ζ^* is $r\,\bar{y}(n)$ where r is a value chosen in the range $(0, 1)$. With this choice, the maximum displacement of the estimate from μ is proportional to the value of the estimate itself. Note that our quality measure can be interpreted as:

$$\frac{\zeta}{\bar{y}(n)} < r$$

The procedure is outlined below and is based entirely on the discussion in Sect. A2.7 [Eq. (A2.36)] has particular relevance):

1. Choose values for r, for the confidence level parameter C and, as well, an initial value for n that is not less than 20.
2. Collect the n observed values $y_1, y_2, \ldots, y_n$ for the random variables Y_k, $k = 1, 2,$ $\ldots, n$, that result from n replications of the simulation model, **M**.
3. From tabulated data for the student's t-distribution, determine $t_{n-1,a}$ where $a = (1 - C)/2$.

4. Compute

$$\bar{y}(n) = \frac{1}{n}\sum_{k=1}^{n} y_k$$

$$s^2(n) = \frac{\sum_{k=1}^{n} (y_k - \bar{y}(n))^2}{n-1} \tag{7.1}$$

$$\zeta(n) = \frac{t_{n-1,\alpha}s(n)}{\sqrt{n}}$$

5. If $\zeta(n)/\bar{y}(n) < r$, then accept $\bar{y}(n)$ as the point estimate of μ and end the procedure; otherwise, continue to step 6.
6. Choose a Δn and carry out a further Δn replications to obtain the additional observations $y_{n+1}, y_{n+2}, \ldots y_{n+\Delta n}$; replace n with $n + \Delta n$, and repeat from step 3.

7.2.3 Output Analysis for the Kojo's Kitchen Project

This section examines how the analysis techniques described in the previous section can be applied to achieving the goal set out in the Kojo's Kitchen project. As outlined in Sect. A1.1 that goal is to investigate the impact on the output variable *PropLongWait* (which represents the proportion of customers waiting longer than 5 minutes) of adding an additional employee during 'busy periods'. The Java ABSmod/J simulation program presented in Sect. 6.4.1 is used to experiment with the simulation model and to generate data for analysis.

Note the following with respect to the Java method in Fig. 7.2 which is used to carry out experiments with the Kojo's Kitchen simulation model.

- The first part of the method generates the random seeds used in all the experiments. The *CERN* Colt Java package offers a Class *RandomSeedGenerator* that provides the means to generate appropriate (uncorrelated) random seeds. This ensures that the replications provide independent values for the *PropLongWait* output variable. Note also that the seeds are stored in an array of *Seeds* objects. Thus, they can be reused when executing the replications for the alternate case. This is important for comparing the two cases as will be discussed in Sect. 7.4. Also note that four seeds make up a *Seeds* object, one for each random number generator that is used in the simulation program.
- For each simulation run, a new *KojoKitchen* object is created using the class constructor. The constructor provides the data necessary for the simulation run, i.e. it specifies the observation interval (the first two arguments specify the right-hand and left-hand boundaries of the observation interval), a value for the *addEmp* parameter (either 0 or 1) and finally a *Seeds* object to seed the random number generators.

```
class KojoExperiment1 {
  public static void main(String[] args) {
    // Constants
    final int NUMRUNS = 20;
    final double CONF_LEVEL = 0.9;  // Confidence levels
    // Local variables
    int i;
    double startTime=0.0, endTime=660.0;
    Seeds[] sds = new Seeds[NUMRUNS];
    KojoKitchen kojo;  // Simulation object
    ConfidenceInterval cfIntCase1, cfIntCase2;
    double [] valuesCase1 = new double[NUMRUNS];
    double [] valuesCase2 = new double[NUMRUNS];
    // Get a set of uncorrelated seeds
    RandomSeedGenerator rsg = new RandomSeedGenerator();
    for(i=0 ; i<NUMRUNS ; i++) sds[i] = new Seeds(rsg);
    // Loop for NUMRUN simulation runs for each case
    // Case 1
    System.out.println("Case 1 - no additional employee");
    for(i=0 ; i < NUMRUNS ; i++) {
      kojo = new KojoKitchen(startT,endT,0,sds[i], false);
      kojo.runSimulation();
      valuesCase1[i] = kojo.getPropLongWait();
    }
    // Case 2
    System.out.println("Case 2 – with additional employee");
    for(i=0 ; i < NUMRUNS ; i++) {
      kojo = new KojoKitchen(startT,endT,1,sds[i], false);
      kojo.runSimulation();
      valuesCase2[i] = kojo.getPropLongWait();
    }
    // Define confidence intervals with 90% confidence level
    cfIntCase1 = new ConfidenceInterval(valuesCase1, CNF_LEVEL);
    cfIntCase2 = new ConfidenceInterval(valuesCase2, CNF_LEVEL);
    /*-- Display the resulting confidence intervals --*/
    // Code for displaying contents of value arrays and the
    // two ConfidenceInterval objects not shown
  }
}
```

Fig. 7.2 Java method for experimentation with the Kojo's Kitchen simulation program

- After each run, the value generated for *PropLongWait* is saved in an array (*valuesCase1* for Case 1 and *valuesCase2* for Case 2). The saved values are subsequently used to instantiate a *ConfidenceInterval* object to compute confidence intervals for the two cases. The method then prints the results for the experiment are shown in Fig. 7.3 (print instructions are not included in Fig. 7.2).

```
Replication Case 1    Case 2
---------------------------------
        1      0.505      0.060
        2      0.457      0.171
        3      0.731      0.118
        4      0.665      0.294
        5      0.634      0.230
        6      0.517      0.062
        7      0.629      0.126
        8      0.701      0.109
        9      0.463      0.017
       10      0.514      0.223
       11      0.570      0.161
       12      0.519      0.212
       13      0.410      0.116
       14      0.609      0.180
       15      0.151      0.014
       16      0.686      0.237
       17      0.572      0.072
       18      0.599      0.204
       19      0.455      0.132
       20      0.659      0.234
---------------------------------
       PE      0.552      0.148
     S(n)      0.131      0.079
     zeta      0.051      0.030
   CI Min      0.502      0.118
   CI Max      0.603      0.179
  zeta/PE      0.092      0.205
---------------------------------
```

Fig. 7.3 Displayed output produced by the Java method for experimentation (see Fig. 7.2)

The values for the point estimate (*PE*), the standard deviation (*S(n)*) and the confidence interval half-length (*zeta*) are shown in Fig. 7.3, where $n = 20$. These were computed within the *ConfidenceInterval* objects using Eq. (7.1) with a 90% confidence level (*CNF_LEVEL*). The left boundary $\bar{y}(n) - \zeta(n)$ and right boundary $\bar{y}(n) + \zeta(n)$ of the confidence interval are shown in the rows labeled CI Min and CI Max, respectively.

Figure 7.4 shows the output produced by an alternate Java method that shows, for each of the two cases, the values of $y(n)$, $s(n)$ and $zeta(n)$ as well as the boundaries of the confidence interval (columns CI Min and CI Max) and the ratio $zeta(n)/y(n)$ when n (the number of replications) is increased up to 10000. Note from the rightmost column how the ratio $zeta(n)/\bar{y}(n)$ decreases as n increases.

```
-----------------------------------------------------------------------
                                  Case 1
-----------------------------------------------------------------------
n         y(n)     s(n)     zeta(n)  CI Min   CI Max   zeta(n)/y(n)
-----------------------------------------------------------------------
    20    0.552    0.131    0.051    0.502    0.603    0.092
    30    0.558    0.114    0.035    0.523    0.594    0.063
    40    0.565    0.113    0.030    0.535    0.595    0.053
    60    0.582    0.108    0.023    0.558    0.605    0.040
    80    0.578    0.109    0.020    0.558    0.599    0.035
   100    0.581    0.108    0.018    0.564    0.599    0.031
  1000    0.587    0.110    0.006    0.581    0.593    0.010
 10000    0.586    0.113    0.002    0.584    0.588    0.003
-----------------------------------------------------------------------


-----------------------------------------------------------------------
                                  Case 2
-----------------------------------------------------------------------
n         y(n)     s(n)     zeta(n)  CI Min   CI Max   zeta(n)/y(n)
-----------------------------------------------------------------------
    20    0.148    0.079    0.030    0.118    0.179    0.205
    30    0.151    0.087    0.027    0.123    0.178    0.180
    40    0.162    0.109    0.029    0.133    0.191    0.180
    60    0.173    0.117    0.025    0.148    0.198    0.145
    80    0.166    0.112    0.021    0.145    0.187    0.126
   100    0.167    0.112    0.019    0.149    0.186    0.111
  1000    0.174    0.121    0.006    0.168    0.180    0.036
 10000    0.174    0.119    0.002    0.172    0.176    0.011
-----------------------------------------------------------------------
```

Fig. 7.4 Impact of number of replications on the confidence interval

This is mainly a consequence of a decreasing value for the confidence interval half-length $zeta(n)$.

Observe that for case 1, with 20 replications, the half-length of the confidence interval is just under 10% of the point estimate (see rightmost column where the value is 0.092). However, with 20 replications, the interval half-length in case 2 is about 20% of the point estimate (value in rightmost column is 0.205). For case 2, 100 replications are required to achieve a confidence interval that is comparable to case 1.

7.3 Steady-State Studies

The fundamental requirement in a steady-state study is the postponement of data collection during a simulation run until it is apparent that the simulation model has reached steady-state conditions, i.e. the stochastic processes associated with the output variables of interest have become stationary. A necessary (but not sufficient) condition for steady-state behaviour of the simulation model is the requirement that the underlying random variables associated with autonomous stochastic processes (i.e. input), such as arrival rates and services rates, are themselves stationary. But even when this is the case, the model's initial conditions usually give rise to circumstances that cause dependent stochastic processes in the simulation model to pass through a transient phase at the start of a simulation run.

The right-hand boundary of the observation interval is not specified for steady state studies. This provides the flexibility to execute a simulation run for as long as necessary in order to first reach steady-state conditions and then acquire sufficient data to permit meaningful conclusions. Consequently, the execution of simulation runs for steady-state studies must address two important issues:

- Determining a warm-up period: A transient period is always present at the beginning of any simulation run. Behaviour data from this interval is (by definition) incompatible with the steady-state requirements of the study. The implication here is that a 'warm-up period' that precedes the collection of data needs to be recognized. The duration of this period cannot be predicted, and hence a mechanism for determining the end of the warm-up period must be incorporated into the experimentation procedure. Data collection can begin only after this transient or warm-up period has come to an end.
- Establishing confidence in the conclusions: If simulation runs are not sufficiently long, credibility of point estimates and confidence intervals may be undermined (on the basis of the law of large numbers).

7.3.1 Determining the Warm-up Period

Considerable research has explored the problem of establishing a suitable warm-up period for a simulation run, i.e. an interval which allows sufficient time for the dependent stochastic process of interest to reach a steady state, e.g. [4, 8, 9, 11]. Welch's moving average method is one of the many available approaches. It is graphically oriented and relatively straightforward and provides reasonable estimates. This section outlines the application of this method (a more extensive presentation can be found in Law [5]).

Welch's moving average method relies on a relatively small number of simulation replications (e.g. 5–10). The duration of each replication needs to be sufficiently long so that it extends beyond the transient period. A typical run is shown in Fig. 7.5 which illustrates a representative transient condition at the start of the simulation run (in this case the simulation model begins without any consumer entities being present). The horizontal axis in this figure corresponds to the (simulated) time which has been compartmentalized into m time cells. The vertical axis shows how the average value for some output variable might change if separate averages were computed within the time cells. The changing shapes of the idealized distribution functions for this output variable are superimposed. The figure is intended to show that starting at time cell D, changes in average value no longer occur and hence steady state can be assumed.

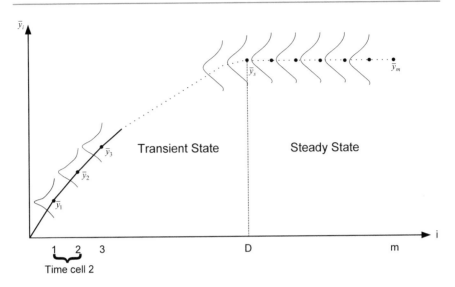

Fig. 7.5 Reaching steady state

Selecting the size of the time cells and the number of time cells (which is equivalent to establishing the length of the simulation run) depends on the underlying nature of the simulation model. The size of the time cell should be large enough to provide reasonable results (i.e. enough data points to compute a credible average within the cell) and yet short enough to be able to detect the existence of the transient.

Replication j generates an output sequence of n_j values, e.g. $\{y_{k,j}: k = 1, 2, \ldots, n_j\}$. The average of those values that fall into time cell i is computed to produce $\bar{y}_{i,j}$ which is the ith cell average for the jth replication. Thus, n replications will produce the set of n averages $\{\bar{y}_{i,j}: j = 1, 2, \ldots n\}$ where i is the time cell index. The following steps are carried out to obtain an estimate of the time cell index where the system transient terminates or, in other words, the system reaches steady state:

1. Obtain the value $\bar{a}_i$ as the average over the n replications of the ith cell averages $(\bar{y}_{i,j})$, i.e.

$$\bar{a}_i = \frac{1}{n}\sum_{j=1}^{n}\bar{y}_{i,j}$$

2. The values $\bar{a}_i$, $i = 1, 2, \ldots, m$, usually vary considerably. If plotted against index i, the resulting graph is 'choppy' and difficult to interpret. A smoothing operation is required in order to smooth out the rapid variations to obtain a smoother curve that captures the long-run trend. For this purpose, the *moving average values* $\bar{a}_i(w)$ are computed using Eq. (7.2). The parameter w represents a window size that controls the smoothing operation. Its selection is by trial and error.

Usually, a number of values for w need to be tried. The objective is to find as small a value as possible that provides the desired smoothing effect.

$$\bar{a}_i(w) = \begin{cases} \dfrac{\displaystyle\sum_{h=-(i-1)}^{i-1} \bar{a}_{i+h}}{2i-1} & i = 1, \ldots, w \\[2em] \dfrac{\displaystyle\sum_{h=-w}^{w} \bar{a}_{i+h}}{2w+1} & i = w+1, \ldots, m-w \end{cases} \tag{7.2}$$

Equation (7.2) is not as complex as it might appear. When $i > w$, there are w cell averages on either side of $\bar{a}_i$ that are averaged to produce the running average value $\bar{a}_i(w)$. When $i \leq w$, there are not enough values preceding time cell i to fill the window. In this case, w is replaced with $(i-1)$. Table 7.2 shows how the running averages are computed for the case where $w = 3$.

3. The values $\bar{a}_i(w)$ are plotted against the cell index i, and it should be apparent from this graph when steady state has been achieved. A good practice is to extend the apparent length of the warm-up period (say, by 30%). The idea here is to err on the safe side by making the warm-up period too long rather than too short.

Table 7.2 Welch's moving average method with $w = 3$

i	$\bar{a}_i(3)$ equation	$\bar{a}_i(3)$ expansion
1	$\dfrac{\sum_{h=0}^{0} \bar{a}_{i+h}}{1}$	$\dfrac{\bar{a}_1}{1}$
2	$\dfrac{\sum_{h=-1}^{1} \bar{a}_{i+h}}{3}$	$\dfrac{\bar{a}_1 + \bar{a}_2 + \bar{a}_3}{3}$
3	$\dfrac{\sum_{h=-2}^{2} \bar{a}_{i+h}}{5}$	$\dfrac{\bar{a}_1 + \bar{a}_2 + \bar{a}_3 + \bar{a}_4 + \bar{a}_5}{5}$
4	$\dfrac{\sum_{h=-3}^{3} \bar{a}_{i+h}}{7}$	$\dfrac{\bar{a}_1 + \bar{a}_2 + \bar{a}_3 + \bar{a}_4 + \bar{a}_5 + \bar{a}_6 + \bar{a}_7}{7}$
5	$\dfrac{\sum_{h=-3}^{3} \bar{a}_{i+h}}{7}$	$\dfrac{\bar{a}_2 + \bar{a}_3 + \bar{a}_4 + \bar{a}_5 + \bar{a}_6 + \bar{a}_7 + \bar{a}_8}{7}$
$\vdots$	$\vdots$	$\vdots$
$m-3$	$\dfrac{\sum_{h=-3}^{3} \bar{a}_{i+h}}{7}$	$\dfrac{\bar{a}_{m-6} + \bar{a}_{m-5} + \bar{a}_{m-4} + \bar{a}_{m-3} + \bar{a}_{m-2} + \bar{a}_{m-1} + \bar{a}_m}{7}$

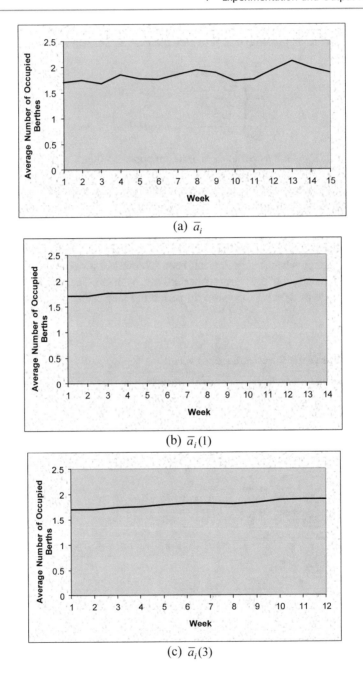

Fig. 7.6 Welch's method applied to the number of occupied berths

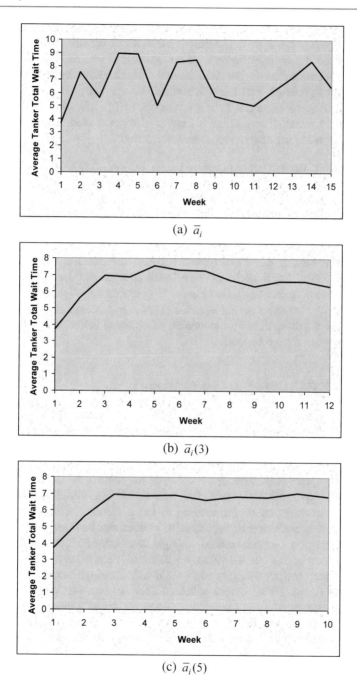

(a) $\bar{a}_i$

(b) $\bar{a}_i(3)$

(c) $\bar{a}_i(5)$

Fig. 7.7 Welch's method applied to tanker total wait time

We illustrate the use of the Welch's moving average method using Version 1 of our Port project (Sect. A1.3). The output variables of interest are the number of occupied berths (i.e. *numOccBerths*) and the tanker total wait time (i.e. *tankerWait*). The class *WelchAverage* is provided in the ABSmod/J library to carry out the computations associated with Welch's method. The output produced was graphed using Excel. Figures 7.6 and 7.7 show the results of 10 replications ($n = 10$), each of duration 15 weeks. The time cells have a width of 1 week which means that $m = 15$. The following observations are noteworthy:

1. In the case of the number of occupied berths, there is no apparent transient. Even without the use of running averages (see Fig. 7.6a), the graph is relatively smooth. This result can be attributed to the small size of the group (namely, 3) which results in the available berths being quickly filled by the first few arrivals of tankers.
2. A transient is certainly apparent for the tanker total wait time as shown in Fig. 7.7 and moving averages are required to smooth out the graph. A window size of five provides a suitable result and shows that the transient lasts for approximately 3 weeks. Either 4 or 5 weeks can be selected as a suitable warm-up period.
3. The warm-up period has relevance for the elimination of the transient in the tanker total wait time output variable. However, because no transient exists during the warm-up time for the number of occupied berths, the data during the warm-up period may be used.

7.3.2 Collection and Analysis of Results

Extending the right-hand boundary of the observation interval allows more data to be collected during a simulation run, and this provides the basis for a number of methods for generating the data that is necessary for analysis (i.e. a set of independent and identically distributed (IID) values for the output variable). We examine two approaches. The replication–deletion method is described and illustrated using experimentation with the port simulation model as presented in Sect. 7.3.3. An overview of the method of batch means is also given. A more comprehensive presentation of the available options can be found in Law [5].

In the replication–deletion method, the right boundary of the observation interval (i.e. t_f) is simply taken to be the value of (simulated) time when a sufficient amount of data has been collected to generate a valid and meaningful collection of output observations (in particular, output collected after the warm-up period). In fact, a sequence of n replications is executed to produce a set of n values for the output variable. The main advantage of this method is that it naturally generates a set of IID values. It does however have the computing time overhead resulting from the repetition of the warm-up period for each of the replications.

The replication-deletion method resembles the experimentation and output analysis previously outlined for a bounded horizon project (see Sect. 7.2). From the output data, a point estimate and confidence interval can be obtained using a number of replications and applying Eq. (7.1). In the discussion of Sect. 7.2.2, it was noted that increasing the number replications reduced the confidence interval half-length $\zeta(n)$ and increased the quality of the point estimate. This equally applies in the replication–deletion approach for a steady-state study. However, in a steady-state study, $\zeta(n)$ can also be reduced by increasing the length of the replications, that is, by adjusting the right-hand boundary, t_f, of the observation interval. Based on these observations, the procedure for the replication–deletion method can be formulated as a straightforward extension of the earlier procedure presented in Sect. 7.2.2. It is as follows:

1. Choose a value for r, for the confidence level parameter C and, as well, an initial reasonable value for t_f and an initial value for n that is not smaller than 20.
2. Collect the n observed values $y_1, y_2, \ldots, y_n$ for the random variables Y_k, $k = 1, 2, \ldots, n$ that result from n replications of the simulation model **M**, that terminate at time t_f (Note that the output sequence for the y_i's is obtained from data collected after the end of the warm-up period.).
3. From the tabulated data for the student's t-distribution, determine $t_{n-1,a}$ where $a = (1 - C)/2$.
4. Compute $\bar{y}(n)$ and $\zeta(n)$ using Eq. (7.1).
5. If $\zeta(n)/\bar{y}(n) < r$, then accept $\bar{y}(n)$ as the estimate of μ and end the procedure; otherwise, continue to step 6.
6. Either choose Δn (≥ 1) and collect the additional observations $y_{n+1}, y_{n+2}, \ldots, y_{n+\Delta n}$ through a further Δn replications, replace n with $n + \Delta n$ and repeat from step 3 or increase the value of t_f (for example, by 50%) and repeat from step 2.[2]

The batch means method, on the other hand, is an entirely different approach that requires only a single (but potentially 'long') simulation run. An advantage of this approach is economy of computing time since the warm-up period only needs to be accommodated once. The end of the observation interval, t_f, is selected to generate all the data necessary for analysis. But possible autocorrelation of the output data must be dealt with in order to generate the necessary IID data.

To generate a set of IID data values, the observation interval beyond the warm-up period is divided into n time cells as shown in Fig. 7.8. The output data that falls into a time cell is then collected into a batch. The end result is a set of

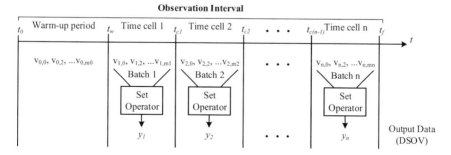

Fig. 7.8 Output values using the method of batch means

---

[2]In programming environments (e.g. Java), it is generally possible to save the state of the simulation model for each run so that simulation runs can be continued from the previously specified t_f.

n batches or sets of values. A DSOV output value is then computed for each batch, providing a set of output values y_i for $i = 1, 2, ..., n$. Equation (7.1) can then be applied to the set of output values to obtain a point estimate and confidence interval for each output variable of interest.

The challenge with this method is to select a large enough time cell so that the set of y_i values are uncorrelated. Correlation of the y_i values needs to be checked to ensure that proper batch size has been selected. Details about this method (and also other methods which avoid using replications and use a single simulation run) can be found in Law [5].

7.3.3 Experimentation and Data Analysis for the Port Project (Version 1)

The Java program given in Fig. 7.9 carries out the experiments for the steady-state study of the Port project and generates the tabular output shown in Fig. 7.10. The replication–deletion method is being used with a warm-up period of 5 weeks (previously determined by using Welch's moving average method). The main aspects are the following:

1. Several constants are created to support the logic that produces the tabular output; these include *NUMRUNS* that defines the number of replications to carry out for each case, *NUM_WEEKS* that gives three lengths for the observation interval, *NUM_BERTHS* that gives the two values for the model parameter, *WEEK* that gives the number of hours in a week, *WARM_UP_PERIOD* that specifies the warm-up period (in hours), *CONF_LEVEL* used as the desired confidence level when calculating the confidence intervals, and *NUM_RUN-S_ARRAY* that contains the different number of runs replications used to calculate the confidence interval.

2. Output values are stored in two arrays: *avgOccBerth* used to store the average number of occupied berths and *avgWaitTime* used to store the average number of tanker waiting times.

3. A set of uncorrelated seeds are generated (to ensure independent replications) and saved into an array. The seeds are reused when carrying out the replications for the alternative case of the Port project (this implements the use of common random numbers described in Sect. 7.4).

4. Two experiments are carried out: one set for the case where the parameter *RG.Berths.numBerths* = 3 (the base case) and the other for the case where *RG.Berths.numBerths* = 4 (the alternate case). This is implemented as a loop that sets *nb* (the value of *RG.Berths.numBerths*) to the values in the *NUM_BERTHS* array.

5. For each experiment (i.e. for each value of the model parameter *RG.Berths.numBerths*), three observations interval lengths are used. The length (in weeks) of each observation interval is given in the array *NUM_WEEKS*. The length in hours is computed and stored in the variable *end*. A loop iterates over these values.

6. A third embedded loop will execute *NUMRUNS* replications for each combination of parameter value/observation interval length.

```
public class PortV1Exp2
{
    // Some experimental constants
    public static final int NUMRUNS = 10000;   // Number of runs
    public static final int [] NUM_WEEKS = { 10, 20, 30 }; // Obs. Int.
    public static final int [] NUM_BERTHS = { 3, 4 }; // Parameter
    public static final double WEEK = 7.0 * 24.0;  // hours/week
    public static final double WARM_UP_PERIOD = 5 * WEEK;  // Warm up
    // Arrays to collect experimental data
    // One output data set per end time for each case
    public static double [][][] avgOccBerths =
        new double[NUM_BERTHS.length][NUM_WEEKS.length][NUMRUNS];
    public static double [][][] avgWaitTime =
        new double[NUM_BERTHS.length][NUM_WEEKS.length][NUMRUNS];
    // For output analysis
    static final double CONF_LEVEL = 0.9;
    static final int [] NUM_RUNS_ARRAY = {20, 30, 40, 60, 80,
                                    100, 1000, 10000};
    public static void main(String[ ] args) {
        double start=0.0; // Observation interval starts at t = 0
        double end; // End of observation interval, see NUM_WEEKS
        Seeds [ ] sds = new Seeds[NUMRUNS];
        int nb;   // parameter, see NUM_BERTHS
        // Get a set of uncorrelated seeds
        RandomSeedGenerator rsg = new RandomSeedGenerator();
        for(int i=0 ; i<NUMRUNS ; i++) sds[i] = new Seeds(rsg);
        for(int ix_nb = 0 ; ix_nb < NUM_BERTHS.length ; ix_nb++)  {
            nb = NUM_BERTHS[ix_nb];
            System.out.println("Number of berths = " + nb);
            for(int ix_nw = 0; ix_nw < NUM_WEEKS.length; ix_nw++)  {
                end=NUM_WEEKS[ix_nw] * WEEK;
                for(int i=0 ; i<NUMRUNS ; i++)  {
                    PortVer1 portSys =
                                new PortVer1(start, nb, sds[i], false);
                    portSys.setTimef(WARM_UP_PERIOD);
                    portSys.runSimulation();
                    portSys.clearPHIiCTankerTotalWait();
                    portSys.setTimef(end);
                    portSys.runSimulation();
                    // Save the results in the arrays
                    avgOccBerths [ix_nb][ix_nw][i] =
                                    portSys.getAvgOccBerths(start, end);
                    avgWaitTime [ix_nb][ix_nw][i] = portSys.getAvgWaitTime();
                }
            }
        }
        displayTable(avgOccBerths, avgWaitTime);
    }
}
```

Fig. 7.9 Experiments with the ABSmod/J Port simulation model (corresponds to the ABCmod conceptual model presented in Sect. A1.3)

```
+-------+----------------------------------+----------------------------------+----------------------------------+
|                                      Number of berths 3                                                       |
+-------+----------------------------------+----------------------------------+----------------------------------+
| tf:   |            10 weeks              |             20 weeks             |             30 weeks             |
+-------+----------------------------------+----------------------------------+----------------------------------+
| n     | yb(n)  s(n)  z(n) z(n)/yb(n)|  yb(n)  s(n)  z(n) z(n)/yb(n)|  yb(n)  s(n)  z(n) z(n)/yb(n)|
+-------+----------------------------------+----------------------------------+----------------------------------+
|                                  Average Number of Occupied Berths                                            |
+-------+----------------------------------+----------------------------------+----------------------------------+
|    20 | 1.845  0.138  0.053 0.0290 | 1.858  0.072  0.028 0.0149 | 1.866  0.066  0.025 0.0137 |
|    30 | 1.820  0.139  0.043 0.0237 | 1.833  0.088  0.027 0.0149 | 1.842  0.078  0.024 0.0132 |
|    40 | 1.824  0.138  0.037 0.0201 | 1.833  0.095  0.025 0.0138 | 1.838  0.071  0.019 0.0103 |
|    60 | 1.827  0.147  0.032 0.0173 | 1.833  0.099  0.021 0.0117 | 1.828  0.077  0.017 0.0090 |
|    80 | 1.828  0.138  0.026 0.0141 | 1.834  0.102  0.019 0.0103 | 1.833  0.076  0.014 0.0077 |
|   100 | 1.823  0.138  0.023 0.0126 | 1.834  0.101  0.017 0.0091 | 1.830  0.078  0.013 0.0071 |
|  1000 | 1.834  0.130  0.007 0.0037 | 1.833  0.094  0.005 0.0027 | 1.833  0.077  0.004 0.0022 |
| 10000 | 1.826  0.132  0.002 0.0012 | 1.832  0.095  0.002 0.0009 | 1.833  0.078  0.001 0.0007 |
+-------+----------------------------------+----------------------------------+----------------------------------+
|                                  Average Tanker Total Wait Time                                               |
+-------+----------------------------------+----------------------------------+----------------------------------+
|    20 | 7.186  2.360  0.913 0.1270 | 7.216  1.931  0.747 0.1035 | 7.269  1.540  0.595 0.0819 |
|    30 | 6.862  2.411  0.748 0.1090 | 7.060  1.690  0.524 0.0743 | 7.280  1.394  0.432 0.0594 |
|    40 | 6.741  2.482  0.661 0.0981 | 7.117  2.018  0.538 0.0755 | 7.331  1.487  0.396 0.0540 |
|    60 | 7.998  5.740  1.238 0.1548 | 7.467  2.625  0.566 0.0758 | 7.334  1.727  0.373 0.0508 |
|    80 | 7.913  5.248  0.977 0.1234 | 7.476  2.435  0.453 0.0606 | 7.422  1.760  0.327 0.0441 |
|   100 | 7.899  5.158  0.856 0.1084 | 7.686  2.496  0.415 0.0539 | 7.469  1.780  0.296 0.0396 |
|  1000 | 7.672  4.538  0.236 0.0308 | 7.701  2.820  0.147 0.0191 | 7.711  2.300  0.120 0.0155 |
| 10000 | 7.504  4.529  0.074 0.0099 | 7.705  2.929  0.048 0.0063 | 7.740  2.295  0.038 0.0049 |
+-------+----------------------------------+----------------------------------+----------------------------------+

+-------+----------------------------------+----------------------------------+----------------------------------+
|                                      Number of berths 4                                                       |
+-------+----------------------------------+----------------------------------+----------------------------------+
| tf:   |            10 weeks              |             20 weeks             |             30 weeks             |
+-------+----------------------------------+----------------------------------+----------------------------------+
| n     | yb(n)  s(n)  z(n) z(n)/yb(n)|  yb(n)  s(n)  z(n) z(n)/yb(n)|  yb(n)  s(n)  z(n) z(n)/yb(n)|
+-------+----------------------------------+----------------------------------+----------------------------------+
|                                  Average Number of Occupied Berths                                            |
+-------+----------------------------------+----------------------------------+----------------------------------+
|    20 | 1.859  0.149  0.058 0.0310 | 1.872  0.075  0.029 0.0154 | 1.877  0.068  0.026 0.0140 |
|    30 | 1.835  0.146  0.045 0.0247 | 1.847  0.090  0.028 0.0152 | 1.854  0.079  0.025 0.0132 |
|    40 | 1.839  0.145  0.039 0.0210 | 1.847  0.096  0.025 0.0138 | 1.850  0.071  0.019 0.0103 |
|    60 | 1.844  0.155  0.033 0.0181 | 1.846  0.101  0.022 0.0118 | 1.840  0.078  0.017 0.0092 |
|    80 | 1.845  0.146  0.027 0.0147 | 1.847  0.105  0.019 0.0105 | 1.845  0.077  0.014 0.0078 |
|   100 | 1.839  0.144  0.024 0.0130 | 1.847  0.104  0.017 0.0094 | 1.841  0.080  0.013 0.0072 |
|  1000 | 1.850  0.135  0.007 0.0038 | 1.846  0.097  0.005 0.0027 | 1.845  0.079  0.004 0.0022 |
| 10000 | 1.842  0.136  0.002 0.0012 | 1.845  0.098  0.002 0.0009 | 1.846  0.080  0.001 0.0007 |
+-------+----------------------------------+----------------------------------+----------------------------------+
|                                  Average Tanker Total Wait Time                                               |
+-------+----------------------------------+----------------------------------+----------------------------------+
|    20 | 2.538  1.045  0.404 0.1593 | 2.536  0.641  0.248 0.0977 | 2.535  0.421  0.163 0.0642 |
|    30 | 2.494  0.939  0.291 0.1168 | 2.558  0.569  0.176 0.0690 | 2.600  0.449  0.139 0.0536 |
|    40 | 2.461  0.886  0.236 0.0959 | 2.557  0.563  0.150 0.0587 | 2.583  0.421  0.112 0.0434 |
|    60 | 2.780  1.554  0.335 0.1206 | 2.604  0.703  0.152 0.0583 | 2.549  0.485  0.105 0.0411 |
|    80 | 2.720  1.424  0.265 0.0974 | 2.586  0.689  0.128 0.0496 | 2.551  0.497  0.093 0.0363 |
|   100 | 2.725  1.382  0.229 0.0842 | 2.630  0.683  0.113 0.0431 | 2.555  0.495  0.082 0.0322 |
|  1000 | 2.555  1.117  0.058 0.0228 | 2.563  0.640  0.033 0.0130 | 2.563  0.509  0.026 0.0103 |
| 10000 | 2.517  1.097  0.018 0.0072 | 2.562  0.662  0.011 0.0043 | 2.567  0.512  0.008 0.0033 |
+-------+----------------------------------+----------------------------------+----------------------------------+
```

Fig. 7.10 Simulation program output showing experimentation results for the Port project (Version 1)

7. A simulation run consists of instantiating the *PortVer1* object that is initialized with the start time, number of berths and random number generator seeds, and a *false* Boolean argument that turns off logging. The termination time of the *PortVer1* object is first set to *WARM_UP_PERIOD* (using the *setTimef* method), and the simulation program executes for the warm-up period. The output sequence for the tanker total wait time is cleared with the execution of the method *clearPHIiCTankerTotalWait()*. The termination time of the *PortVer1* object is now set to *end* and continues execution until the end of the observation interval.

8. The output data values are then obtained with the *getAvgOccBerths* and *getAvgWaitTime* methods and stored respectively in the arrays *avgOccBerths* and *avgWaitTime*. The values stored in these arrays contain the output from 10000 (*NUMRUNS*) replications for each of the two cases (number of berths equals 3 and 4 as stored in *NUM_BERTHS*). For each parameter value assignment, three observation interval lengths are used; namely 10, 20 and 30 weeks (values in *NUM_WEEKS*).

9. The *displayTable* method generates the table shown in Fig. 7.10 using the data in the arrays *avgOccBerths* and *avgWaitTime*. It will determine confidence intervals for a sequence of increasing values of n (number of replications specified in *NUM_RUNS_ARRAY*). The table shows the point estimate ($yb(n)$), the standard deviation ($s(n)$), the confidence interval half-length ($z(n)$) values and the ratio of the half-length to the point estimate ($z(n)/yb(n)$) computed using the ABSmod/J class *ObservationInterval*.

As expected, increasing the number of replications (n) reduces the confidence interval half-length, $\zeta(n)$. Increasing the simulation run length also improves the quality of the results. An evaluation of the difference between the 3 and 4 berth options is undertaken in Sect. 7.4.1 using an appropriate statistical framework.

7.4 Comparing Alternatives

A frequently occurring goal of a modelling and simulation project is the evaluation of several alternate 'system designs'. For example, what reduction in maximum patient waiting time could be expected in the emergency admitting area of a large hospital if an additional orthopaedic specialist was hired or what might be the impact on traffic flow in the downtown core of a large city if a network of one-way streets was implemented? There can be a large number of such design alternatives that need to be evaluated, but we first examine the case where there are only two.

In principle, the steps involved in obtaining a problem solution are straightforward: (1) develop a simulation program which incorporates the means to invoke the model for each of the scenarios (alternatives of interest), (2) obtain a point estimate (and confidence interval) for a designated performance measure that provides a meaningful evaluation of each scenario (e.g. a mean value estimate for some DSOV) and (3) compare the point estimates obtained. There is, however, a serious complication that emerges, namely, what assurance is there that any observed difference between the performance measure values is a consequence of the design difference being studied and not simply a consequence of the inherent random behaviour within the model?

A number of different approaches have emerged for dealing with this problem and comprehensive discussions can be found in the literature (e.g. Banks et al. [1], Goldsman and Nelson [2]). One of the most straightforward is called the *paired-t confidence interval method* (see Law [5]). The objective here is to first establish a confidence interval for an estimate of the difference between the performance measure values associated with each of the scenarios. A decision about relative superiority is then based on the position of the confidence interval relative to zero. Some details are provided below.

7.4.1 Comparing Two Alternatives

Suppose that the variable Y is the performance measure used for the evaluation and let's assume that we seek as large a value as possible for this variable. The simulation program for each of the design alternatives is executed n times (i.e., n replications). Suppose that y_{1k} is the value of Y obtained for case 1 on the kth replication and suppose that y_{2k} is the value for case 2 on the kth replication. Let

$$d_k = y_{2k} - y_{1k} \quad k = 1, 2, \ldots n$$

$$\bar{d}(n) = \frac{1}{n} \sum_{k=1}^{n} d_k$$

$$s^2(n) = \frac{\sum_{k=1}^{n} (d_k - \bar{d}(n))^2}{n - 1} \quad (7.3)$$

$$\zeta = \frac{t_{n-1,a} s(n)}{\sqrt{n}}$$

where $t_{n-1,a}$ is a value from the student's t-distribution (see Table A2.4) that corresponds to $(n - 1)$ degrees of freedom and $a = (1 - C)/2$ with C the confidence level parameter. Here, $\bar{d}(n)$ is a point estimate of the mean of the differences and $s^2(n)$ is the sample variance. (The similarity of these results with those given in Eq. (7.1) is worth noting.) The associated confidence interval is $CI(n) = [\bar{d}(n) - \zeta, \bar{d}(n) - \zeta]$.

There are three possible outcomes based on CI(n), namely:

(a) If CI(n) lies entirely to the right of zero, then the result of case 2 exceeds the result of case 1 with a level of confidence given by $100C\%$.
(b) If CI(n) lies entirely to the left of zero, then the result of case 1 exceeds the result of case 2 with a level of confidence given by $100C\%$.
(c) If CI(n) includes zero, then with a level of confidence given by $100C\%$, there is no meaningful difference between the two cases.

The procedure outlined above is best carried out in conjunction with a technique called *common random numbers* in order to produce narrow confidence intervals (see [1, 3, 5] for details on this technique).

Applying Eq. A2.15 to finding the $\mathrm{var}[\overline{D}]$ gives

$$\mathrm{var}[\overline{D}] = \mathrm{var}[\overline{Y}_2 - \overline{Y}_1] = \mathrm{var}[\overline{Y}_2] + \mathrm{var}[\overline{Y}_1] - 2\mathrm{cov}[\overline{Y}_2, \overline{Y}_1] \qquad (7.4)$$

where $\overline{D}, \overline{Y}_1, \overline{Y}_2$ are estimated by $\overline{d}(n), \overline{y}_1(n)$ and $\overline{y}_2(n)$ given in Eq. (7.3). The objective of applying common random numbers is to increase correlation between the output values y_{2k} and y_{1k}; thus, the covariance term in Eq. (7.4) is increased and the confidence interval for $\overline{d}(n)$ is reduced.

The application of the technique corresponds to endeavouring to ensure that, insofar as possible, the random phenomena within the two alternatives are coordinated, e.g. comparable entities flowing in the model for the two alternatives have the same arrival times. In principle, this can be achieved by the strict management of the random variate generation procedures for the two alternatives of the model using different random number generator streams and the seeds used with them. This coordination depends on the model and is straightforward for input data models. The coordination task can also be easily achieved with all data models when the simulation model is relatively simple.

This coordination is achieved by applying the following guidelines:

1. Use different random number generators (recall that random variate generators use random number generators) for specific purposes, that is, for each different data model. Thus, different streams are used for interarrival times, service times, etc. The Colt Java package used with ABSmod/J provides the class *RandomNumberSeedGenerator* (see Fig. 7.10) to provide uncorrelated seeds. Different seeds are used in each new replication during an experiment with each case. The same set of seeds is used for each experiment for the two alternate cases.
2. For transient entities, store at the time of arrival in entity attributes all random values required during the lifetime of the entity in the model. For example, save into entity attributes all random service times required for services the entity is to undergo during its time in the model.
3. When a non-transient entity is involved in repeated activity instances, a single random number stream is assigned to the entity. For example, in the case where a resource cycles between being available and unavailable for random periods of time, a dedicated stream is used to generate the available and unavailable periods of time for the resource.
4. Where coordination of random numbers is not feasible between the alternate cases, independent random number streams should be used for the random variates (e.g. the situation where additional random behaviour is inherent in one of the cases).

We return now to our experiments with the Port project (Version 1) as outlined in Sect. 7.3.3. For the two cases (the number of berths is either 3 or 4) Table 7.3 shows the output data for each of the output variables (the average group size and tanker total wait time) from a sequence of experiments with $t_f = 30$ weeks and $n = 30$. The 'Difference' column is obtained as the difference between the value in column (RG.Berths.numBerths = 4) and the column (RG.Berths.numBerths = 3).

Table 7.3 Comparing alternative cases in the Port project (with coordinated CRNs and $n = 30$)

| | Average Number of Occupied Berths | | | Average Tanker Total Wait Time | | |
| | RG.Berths.numBerths | | | RG.Berths.numBerths | | |
Replication	3	4	Difference	3	4	Difference
1	1.683	1.691	0.007	5.035	1.992	-3.043
2	1.863	1.874	0.011	6.905	2.775	-4.130
3	1.846	1.858	0.012	6.513	2.447	-4.066
4	1.893	1.904	0.011	6.956	2.249	-4.707
5	1.906	1.919	0.013	6.527	2.300	-4.227
6	1.872	1.880	0.008	5.301	1.876	-3.424
7	1.876	1.887	0.012	7.367	2.928	-4.439
8	1.958	1.983	0.025	10.830	3.379	-7.451
9	1.769	1.778	0.009	6.542	2.476	-4.066
10	1.849	1.858	0.009	7.182	2.522	-4.659
11	1.909	1.914	0.005	7.349	2.480	-4.869
12	1.864	1.878	0.014	7.806	2.587	-5.220
13	1.772	1.789	0.017	7.870	2.323	-5.548
14	1.945	1.960	0.015	11.151	3.340	-7.811
15	1.865	1.863	-0.001	5.435	1.873	-3.562
16	1.905	1.926	0.022	6.989	2.929	-4.060
17	1.950	1.959	0.009	8.651	2.954	-5.697
18	1.905	1.916	0.011	7.096	2.363	-4.732
19	1.849	1.856	0.007	6.488	2.270	-4.219
20	1.840	1.851	0.012	7.364	2.629	-4.735
21	1.942	1.952	0.010	8.516	3.025	-5.491
22	1.806	1.815	0.009	6.472	2.529	-3.943
23	1.744	1.766	0.021	6.103	2.427	-3.675
24	1.820	1.847	0.027	8.815	3.856	-4.959
25	1.709	1.706	-0.003	6.607	2.315	-4.291
26	1.814	1.821	0.007	6.444	2.303	-4.140
27	1.708	1.727	0.019	6.903	2.268	-4.635
28	1.689	1.705	0.016	6.921	2.800	-4.121
29	1.889	1.907	0.018	9.249	3.138	-6.111
30	1.833	1.842	0.009	6.989	2.631	-4.357
$\bar{y}(n)$	1.842	1.854	0.012	7.279	2.599	-4.680
$s(n)$	0.078	0.079	0.007	1.396	0.450	1.054
$\zeta(n)$	0.024	0.024	0.002	0.433	0.140	0.327
CI Min	1.818	1.830	0.010	6.846	2.460	-5.007
CI Max	1.867	1.879	0.014	7.712	2.739	-4.353

Table 7.4 Comparing alternative cases in the Port project (with single RNG and $n = 30$)

| | Average Number of Occupied Berths | | | Average Tanker Total Wait Time | | |
| | RG.Berths.numBerths | | | RG.Berths.numBerths | | |
Replication	3	4	Difference	3	4	Difference
1	1.916	1.822	-0.094	8.329	2.465	-5.865
2	1.832	1.808	-0.024	6.485	2.091	-4.394
3	1.832	1.731	-0.101	7.748	2.064	-5.684
4	1.776	1.940	0.164	5.737	2.306	-3.431
5	1.828	1.848	0.020	7.514	3.090	-4.424
6	1.872	1.856	-0.016	7.633	2.831	-4.802
7	1.703	1.890	0.187	5.828	2.780	-3.048
8	1.892	1.907	0.015	12.773	2.452	-10.321
9	1.722	1.747	0.025	5.701	1.994	-3.706
10	1.630	1.762	0.132	5.666	2.482	-3.183
11	1.878	1.789	-0.089	6.116	2.251	-3.865
12	1.951	1.765	-0.187	8.846	2.890	-5.956
13	1.741	1.842	0.101	7.108	2.759	-4.349
14	1.768	1.676	-0.092	7.100	1.470	-5.629
15	1.800	1.916	0.116	7.740	2.940	-4.800
16	2.001	1.850	-0.151	11.351	3.472	-7.879
17	1.910	1.943	0.034	7.764	2.501	-5.263
18	1.742	1.686	-0.056	6.379	2.386	-3.993
19	1.779	1.908	0.128	5.587	3.111	-2.475
20	1.843	1.816	-0.027	7.249	1.952	-5.297
21	1.752	1.800	0.048	6.153	2.023	-4.131
22	1.913	1.870	-0.043	10.230	2.650	-7.580
23	1.667	1.921	0.254	4.647	2.637	-2.010
24	1.826	1.820	-0.006	10.078	2.095	-7.983
25	1.738	1.899	0.162	5.827	2.728	-3.099
26	1.818	1.911	0.093	4.225	2.680	-1.544
27	1.673	1.846	0.173	6.512	3.342	-3.170
28	1.716	1.778	0.062	6.595	2.186	-4.409
29	1.875	1.769	-0.106	7.106	3.431	-3.675
30	1.940	1.823	-0.117	8.612	2.230	-6.383
$\bar{y}(n)$	1.811	1.831	0.020	7.288	2.543	-4.745
$s(n)$	0.093	0.072	0.113	1.907	0.476	1.918
$\zeta(n)$	0.029	0.022	0.035	0.592	0.148	0.595
CI Min	1.782	1.809	-0.015	6.696	2.395	-5.340
CI Max	1.840	1.854	0.055	7.880	2.691	-4.150

The comparison of the two alternatives is carried out using Eq. (7.3). The summary information for the confidence intervals is provided at the bottom of Table 7.3.

Some interpretation of the data is as follows:

1. It is clear that increasing the number of berths to four does decrease the average tanker total wait time (by almost 4.5 hours).
2. Although the confidence interval for the difference in the average number of occupied berths is to the right of zero, the point estimate of the difference is so small relative to the individual point estimates we are obliged to conclude that increasing the berth size has no effect on this output variable. This is somewhat counter-intuitive but is a consequence of the relative values of tanker arrival rate, the tug's cycle time (time to deberth and berth a tanker) and the tanker loading times. For example, if the loading times are increased, then the average number of occupied berths does increase and vice versa. An alternate measure that would be interesting is the percentage of time that all available berths are occupied. The interested reader is encouraged to experiment with the simulation program by exploring the effects of changing these various times.

Table 7.4 shows the data obtained from equivalent replications which do not use common random numbers (CRNs) for the two cases of interest, i.e. *RG.Berths.numBerths* = 3 and *RG.Berths.numBerths* = 4. This was achieved by using a single random number generator for all random variate generators used to implement the data modules in the replications. Different seeds were used for all replications for both cases (i.e. seeds are not coordinated between the two cases). Note that the confidence interval half-length $\zeta(n)$ increases for both output variables when compared to the results in Table 7.3. Note also that the covariance term in Eq. 7.4) increases when the CRN procedure is applied. This decreases the variance and thus the width of the confidence interval).

7.4.2 Comparing Three or More Alternatives

The paired-t confidence interval method described above can be extended to the case where multiple comparisons need to be carried out. The basis for carrying out this extension is provided by the Bonferroni inequality (sometimes called the Boole inequality). It states that

$$P\left[\bigcap_{k=1}^{K} A_k\right] \geq (1 - K) + \sum_{k=1}^{K} P[A_k]$$

In our context, the A_k can be interpreted as the event (in a probability context) that the kth confidence interval contains the kth mean in a designated collection of K (pairwise) comparisons; note that $P[A_k] = C_k$. The Bonferroni inequality, in effect, places constraints on the individual comparisons in order to achieve an overall result that has a prescribed level of confidence, $100C\%$. In other words, with $100C\%$ confidence, the mean differences *all* fall into their respective confidence intervals; note that $C = P\left[\bigcap_{k=1}^{K} A_k\right]$. The (simplified) result that flows from the Bonferroni inequality is that each of the K comparisons should be carried out with a confidence level parameter value of

$$C_K = 1 - \left(\frac{1-C}{K}\right) \tag{7.5}$$

Note that the result given in Eq. (7.5) is overly restrictive because it has imposed the unnecessary (but simplifying) constraint that the confidence level parameter for all pairwise comparisons has the same value (namely C_k).

The following is a typical scenario. There exists a 'base case' which normally corresponds to the current status of the SUI. The project goal introduces M alternate 'designs' together with the requirement to identify the best of the alternate designs by comparing each alternative to the base case. Thus, $K = M$ comparisons need to be made. If an overall confidence level of $100C\%$ is stipulated, then the K individual comparisons have to be carried out with a confidence level parameter of C_K as given in Eq. (7.5).

It may, on the other hand, be stipulated in the project goal that the M alternative designs not only be compared to the base case but also be pairwise compared to each other. In this case, there is a requirement for $K = M(M - 1)/2$ comparisons. The number of comparisons can easily rise quickly and the reliability of the procedure deteriorates. In addition, of course, the computational overhead can become overwhelming.

Some illustrative results obtained using the multiple alternatives procedure outlined above are given in Fig. 7.11. The results relate to the Kojo's Kitchen project (version 2, Sect. A1.2). We consider a base case (case 1) which corresponds to the two employees working over the entire business day (10h00 to 21h00) and three alternative employee scheduling options (cases 2, 3, 4). These options allocate different numbers of employees to various segments of the day. The employee scheduling schemes are summarized in Table 7.5. The rightmost column of this table provides the total number of employee hours associated with each option. This is relevant in the ultimate selection decision because it represents the 'cost' of the option. The basic output continues to be the percentage of customers who wait more than 5 minutes before receiving service.

Figure 7.11 shows the output from an experimentation program (similar to the one discussed in Sect. 7.2.3). It provides the confidence intervals for each of the three comparisons. The results shown for Diff21 are obtained by subtracting the

```
-------------------------------------------------------------------
Comparison   PE(yb(n))  s(n)     zeta   CI Min   CI Max  zeta/ybar(n)
-------------------------------------------------------------------
   Diff21     -0.414    0.011   0.023   -0.437   -0.391     0.056
   Diff31     -0.289    0.014   0.026   -0.315   -0.263     0.090
   Diff41     -0.332    0.013   0.025   -0.357   -0.307     0.074
-------------------------------------------------------------------
```

Fig. 7.11 Results for multiple scheduling alternatives

Table 7.5 Multiple scheduling alternatives for Kojo's Kitchen

	Slow (10h00- 11h30)	Busy (11h30- 13h30)	Slow (13h30- 17h00)	Busy (17h00- 19h:00)	Slow (19h00- 21h00)	Emp- Hours
Case 1 (Base Case)	2	2	2	2	2	22
Case 2	2	3	2	3	2	26
Case 3	1	3	1	3	1	19
Case 4	1	3	2	3	1	22.5

results of the base case (case 1) from case 2 and applying Eq. (7.3). Results for Diff31 and Diff41 are obtained in the same fashion. In each of the comparisons, the results are based on data from 100 replications ($n = 100$) and use a confidence level parameter value of $C_k = 0.968$ in the determination of the confidence interval for the individual comparisons. This gives a value of $C = 0.904$ using Eq. (7.5), i.e. a confidence level of 90.4% in the conclusions from the comparison. Figure 7.11 suggests that the scheduling alternative of case 2 provides the best improvement over the base case. Unfortunately, it is also the most expensive (see Table 7.5)! Note that scheduling in Case 4 provides a significant improvement at very little additional cost and that Case 3 provides an improvement with a reduction in cost.

7.5 Design of Experiments

7.5.1 Introduction

The design of experiments (DoE) is a collection of concepts that was introduced more than a century ago [7]. The concepts have been widely discussed in the literature [5, 6] and the topic continues to be of current interest as indicated by discussion at recent conferences [10]. The objective of this body of work is to provide help in determining the most promising experimental alternatives to explore when a large number of options are available. The basic underlying procedure assumes the existence of a set of numerical parameters[3] that are available for adjustment, each of which has specified upper and lower bounds. The thrust of the procedure focuses on the determination of how a defined system performance measure (SPM) is affected when any one of the parameters changes in value from its maximum to its minimum. Included as well are procedures for investigating how parameters interact with each other and how such interactions affect the SPM. Such analysis provides valuable insight into the identification of those parameters that

---

[3]The procedure is extendable to the case where the parameters have non-numeric values.

have more significant impact upon the SPM thereby allowing a more focused search for conditions that yield best system performance. This feature can also contribute to formal optimization studies by enabling a reduction in the number of parameters that are within the search space (see Sect. 12.3).

An overview of the topic called 2^m factorial design[4] is provided below to illustrate some of the basic concepts associated with the DoE (here m represents the number of parameters under consideration). The approach is especially helpful when m is relatively small (e.g. less 8). When m becomes large the practicality of the approach is challenged but an alternate approach called 2^{m-p} factorial design can be used. Other 'screening' methods exist when the number of parameters is much larger. An introduction to 2^{m-p} factorial design and similar methods can be found in [5, 6].

The simplest case of 2^m factorial design is outlined below. We introduce the m-dimensional parameter vector $\mathbf{p}$ and assume that each component of $\mathbf{p}$ has a range constraint, that is, $L_j \leq p_j \leq U_j$ for $1 \leq j \leq m$ where L_j and U_j are known constants. Let Ω be the set of distinct parameter vectors $\mathbf{p}$ such that $\mathbf{p} \in \Omega$ if and only if either $p_j = L_j$ or $p_j = U_j$ for $1 \leq j \leq m$. Note that the size (cardinality) of Ω is 2^m. A useful way of ordering the parameter vectors within Ω is provided by the matrix $\mathbf{P}$ shown in Eq. (7.6). Here each row is an admissible value of the parameter vector $\mathbf{p}$.

$$
\mathbf{P} = \begin{bmatrix}
\overset{p_1}{L_1} & \overset{p_2}{L_2} & \overset{p_3}{L_3} & \overset{\cdots}{\cdots} & \overset{p_m}{L_m} \\
U_1 & L_2 & L_3 & \cdots & L_m \\
L_1 & U_2 & L_3 & \cdots & L_m \\
U_1 & U_2 & L_3 & \cdots & L_m \\
\vdots & \vdots & \vdots & \ddots & \vdots \\
U_1 & U_2 & U_3 & \cdots & U_m
\end{bmatrix}
\tag{7.6}
$$

The simulation model is now executed for each parameter vector $\mathbf{p} \in \Omega$ (or equivalently, the parameter vectors as specified in the 2^m rows of $\mathbf{P}$), and the resulting values for the SPM are stored in the vector $\mathbf{R}$.

$$
\mathbf{R} = \begin{bmatrix}
R_1 \\
R_2 \\
\vdots \\
R_{2^m}
\end{bmatrix}
\tag{7.7}
$$

Consider now the ith component p_i of the parameter vector $\mathbf{p}$ (in effect, the ith column of $\mathbf{P}$). There are 2^{m-1} occurrences of $\mathbf{p}$ where p_i has the value U_i. Choose one such occurrence, e.g. $\mathbf{p}^+$, and denote its SPM value (taken from $\mathbf{R}$) by R_i^+. Note

---

[4]In the DoE literature, this method is generally referred to as 2^k factorial design, where k is the number of factors (i.e. parameters). We use 2^m to be consistent with our convention of using m as the number parameters.

that there necessarily is a parameter vector $\mathbf{p}^-$ in Ω whose ith component, p_i, is equal to L_i and for which all other components are the identical to those of $\mathbf{p}^+$. Let R_i^- denote its SPM value (again taken from $\mathbf{R}$).

The difference $R_i^+ - R_i^-$ indicates how, for the specific parameter vector pair selected, the SPM is affected by changing only the ith component of the parameter vector from the value L_i to the value U_i when all other components are fixed in value. There are 2^{m-1} such parameter pairs and consequently 2^{m-1} such differences. The 'main effect', e_i, of the ith component, p_i, of the parameter vector $\mathbf{p}$ is defined as the average of these 2^{m-1} differences.

The vector $\mathbf{E}$ shown in Eq. (7.8) provides a convenient computational means for obtaining the main effect for each of the components of the parameter vector $\mathbf{p}$. In this equation, the matrix $\mathbf{D}$ is called the 'design matrix' and is obtained from $\mathbf{P}$ by replacing each occurrence of U_j by $+1$ and each occurrence of L_j by -1 as shown in Eq. (7.9).

$$\mathbf{E} = \begin{bmatrix} e_1 \\ e_2 \\ e_3 \\ \vdots \\ e_m \end{bmatrix} = \frac{1}{2^{m-1}} \mathbf{D}^\mathsf{T} \mathbf{R} \tag{7.8}$$

$$\mathbf{D} = \begin{matrix} \begin{matrix} p_1 & p_2 & p_3 & \cdots & p_m \end{matrix} \\ \begin{bmatrix} -1 & -1 & -1 & \cdots & -1 \\ +1 & -1 & -1 & \cdots & -1 \\ -1 & +1 & -1 & \cdots & -1 \\ +1 & +1 & -1 & \cdots & -1 \\ \vdots & \vdots & \vdots & \ddots & \vdots \\ +1 & +1 & +1 & \cdots & +1 \end{bmatrix} \end{matrix} \tag{7.9}$$

Each of the main effects, e_i, is a random variable because the components of $\mathbf{R}$ are random variables. Therefore, a point estimate and confidence interval must be determined for each component of the vector $\mathbf{E}$. From the principles discussed in Sects. 7.2.2 and 7.3.2, it follows that many runs are required to find a meaningful set of values for $\mathbf{R}$ in order to determine a set of values for each component in $\mathbf{E}$ which in turn make possible the derivation of a confidence interval for each of the main effects in $\mathbf{E}$ [see Eq. (7.1)].

The main effect of a parameter provides a value that reflects upon the extent to which that parameter affects the SPM. More specifically, a low value of the main effect provides an indication that the corresponding parameter may not be of interest to achieve the project goal, while a high value indicates that the parameter likely has direct relevance. However, even though the main effect of a parameter is low, it may not be appropriate to eliminate it outright. Its interaction with other parameters needs to be considered. We now explore this interaction effect (or simply interaction).

Suppose we wish to determine the interaction between the two parameters p_i and p_j. Let ε_i^+ be the average of all difference terms used to compute the main effect for p_i where p_j is set to its upper value, U_j and let ε_i^- be the average of all difference terms used to compute the main effect for p_i where p_j is set to its lower value, L_j. The difference between these two values, i.e. $\varepsilon_i^+ - \varepsilon_i^-$ gives a measure of the interaction between p_i and p_j, that is, a measure of how changing the value of p_j from its lower value to its upper value will influence the effect of the parameter p_i on the SPM. As pointed out by Law [5], one half of this difference is, by convention, called the 'two-factor interaction effect' and denoted by e_{ij}.

The design matrix can be used to compute the interactions between parameters. The elements of each column associated with parameters p_i and p_j are multiplied to create a vector of dimension 2^m, which we denote by $\mathbf{I}_{ij}$ and is given in Eq. (7.10).

$$\mathbf{I}_{ij} = \begin{bmatrix} D_{1,i} \times D_{1,j} \\ D_{2,i} \times D_{2,j} \\ D_{3,i} \times D_{3,j} \\ \\ D_{2^m,i} \times D_{2^m,j} \end{bmatrix} \tag{7.10}$$

Equations (7.10) and (7.11) provide a convenient computational means for determining the interaction from the design matrix $\mathbf{D}$ and the result vector $\mathbf{R}$. This is repeated for each combination of interactions. It is easily seen that $e_{ij} = e_{ji}$. Notice that there are $m(m-1)/2$ distinct two-way interactions. Since a confidence interval is required for each interaction, the same set of values for $\mathbf{R}$ that is used in determining the confidence intervals for the main effects can be applied.

$$e_{ij} = \frac{1}{2^{m-1}} \mathbf{I}_{ij}^{\mathrm{T}} \mathbf{R} \tag{7.11}$$

The above process can also be applied to evaluate interactions for three or more parameters. For example, in a three parameter interaction, say parameter p_h, p_i and p_j, the interaction $e_{hij} = (1/2^{m-1})\mathbf{I}_{hij}\mathbf{R}$ where $\mathbf{I}_{hij}$ is formulated from the design matrix entries as shown in Eq. (7.12).

$$\mathbf{I}_{hij} = \begin{bmatrix} D_{1,h} \times D_{1,i} \times D_{1,j} \\ D_{2,h} \times D_{2,i} \times D_{2,j} \\ D_{3,h} \times D_{3,i} \times D_{3,j} \\ \vdots \\ D_{2^m,h} \times D_{2^m,i} \times D_{2^m,j} \end{bmatrix} \tag{7.12}$$

By way of illustration, consider the case where $m = 3$. The matrix $\mathbf{P}$ in Eq. (7.13) lists the eight (2^3) members of the set Ω and the vector $\mathbf{R}$ provides the corresponding SPM values obtained from the execution of the simulation model for each of the parameter vectors in the rows of $\mathbf{P}$. Equation (7.13) also shows the corresponding design matrix.

$$
\mathbf{P} =
\begin{array}{c}
\begin{array}{ccc} p_1 & p_2 & p_3 \end{array} \\
\begin{bmatrix}
L_1 & L_2 & L_3 \\
U_1 & L_2 & L_3 \\
L_1 & U_2 & L_3 \\
U_1 & U_2 & L_3 \\
L_1 & L_2 & U_3 \\
U_1 & L_2 & U_3 \\
L_1 & U_2 & U_3 \\
U_1 & U_2 & U_3
\end{bmatrix}
\end{array}
\quad
\mathbf{R} =
\begin{bmatrix}
R_1 \\
R_2 \\
R_3 \\
R_4 \\
R_5 \\
R_6 \\
R_7 \\
R_8
\end{bmatrix}
\quad
\mathbf{D} =
\begin{array}{c}
\begin{array}{ccc} p_1 & p_2 & p_3 \end{array} \\
\begin{bmatrix}
-1 & -1 & -1 \\
+1 & -1 & -1 \\
-1 & +1 & -1 \\
+1 & +1 & -1 \\
-1 & -1 & +1 \\
+1 & -1 & +1 \\
-1 & +1 & +1 \\
+1 & +1 & +1
\end{bmatrix}
\end{array}
\tag{7.13}
$$

Equation (7.14) shows how the main effects are computed by averaging the difference terms, $R_i^+ - R_i^-$, using the values in $\mathbf{P}$ and $\mathbf{R}$. Rearrangement of the terms in this equation and reformatting into a vector format give Eq. (7.15). Note that Eq. (7.15) is the outcome of Eq. (7.8).

$$
\begin{aligned}
e_1 &= \frac{(R_2 - R_1) + (R_4 - R_3) + (R_6 - R_5) + (R_8 - R_7)}{4} \\
e_2 &= \frac{(R_3 - R_1) + (R_4 - R_2) + (R_7 - R_5) + (R_8 - R_6)}{4} \\
e_3 &= \frac{(R_5 - R_1) + (R_6 - R_2) + (R_7 - R_3) + (R_8 - R_4)}{4}
\end{aligned}
\tag{7.14}
$$

$$
\begin{bmatrix}
e_1 \\
e_2 \\
e_3
\end{bmatrix}
= \frac{1}{4}
\begin{bmatrix}
-R_1 + R_2 - R_3 + R_4 - R_5 + R_6 - R_7 + R_8 \\
-R_1 - R_2 + R_3 + R_4 - R_5 - R_6 + R_7 + R_8 \\
-R_1 - R_2 - R_3 - R_4 + R_5 + R_6 + R_7 + R_8
\end{bmatrix}
\tag{7.15}
$$

In a similar fashion, Eqs. (7.16) and (7.17) illustrate how the two-way interaction effects e_{12}, e_{13} and e_{23} can be found using Eq. (7.11)

$$
\begin{aligned}
e_{12} &= \frac{1}{2} \left[\frac{[(R_4 - R_3) + (R_8 - R_7)]}{2} - \frac{[(R_2 - R_1) + (R_6 - R_5)]}{2} \right] \\
e_{13} &= \frac{1}{2} \left[\frac{[(R_6 - R_5) + (R_8 - R_7)]}{2} - \frac{[(R_2 - R_1) + (R_4 - R_3)]}{2} \right] \\
e_{23} &= \frac{1}{2} \left[\frac{[(R_7 - R_5) + (R_8 - R_6)]}{2} - \frac{[(R_3 - R_1) + (R_4 - R_2)]}{2} \right]
\end{aligned}
\tag{7.16}
$$

$$e_{12} = \frac{R_1 - R_2 - R_3 + R_4 + R_5 - R_6 - R_7 + R_8}{4}$$

$$e_{13} = \frac{R_1 - R_2 + R_3 - R_4 - R_5 + R_6 - R_7 + R_8}{4} \qquad (7.17)$$

$$e_{23} = \frac{R_1 + R_2 - R_3 - R_4 - R_5 - R_6 + R_7 + R_8}{4}$$

7.5.2 Examples of the Design of Experiment (DoE) Methodology

7.5.2.1 Kojo's Kitchen

We consider here a variation of the Kojo's Kitchen (Version 2) example, (see Sect. A1.2) which allows the number of employees during each of the five time slots of the schedule to vary between 1 and 5 employees.

The intent is to determine the impact of employee assignments in the five time slots on the SPM, namely, the percentage of customers that wait more than 5 minutes for service. We use the DoE methodology to evaluate the extent to which employee assignments in the various time slots affect the SPM. The results obtained provide insights into which time slots merit more careful evaluation during further experimentation that is focused on achieving the project goal.

The components of the parameter vector $\mathbf{p}$ correspond to the number of employees assigned to each of the five time slots (s1 to s5). Equation (7.18) shows the matrix $\mathbf{P}$ and the results obtained for the vector $\mathbf{R}$ when a single run is executed for each of the 2^5 parameter vector values. The design matrix, $\mathbf{D}$, is also shown.

The main effect (e_{s1} to e_{s5}) of each of the parameters on the SPM (percentage of customers that wait longer than 5 minutes) is provided by Eq. (7.19). To obtain confidence intervals for each of the main effects, 20 replications were carried out for each of the 32 components of $\mathbf{R}$. The confidence intervals for the main effects are shown in Table 7.6; the four columns of table provide the point estimate (PE), the confidence interval half-length (Zeta), the minimum (CI Min) and the maximum (CI Max) values of the confidence intervals. Clearly, time slots s2 and s4 have the greatest effect on the SPM (47% and 30% reduction, respectively); in other words, the greatest reduction on the percentage of customers that wait longer than 5 minutes. Parameter s1 has very little effect (only 1% reduction). Since increasing the value from its minimum to its maximum value has virtually no effect on the SPM of interest, this parameter can be fixed at 1 (its lowest value) and not varied in any future experimentation.

$$
\mathbf{P} =
\begin{bmatrix}
1 & 1 & 1 & 1 & 1 \\
5 & 1 & 1 & 1 & 1 \\
1 & 5 & 1 & 1 & 1 \\
5 & 5 & 1 & 1 & 1 \\
1 & 1 & 5 & 1 & 1 \\
5 & 1 & 5 & 1 & 1 \\
1 & 5 & 5 & 1 & 1 \\
5 & 5 & 5 & 1 & 1 \\
1 & 1 & 1 & 5 & 1 \\
5 & 1 & 1 & 5 & 1 \\
1 & 5 & 1 & 5 & 1 \\
5 & 5 & 1 & 5 & 1 \\
1 & 1 & 5 & 5 & 1 \\
5 & 1 & 5 & 5 & 1 \\
1 & 5 & 5 & 5 & 1 \\
5 & 5 & 5 & 5 & 1 \\
1 & 1 & 1 & 1 & 5 \\
5 & 1 & 1 & 1 & 5 \\
1 & 5 & 1 & 1 & 5 \\
5 & 5 & 1 & 1 & 5 \\
1 & 1 & 5 & 1 & 5 \\
5 & 1 & 5 & 1 & 5 \\
1 & 5 & 5 & 1 & 5 \\
5 & 5 & 5 & 1 & 5 \\
1 & 1 & 1 & 5 & 5 \\
5 & 1 & 1 & 5 & 5 \\
1 & 5 & 1 & 5 & 5 \\
5 & 5 & 1 & 5 & 5 \\
1 & 1 & 5 & 5 & 5 \\
5 & 1 & 5 & 5 & 5 \\
1 & 5 & 5 & 5 & 5 \\
5 & 5 & 5 & 5 & 5
\end{bmatrix}
\quad
\mathbf{R} =
\begin{bmatrix}
0.922819 \\
0.909358 \\
0.411249 \\
0.400499 \\
0.802096 \\
0.790124 \\
0.376229 \\
0.365487 \\
0.639692 \\
0.628376 \\
0.109047 \\
0.099103 \\
0.473525 \\
0.462484 \\
0.077314 \\
0.067370 \\
0.875906 \\
0.864865 \\
0.368739 \\
0.358795 \\
0.731869 \\
0.720829 \\
0.335659 \\
0.325714 \\
0.57286 \\
0.561545 \\
0.042216 \\
0.032271 \\
0.406693 \\
0.395653 \\
0.010483 \\
0.000538
\end{bmatrix}
\quad
\mathbf{D} =
\begin{bmatrix}
-1 & -1 & -1 & -1 & -1 \\
+1 & -1 & -1 & -1 & -1 \\
-1 & +1 & -1 & -1 & -1 \\
+1 & +1 & -1 & -1 & -1 \\
-1 & -1 & +1 & -1 & -1 \\
+1 & -1 & +1 & -1 & -1 \\
-1 & +1 & +1 & -1 & -1 \\
+1 & +1 & +1 & -1 & -1 \\
-1 & -1 & -1 & +1 & -1 \\
+1 & -1 & -1 & +1 & -1 \\
-1 & +1 & -1 & +1 & -1 \\
+1 & +1 & -1 & +1 & -1 \\
-1 & -1 & +1 & +1 & -1 \\
+1 & -1 & +1 & +1 & -1 \\
-1 & +1 & +1 & +1 & -1 \\
+1 & +1 & +1 & +1 & -1 \\
-1 & -1 & -1 & -1 & +1 \\
+1 & -1 & -1 & -1 & +1 \\
-1 & +1 & -1 & -1 & +1 \\
+1 & +1 & -1 & -1 & +1 \\
-1 & -1 & +1 & -1 & +1 \\
+1 & -1 & +1 & -1 & +1 \\
-1 & +1 & +1 & -1 & +1 \\
+1 & +1 & +1 & -1 & +1 \\
-1 & -1 & -1 & +1 & +1 \\
+1 & -1 & -1 & +1 & +1 \\
-1 & +1 & -1 & +1 & +1 \\
+1 & +1 & -1 & +1 & +1 \\
-1 & -1 & +1 & +1 & +1 \\
+1 & -1 & +1 & +1 & +1 \\
-1 & +1 & +1 & +1 & +1 \\
+1 & +1 & +1 & +1 & +1
\end{bmatrix}
\tag{7.18}
$$

$$
\mathbf{E} =
\begin{bmatrix}
e_{s1} \\
e_{s2} \\
e_{s3} \\
e_{s4} \\
e_{s5}
\end{bmatrix}
= \frac{1}{16} \mathbf{D}^T
\begin{bmatrix}
R_1 \\
R_2 \\
R_3 \\
R_4 \\
R_5 \\
\vdots \\
R_{32}
\end{bmatrix}
\tag{7.19}
$$

Table 7.6 Confidence intervals of the main effects

Main effect	PE	Zeta	CI Min	CI Max
e_{s1}	−0.01	0.001	−0.011	−0.009
e_{s2}	−0.466	0.006	−0.473	−0.46
e_{s3}	−0.097	0.003	−0.101	−0.094
e_{s4}	−0.303	0.005	−0.308	−0.299
e_{s5}	−0.062	0.003	−0.065	−0.059

Table 7.7 Interaction between parameters

Interaction	PE	Zeta	CI Min	CI Max
e_{s1s2}	0.00081	0.00021	0.0006	0.00102
e_{s1s3}	0.00012	0.00005	0.00007	0.00016
e_{s1s4}	0.00023	0.00005	0.00019	0.00028
e_{s1s5}	0.00028	0.00003	0.00024	0.00031
e_{s2s3}	0.0586	0.00148	0.05712	0.06007
e_{s2s4}	−0.00434	0.0017	−0.00604	−0.00264
e_{s2s5}	0.00359	0.0006	0.00299	0.00419
e_{s3s4}	−0.00959	0.00147	−0.01106	−0.00812
e_{s3s5}	−0.00277	0.00032	−0.0031	−0.00245
e_{s4s5}	−0.00483	0.00266	−0.00749	−0.00217

However, before actually abandoning the adjustment of parameters whose main effect is small, it is important to determine if there are significant interactions between parameters. It needs to be recognized that although a parameter's main effect is minimal, it may influence other parameters.

Table 7.7 shows the confidence intervals for the two-way interactions between the parameters. Clearly, interactions are essentially nonexistent (less than 1% change in the output) for all parameters except for interaction e_{s2s3} where there is almost 6% change in the percentage of customers waiting more than 5 minutes. Thus, from the main effect and interaction results, we can conclude that the first and fifth components of the parameter vector (slots s_1 and s_5) do not have a significant effect and can be set to 1 with continuing experimentation focusing on the other three parameters.

7.5.2.2 Acme Manufacturing Example

The modelling and simulation project that is considered in the discussion that follows is outlined in detail in Sect. A1.7. The project is also used as a vehicle for exploring optimization within the modelling and simulation context in Sect. 12.4. In order to fully appreciate the discussion below, the reader is urged to acquire some familiarity with the project by referring to Sect. A1.7.

To simplify the discussion that follows the vector **M** is introduced as a surrogate for the first four components of the parameter vector **p** (i.e. the number of machines in each of the four workstations) and the vector **S** is introduced as a surrogate for the last three components of the parameter vector **p** (i.e. the number of spaces in each of the three buffers). It should be noted, as well, that the project includes two SPM's; namely, profit and number of parts produced.

The main effects of each of the seven parameters and the interaction between parameters were determined using the concepts presented earlier in Sect. 7.5.1. The confidence intervals of the main effect upon profit for each of the four components of **M** are shown in Table 7.8 and similarly the confidence intervals for the main effect upon profit of the three components of **S** are shown in Table 7.9 (the point estimates are plotted in Fig. 7.12). The main effect of the parameters upon number of parts produced is shown in Tables 7.10 and 7.11 (the point estimates are plotted in Fig. 7.13). All confidence interval results presented in these four tables are based on 20 replications of the simulation model execution for each value R_j in Eq. (7.7). The four columns of tables provide the point estimate (PE), the confidence interval half-length (Zeta) the minimum (CI Min) and the maximum (CI Max) values of the confidence intervals.

With respect to the main effect values, note that

- The number of machines in Workstations 1 and 2 (components 1 and 2 of the parameter vector **M**) has a significant effect on both profit and the number of parts produced while the number of spaces in the buffers (the components of the parameter vector **S**) is minimal.
- Negative values in the profit tables (Tables 7.8 and 7.9) are the consequence of insufficient parts being produced to compensate for the increase in maintenance costs for the machines and buffers. By way of clarification, recall that the cost of adding a machine to a workstation is $25,000 and that a produced part generates an income of $200. Table 7.10 shows that increasing the number of machines from 1 to 3 increases part production by about 164 parts which increments profit

Table 7.8 Main effect of the parameter vector **M** on profit

Main effect	PE	Zeta	CI Min	CI Max
e_{M1}	70699.73	680.11	70019.62	71379.83
e_{M2}	208800.20	901.00	207899.20	209701.20
e_{M3}	-17283.00	193.28	-17476.30	-17089.70
e_{M4}	2322.48	289.87	2032.60	2612.35

Table 7.9 Main effect of the parameter vector **S** on profit

Main effect	PE	Zeta	CI Min	CI Max
e_{S1}	6195.31	145.74	6049.57	6341.04
e_{S2}	1127.79	96.77	1031.02	1224.56
e_{S3}	-1551.81	51.64	-1603.44	-1500.17

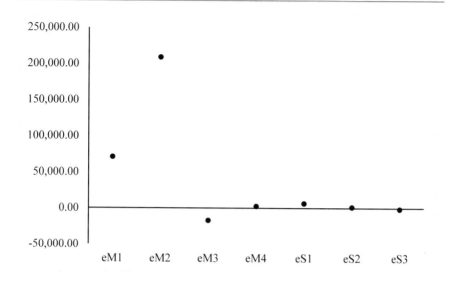

Fig. 7.12 Main effects on profit

Table 7.10 Main effect of parameter vector **M** on number of parts

Main effect	PE	Zeta	CI Min	CI Max
e_{M1}	603.50	3.40	600.10	606.90
e_{M2}	1294.00	4.51	1289.50	1298.51
e_{M3}	163.59	0.97	162.62	164.55
e_{M4}	261.61	1.45	260.16	263.06

Table 7.11 Main effect of parameter vector **S** on number of parts

Main effect	PE	Zeta	CI Min	CI Max
e_{S1}	75.98	0.73	75.25	76.71
e_{S2}	50.64	0.48	50.16	51.12
e_{S3}	37.24	0.26	36.98	37.50

by about \$32,800. However, the addition of two machines incurs a cost of \$50,000 which results in a profit decrease of about \$17,200. This corresponds to the main effect value e_{M3} shown in Table 7.8.

The effect of parameter interactions with respect to profit is shown in Table 7.12 and with respect to number of parts is shown in Table 7.13. The values of the point estimate for these interaction effects are displayed in Fig. 7.14 and Fig. 7.15, respectively.

Table 7.12 Effect of parameter interaction on profit

Parameter interaction	PE	Zeta	CI Min	CI Max
$e_{M1,M2}$	103,935.06	724.90	103,210.16	104,659.95
$e_{M1,M3}$	28,727.29	174.62	28,552.67	28,901.91
$e_{M1,M4}$	50,707.35	301.44	50,405.91	51,008.79
$e_{M2,M3}$	27,825.81	170.24	27,655.56	27,996.05
$e_{M2,M4}$	51,636.68	294.17	51,342.51	51,930.85
$e_{M3,M4}$	14,708.63	184.63	14,524.01	14,893.26
$e_{M1,S1}$	−9,296.54	129.94	−9,426.47	−9,166.60
$e_{M1,S2}$	5,997.85	58.78	5,939.07	6,056.64
$e_{M1,S3}$	5,886.66	44.96	5,841.70	5,931.63
$e_{M2,S1}$	−1,277.52	116.10	−1,393.63	−1,161.42
$e_{M2,S2}$	4,918.81	88.01	4,830.80	5,006.81
$e_{M2,S3}$	6,757.84	52.40	6,705.44	6,810.23
$e_{M3,S1}$	2,338.65	48.74	2,289.91	2,387.39
$e_{M3,S2}$	−8,408.37	82.53	−8,490.90	−8,325.84
$e_{M3,S3}$	−6,052.59	40.42	−6,093.00	−6,012.17
$e_{M4,S1}$	2,649.74	38.35	2,611.39	2,688.09
$e_{M4,S2}$	3,685.60	51.82	3,633.78	3,737.42
$e_{M4,S3}$	−6,763.96	60.08	−6,824.04	−6,703.89
$e_{S1,S2}$	−78.82	30.32	−109.14	−48.50
$e_{S1,S3}$	−131.32	13.19	−144.51	−118.13
$e_{S2,S3}$	−895.46	20.65	−916.11	−874.81

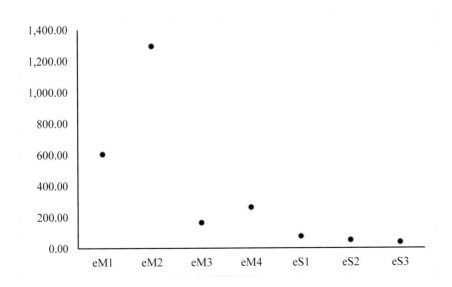

Fig. 7.13 Main effects on number of parts

Table 7.13 Effect of parameter interaction on number of parts

Parameter interaction	PE	Zeta	CI Min	CI Max
$e_{M1,M2}$	519.68	3.62	516.05	523.30
$e_{M1,M3}$	143.64	0.87	142.76	144.51
$e_{M1,M4}$	253.54	1.51	252.03	255.04
$e_{M2,M3}$	139.13	0.85	138.28	139.98
$e_{M2,M4}$	258.18	1.47	256.71	259.65
$e_{M3,M4}$	73.54	0.92	72.62	74.47
$e_{M1,S1}$	-46.48	0.65	-47.13	-45.83
$e_{M1,S2}$	29.99	0.29	29.70	30.28
$e_{M1,S3}$	29.43	0.23	29.21	29.66
$e_{M2,S1}$	-6.39	0.58	-6.97	-5.81
$e_{M2,S2}$	24.59	0.44	24.15	25.03
$e_{M2,S3}$	33.79	0.26	33.53	34.05
$e_{M3,S1}$	11.69	0.24	11.45	11.94
$e_{M3,S2}$	-42.04	0.41	-42.45	-41.63
$e_{M3,S3}$	-30.26	0.20	-30.47	-30.06
$e_{M4,S1}$	13.25	0.19	13.06	13.44
$e_{M4,S2}$	18.43	0.26	18.17	18.69
$e_{M4,S3}$	-33.82	0.30	-34.12	-33.52
$e_{S1,S2}$	-0.39	0.15	-0.55	-0.24
$e_{S1,S3}$	-0.66	0.07	-0.72	-0.59
$e_{S2,S3}$	-4.48	0.10	-4.58	-4.37

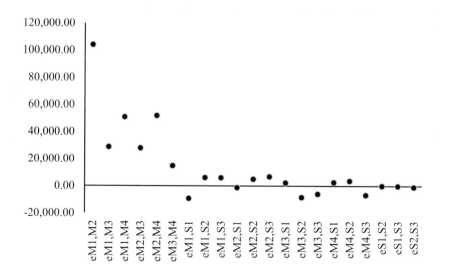

Fig. 7.14 Effect of parameter interaction on profit

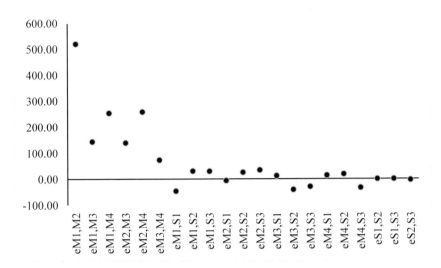

Fig. 7.15 Effect of parameter interaction on number of parts

The following observations relating to the values of the parameter interaction effects are especially noteworthy:

- Interactions are most significant between parameters in vector M, particularly the parameters M_1 and M_2.
- Interactions between number of machines (parameters in vector **M**) and buffer sizes (parameters in vector **S**) are much less significant.
- Interactions between buffers sizes are relatively insignificant.
- The interaction between M_1 and S_1 shows that increasing the value of S_1 (buffer size) from 1 to 10 actually reduces the effect of increasing the number of machines in Workstation 1 from 1 to 3.

These observations are used in developing heuristic experimentation strategy for the ACME Manufacturing Project in the next section.

7.5.2.3 Heuristic Experimentation

The values for the main effects and the interaction effects for the ACME Manufacturing Project demonstrates that the number of machines in each workstation has a considerable effect upon both the monthly profit and the number of parts produced. On the other hand, the size of the buffers has very little impact. In this section, we use this knowledge to develop a heuristic experimentation procedure whose purpose is the achievement of the project goal (the maximization of profit).

The main effect of the parameters (see Figs. 7.12 and 7.13) shows that the number of machines in Workstations 1 and 2 has a major impact on producing parts and generating profit while the number of machines in Workstations 3 and 4 has little direct effect. From the parameter interaction effect (see Figs. 7.14 and 7.15), it's clear that there is important interaction between number of machines in Workstations 3 and 4 (particularly 4) and those in Workstations 1 and 2. From these results, clearly the parameters from the vector **M** must be considered during experimentation for achieving the project goal. The buffer sizes (parameters in the **S** vector) have little direct effect on profit and virtually no interaction with each other, but have some small interaction with the machines in the workstations. Less extensive experimentation with buffers sizes may be appropriate.

Using our understanding of the system operation and these results from the DoE experiments, the following two-part approach for maximizing profit can be formulated. The first part undertakes to determine the number of machines in each workstation which maximizes profit under the condition that the number of buffers between workstations is set to the maximum value of 10 to eliminate, as far as possible, the occurrence of bottlenecks.

The procedure for part 1 is as follows (recall that each part generates $200 of income).

1. Set the number of machines in all workstations to 1.
2. Increment by 1 the number of machines in Workstation 1 and determine if the income increase exceeds the maintenance cost of that machine. If so, further increase the number of machines in Workstation 1 (up to the limit of 3), if not remove the added machine and proceed to step 3.
3. Increment by 1 the number of machines in Workstation 2 (up to the limit of 3) and determine if the income increase exceeds the maintenance cost of that machine. If so, return to step 2 on the assumption that it may be beneficial to increase the number of machines in Workstation 1, if not remove the added machine and proceed to step 4.
4. Repeat step 3 for Workstations 3 and 4. After a new machine is successfully added to either Workstation 3 or 4, check for the possible usefulness of the addition of a machine in previous workstations starting with Workstation 1 (i.e. return to step 2).
5. Stop the process when the profit increase from adding a machine in Workstation 4 does not exceed the maintenance cost of the machine.

Table 7.14 shows the progress in applying the above procedure for establishing the number of machines in each workstation. Each row represents the results of experiments with the parameter values shown in the first two columns. The third column shows the average profit found and the last column the average number of parts produced. The fourth column provides the increase in profit which is the increase in income minus the cost of the machine maintenance ($25,000 for the added machine); note that the increased income is computed from the increase in the number of parts. After 16 experiments, the procedure terminates with a profit value equal to $584,290 and a machine allocation of [3, 3, 2, 2].

Step 2 is concerned with selecting the sizes of the buffers. Recall that the results from the DoE experiments show that the size of buffers has only a small effect on

Table 7.14 Experimentation steps for finding number of machines in each workstation

Step	Number machines	Number buffers	Average profit	Increase in average profit − $25,000	Average number of parts
1	[1, 1, 1, 1]	[10, 10, 10]	159,670		1,448
2	[2, 1, 1, 1]	[10, 10, 10]	135,760	−23,910	1,454
3	[1, 2, 1, 1]	[10, 10, 10]	274,390	114,720	2,147
4	[2, 2, 1, 1]	[10, 10, 10]	359,590	85,200	2,698
5	[3, 2, 1, 1]	[10, 10, 10]	334,880	−24,710	2,699
6	[2, 3, 1, 1]	[10, 10, 10]	357,750	−1,840	2,814
7	[2, 2, 2, 1]	[10, 10, 10]	343,910	−15,680	2,745
8	[2, 2, 1, 2]	[10, 10, 10]	363,080	3,490	2,840
9	[3, 2, 1, 2]	[10, 10, 10]	338,520	−24,560	2,843
10	[2, 3, 1, 2]	[10, 10, 10]	468,060	104,980	3,490
11	[3, 3, 1, 2]	[10, 10, 10]	445,410	−22,650	3,502
12	[2, 3, 2, 2]	[10, 10, 10]	551,910	83,850	4,035
13	[3, 3, 2, 2]	[10, 10, 10]	584,290	32,380	4,321
14	[3, 3, 3, 2]	[10, 10, 10]	559,880	−24,410	4,324
15	[3, 3, 2, 3]	[10, 10, 10]	559,510	−24,780	4,323
16	[3, 3, 2, 2]	[10, 10, 10]	584,290		4,321

profits but some small interaction with the number of machines in the workstations. Thus, we adopt a simple approach to explore the size of buffers, that is, explore three cases, one where all buffer sizes are set to 1, one where all buffer sizes are set to 5 and one where all buffer sizes are set to 10. Table 7.15 shows that setting the size of all buffers to five produces the most profit.

A second possible strategy is to determine "best" values for the buffer sizes using the same algorithmic approach used earlier in determining the number of machines in the workstations. Table 7.16 shows the results of applying such an algorithm. The approach finds an option that generates more profit ($589,140) than the $584,760 found in the previous simple approach with the three cases as shown in Table 7.15. This increase of $4,380 represents an increase of 0.7% in profit but which requires an $8000 increase in cost for the larger buffer sizes. Considering that the value of profit is a point estimate within a confidence interval, the difference in profit from both options is not statistically significant. Inasmuch as the cost is a fixed deterministic value, while the profit value is a random variable, it is reasonable to consider the case with the smaller buffer sizes as superior.

An effective modification of the earlier algorithmic approach is to increase buffer size only if it increases the profit by twice the cost of increasing the buffer size (namely, $2000). This is an attempt to partially accommodate the width of the profit's confidence interval which was found to be approximately $12,000. The result of this experimentation is $S = [6, 10, 2]$ (with $M = [3, 3, 2, 2]$), which generates an average profit of $587,840 and which produces an average of 4,279

Table 7.15 Finding size of buffers using the three case simple approaches

Case	Number machines	Number buffers	Average profit	Average number of parts
1	[3, 3, 2, 2]	[1, 1, 1]	518,160	3,856
2	[3, 3, 2, 2]	[5, 5, 5]	584,760	4,249
3	[3, 3, 2, 2]	[10, 10, 10]	584,290	4,321

Table 7.16 Finding size of buffers using algorithmic approach

Step	Number machines	Number buffers	Average profit	Increase in average profit − $1000	Average number of parts
1	[3, 3, 2, 2]	[1, 1, 1]	518,160	0	3,856
2	[3, 3, 2, 2]	[2, 1, 1]	524,950	6,790	3,895
3	[3, 3, 2, 2]	[3, 1, 1]	527,720	2,770	3,914
4	[3, 3, 2, 2]	[4, 1, 1]	528,560	840	3,923
5	[3, 3, 2, 2]	[5, 1, 1]	528,480	−80	3,927
6	[3, 3, 2, 2]	[4, 2, 1]	546,190	17,630	4,016
7	[3, 3, 2, 2]	[5, 2, 1]	546,420	230	4,022
8	[3, 3, 2, 2]	[6, 2, 1]	545,970	−450	4,025
9	[3, 3, 2, 2]	[5, 3, 1]	558,080	11,660	4,085
10	[3, 3, 2, 2]	[6, 3, 1]	557,840	−240	4,089
11	[3, 3, 2, 2]	[5, 4, 1]	566,440	8,360	4,132
12	[3, 3, 2, 2]	[6, 4, 1]	566,470	30	4,137
13	[3, 3, 2, 2]	[7, 4, 1]	566,080	−390	4,140
14	[3, 3, 2, 2]	[6, 5, 1]	572,520	6,050	4,173
15	[3, 3, 2, 2]	[7, 5, 1]	572,170	−350	4,176
16	[3, 3, 2, 2]	[6, 6, 1]	576,970	4,450	4,200
17	[3, 3, 2, 2]	[7, 6, 1]	576,670	−300	4,203
18	[3, 3, 2, 2]	[6, 7, 1]	580,000	3,030	4,220
19	[3, 3, 2, 2]	[7, 7, 1]	580,040	40	4,225
20	[3, 3, 2, 2]	[8, 7, 1]	579,680	−360	4,228
21	[3, 3, 2, 2]	[7, 8, 1]	582,490	2,450	4,242
22	[3, 3, 2, 2]	[8, 8, 1]	582,160	−330	4,246
23	[3, 3, 2, 2]	[7, 9, 1]	583,980	1,490	4,255
24	[3, 3, 2, 2]	[8, 9, 1]	583,750	−230	4,259
25	[3, 3, 2, 2]	[7, 10, 1]	585,140	1,160	4,266
26	[3, 3, 2, 2]	[8, 10, 1]	584,970	−170	4,270
27	[3, 3, 2, 2]	[7, 10, 2]	588,070	2,930	4,285
28	[3, 3, 2, 2]	[8, 10, 2]	587,940	−130	4,290
29	[3, 3, 2, 2]	[7, 10, 3]	588,880	810	4,294
30	[3, 3, 2, 2]	[8, 10, 3]	588,880	0	4,299
31	[3, 3, 2, 2]	[9, 10, 3]	588,670	−210	4,303
32	[3, 3, 2, 2]	[8, 10, 4]	589,140	260	4,306
33	[3, 3, 2, 2]	[9, 10, 4]	588,630	−510	4,308
34	[3, 3, 2, 2]	[8, 10, 5]	588,730	−410	4,309
35	[3, 3, 2, 2]	[8, 10, 4]	589,140	0	4,306

parts. As with the other options found, essentially the same profit is generated (no more than a fraction of a percent difference among the three alternate approaches).

An exhaustive search over all possible parameter combinations requires 81,000 experiments each involving 20 replications (see Sect. 12.4.2.2 for a detailed presentation). We note that the solution provided in Table 7.16 coincides with the maximum profit found in the exhaustive search. It is also pointed out in Sect. 12.4.2.2 that approximately 350 solution options provide a profit that is less than 1% from the maximum profit. All these solutions have $\mathbf{M} = [\ 3, 3, 2, 2]$, but do utilize different buffer sizes (values of the vector $\mathbf{S}$). This clearly shows that indeed the impact of buffer size on the profit is small.

Thus of the three approaches for determining the buffer sizes, the simple approach (where only three cases are considered) is adequate. It finds a good solution (if not the best) with only 18 experiments compared to 51 experiments for the other two approaches. However, this does not include the number of experiments required for the DoE experiments. Nevertheless, it needs to be emphasized that DoE results provide insights which suggest an experimental approach that finds a solution with a relatively small number of experiments.

7.6 Exercises and Projects

7.1 Use the program developed in Problem 6.2 to carry out experiments that provide the values required for the graphs that are stipulated in the goal of the project outlined in Problem 4.2. Write a short report that outlines the problem, the goals of the modelling and simulation project and the conclusions obtained from the study.

7.2 Use the program developed in Problem 6.3 to carry out experiments that provide values for the proposed performance measures referred to in part (a) of Problem 4.4. Write a short report that outlines the problem, the goals of the modelling and simulation project and the conclusions obtained from the study.

7.3 Use the program developed in Problem 6.5 to carry out experiments that provide values for the proposed performance measures referred to in part (b) of Problem 4.5. Write a short report that outlines the problem, the goals of the modelling and simulation project and the conclusions obtained from the study.

7.4 Use the program developed in Problem 6.6 to carry out experiments to evaluate the effects of reneging introduced in Problem 4.6. Write a short report that outlines the problem, the goals of the modelling and simulation project and the conclusions obtained from the study.

References

1. Banks J, Carson JS II, Nelson BL, Nicol DM (2005) Discrete-event system simulation, 4th edn. Pearson Prentice Hall, Upper Saddle River
2. Goldsman D, Nelson BL (1998) Comparing systems via simulation. In: Banks J (ed) Handbook of simulation. Wiley, New York, pp 273–306
3. Goldsman D, Nelson BL (2001) Statistical selection of the best system. In: Peters BA, Smith JS, Medeiros DJ, Rohrer MW (eds) Proceeding of the 2001 winter simulation conference. IEEE Press, Piscataway, pp 139–146
4. Goldsman D, Schruben LW, Swain JJ (1994) Test for transient means in simulation time series. Nav Res Logist Q 41:171–187
5. Law A (2015) Simulation model and analysis, 5th edn. McGraw-Hill Education, New York, NY
6. Montgomery DC (2017) Design and analysis of experiments, 9th edn. Wiley & Sons Inc
7. Niedz RP, Evens TJ (2016) Vitro Cell Dev Biol Plant 52:547. https://doi.org/10.1007/s11627-016-9786-1
8. Robinson S (2002) A statistical process control approach for estimating the warm-up period. Proceeding of the 2002 winter simulation conference. IEEE Press, Piscataway, pp 439–446
9. Roth E (1994) The relaxation time heuristic for the initial transient problem in M/M/K queuing systems. Eur J Oper Res 72:376–386
10. Sanchez SM, Wan H (2015) Work smarter, not harder: a tutorial in designing and conducting simulation experiments. In: Proceedings of the 2015 winter simulation conference. IEEE Press, Piscataway, pp 1795–1809
11. Welch P (1983) The statistical analysis of simulation results. In: Lavenberg S (ed) The computer performance modeling handbook. Academic, New York, pp 268–328

Part III

CTDS Modelling and Simulation

There are several features that distinguish the modelling and simulation activity within the continuous time dynamic system (CTDS) domain. Perhaps one of the most important is the dependence of the project's success upon the selection of the behaviour generation tool that is best suited to the nature of the conceptual model. Because the conceptual model in this domain always includes a set of ordinary differential equations (ode's), the tools in question relates to the numerical procedures for solving these equations.[1]

Many families of approaches for the solution of ode's can be found in the literature, and within each family there generally are numerous specific options. The methods in these families have their characteristic strengths and weaknesses and are often best suited for specific categories of problems. Furthermore, the use of any of these methods usually involves the specification of values for embedded parameters. The range of options is indeed wide and can even become daunting. To embark on a modelling and simulation project in this environment without some appreciation for the issues involved can be foolhardy. Our objective in Part III of this book is to provide a basic foundation for dealing with these issues.

In Chap. 8, we establish a context for the discussion by formulating a range of simple CTDS conceptual models. For the most part, these have their origins in the portions of the physical world where behaviour can be readily characterized by familiar laws of physics. This central role of the laws of physics is a typical circumstance in the CTDS domain and should not be interpreted as a biased perspective. However, this is not to suggest that CTDS models cannot be formulated in the absence of directly applicable physical laws, and we illustrate this point by providing an example of the formulation of a credible CTDS model based entirely on intuitive arguments. The final topic in Chap. 8 is a brief examination of the

---

[1]Consideration of CTDS models whose formulation includes partial differential equations (pde's) is beyond the scope of our discussions in this book.

problem of transforming a conceptual model that has evolved with higher order differential equations, into an equivalent set of first-order differential equations. Such a format is usually required by numerical software.

In Chap. 9, we provide an overview of some of the basic numerical tools for solving the ode's of the CTDS conceptual model. The presentation is relatively informal and is at an introductory level. Features that have practical relevance, especially those that can lead to numerical difficulties, are emphasized.

As a concluding comment in this synopsis, we encourage the reader to examine Annex 4 where we have provided a brief introduction to MATLAB, a comprehensive software tool for a wide range of numerical problems. Our emphasis in Annex 4 is, however, very much restricted to features that have relevance to simulation experiments with CTDS conceptual models.

Modelling of Continuous Time Dynamic Systems

8

8.1 Introduction

Our concern in this chapter is with exploring the modelling process within the context of continuous time dynamic systems (CTDS). From our perspective, the essential distinguishing feature of this category of system is the fact that a conceptual model can be formulated as a set of differential equations, possibly augmented with a set of algebraic equations. For the most part, such models emerge from a *deductive process* that has its basis in physical laws that are known to govern the behaviour we seek to explore; i.e. the behaviour of the SUI. This is in contrast to an *inductive process*, whereby a model is developed on the basis of observed (or hypothesized) behaviour, as is the case in the development of almost all models in the realm of discrete event dynamic systems (DEDS). The deductive model building process is generally associated with systems that have their origins in engineering or in the physical sciences. Because of the availability of 'deep' knowledge provided by relevant physical laws, such models can incorporate subtleties and a level of detail that is not usually possible within the DEDS context. This enhances the scope of project goals that are realistically achievable.

For convenience, we shall refer to conceptual models that have a differential equation format as *CTDS models*. While such models arise most commonly from a deductive process, it needs to be stressed that this is not a prerequisite. It is most definitely possible to develop credible and useful CTDS models via an inductive process in certain cases where the SUI falls outside the realm of established physical laws. The fields of biology and economics provide many examples of such an approach.

CTDS models can be formulated in terms of either ordinary or partial differential equations (or both). When the modelling power of partial differential equations is required, the SUI is usually called a *distributed parameter system*. Such systems arise in a wide variety of domains. Included here are heat transfer, hydrodynamics, electromagnetics and elasticity. The treatment of models that depend on this

© Springer Nature Switzerland AG 2019
L. G. Birta and G. Arbez, *Modelling and Simulation*, Simulation Foundations, Methods and Applications, https://doi.org/10.1007/978-3-030-18869-6_8

formalism is, however, beyond the scope of this book. Our considerations are restricted to CTDS models that can be formulated exclusively within the framework of ordinary differential equations.

Frequently random effects are absent in continuous models. Although this is not an essential property, we shall limit our considerations in this chapter to this restricted (i.e. deterministic) case. One especially significant feature associated with the deterministic context is that a search for operating conditions that yield some prescribed behaviour for the SUI becomes a significantly simpler task because there is no requirement to assess the efficacy of a candidate solution over some potentially vast stochastic environment.

The nature of CTDS models as outlined above suggests a number of differences from the class of DEDS models examined earlier in this book. In effect, CTDS models exhibit a 'smoothness' property in the sense that the time trajectories of the variables within the model tend to undergo only small changes in response to small changes in parameters or in operating conditions. This feature, combined with the characterization of system behaviour at a relatively detailed level (resulting from the underlying deep knowledge that is typically available) and the absence of stochastic effects, permits the formulation of a project goal that can be more demanding in terms of expected precision and reliability. This, in particular, makes feasible credible optimization studies whose outcome can provide the basis for system implementation. We examine this topic in some detail in Part IV of this book.

Another important difference between CTDS and DEDS models relates to the nature of the time advance mechanism required in the simulation program. In the case of CTDS models, the fundamental requirement is that of solving the underlying differential equations within the conceptual model and that process intrinsically incorporates a time advance procedure. The mechanisms in question are explored in Chap. 9.

8.2 Some Examples of CTDS Conceptual Models

8.2.1 Simple Electrical Circuit

An electrical circuit consisting of a resistor (R), capacitor (C), inductor (L) and a voltage source $(E(t))$ connected in series (see Fig. 8.1) provides an archetypical example of a system whose dynamics can be represented using a CTDS model. An analysis of the circuit based on the application of Kirchoff's voltage law yields the equation:

$$Lq''(t) + Rq'(t) + \frac{q(t)}{C} = E(t) \tag{8.1}$$

where $q(t)$ is the charge on the capacitor, C, and $q'(t)$ is the current in the circuit. If we denote by t_0 the left-hand boundary of the observation interval, then it is

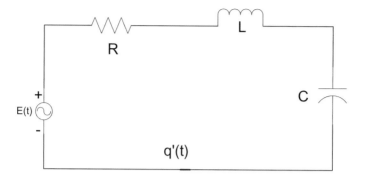

Fig. 8.1 A simple electrical circuit

important to observe here that the solution of Eq. (8.1) (i.e. the behaviour generation process) requires the specification of two initial conditions, namely, $q(t_0)$ and
$q'(t_0)$, as well as the explicit specification of the function $E(t)$. In other words,
$q(t)$ and $q'(t)$ are state variables for the model and $E(t)$ is an input variable.

8.2.2 Automobile Suspension System

A vehicle of mass 4M (mass M on each of the four wheels) is travelling forward at
constant velocity over a road which is initially smooth and horizontal. It is in an
equilibrium condition, and any particular point on the body of the vehicle has a
constant vertical displacement from the road surface. The body is connected to each
of the four wheels through a spring/shock absorber system, and each wheel supports
one quarter of the total mass.

At time $t = t_0$, the vehicle begins to travel over a section of the road which has an
irregular surface (see Fig. 8.2). This causes vertical motion of the vehicle about its
equilibrium position. If we use $y(t)$ to represent this vertical displacement, then from
the application of Newton's second law, the trajectory of $y(t)$ is defined by

$$M y''(t) + f_b(t) + f_a(t) = 0 \qquad (8.2)$$

where $f_a(t)$ and $f_b(t)$ represents the forces associated with the spring and the shock
absorber, respectively. We choose the variable u to represent the vertical irregularities in the road surface, taken with respect to the road's smooth, horizontal
(equilibrium) condition. Although u is a function of horizontal displacement from
some reference point, it can, from the perspective of the vehicle moving over it at a
constant speed, be treated as a function of time, i.e. $u = u(t)$. This time function $u(t)$,
in fact, represents an input to the CTDS model being formulated. A particular
choice for $u(t)$ that matches the presentation in Fig. 8.2 is

$$u(t) = \begin{cases} 0 & \text{for } t < t_0 \\ \left(\frac{u_{max}}{2}\right)(1 - \cos(\omega(t - t_0))) & \text{for } t \geq t_0 \end{cases} \tag{8.3}$$

where ω is proportional to the vehicle's horizontal velocity.

For definiteness, we assume that the spring is linear; hence, $f_a(t) = k \, (y(t) - u(t))$ where k is the spring constant. On the other hand, let's assume that the shock absorber is non-linear and that the associated force is

$$f_b(t) = \psi |v(t)| v(t) \tag{8.4}$$

where $v(t) = (y'(t) - u'(t))$ and ψ is the shock absorber constant.

If we choose $y(t)$ and $y'(t)$ to be the state variables for the model, then the solution of the second-order differential equation Eq. (8.2) requires the two initial conditions: $y(t_0)$ and $y'(t_0)$. From the definition of $y(t)$ and as a direct consequence of the equilibrium assumption prior to $t = t_0$, both of these values are zero.

A possible goal for a modelling and simulation study associated with the above model could be the determination of values for the spring and shock absorber constants, which yield the best value for some prescribed measure of ride quality.

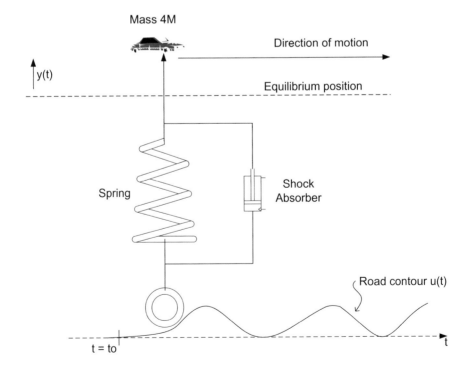

Fig. 8.2 Automobile suspension system

8.2.3 Fluid Level Control

The cleaning solution required in an industrial process passes through two holding tanks (see Fig. 8.3). Valves control the inflow into each of the tanks, and the position of these valves is established by a control strategy based on the height of the liquid in the respective tanks. The rate of change of the volume of liquid in each tank is equal to the difference between the inflow rate and the outflow rate. If we let A_1 and A_2 represent the cross-sectional areas of Tank 1 and Tank 2, respectively, and let h_1 and h_2 represent the height of the liquid in Tank 1 and Tank 2, respectively, then

$$A_1 h_1'(t) = w_0(t) - w_1(t)$$
$$A_2 h_2'(t) = w_1(t) - w_2(t) \tag{8.5}$$

where $w_0(t)$, $w_1(t)$ and $w_2(t)$ are the volume flow rates (e.g. cubic meters per second) into and out of the tanks as shown in Fig. 8.3 (Note that the solution of Eq. (8.5) requires initial conditions for h_1 and h_2.).

Flow Control Specifications
Tank 1: The valve V_0 opens when the level in Tank 1 is decreasing and falls below a half-full tank. More precisely, V_0 moves from a closed to an open position at time t_a where $h'_1(t_a) < 0$ and $h_1(t_a) < H_1/2$. Once open, V_0 stays open until $h_1(t)$ reaches the level H_1 which is the full-tank condition. When V_0 is open, the inflow rate is constant, i.e. $w_0(t) = K$.

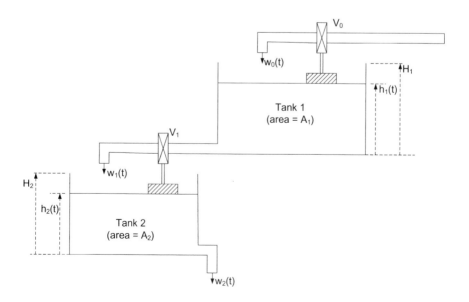

Fig. 8.3 Fluid level control

Tank 2: The control policy for valve V_1 is analogous, i.e. V_1 moves from a closed to an open position at time t_b, where $h'_2(t_b) < 0$ and $h_2(t_b) < H_2/2$. Once open, V_1 stays open until $h_2(t)$ reaches the level H_2 (the full-tank condition). However, when valve V_1 is open, the inflow rate is given by $w_1(t) = k\, h_1(t)$ where k is a constant. The outflow rate from Tank 2, $w_2(t)$, is given by $w_2(t) = u(t)\, h_2(t)$ where $u(t)$ is a control input to valve V_2 (not shown), which reflects the external demand for cleaning solution. Because of physical limitations of the piping system, $u(t)$ is constrained, i.e. $0 \le u(t) \le u_{max}$.

The SUI outlined above clearly has a control system context. The conceptual model for the SUI is given by Eq. (8.5) together with the (algebraic) equations implicit in the control strategy. A likely project goal here could be the resolution of the design problem of choosing appropriate values for the various parameters (e.g. K, k, H_1, H_2, A_1, A_2) within the model based on assumptions about the external demand, $u(t)$ and some criterion for evaluating performance.

There is a distinctive aspect of this example that is worth noting. In Example 1, the conceptual model for the SUI was formulated entirely on the basis of underlying physical laws. In this example, only part of the conceptual model has such 'natural' origins (i.e. Eq. (8.5)). The remainder of the model relates to behaviour that is superimposed by the technological artefact of the control policy (see Flow Control Features above). This latter behaviour can be readily altered by the control policy's developer. In fact, its possible modification is likely implicit in the project goal.

8.2.4 Population Dynamics

Often the model associated with a modelling and simulation project in the domain of environmental studies must incorporate a representation of the manner in which the population of various species evolves over time. In many cases, the model must reflect the interdependence of several species. Perhaps, the best example of the case of interacting populations is given by predator/prey (or host/parasite) situations, e.g. wolf/caribou or lynx/hare populations. The characterization of the behaviour of such populations with a CTDS model implies that the variables representing population values will acquire 'real' (i.e. fractional) values rather than values that are strictly integer. This may appear counter-intuitive, but with the assumption that the populations are 'large', the fractional parts of real values have little consequence on the general features of the results obtained.

There are no underlying physical laws upon which to base the development of such population models (unlike the circumstances in the examples discussed in Sects. 8.2.1, 8.2.2 and 8.2.3). Consequently, the development is based on essentially intuitive arguments. As demonstrated below, a credible structure for such models can be formulated in a reasonably straightforward manner. However, accommodating the associated data requirements (parameter values) can present a challenge.

We consider first a 'single population' model and let $P(t)$ represent the population at time t. A natural assumption is that the rate of change of population is dependent on two effects, namely, the birth rate, $b(t)$ (births per unit time) and the death rate, $d(t)$ (deaths per unit time). This yields the basic equation:

$$P'(t) = b(t) - d(t) \tag{8.6}$$

It is reasonable to assume that both $b(t)$ and $d(t)$ are dependent on the current population. If this dependence is linear, i.e. $b(t) = k_b P(t)$ and $d(t) = k_d P(t)$, then the model becomes

$$P'(t) = k\, P(t) \tag{8.7}$$

where $k = (k_b - k_d)$. The solution to Eq. (8.7) can be easily verified to be

$$P(t) = \exp(kt)P_0$$

where P_0 is the population at some (initial) time t_0. Clearly, if $k > 0$, the population will grow without bound while if $k < 0$, the population will eventually vanish; hence, the model is relatively rudimentary.

A possible refinement is to conjecture that k is indeed positive but that there are 'external effects' that prevent the population from exceeding a value of $P_{\max}$. This behaviour can be achieved with a simple modification to the model of Eq. (8.7), i.e.

$$P'(t) = k\left[1 - \left(\frac{P(t)}{P_{\max}}\right)\right]P(t) \tag{8.8}$$

Now, as $P(t)$ approaches $P_{\max}$, the growth rate approaches zero.

As an alternative, suppose we choose the dependence in the case of $b(t)$ to be linear but non-linear in the case of $d(t)$. Specifically, let's choose

$$b(t) = \alpha P(t)$$
$$d(t) = \beta P^2(t)$$

where α and β are constants whose values (necessarily positive) remain to be determined as part of the data modelling phase. With the substitution of these relations in Eq. (8.6) and with some straightforward manipulation, we obtain

$$P'(t) = \alpha P(t)[1 - KP(t)] \tag{8.9}$$

Here, $1/K = \alpha/\beta$ plays the role of an equilibrium value for the population, $P(t)$. In other words, the solution of Eq. (8.9) approaches the value $1/K$ from any initial condition $P_0 = P(t_0)$.

We now extend our considerations to the case of two populations that function in a predator/prey framework. We use P_1 and P_2 to represent the predator and the prey populations, respectively. The behaviour of each of these populations can be assumed to be represented by an equation of the form of Eq. (8.9) but suitably augmented by some reflection of the mutual interaction. We assume that the interaction can be characterized by a term that is proportional to the product of the two population sizes. Furthermore, it is reasonable to assume that the interaction is beneficial to the predator population growth rate but is detrimental to the growth rate of the prey population. Under these circumstances, we obtain the following CTDS model:

$$P'_1(t) = \alpha_1 P_1(t)[1 - K_1 P_1(t)] + \lambda_1 P_1(t)P_2(t)$$
$$P'_2(t) = \alpha_2 P_2(t)[1 - K_2 P_2(t)] - \lambda_2 P_1(t)P_2(t)$$

(8.10)

where the positive constants λ_1 and λ_2 reflect the 'strength' of the interactions.

A common simplification to the model given in Eq. (8.10) is to ignore the effect of 'natural' death rates by setting $\beta_1 = \beta_2 = 0$, which then results in $K_1 = K_2 = 0$. This gives

$$P'_1(t) = \alpha_1 P_1(t) + \lambda_1 P_1(t)P_2(t)$$
$$P'_2(t) = \alpha_2 P_2(t) - \lambda_2 P_1(t)P_2(t)$$

(8.11)

From a validation point of view, it is reasonable to require that the predator population, $P_1(t)$, approach zero if the prey population vanishes (i.e. if $P_2 = 0$). This requirement can be achieved only if α_1 has a negative value. An equivalent effect can be achieved by replacing α_1 with $-\alpha_1$ (and then taking both α_1 and α_2 to be positive). Our model then becomes

$$P'_1(t) = -\alpha_1 P_1(t) + \lambda_1 P_1(t)P_2(t) = -\alpha_1 P_1(t)[1 - (\lambda_1/\alpha_1)P_2(t)]$$
$$P'_2(t) = \alpha_2 P_2(t) - \lambda_2 P_1(t)P_2(t) = \alpha_2 P_2(t)[1 - (\lambda_2/\alpha_2)P_1(t)]$$

(8.12)

Equation (8.12) has an equilibrium point given by $P^*_1 = \alpha_2/\lambda_2$ and $P^*_2 = \alpha_1/\lambda_1$. These values correspond to the case where both $P'_1(t)$ and $P'_2(t)$ are zero. The equilibrium point, however, is unstable, and any small perturbation from it leads to an oscillatory trajectory for both $P_1(t)$ and $P_2(t)$ about their respective equilibrium values. Representative trajectories are shown in Fig. 8.4.

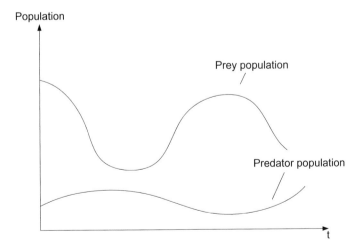

Population

Prey population

Predator population

t

Fig. 8.4 Predator/prey population

The CTDS model of Eq. (8.12) has been extensively studied, and the equations are known as the Lotka–Volterra equations (see e.g. [2]). An interesting study of predator/prey behaviour when harvesting is introduced can be found in [1].

8.3   Safe Ejection Envelope: A Case Study

Several CTDS models have been presented in the preceding section to illustrate the nature of this family of conceptual models. In this section, our focus is on another SUI which gives rise (via a deductive approach) to an CTDS model. However, in this case, we identify a specific goal, and in effect, we formulate a modelling and simulation project.

The problem is one that has been frequently used in the modelling and simulation literature in relation to continuous time dynamic systems. It concerns the safe ejection of a pilot from the cockpit of a disabled fighter aircraft. The specific situation we investigate concerns an aircraft that is flying horizontally at an altitude, H, with a constant speed of V_a, when an emergency situation arises and the pilot is obliged to activate the on-board ejection mechanism and abandon the aircraft. Figure 8.5 shows the pilot's general trajectory following ejection.

The ejection mechanism ensures that the pilot safely leaves the cockpit, and once disconnected from the aircraft, the pilot[1] follows a ballistic trajectory that is governed by two forces. One of these is a drag force and the other is the force of gravity

Fig. 8.5 Trajectory of the ejected pilot

---

[1]For convenience, we shall usually refer simply to the trajectory of the pilot, but it should be recognized that upon leaving the aircraft, the pilot remains connected to the seat and it is the trajectory of the pilot plus seat that is, in reality, being studied. We assume that the seat is jettisoned at some point in time that is beyond the observation interval of interest.

which will ultimately return the pilot to the surface of the earth. Notice, however, that once the pilot leaves the aircraft, the aircraft's tail section becomes a projectile that can potentially strike the pilot and cause serious injury. Our concern is with exploring the circumstances that cause such a collision.

A prerequisite for achieving this objective is a model of the dynamic behaviour of the pilot and the aircraft. The modelling perspective which we adopt incorporates two important assumptions, namely:

- The motion is restricted to two dimensions; more specifically, the pilot's trajectory stays in the plane defined by the cockpit and tail section (in other words, wind forces that might alter this planar motion are ignored).
- During a free flight (ballistic) trajectory any object (in this case the pilot) is subjected to a drag force, $D(t)$, which results from the resistance introduced by air friction. This force acts in a direction opposite to the velocity vector, and we adopt the usual assumption that it can be expressed as

$$D(t) = \mu V^2(t) \tag{8.13}$$

where $\mu = \hat{C}_D \rho$. Here, $\hat{C}_D$ is a constant that depends on the physical shape of the moving object and ρ is the local air density which is dependent on altitude. This relationship is known only in terms of a number of data points as provided in Table 8.1.

There is a variety of factors that influence the form of the pilot's trajectory and hence, the possibility of a collision with the tail section; e.g. the orientation, θ_r, of the ejection rail, the ejection velocity V_r, the position of the tail assembly, the velocity, V_a, of the aircraft and the altitude, H, at which the aircraft is flying. Note that the latter is a consequence of the dependence of drag, $D(t)$, on air density which, in turn, depends on H.

Table 8.1 The altitude/air density relationship

Altitude (ft)	Air density (ρ) (lbs/ft^3)
0	2.3777×10^{-3}
1000	2.208×10^{-3}
200	2.241×10^{-3}
4000	2.117×10^{-3}
6000	1.987×10^{-3}
10,000	1.755×10^{-3}
15,000	1.497×10^{-3}
20,000	1.267×10^{-3}
30,000	0.891×10^{-3}
40,000	0.587×10^{-3}
50,000	0.364×10^{-3}
60,000	0.2238×10^{-3}

The specific relationship we undertake to investigate in this project is the one that exists between the constant horizontal velocity of the aircraft (V_a) and a variable we call H_{min}. As is apparent from Eq. (8.13), the drag force $D(t)$ acting on the pilot is dependent on the altitude at which the aircraft is flying (indirectly via the air density relationship). Suppose the aircraft is flying at some altitude with horizontal velocity $V_a = \alpha$. A collision will result if the drag force is too high. To avoid a collision at the velocity α, the altitude needs to be increased (air density and hence drag force, both decrease as altitude increases (see Table 8.1), and the least altitude (say β) at which a collision is avoided is the H_{min} value associated with the velocity α. Our project goal is to determine a value of H_{min}, corresponding to each of a selected sequence of values of V_a. A graph of the form shown in Fig. 8.6 could be a possible means for presenting the data thus acquired.

The ejection mechanism, once activated (at time $t = 0$), propels the pilot over a short length of rail at a constant velocity V_r. This rail is inclined at an angle θ_r from the vertical (see Fig. 8.7). The seat becomes disengaged from the rail after it has risen a vertical distance of Y_r. At that moment (time $t = t_E$), the pilot (and seat) begins a ballistic trajectory that may either pass over or strike the tail section.

There is a variety of ways in which the conceptual model for the dynamic behaviour of interest can be formulated. In our approach, we choose $X_p(t)$ and $Y_p(t)$ to represent the horizontal and vertical displacements, respectively, of the pilot measured relative to a reference point in space, A°, whose location is fixed in time. A convenient choice for the A° is the point on the aircraft where the seat is located prior to ejection. If we assume that the ejection process begins at time $t = t_0 = 0$, then $X_p(0) = Y_p(0) = 0$, namely at point A°.

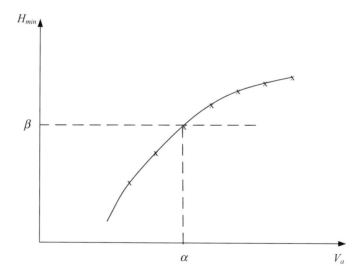

Fig. 8.6 Generic form of the safe ejection envelope

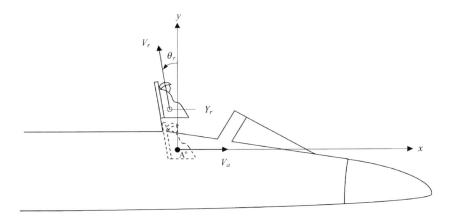

Fig. 8.7 Initial phase of the ejection trajectory

We make the simplifying assumption that the leading edge of the tail section is vertical, and we let $(X_T(t), Y_T(t))$ be the coordinates of the point at the top of the leading edge. This particular point is of interest because it is a reference point for our safe ejection study. We assume that the leading edge of the tail is located a distance BT units from A°. Since the aircraft is moving with a constant horizontal velocity of V_a, it follows that (relative to the fixed point A°) $X_T(t) = (V_a t - \text{BT})$ for $t \geq 0$. Similarly, we assume that the top point of the tail section is displaced a distance of HT above the anchor point; thus, $Y_T(t) = \text{HT}$ for $t \geq 0$. Both BT and HT are positive constants yet to be specified.

We use t^* to denote the value of time when the pilot is located at the leading edge of the tail section. The value of t^* is implicitly defined by the relation:

$$X_p(t^*) = X_T(t^*) = V_a t^* - \text{BT} \tag{8.14}$$

At $t = t^*$, the pilot is either passing over the leading edge of the tail section $(Y_p(t^*) > Y_T(t^*) = \text{HT})$ or is striking it $Y_p(t^*) \leq \text{HT}$. It should also be observed that Eq. (8.14), in fact, provides an implicit definition of the right-hand end of the observation interval.

Although, in principle, the collision boundary corresponds to $Y_p(t^*) = \text{HT}$, it is realistic to adopt a more conservative criterion (a 'safe miss') which we define to be one where the trajectory passes over the tail section with a vertical displacement of a least $(\text{HT} + S_f)$ where S_f is a 'safety factor'. The intent here is to accommodate inherent uncertainties in many of the constants embedded in the dynamic model. Throughout the remaining discussion, references to 'missing the tail' will therefore imply $Y_p(t^*) > (\text{HT} + S_f)$.

If we denote by $V(t)$ the pilot's velocity vector, then the generic form of the pilot's motion can be represented as shown in Fig. 8.8, from which it follows that:

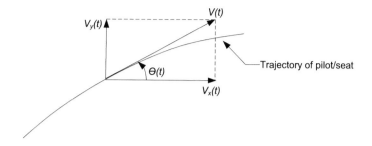

Fig. 8.8 Generic trajectory of the pilot/seat

$$V_x(t) = V(t)\cos\theta(t)$$
$$V_y(t) = V(t)\sin\theta(t) \tag{8.15}$$

While on the rails, the pilot's velocity vector $V(t)$ is the sum of the constant horizontal velocity of the aircraft V_a and the constant ejection velocity V_r. The configuration is shown in Fig. 8.9 from which it follows directly that:

$$X'_p(t) = V_x(t) = V_a - V_r\sin\theta_r$$
$$Y'_p(t) = V_y(t) = V_r\cos\theta_r \tag{8.16}$$

Furthermore, because both the magnitude and the orientation of $V(t)$ are constant while the pilot is on the rails, we have that $V'_x(t) = V'_y(t) = 0$.

Suppose we assume that the pilot/seat leaves the rails at time $t = t_E$. It is straightforward to establish that:

$$t_E = Y_r/(V_r\cos\theta_r)$$
$$X_p(t_E) = (V_a - V_r\sin\theta_r)t_E$$
$$Y_p(t_E) = Y_r$$
$$X'_p(t_E) = V_x(t_E) = V_a - V_r\sin\theta_r \tag{8.17}$$
$$Y'_p(t_E) = V_y(t_E) = V_r\cos\theta_r$$

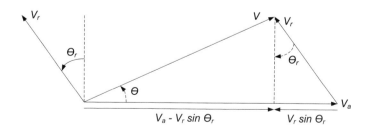

Fig. 8.9 Constrained motion on rails ($Y_p \le Y_r$)

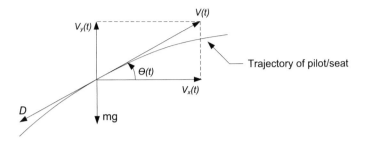

Fig. 8.10 Free fall motion (ballistic trajectory)

Once the pilot/seat is 'disconnected' from the aircraft (i.e. leaves the rails), its motion is governed by two forces, namely, the force of gravity and the drag force $D(t)$ as shown in Fig. 8.10. Together, these forces create a trajectory that (from the perspective of an observer moving with a horizontal velocity of V_a) arcs backwards over the rear of the aircraft (see Fig. 8.5).

Because two forces now act upon the pilot, there are acceleration effects introduced as a consequence of Newton's second law. In other words, $V_x(t)$ and $V_y(t)$ are no longer constant. The dynamic model becomes (see Fig. 8.10)

$$X'_p(t) = V_x(t) \tag{8.18a}$$

$$Y'_p(t) = V_y(t) \tag{8.18b}$$

$$V'_x(t) = -(D(t)/m)\cos\theta(t) \tag{8.18c}$$

$$V'_y(t) = -(D(t)/m)\sin\theta(t) - g \tag{8.18d}$$

The conceptual model we seek is provided, in its most fundamental form, by Eqs. (8.18a, 8.18b, 8.18c, 8.18d). One shortcoming, however, is the dependence on $V(t)$ (through $D(t)$) and on $\theta(t)$. Two approaches are possible for dealing with this. In the approach we adopt, this explicit dependence is eliminated with some algebraic manipulation that incorporates Eq. (8.15) and the specification for $D(t)$ (see Eq. (8.13)). Equations (8.18c) and (8.18d) then become

$$V'_x(t) = -\psi(t)V_x(t)$$
$$V'_y(t) = -\psi(t)V_y(t) - g$$
$$\text{where } \psi(t) = \frac{\left[\hat{C}_D\rho(h)\left(V_x^2(t) + V_y^2(t)\right)^{0.5}\right]}{m}$$
$$h = H + Y_p(t)$$

There now remains the requirement of specifying the observation interval, I_O, that is pertinent to the project goal. The right-hand end of I_O has, in fact, been established earlier (see Eq. (8.14)). The nominal left-hand end of I_O is the moment when the pilot initiates the ejection process, and we have previously associated this with $t = 0$. The values of the four state variables (X_p, Y_p, V_x, V_y) are certainly known at $t = 0$. Notice, however, that values for the state variables are also known at the later time $t = t_E$ (see Eq. (8.17)). The fact that there is a severe discontinuity in the derivatives $V'_x(t)$ and $V'_y(t)$ as t passes over the point $t = t_E$ suggests that $t = t_E$ is a more practical choice for the left-hand boundary of I_O (see Sect. 8.4.2). In view of this, we choose our conceptual model to be the set of equations given in Eq. (8.19).

$$X'_p(t) = V_x(t)$$
$$Y'_p(t) = V_y(t)$$
$$V'_x(t) = -\psi(t)V_x(t)$$
$$V'_y(t) = -\psi(t)V_y(t) - g$$
$$\text{where} \quad \psi(t) = \frac{[\hat{C}_D\rho(h)V(t)]}{m}$$
$$V(t) = \left(V_x^2(t) + V_y^2(t)\right)^{0.5}$$
$$h = H + Y_p(t)$$

(8.19)

where the corresponding 'initial' conditions are at $t = t_E$ as prescribed in Eq. (8.17). A summary of the various constants associated with the model is given in Table 8.2.

It is interesting to also formulate an alternate elaboration of Eqs. (8.18c) and (8.18d). In this approach, we begin with Eq. (8.15) from which it follows that:

$$V'_x(t) = V'(t)\cos\theta(t) - \theta'(t)V(t)\sin\theta(t) \tag{8.20a}$$

$$V'_y(t) = V'(t)\sin\theta(t) + \theta'(t)V(t)\cos\theta(t) \tag{8.20b}$$

Table 8.2 Summary of constants

Constant	Numerical value	Units	Role
BT	30	ft	horizontal displacement of tail section behind origin
$\hat{C}_D$	5		drag factor
g	32.2	ft/sec^2	acceleration due to gravity
HT	12	ft	vertical height of tail section
m	7	slugs	mass of the pilot/seat combination
S_f	8	ft	safety factor for avoiding tail section
θ_r	15	degrees	displacement angle of ejection rails from vertical
V_r	40	ft/sec	seat velocity while on rails
Y_r	4	ft	vertical height of rails

Multiplication of (8.20a) by $\cos \theta(t)$ and (8.20b) by $\sin \theta(t)$, then addition and substitution of Eqs. (8.18c) and (8.18d) yield

$$V'(t) = -(D(t)/m) - g \sin \theta(t)$$

Similarly, multiplication of (8.20a) by $\sin \theta(t)$ and (8.20b) by $\cos \theta(t)$, addition and then substitution of Eqs. (8.18c) and (8.18d) yield

$$\theta'(t) = -\frac{(g \cos \theta(t))}{V(t)}$$

Thus, an alternate conceptual model for the ballistic trajectory $(t > t_E)$ is

$$X'_p(t) = V(t) \cos \theta(t)$$
$$Y'_p(t) = V(t) \sin \theta(t)$$
$$V'(t) = -(D(t)/m) - g \sin \theta(t)$$
$$\theta'(t) = -\frac{(g \cos \theta(t))}{V(t)}$$
$$\text{where} \quad D(t) = \hat{C}_D \rho(h) V^2(t)$$
$$\text{and} \quad h = H + Y_p(t)$$

The safe ejection envelope project is revisited in Sect. 9.6 where a procedure for its completion is presented together with a MATLAB simulation program, which carries out the procedure.

8.4 State-Space Representation

8.4.1 The Canonical Form

The differential equations that evolve in the development of a conceptual model for a CTDS can have a variety of formats, for example, they may be linear or non-linear, they may be a set of first-order equations or they may be equations of higher order, they may be autonomous or they may instead have input functions that reflect pertinent interaction with their environment. Illustrations of these various alternatives can be found in the examples of the previous discussion. The model developed for the electric circuit (Eq. (8.1)) is linear, of second order and is non-autonomous (the voltage source, E, represents an input). The suspension system model of Eq. (8.2) is also a second-order equation but is non-linear; it also is non-autonomous (the irregular road surface provides the input). The fluid level control model of Eq. (8.5) is a pair of first-order equations which are non-linear (because of the non-linear dependence of $w_0(t)$ on $h_1(t)$ and $w_1(t)$ on $h_2(t)$) and non-autonomous (the outflow demand represents an input to the model). In the final example, the population model presented in Eq. (8.12) is a pair of first-order, non-linear equations that are autonomous.

The above discussion illustrates the wide range of formats in which CTDS conceptual models can evolve. This same variability is certainly present in the realm of DEDS models and it is not surprising to encounter it again. However, continuous change models do have a particularly important feature in this regard, namely, that it is possible to transform all of these formats into a standard (canonical) form. This can be written as

$$
\begin{aligned}
\mathbf{x}'(t) &= \mathbf{f}(\mathbf{x}(t),\ \mathbf{u}(t), t) \\
\text{with } \mathbf{x}(t_0) &= \mathbf{x_0}
\end{aligned}
\tag{8.21a}
$$

$$
\text{and } \mathbf{y}(t) = \mathbf{g}(\mathbf{x}(t))
\tag{8.21b}
$$

Here, $\mathbf{x}(t)$, $\mathbf{u}(t)$ and $\mathbf{y}(t)$ are vectors of dimension N, p and q respectively, and represent the state, the input and the output variables respectively, of the CTDS model. The functions $\mathbf{f}$ and $\mathbf{g}$ are likewise vectors with dimensions that are consistent with usage. Equation (8.21a) represents a set of N first-order differential equations, and as noted, the initial conditions required for the solution of Eq. (8.21a) are assumed to be given. Equation (8.21b) makes provision for the situation where the output variables of the model do not correspond directly to any of the state variables but rather are prescribed functions of the state variables.

The representation given in Eqs. (8.21a, 8.21b) is called the *state-space* representation for the particular CTDS model that is under consideration. This representation has two components, the first is the *state equations*, given by Eq. (8.21a), and the second component, given by Eq. (8.21b), is called the output equation of the model. Neither, however, is unique. Nevertheless, as we shall outline below, there often are natural choices for the state variables, $x_i(t)$, which are the members of the state vector, $\mathbf{x}(t)$.

The state-space representation for any CTDS model has several important aspects. Among these is the fact that a very substantial body of knowledge about equations of the form of Eq. (8.21a) has evolved within the domain of applied mathematics. This knowledge is therefore applicable for investigating the properties of CTDS models. Included here are issues that range from the very fundamental, e.g. the question of the existence of solutions to the equations that comprise the model, to issues that characterize the properties of the solution, e.g. stability. Exploration of these topics is, however, beyond the scope of the considerations in this book. The interested reader is encouraged to explore these topics in references such as [3] and [4].

In addition to the important behavioural properties of a CTDS conceptual model that can be explored via its state-space representation, there is one very practical benefit also associated with it. Recall that experimentation with any continuous change model requires the means to generate the numerical solution of differential equations. This is a problem that has been extensively studied in the applied mathematics literature and an extensive body of relevant knowledge about the problem exists. But with few exceptions, this body of knowledge addresses the problem of solving a set of differential equation that is of the form of Eq. (8.21a) and likewise, the available solution methods apply to this case. Thus, the transformation of a CTDS conceptual model into its state-space representation is an essential step for purposes of harnessing the numerical tools for solution generation or, more specifically, for carrying out simulation studies.

8.4.2   The Transformation Process

If any CTDS conceptual model has a state-space representation, i.e. can be transformed into the form of Eqs. (8.21a, 8.21b), then this must certainly be true for a linear model of the form

$$
\begin{aligned}
y^{(N)}(t) &+ a_{N-1}y^{(N-1)}(t) + a_{N-2}y^{(N-2)}(t) + \cdots + a_1 y'(t) + a_0 y(t) \\
&= b_m u^{(m)}(t) + b_{m-1}u^{(m-1)}(t) + b_{m-2}u^{(m-2)}(t) + \cdots + b_1 u'(t) + b_0 u(t)
\end{aligned}
\tag{8.22}
$$

where we assume $m \leq N$ and that $u(t)$ and $y(t_0)$, $y'(t_0), \ldots y^{(N-1)}(t_0)$ are given. The implicit assumption here is that the model has a single input variable, $u(t)$, and a single output variable, $y(t)$. This general linear case is used to illustrate some features of the transformation process.

First, we consider the special case where $m = 0$, i.e. the right-hand side of Eq. (8.22), contains no derivatives of the input function $u(t)$ (an example of this case is provided by the electrical circuit example, specifically Eq. (8.1)). The transformation here is particularly straightforward. Let

$$
\begin{aligned}
x_1(t) &= y(t) \\
x_2(t) &= y'(t) \\
&\;\vdots \\
x_N(t) &= y^{(N-1)}(t)
\end{aligned}
\tag{8.23}
$$

The state equations are then

$$
\begin{aligned}
x_1'(t) &= x_2(t) \\
x_2'(t) &= x_3(t) \\
&\;\vdots \\
x_N'(t) &= -a_0 x_1(t) - a_1 x_2(t)\ldots - a_{N-1}x_N(t) + b_0 u(t)
\end{aligned}
\tag{8.24a}
$$

$$
\text{with } y(t) = x_1(t)
\tag{8.24b}
$$

The more conventional compact form for Eqs. (8.24a, 8.24b) is[2]

$$\mathbf{x}'(t) = \mathbf{A}\mathbf{x}(t) + \mathbf{b}u(t)$$
$$y(t) = \mathbf{c}^T\mathbf{x}(t)$$

where

$$A = \begin{bmatrix} 0 & 1 & 0 & \cdots & 0 \\ 0 & 0 & 1 & \cdots & 0 \\ \cdot & & \cdot & \cdot & \cdot \\ \cdot & & \cdot & \cdot & \cdot \\ \cdot & & \cdot & \cdot & \cdot \\ 0 & 0 & 0 & \cdots & 1 \\ -a_0 & -a_1 & -a_2 & \cdots & -a_{N-1} \end{bmatrix}$$

$$\mathbf{b}^T = [0, 0, 0, \ldots 0, 1]$$
$$\mathbf{c}^T = [1, 0, 0, \ldots 0, 0]$$
$$\mathbf{x}^T = [x_1(t), x_2(t), \ldots x_N(t)]$$

The initial conditions for the state equations of (8.24a) follow directly from the assumptions following Eq. (8.22) and the definitions of Eq. (8.23).

Let's now consider the case where $m > 0$ in Eq. (8.22). A specific example (with $m = 1$) can be obtained from the automobile suspension system model developed earlier if the non-linear shock absorber is replaced with a linear device. In other words, if we replace the earlier specification for $f_b(t)$ with simply

$$f_b(t) = \psi v(t), \quad \text{where } v(t) = y'(t) - u'(t)$$

then Eq. (8.2) can be written as

$$y''(t) + a_1 y'(t) + a_0 y(t) = b_1 u'(t) + b_0 u(t) \tag{8.25}$$

where $a_1 = b_1 = \psi/M$, $a_0 = b_0 = k/M$.

Suppose the procedure we outlined above is applied, i.e. we let $x_1(t) = y(t)$ and $x_2(t) = y'(t)$. The state-space representation then becomes

$$x_1'(t) = x_2(t) \tag{8.26a}$$

$$x_2'(t) = -a_0 x_1(t) - a_1 x_2(t) + b_1 u'(t) + b_0 u(t) \tag{8.26b}$$

$$\text{with } y(t) = x_1(t) \tag{8.26c}$$

---

[2]We use the superscript T to denote the transpose of a vector or matrix.

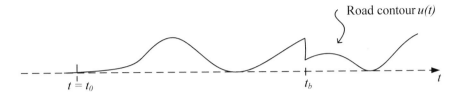

Fig. 8.11 Discontinuous road surface

The perplexing outcome here is the explicit reference to the derivative of the input function that appears on the right-hand side of Eq. (8.26b). It is not unreasonable to imagine cases of interest, where $u(t)$ is not differentiable at all values of t in the observation interval. Recall that in the example that is under consideration, $u(t)$ corresponds to the road surface over which the automobile is travelling. A discontinuity in the road surface could correspond to a pothole in the road as shown in Fig. 8.11. Because of this discontinuity in $u(t)$, the derivative of $u'(t)$ does not exist at $t = t_b$. Does this mean that Eq. (8.25) cannot be solved? Fortunately, the answer is 'no'. The dilemma that we have encountered arises because of a poor choice of state variables.

As an alternative candidate for the state-space representation, consider

$$x_1'(t) = x_2(t) \tag{8.27a}$$

$$x_2'(t) = -a_0 x_1(t) - a_1 x_2(t) + u(t) \tag{8.27b}$$

$$\text{with } y(t) = b_0 x_1(t) + b_1 x_2(t) \tag{8.27c}$$

This representation certainly has the desired feature of being independent of any derivative of the input function, $u(t)$. But is it a valid representation? To confirm that it is, it must be possible to reconstruct the original continuous system model of Eq. (8.25) from Eqs. (8.27a, 8.27b, 8.27c), and this can, in fact, be achieved. The process involves straightforward mathematical manipulation that includes successively differentiating Eq. (8.27c) (to obtain $y'(t)$ and $y''(t)$) and substitutions from Eqs. (8.27a) and (8.27b) to eliminate derivatives of the state variables, $x_1(t)$ and $x_2(t)$.

There is, however, one further issue that needs to be addressed before Eq. (8.27a, 8.27b, 8.27c) can be accepted as a useful state-space representation. This is the matter of initial conditions. Values are provided for $y(t_0)$ and $y'(t_0)$, and these have to be transformed into initial conditions for the state variables x_1 and x_2 so that Eq. (8.27a) and Eq. (8.27b) can be solved. The necessary transformation can be developed using Eq. (8.27c) together with the result obtained by differentiating Eq. (8.27c) and substituting from Eq. (8.27a) and Eq. (8.27b). With t set to t_0 in the resulting equations, we get

$$y(t_0) = b_0 x_1(t_0) + b_1 x_2(t_0) \tag{8.28a}$$

$$y'(t_0) - b_1 u(t_0) = -a_0 b_1 x_1(t_0) + (b_0 - a_1 b_1) x_2(t_0) \tag{8.28b}$$

Equations (8.28a, 8.28b) provide two linear algebraic equations for the two unknowns $x_1(t_0)$ and $x_2(t_0)$. A sufficient condition for the existence of a solution to these equations is that the determinant, det, of the coefficient matrix on the right-hand side be non-zero. The value of the determinant is

$$det = b_0^2 - a_1 b_0 b_1 + a_0 b_1^2 \qquad (8.29)$$

For the specific case of the (linearized) automobile suspension system, a_0, a_1, b_0 and b_1 have values previously specified (see Eq. (8.25)). With these values substituted, $det = (k/M)^2$ and hence is non-zero. Consequently, we can conclude that Eq. (8.27a, 8.27b, 8.27c) is a satisfactory state-space representation for Eq. (8.25) in the particular context of the specified parameter values.

In general however, there is no guarantee that the value of det as given in Eq. (8.29) is non-zero which means that there is a possibility that the state-space representation of Eq. (8.25) given by Eq. (8.27a, 8.27b, 8.27c) may not be acceptable. It can, for example, be easily shown that if $a_1^2 = 4 a_0$ and $a_1 b_1 = 2 b_0$, then det is identically zero. It is reasonable therefore to wonder about the existence of another state-space representation that circumvents this possible flaw. Such an alternative does exist and is given by

$$x_1'(t) = -a_0 x_2(t) + b_0 u(t) \qquad (8.30a)$$

$$x_2'(t) = x_1(t) - a_1 x_2(t) + b_1 u(t) \qquad (8.30b)$$

$$\text{with } y(t) = x_2(t) \qquad (8.30c)$$

Using the same procedure outlined earlier, it can be demonstrated that Eq. (8.25) can be reconstructed from Eqs. (8.30a, 8.30b, 8.30c), and hence, Eqs. (8.30a, 8.30b, 8.30c) is a valid representation for Eq. (8.25). The equations for the initial conditions follow from Eqs. (8.30c) and (8.30b) (setting $t = t_0$):

$$y(t_0) = x_2(t_0)$$
$$y'(t_0) - b_1 u(t_0) = x_1(t_0) - a_1 x_2(t_0)$$

The determinant of the coefficient matrix for these two algebraic equations has the value -1, and consequently, a solution for $x_1(t_0)$ and $x_2(t_0)$ always exists. Specifically

$$x_1(t_0) = y'(t_0) + a_1 y(t_0) - b_1 u(t_0)$$
$$x_2(t_0) = y(t_0)$$

The state-space representation given in Eq. (8.30a) can be extended to the general case of Eq. (8.22). The form of this representation is given below

$$\mathbf{x}'(t) = \mathbf{F}\mathbf{x}(t) + \mathbf{g}u(t)$$
$$y(t) = \mathbf{h}^T\mathbf{x}(t) + b_N u(t)$$

where

$$\mathbf{F} = \begin{bmatrix} 0 & 0 & 0 & \cdots & 0 & -a_0 \\ 1 & 0 & 0 & \cdots & 0 & -a_1 \\ 0 & 1 & 0 & \cdots & 0 & -a_2 \\ \cdot & \cdot & \cdot & \cdot & \cdot & \cdot \\ \cdot & \cdot & \cdot & \cdot & \cdot & \cdot \\ \cdot & \cdot & \cdot & \cdot & \cdot & \cdot \\ 0 & 0 & 0 & \cdots & 0 & -a_{N-2} \\ 0 & 0 & 0 & \cdots & 1 & -a_{N-1} \end{bmatrix}$$

$$\mathbf{g}^T = [b_0 - a_0 b_N, b_1 - a_1 b_N, b_2 - a_2 b_N, \ldots, b_{N-1} - a_{N-1} b_N]$$
$$\mathbf{h}^T = [0, 0, 0, \ldots, 0, 1]$$
$$\mathbf{x}^T(t) = [x_1(t), x_2(t), \ldots, x_N(t)]$$

The vector $\mathbf{g}$ is shown for the case where $m = N$ in Eq. (8.22). The case where $m < N$ is accommodated by setting $b_k = 0$ for $k = (m + 1), (m + 2) \ldots N$.

Our discussion in this section about the formulation of state-space representations for continuous change models has been somewhat limited in scope. Nevertheless, many of the key issues have been pointed out, and a basis for dealing with them in a broader context has been provided.

References

1. Brauer F, Soudack AC (1979) Stability regions and transition phenomena for harvested predator-prey systems. J Math Biol 8:55–71
2. Hall CAS, Day JW (1977) Ecosystem modelling in theory and practice. Wiley, New York
3. Iserles A (1996) A first course in the numerical analysis of differential equations. Cambridge University Press, Cambridge
4. Lambert JD (1991) Numerical methods for ordinary differential equations. Wiley, London

Simulation with CTDS Models

9

9.1 Overview of the Numerical Solution Process

9.1.1 The Initial Value Problem

An implicit requirement associated with modelling and simulation projects within the realm of continuous change models is a means for solving the differential equations embedded in the conceptual model. In very special cases, these equations can fall into a category for which closed-form analytic solutions can be developed and this certainly has many advantages. Far more common, however, is the case where the features of the equations preclude such a solution approach. In such situations, numerical approximation procedures provide the only solution alternative. Our concern in this section is with exploring some of these numerical procedures. More specifically, our interest focuses on the means for solving the generic Eq. (8.8a) of Chap. 8 (the companion Eq. (8.8b) of Chap. 8 is not relevant here because it simply represents a functional relationship defined on the state vector, $\mathbf{x}(t)$).

Our concern, therefore, is with numerical procedures for generating the solution of the equation

$$\mathbf{x}'(t) = \mathbf{f}(\mathbf{x}(t), t), \tag{9.1}$$

over the observation interval $I_O = [t_0, t_f]$ where t_0, t_f and $\mathbf{x}(t_0) = \mathbf{x}_0$ are assumed to be explicitly given. (Note that the explicit dependence of the derivative function, f, on $u(t)$ that appears in Eq. (8.21a) has been suppressed in this representation; the role of $u(t)$ has been merged into the explicit dependence on t.) In general, $\mathbf{x}$ and $\mathbf{f}$ in Eq. (9.1) are vectors of dimension N.

The problem stated above is commonly called the initial value problem (IVP). It is distinct from a closely related problem called the boundary value problem (BVP) . In both problems, at least N pieces of data about the solution are known. In the case of the IVP, these are the N components of the N-vector $\mathbf{x}_0$. The situation in

© Springer Nature Switzerland AG 2019
L. G. Birta and G. Arbez, *Modelling and Simulation*, Simulation Foundations,
Methods and Applications, https://doi.org/10.1007/978-3-030-18869-6_9

the case of the BVP is different because the known values do not occur at the same value of t. The available data could, for example, be

$$x_1(t_0), x_2(t_0), x_3(t_0), \ldots x_\eta(t_0), x_{\eta+1}(t_f), \ldots x_N(t_f)$$

where $1 < \eta < N$.

Although a continuous change model almost always incorporates more than one first-order differential equation (i.e. the dimension of the state vector, $\mathbf{x}(t)$, is greater than 1), this higher dimensionality introduces unnecessary notational complexity when examining numerical solution methods. Consequently, without loss of generality, we take $N = 1$ throughout most of the discussion that follows.

9.1.2   Existence Theorem for the IVP

The search for the solution of any problem can be undertaken with considerably more confidence when there is an assurance that a solution to the problem does indeed exist. With respect to the solution of Eq. (9.1), this issue has been extensively studied and substantial knowledge is available. We summarize here some of the most significant results in this regard.

As might be expected, it is the characteristics of the derivative function $f(x, t)$, which play a pivotal role in the identification of existence conditions for the solution of Eq. (9.1). Our focus, therefore, is restricted to a function $f(x, t)$ that has two particular features. These are as follows:

(a) $f(x, t)$ is defined and continuous in the strip $-\infty < x < \infty$, $t_0 \leq t \leq t_f$, with t_0 and t_f finite.
(b)   There exists a constant, L, such that for any $t \in [t_0, t_f]$ and any two numbers α and β,

$$|f(\alpha, t) - f(\beta, t)| \leq L|\alpha - \beta|$$

Here (a) and (b) are called the *Lipschitz conditions* and L is called the *Lipschitz constant*.

Theorem A   Let $f(x, t)$ satisfy (a) and (b) and let x_0 be any number. Then there exists exactly one function $X(t)$ with the following properties:

(i) $X(t)$ is continuous and differentiable for $t \in [t_0, t_f]$.
(ii) $X'(t) = f(X(t), t)$ for $t \in [t_0, t_f]$.
(iii) $X(t_0) = x_0$.

Remark 1   Theorem A states that under the assumed conditions on the derivative function, $f(x, t)$, the IVP of Eq. (9.1) not only has a solution, but it is unique.

Remark 2 Suppose $f(x, t)$ has a continuous derivative with respect to x which is bounded in the strip in question (see above), then assumption (a) of Theorem A follows directly, while (b) follows as a consequence of the mean value theorem; hence the two requirements of Theorem A are satisfied.

Remark 3 Unless otherwise noted, we assume through the remainder of this chapter that the conditions of Theorem A hold. Furthermore, we shall call the function $X(t)$ referred to in Theorem A the *true solution* to the IVP under consideration. In some limited circumstances, the true solution may be available as an explicit analytic function. In such circumstances, an 'exact' value can be obtained for the true solution at any value of t within the observation interval I_O, at least to the extent of the precision limitations inherent in the evaluation of the function in question.

9.1.3 What Is the Numerical Solution to an IVP?

A numerical solution to an IVP is a finite set of points, i.e.

$$\{(t_n, x_n); \; n = 0, 2, \ldots, M\}$$

where

- (t_0, x_0) is the given initial condition.
- x_n is a generated numerical approximation for the true solution value at $t = t_n$; i.e. an approximation to $X(t_n)$.
- $t_{n+1} = t_n + h_n$ for $0 \le n \le M - 1$ and $t_M = t_f$.

Here, h_n is called the step size at t_n. If h_n remains invariant for all values of n, then the solution process is said to be of fixed step size; otherwise it is of variable step size. As will become apparent in the discussion below, the step size is a critical parameter in the solution process. As might be expected, its value plays a central role in the accuracy of the results obtained. The issues associated with step-size selection are the following:

- If the step size is to be fixed, then how should the value be selected?
- If it is to be variable, then what is the procedure for making changes?

These are not easy questions to answer but some insight will be provided in the discussion that follows. Decisions relating to step size have to be made by the user of most simulation environments for continuous change models; consequently, some familiarity with the underlying issues is essential. The various notions discussed above are illustrated in Fig. 9.1.

An important feature implicit in Fig. 9.1 is that the numerical solution rarely coincides with the true solution. In fact, it is only at the starting point, $t = t_0$, where

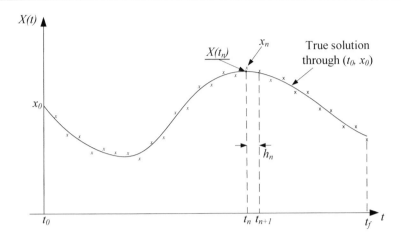

Fig. 9.1 Numerical solution to an IVP

there is certainty that the numerical value is identical to the true value. All other numerical values are, in general, different from the true value. This difference; i.e. the error, has two basic origins:

(a) Truncation (or discretization) error:

    – This is a property of the numercial solution method.

    – If all arithmetic operations could be performed with infinite precision, then this would be the only source of error.

(b) Round-off error:

    – This is a property of the number of calculations carried out by the computer program used to implement the solution method.
    – It arises because of the finite precision in number representation within the computer.

Although not apparent from Fig. 9.1, it is important to appreciate that with all numerical solution methods, each new solution estimate is generated using information from previously generated solution values; in other words, it is constructed from data that may already have significant error. This somewhat disturbing fact sets the stage for the propagation of error that, in turn, can lead to instability. In other words, there is the possibility that the size of the error will grow, in an unbounded manner, as new solution values are generated.

Stability is one of several important attributes that can be associated with any numerical solution method. Others are as follows:

- Order (as discussed in the following sections it is closely related to the notion of truncation error introduced earlier).
- Accuracy (this is a reference to the correspondence between the true solution and the numerical solution).
- Local efficiency (this is a measure of the computational effort required to move the generated solution forward from $t = t_n$ to $t = t_{n+1}$; it is typically measured in terms of the number of evaluation of the derivative function, f).

In the discussion that follows, we'll explore these various matters that have vital importance to the simulation phase of a modelling and simulation study within the CTDS realm.

9.1.4   Comparison of Two Preliminary Methods

(a) The Euler method

The Euler method is the most fundamental of the wide range of approaches that are available for the numerical solution of the IVP. The underlying concept is shown in Fig. 9.2.

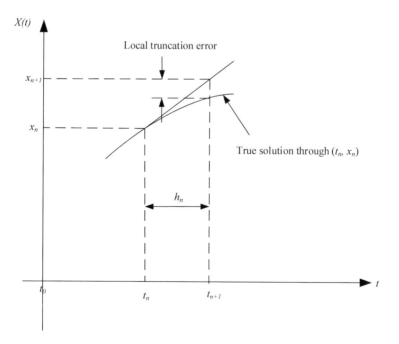

Fig. 9.2 The Euler method

The assumption in Fig. 9.2 is that the solution process has progressed to $t = t_n$ and the solution value generated at $t = t_n$ is x_n. We denote by f_n the slope of the true solution through the point (t_n, x_n), i.e.

$$f_n = f(x_n, t_n)$$

The solution approximation at $t = t_{n+1} = t_n + h_n$ associated with the Euler method is

$$x_{n+1} = x_n + h_n f_n \tag{9.2}$$

While the approach here is intuitively appealing, it can also be viewed as an approximation arising from the definition of a derivative; namely, from the definition:

$$x'(t) = \frac{dx}{dt} = \lim_{\Delta \to 0} \frac{x(t + \Delta) - x(t)}{\Delta}$$

The update formula of Eq. (9.2) is then obtained by ignoring the requirement for Δ to approach zero and by making the following associations: $t = t_n$, $\Delta = h_n$, $x(t_n) = x_n$, $x(t_n + h_n) = x_{n+1}$ and $x'(t_n) = f_n$.

(b)   The modified Euler method (or trapezoidal rule)

The Euler method 'moves forward' on the basis of a single-derivative function evaluation. This value (namely, $f(x_n, t_n)$) is the slope of the true solution that passes through the solution estimate (t_n, x_n). But because (t_n, x_n) is not generally on the 'true' solution (the one through (t_0, x_0)), it is reasonable to conjecture that some other slope value might be a better choice. The modified Euler method creates such an alternate choice by first evaluating the derivative function at the solution estimate produced by the Euler method and then taking an average of two slope values that are thus available. More specifically:

- Take an Euler step to produce the value $p_{n+1} = (x_n + h_n f_n)$ at $t = t_{n+1}$.
- Let $F_{n+1} = f(p_{n+1}, t_{n+1})$.
- Choose the solution estimate at $t = t_{n+1}$ to be

$$x_{n+1} = x_n + h_n(f_n + F_{n+1})/2 \tag{9.3}$$

One difference between solution estimates from the Euler and the Modified Euler methods (given by Eqs. (9.2) and (9.3), respectively) is that the former requires only one derivative function evaluation, while the latter requires two. It is natural, therefore, to expect some advantage from the added effort. An advantage is certainly present and it is realized in terms of superior error performance, at least at the level of local behaviour. This feature can be explored by examining the Taylor series expansion of the true solution of Eq. (9.1) that goes through the point (t_n, x_n). We denote this particular solution by $X_n(t)$. A Taylor series expansion gives

$$X_n(t_n + \delta) = X_n(t_n) + \delta X_n'(t_n) + 0.5\delta^2 X_n''(t_n) + O(\delta^3) \tag{9.4}$$

But

$$X_n(t_n) = x_n$$

and

$$X'_n(t_n) = f(X_n(t_n), t_n) = f(x_n, t_n) = f_n$$

Then, if we set $\delta = h_n$ in Eq. (9.4), we get

$$X_n(t_{n+1}) = x^E_{n+1} + O(h^2_n)$$

where x^E_{n+1} is the Euler solution estimate given by Eq. (9.2). This result demonstrates that the Euler method has a local truncation error that is of order h^2_n, which in turn implies that the Euler method is a first-order method.

A similar analysis with the Modified Euler method gives the result that

$$X_n(t_{n+1}) = x^{ME}_{n+1} + O(h^3_n)$$

where x^{ME}_{n+1} is the solution estimate provided by Eq. (9.3). Thus, the local truncation error of the Modified Euler method is of order h^3_n and the method is a second-order method. This, in particular, demonstrates that the additional derivative function evaluation required by the Modified Euler method provides the benefit of realizing a higher order method.

The order of a solution method is one of its most important characterizing features. As illustrated above, this feature relates to the nature of the error between the solution value produced by the method over a single step relative to the true solution, when both begin at the same starting point. This error estimate evolves from the Taylor series expansion of the true solution around the starting point and reflects the degree of correspondence between the series expansion and the generated solution value.

A practical interpretation of the meaning of a solution method of order r is that such a method generates solution values that have zero error for the case where the true solution is a polynomial of order r or less. Thus, a second-order method will produce exact solution values for the IVP

$$x'(t) = a_1 + 2a_2(t - t_0); \quad x(t_0) = a_0$$

because the solution to this equation is the quadratic function

$$x(t) = a_0 + a_1(t - t_0) + a_2(t - t_0)^2$$

9.2   Some Families of Solution Methods

The most common numerical solution methods for the IVP fall into two broad classes. We examine each of these in turn, beginning with the Runge–Kutta family.

9.2.1   The Runge–Kutta Family

There are two representations for the Runge–Kutta family and these are referred to as explicit and implicit representations. We will restrict our considerations to the explicit representation. The explicit s-stage Runge–Kutta formula is given in Eq. (9.5).

$$x_{n+1} = x_n + h \sum_{i=1}^{s} b_i g_i \tag{9.5}$$

where

$$g_1 = f(x_n, t_n)$$
$$g_2 = f(x_n + h\, a_{2,1} g_1, t_n + c_2 h)$$
$$g_3 = f(x_n + h(a_{3,1} g_1 + a_{3,2} g_2), t_n + c_3 h)$$
$$\vdots$$
$$g_s = f(x_n + h(a_{s,1} g_1 + a_{s,2} g_2 + \cdots a_{s,s-1} g_{s-1}), t_n + c_s h)$$

Remarks

- The s-stage formula requires s evaluations of the derivative function, f, to advance one step (of length h) along the t-axis.
- The s-stage formula has $S = (s^2 + 3s - 2)/2$ free parameters, namely, the collection of b_i's, a_{ij}'s and c_i's. Numerical values for these parameters are determined by a procedure that undertakes to establish an equivalence between the computed value, x_{n+1}, and the first r terms in a Taylor series expansion for the true solution, $X_n(t)$, passing through (t_n, x_n). This creates a formula of order r. It is always true that $r \leq s$. In essentially all cases, there are many ways to select values for the S parameters in order to achieve a formula of order $r \leq s$.
- The general formula given in Eq. (9.5) is explicit because the solution value x_{n+1} evolves directly without the need for the resolution of further numerical issues Observe also that no past solution information is needed to generate x_{n+1}. As we will see in the discussion that follows, these features are not always provided by other methods.

   It is interesting to observe that both the methods introduced earlier in Sect. 9.1.4 are members of the Runge–Kutta family. The first-order Euler method corresponds to the case where $s = 1$, and $b_1 = 1$ and the second-order Modified Euler method correspond to the case where $s = 2$, $b_1 = b_2 = 1/2$, $a_{21} = c_2 = 1$. An alternate second-order method, frequently called the Heun form, is given by $s = 2$, $b_1 = 1/4$, $b_2 = 3/4$, $a_{21} = c_2 = 2/3$.

Third- and fourth-order Runge–Kutta formulas are often used and a representative of each of these classes is provided below. The third-order formula given below is often called the Heun form:

$$x_{n+1} = x_n + \frac{1}{4}h[g_1 + 3g_3]$$
$$g_1 = f(x_n, t_n)$$
$$g_2 = f\left(x_n + \frac{1}{3}hg_1, t_n + \frac{1}{3}h\right)$$
$$g_3 = f\left(x_n + \frac{2}{3}hg_2, t_n + \frac{2}{3}h\right)$$

The fourth-order formula given below is often called the Kutta (or the 'classic') form:

$$x_{n+1} = x_n + \frac{1}{6}h[g_1 + 2g_2 + 2g_3 + g_4]$$

$$g_1 = f(x_n, t_n)$$
$$g_2 = f\left(x_n + \frac{1}{2}hg_1, t_n + \frac{1}{2}h\right)$$
$$g_3 = f\left(x_n + \frac{1}{2}hg_2, t_n + \frac{1}{2}h\right)$$
$$g_4 = f(x_n + hg_3, t_n + h)$$

9.2.2 The Linear Multistep Family

Specific methods in this family are constructed from the following generic formula:

$$x_{n+1} = \sum_{i=1}^{k} \alpha_i x_{n+1-i} + h \sum_{i=0}^{k} \beta_i f_{n+1-i} \tag{9.6}$$

where $f_j = f(x_j, t_j)$. Notice that an essential difference from the Runge–Kutta family is the reliance on past values of the numerical solution and on the slope of the solution at those values, i.e. on the derivative function, f, evaluated at those past values. Several special cases can be identified:

(a) If $k = 1$, then we have a single-step method (reliance on past values is restricted to values at the current time point, t_n).

(b) If $\beta_0 = 0$, then we have an explicit/open/predictor method; if $\beta_0 \neq 0$, then we have an implicit/closed/corrector method.

The implicit case ($\beta_0 \neq 0$) gives rise to a 'circular' situation where the generation of the solution value, i.e. x_{n+1}, requires data that directly depends on x_{n+1}, namely, f_{n+1}. This introduces an accessory problem that needs to be addressed before a practical solution procedure is realized (see Sect. 9.2.2.1).

It is important also to observe here that the dependence of linear multistep methods on past solution values implies a fundamental shortcoming, namely, a 'start-up' problem. Past solution values are needed to initiate the solution procedure, and these can only be obtained by reliance on some ancillary method that is not similarly constrained. Typically, Runge–Kutta methods are used in practice to provide these preliminary values.

9.2.2.1 Predictor–Corrector Methods

The predictor–corrector methods represent the standard implementation approach for the linear multistep family. The underlying idea is to first use an explicit formula to project forward (i.e. to 'predict') a solution value estimate, and then as a second 'refinement' (or corrector) step, an implicit formula is used to create a tentative solution value. This tentative value may or may not be accepted; in the latter case, one or more iterations may follow. This procedure, in effect, deals with the underlying issue introduced by the implicit formula.

Values for the coefficients in a linear multistep formula of order r are established via the same approach used to develop specific members of the Runge–Kutta family, namely, by establishing an equivalence between the computed value, x_{n+1}, and the first r terms in a Taylor series expansion for the true solution, $X_n(t)$, passing through (t_n, x_n). As an example, we give the formulas for the Adams fourth-order predictor–corrector process.

Predictor (Adams–Bashforth)

$$x_{n+1} = x_n + h(55f_n - 59f_{n-1} + 37f_{n-2} - 9f_{n-3})/24 \tag{9.7a}$$

Corrector (Adams–Moulton)

$$x_{n+1} = x_n + h(9f_{n+1} + 19f_n - 5f_{n-1} + f_{n-2})/24 \tag{9.7b}$$

Notice that the predictor is an explicit formula, while the corrector is implicit.

The associated procedure is summarized below

(a) Use Eq. (9.7a) to generate $x_{n+1}^{(p)}$

(b) Use Eq. (9.7b) to generate $x_{n+1}^{(c)}$ (with $x_{n+1}^{(p)}$ used to compute f_{n+1}).

(c) if $\left| x_{n+1}^{(c)} - x_{n+1}^{(p)} \right| < \epsilon$, then $x_{n+1} = x_{n+1}^{(c)}$; otherwise do step (d).

(d) Replace $x_{n+1}^{(p)}$ with $x_{n+1}^{(c)}$ and repeat from step (b).

Here, $\mathcal{E}$ is a predefined operational parameter that provides accuracy control; its value is typically set by the user. Note also that if the error check at step (c) is successful on the first iteration, then the new solution value x_{n+1} is generated after only two derivative function evaluations (assuming past derivative values have been stored). This is a significant improvement over the four evaluations required by a fourth-order Runge–Kutta method. In other words, this predictor–corrector method is potentially significantly more efficient (in terms of derivative function evaluations) than a Runge–Kutta method of like order. This can have important consequences in a simulation project where the conceptual model has many equations and/or the derivative functions are particularly complex.

9.3 The Variable Step-Size Process

Thus far, our discussion has implicitly assumed that the step size, h, used in the numerical solution procedure for the IVP remains invariant. Such solution procedures are certainly widely used in the simulation phase of modelling and simulation projects that have a CTDS context. Such procedures can, however, be inefficient because the nature of the solution may be such that a small value of h is required only over a minor portion of the observation interval, while larger values can be used elsewhere without the danger of compromised solution quality. This gives rise to the need for automatic step-size adjustment.

The realization of such a variable step-size procedure needs to address two basic issues: how to determine when a step-size change is needed (either increase or decrease) and how to carry out a meaningful change in the value of the step size. It is reasonable to assume that the criterion for step-size change ought to be based on the size of the local truncation error (or an estimate of this error, say, E_{est}) relative to some (user-specified) error tolerance, E_{tol}.

The specification of a variable step-size process within such a context can be summarized as shown in Fig. 9.3. Each repetition of the process moves the solution forward by one time step, and each begins with the current solution value, (t_n, x_n), and a nominal step size, h_n.

Several of the steps in Fig. 9.3 require some elaboration and this is provided below.

Step 2: Obtaining an estimate for the local truncation error is a key aspect of the variable step-size process. A variety of approaches have emerged, but for the most part, their comprehensive development depends on the exploration of issues in numerical mathematics that are beyond the scope of this book (the interested reader can find the relevant discussion in [5, 15]). The general nature of a few of these approaches is provided in the brief summaries given in the discussion that follows:

0. set n = 0 and provide initial values for t_0 and h_0
1. compute a solution estimate x_{n+1} at $t_{n+1} = t_n + h_n$
2. compute E_{est}, an estimate of the magnitude of the local truncation error at x_{n+1}
3. compute E_{tol}, the upper bound for the admissible value for E_{est}
4. if $E_{est} > E_{tol}$, then
      4.1 reduce the value of h_n
    else
      4.2 accept the solution estimate x_{n+1} and set $t_{n+1} = t_n + h_n$
      4.3 compute a "best estimate" for the next step-size, h_{n+1}
      4.4 increment n
5. repeat from step 1

Fig. 9.3 The variable step-size process

(a) The half-step approach

The idea here is to obtain two estimates for the solution at time t_{n+1}; the first obtained on a single step with step size, h_n, and the other obtained using two half steps, each of size $h_n/2$. If we denote these two solution estimates by x_{n+1} and x^*_{n+1}, respectively, then (with certain assumptions) it can be shown that a reasonable estimate of the local truncation error at t_{n+1} is

$$E_{est} = \lambda_r |x^*_{n+1} - x_{n+1}|$$

where $\lambda_r = 2^r/(2^r - 1)$ and r is the order of the solution method. A notable feature of this approach is that it has general applicability inasmuch as it is not linked to any particular solution method. Clearly, a significant disadvantage is a substantial efficiency penalty because there is a threefold increase in the number of derivative function evaluations that would otherwise be required to advance the solution by one step.

(b) The embedded approach

A good illustration of this approach is provided by the Runge–Kutta–Fehlberg method that is given in Eq. (9.8). The underlying idea here is the development of two Runge–Kutta formulas that differ in order by one and can be constructed from a shared collection of derivative function evaluations. In the Runge–Kutta–Fehlberg method, two solution estimates x_{n+1} and x^*_{n+1} of order 4 and 5, respectively, are produced at each step. Their difference $|x^*_{n+1} - x_{n+1}|$ provides a good estimate of the local truncation error in the lower order result. Notice that six derivative function evaluations are required, and if the fourth-order result is used, then there is a 50% overhead incurred here in obtaining the error estimate, relative to the 'classic' fourth-order Runge–Kutta formula given earlier:

$$x_{n+1} = x_n + h \left[\frac{25}{216} g_1 + \frac{1,408}{2,565} g_3 + \frac{2,197}{4,104} g_4 - \frac{1}{5} g_5 \right]$$

$$x_{n+1}^* = x_n + h \left[\frac{16}{135} g_1 + \frac{6,656}{12,825} g_3 + \frac{28,561}{56,430} g_4 - \frac{9}{50} g_5 + \frac{2}{55} g_6 \right]$$

$$g_1 = f(x_n, t_n)$$

$$g_2 = f\left(x_n + \frac{h}{4} g_1, t_n + \frac{h}{4} \right)$$

$$g_3 = f\left(x_n + \frac{h}{32}(3g_1 + 9g_2), t_n + \frac{3h}{8} \right) \qquad (9.8)$$

$$g_4 = f\left(x_n + \frac{h}{2,197}(1,932g_1 - 7,200g_2 + 7,296g_3), t_n + \frac{12}{13}h \right)$$

$$g_5 = f\left(x_n + h\left(\frac{439}{216} g_1 - 8g_2 + \frac{3,680}{513} g_3 - \frac{845}{4,104} g_4 \right), t_n + h \right)$$

$$g_6 = f\left(x_n + h\left(-\frac{8}{27} g_1 + 2g_2 - \frac{3,544}{2,565} g_3 + \frac{1,859}{4,104} g_4 - \frac{11}{40} g_5 \right), t_n + \frac{1}{2}h \right)$$

$$E_{est} = |x_{n+1}^* - x_{n+1}| = h \left| \frac{1}{360} g_1 - \frac{128}{4,275} g_3 - \frac{2,197}{75,240} g_4 + \frac{1}{50} g_5 + \frac{2}{55} g_6 \right|$$

(c) A predictor–corrector approach

As the name suggests, this approach is specific to predictor–corrector methods. With suitable assumptions, the underlying analysis shows that a reasonable estimate of the local truncation error at t_{n+1} has the form

$$E_{est} = \lambda_r |x_{n+1}^* - x_{n+1}|$$

where x_{n+1} and x_{n+1}^* are the corrector and predictor values, respectively (necessarily of the same order), and the positive constant λ_r is dependent on the order of the method.

Step 3: A standard format for the bound on the local truncation error is $E_{tol} = (K_1 + K_2 |x_{n+1}|)$ where K_1 and K_2 are user-specified (positive) parameters. The first term (K_1) provides an 'absolute' contribution to the tolerance bound, while the second term ($K_2 |x_{n+1}|$) provides a 'relative contribution'; i.e. if the solution value itself is large, then the error tolerance increases.

Step 4.1: The result of the analysis leading to a meaningful formula for reducing the value of h_n is surprising simple (the analysis itself, however, is outside the scope of our present interest; relevant discussion can be found in [15]). The general form of the update formula is

$$h_n \Leftarrow c \left(\frac{E_{tol}}{E_{est}} \right)^{\frac{1}{r+1}} h_n \qquad (9.9)$$

where r is the order of the solution value x_{n+1} and c is a 'safety factor' that is typically incorporated (a common value is 0.9). A reduction in size results because $E_{est} > E_{tol}$ at Step 4.1.

Step 4.3: The situation represented at this step corresponds to the case where $E_{est} \leq E_{tol}$. This can be interpreted as reflecting a step size that is overly conservative and therefore could possibly be increased on the subsequent phase of the solution process. The underlying analysis shows that the appropriate update formula for h_n is again given by Eq. (9.9).

9.4 Circumstances Requiring Special Care

Thus far in this chapter, we have explored features of the most important numerical tools commonly used to solve the IVP and hence to carry out simulation studies with CTDS models. Like all tools, these likewise have inherent limitations and restrictions on their applicability, and it is prudent for tool users to be aware of these. Our goal in this section is to provide some insight into this important topic.

9.4.1 Stability

The notion of stability is concerned with the existence of upper bounds on the magnitude of the step size, h, used in the solution generating process. In-depth investigation of this important feature is, of necessity, carried out in the context of linear systems because extensive analysis is possible only in this restricted context. Nevertheless, these results can often be extended to the general case of non-linear models by observing that linear approximations can be constructed for non-linear models around any particular point on the solution trajectory. Although relevance of such approximations is restricted to a small region about the chosen point, useful insights into behaviour can nevertheless be obtained.

The essential point can be illustrated by considering the following simple linear IVP:

$$\begin{aligned} u'(t) &= -c_1 u(t) + v(t); & u(0) &= \alpha_1 \\ v'(t) &= -c_2 v(t); & v(0) &= \alpha_2 \end{aligned} \qquad (9.10)$$

where c_1 and c_2 are positive constants. It can be easily verified (e.g. by direct substitution) that the true solution of Eq. (9.10) is

$$u(t) = (\alpha_1 + \gamma)\exp(-c_1 t) - \gamma\exp(-c_2 t)$$
$$v(t) = -\alpha_2\exp(-c_2 t)$$

where $\gamma = \alpha_2/(c_1 + c_2)$. Observe that both $u(t)$ and $v(t)$ approach 0 as $t \to \infty$ independent of the specific values chosen for c_1, c_2, α_1 and α_2.

Suppose now that a fixed step-size Euler method is applied to generate a numerical solution to Eq. (9.10). The iterative process that results can be expressed as

$$u_{n+1} = u_n + h(-c_1 u_n + v_n) = (1 - c_1 h)u_n + h v_n$$
$$v_{n+1} = v_n + h(-c_2 v_n) = (1 - c_2 h)v_n$$

Clearly if the numerical solution is to have any credibility whatsoever, a fundamental requirement is that both $u_n \to 0$ and $v_n \to 0$ as $n \to \infty$. The necessary and sufficient conditions for this to occur are

$$|1 - c_1 h| < 1; \quad \text{that is} - 1 < (1 - c_1 h) < 1$$
$$\text{and} \quad |1 - c_2 h| < 1; \quad \text{that is} - 1 < (1 - c_2 h) < 1$$

which, in turn, implies $h < \min[2/c_1, 2/c_2]$. In other words, there is a very practical constraint on how large a value can be assigned to the step size, h. If this upper bound is exceeded, then the numerical solution is simply 'unstable' and has no relationship to the true solution.

This result clearly raises several important questions; e.g. are all solution methods subject to such step-size constraints and is there anything special (generalizable) about the nature of the specific constraint obtained above? With respect to the first of these questions, it is certainly true that such a constraint does exist for all members of the Runge–Kutta family. However, the constraint does not apply to all solution methods. This can be illustrated by considering a method called the backward Euler method which is a special case of the linear multistep family given in Eq. (9.6). This method is a single-step implicit method ($k = 1$ and $\beta_0 \neq 0$). The updating formula for the backward Euler method is

$$x_{n+1} = x_n + h f(x_{n+1}, t_{n+1})$$

When this formula is applied to our 'test case' of Eq. (9.10), the iterative process that results is

$$u_{n+1} = \frac{u_n}{(1 + c_1 h)} + \frac{h v_n}{(1 + c_1 h)(1 + c_2 h)}$$
$$v_{n+1} = \frac{v_n}{(1 + c_2 h)}$$

It is easy to conclude here that the necessary and sufficient conditions to ensure that both $u_n \to 0$ and $v_n \to 0$ as $n \to \infty$ are

$$|1 + c_1 h| > 1 \text{ and } |1 + c_2 h| > 1$$

Both these conditions are satisfied for any (positive) value of h (recall our original assumption that both c_1 and c_2 are positive). Hence we have an example of a method that does not place an upper bound on the size of the step size, h, from the perspective of solution stability.

We return now to our earlier observation of the instability that results when an unacceptably large value of step size is used to solve Eq. (9.10) with the Euler method. Are there more general conclusions that can be identified? The answer most certainly is 'yes'. To proceed, we generalize our 'test case' to an IVP that is set of N linear first-order equations, i.e.

$$\mathbf{x}'(t) = \mathbf{A}\mathbf{x}(t) \tag{9.11}$$

with $\mathbf{x}(t_0) = \mathbf{x}_0$. We assume here the simplest case where the $N \times N$ coefficient matrix $\mathbf{A}$ has real, distinct and negative eigenvalues.[1] In this circumstance, it can be shown that the true solution of Eq. (9.11) approaches zero independent of the initial value, $\mathbf{x}_0$. If the Euler method is used to generate the solution of Eq. (9.11), then it can be demonstrated that the stability requirement (namely, the requirement that the computed solution likewise approaches zero) is $h < 2/\lambda_{max}$ where λ_{max} is the largest in magnitude of the eigenvalues of $\mathbf{A}$. We leave as an exercise for the reader to confirm that our earlier stability conclusion for the special case of Eq. (9.10) is entirely consistent with this general result (Hint: show that the eigenvalues of the coefficient matrix in Eq. (9.10) are $-c_1$ and $-c_2$.)

The general result above is restricted to the most fundamental of the methods in the Runge–Kutta family. One might reasonably wonder about the nature of the stability requirement for other members of this family. This is a topic that has been extensively investigated in the numerical mathematics literature, and information can be found in books such as [8, 9, 12]. In this regard, we note that the stability bound for the fourth-order Kutta form given earlier is $h < 2.78/\lambda_{max}$ under the assumed conditions on the coefficient matrix, $\mathbf{A}$, in Eq. (9.11).

9.4.2 Stiffness

Stiffness is a property of some CTDS models. It is of particular importance because it interacts with the step-size constraint that is intrinsic to many numerical solution methods in a manner that can seriously undermine the efficiency of the solution process. The background prerequisites for a comprehensive presentation of the

---

[1]The eigenvalues of the $N \times N$ matrix A are the N solutions, $\lambda_1, \lambda_2, \ldots \lambda_N$ to the equation $\det(\lambda\mathbf{I}\text{-}\mathbf{A}) = 0$ where $\det()$ represents the determinant.

topic are substantial, and hence its treatment is beyond the scope of this book. Nevertheless, the essential nature of the problem can be readily illustrated by examining a straightforward example. (The interested reader is encouraged to explore the issue in the numerical mathematics literature, e.g. [5].)

Consider the following two simple linear CTDS models:

Model A

$$u'(t) = -u(t) + 2; \; u(0) = 0$$
$$v'(t) = -v(t) + 2; \; v(0) = 0 \tag{9.12}$$

Model B

$$p'(t) = -500.5p(t) + 499.5q(t) + 2; p(0) = -0.1$$
$$q'(t) = -499.5p(t) - 500.5q(t) + 2; q(0) = 1 \tag{9.13}$$

It is easy to confirm (e.g. by direct substitution) that the solution to Eq. (9.12) is

$$u(t) = v(t) = 2(1 - e^{-t}) \tag{9.14}$$

and that the solution to Eq. (9.13) is

$$p(t) = u(t) - \delta(t)$$
$$q(t) = v(t) + \delta(t)$$

where $\delta(t) = 0.1 \; e^{-1000t}$. Observe that the solutions to Eqs. (9.12) and (9.13) are essentially identical for $t > 0.02$ because $\delta(t)$ has almost vanished.

It's now important to consider what might constitute a reasonable value for the right boundary of the observation interval, I_O, for these two simple models (the left boundary has already been set to 0). This can easily be inferred from Eq. (9.14) from which it is apparent that the solution in both cases is dominated by the term e^{-t} which tends towards zero as t increases. Inasmuch as this term has effectively vanished after $t = 10$, a reasonable choice for the right boundary of I_O is 10. In other words, it is unlikely that an interest in the behaviour of either of these models would extend beyond $t = 10$.

Let's now examine what impact the stability constraint of a numerical solution method would have. On the basis of our earlier considerations, we assume a constraint of the form $h < K/\lambda_{max}$ where K could, for example, be in the range between 2 and 3. To proceed we require the eigenvalues of the two linear models given in Eqs. (9.12) and (9.13). For model A, it is easily seen that the two eigenvalues of the coefficient matrix are both equal to -1. For model B, it can be demonstrated that the eigenvalues are -1 and -1000. The surprising result that now flows from the stability constraint is that even though the true solutions for both models is 'almost' identical (at least for $t > 0.02$), a maximum step size of K would be permitted in studying model A, while the step size would have to be restricted to less than $K/1000$ when studying model B! Apart from the computational burden that is thus imposed upon the investigation of model B, the unavoidable round-off error that could accumulate during the relatively large number of steps needed to traverse the observation interval could seriously deteriorate the solution quality. The study of model A would not encounter either of these difficulties.

This rather unexpected result has its origins in the wide separation between the largest and smallest eigenvalues of model B. This property is called *stiffness*. As might be expected, it has been extensively studied in the numerical mathematics literature and a considerable body of knowledge about it has been emerged, e.g. [5, 7]. These studies are often in the context of linear systems because of the convenience of analysis that linearity provides. The phenomenon nevertheless does arise in non-linear systems which can always be linearly approximated in suitably small regions. The underlying difficult arises simply because the smallest (in magnitude) eigenvalue generally determines the right boundary of the observation interval, while the largest (in magnitude) eigenvalue can introduce a size constraint on the step size, h. As we have illustrated above, these two affects have conflicting and undesirable impacts on the numerical solution process.

It needs to be stressed, however, that solution methods specifically designed to accommodate stiffness have been developed and should be used in any simulation experiment, where there is a possibility that the CTDS model may exhibit stiffness (see e.g. [5]). These methods do involve additional computational overhead and are not recommended for general usage.

One might be tempted to conjecture that stiff systems are no more than curiosities intended mainly to provide a platform for mathematical analysis. It is easy to demonstrate that this is not the case. Consider, for example, the automobile suspension system that was introduced in Sect. 8.2.2 and subsequently linearized in Eq. (8.25). Suppose we assign the specific values $k = 0.5$ and $M = 0.5$ to the spring constant and the mass, respectively, then Eq. (8.25) becomes

$$y''(t) + 2\psi y'(t) + y(t) = 2\psi u'(t) + u(t)$$

where ψ is the stiffness parameter of the shock absorber. With transformation to the state variable representation of Eq. (8.30), we obtain

$$x_1'(t) = -x_2(t) + u(t)$$
$$x_2'(t) = x_1(t) - 2\psi x_2(t) + 2\psi u(t)$$
$$\text{with } y(t) = x_2(t)$$

It can be easily established that the two eigenvalues, λ_1 and λ_2, of the coefficient matrix are the solutions to the algebraic equation

$$\lambda^2 + 2\psi \lambda + 1 = 0$$

i.e. $\lambda_1 = -\psi + \mathrm{sqrt}(\psi^2 - 1)$ and $\lambda_2 = -\psi - \mathrm{sqrt}(\psi^2 - 1)$. Now assume that ψ is large, in particular that it is much greater than 1. With this assumption, the value -2ψ is a reasonable approximation for λ_2. To obtain a helpful approximation for λ_1, we note that for small δ, a first-order Taylor series approximation for the function $R(z) = \mathrm{sqrt}(z)$ is

$$R(z + \delta) = R(z) + 0.5 \ \delta/R(z)$$

Consequently (bearing in mind the assumption that ψ is much larger than 1)

$$\mathrm{sqrt}(\psi^2 - 1) = \psi\,\mathrm{sqrt}\left(1 - \left(\frac{1}{\psi^2}\right)\right) \approx \psi\left[\frac{\mathrm{sqrt}(1) - 0.5}{(\psi^2\mathrm{sqrt}(1))}\right] = \psi - \left(\frac{0.5}{\psi}\right)$$

and so an approximate value for λ_1 is $-0.5/\psi$. Thus, when the shock absorber constant, ψ, is large (relative to the spring constant, k), there is a significant spread between the two eigenvalues; in particular, $|\lambda_2/\lambda_1| = 4\psi^2$ (which, for example, equals 900 when $\psi = 15$).

In practical terms, a large value for ψ (relative to k) means that a ride over an uneven road surface would be very bumpy for the passengers in the automobile because the suspension system would appear to be very 'stiff'. The need to investigate such a circumstance could very well arise if the project goal included assessment of an automobile's dynamic behaviour in extreme conditions, e.g. evaluation of the impact of a shock absorber failure which could correspond to ψ becoming very large. Hence the need to deal with a CTDS model that has the stiffness property.

A meaningful and generally accepted formal definition of stiffness has proved to be elusive. Instead, it is simply regarded as a property of CTDS models that imposes upon *some* numerical solution procedures the requirement for an unusually small step size over a substantial portion of the observation interval. As we have demonstrated above, in the special case of a linear system whose coefficient matrix has distinct real eigenvalues, this property is present when there is a significant spread between the smallest and the largest eigenvalues.

9.4.3 Discontinuity

CTDS models frequently incorporate discontinuities. Unless special precautions are taken in handling these, it is almost certain that the solution trajectories that are obtained will be flawed. In some cases, these flawed solutions may still be adequate within the context of the goal of the modelling and simulation project, while in other cases, these flaws cannot be tolerated and specialized numerical procedures need to be used.

Two of the examples previously considered have embedded discontinuities; namely, the bouncing ball project (Sect. 2.2.6) and the pilot ejection project

(Sect. 8.3). In the case of the bouncing ball, the discontinuity occurs each time the ball strikes the ice surface and 'bounces'. The bounce corresponds to an instantaneous change in both the horizontal and the vertical velocities of the ball. The latter case is especially severe inasmuch as both the direction and magnitude of the ball's vertical velocity change. In the case of the model for the pilot ejection project, the discontinuity occurs at the moment when the pilot/seat leaves the rails. At that moment there is an instantaneous change in the rate of change of both the horizontal and vertical velocities of the pilot/seat (while on the rails both $V'_x(t)$ and $V'_y(t)$ are zero but this changes instantaneously at the moment when the pilot/seat leaves the rails).

A discontinuity occurs when one or more state variables or the derivatives of state variables undergo an instantaneous change. Such an occurrence is usually called an 'event'. Events fall into two categories; namely, time events and state events. The distinguishing feature of a time event is that the time at which it occurs is explicitly known. The time of occurrence of a state event is known only implicitly through some functional specification that involves the state variables. For example, in the case of the bouncing ball, there is a sequence of state events, and the time of occurrence of each corresponds to the condition $y = 0$ (vertical displacement is zero; i.e. the ball is striking the ice surface).

The fact that the time of occurrence of a time event is known is very significant because it enables a simple circumvention of the numerical difficulty that, as we will outline below, is otherwise present. More specifically, if it is known that a time event occurs at $t = t^*$, then the obvious practical approach is simply to execute the solution procedure up to $t = t^*$, carry out the change(s) associated with the event and then continue the solution either to the next time event or to the right boundary of the observation interval, whichever occurs first. This approach preserves the integrity of the solution and requires only a minor disruption in the normal flow of the solution procedure. Handling time events, therefore, is relatively straightforward.

It is interesting to observe that in the case of the pilot/seat model, the simple analysis that yields Eq. (8.17) effectively transforms the apparent state event into a time event. Because of the constant velocity that prevails while the seat is on the rails, the time when the seat leaves the rails is easily determined to be $t_E = Y_r/(V_r \cos \theta_r)$. Furthermore, there is nothing in the goal of the project that necessitates trajectory information prior to t_E, and consequently, the situation becomes even more straightforward; i.e. simply initiate the numerical solution at the event time t_E (or, more precisely, incrementally beyond the event time).

The circumstances in the case of the bouncing ball are quite different; the state events that occur at the bounces cannot be circumvented. What then is the numerical issue that emerges? To address this question, we need to reflect on the program code requirements that are necessitated by the discontinuity. As the following discussion points out, to deal with the state event, the simulation model itself must now acquire a facet that is beyond the simple programming of the algebraic expressions that constitute the derivative functions of the model.

The basic requirement here is clearly a means for locating the occurrence of the state event so that the changes associated with it can be carried out. This is usually achieved by the introduction of *switch functions*. One such function is created for each state event that needs to be accommodated in the CTDS model. The key requirement in defining these switch functions is to capture, in a simple way, the implicit specification of the time of occurrence of the state event. A standard approach is to define the switch function so that its algebraic sign changes when the state event occurs. In other words, a zero value for the switch function signals the occurrence of the state event. For example, in the case of the bouncing ball model, an appropriate switch function is $\varphi_1(t) = y_1(t)$ (recall that $y_1(t)$ represents the vertical position of the ball above the ice surface). In the general case, we assume the existence of m such switch functions associated with the CTDS model being studied, e.g. $\varphi_1(t)$, $\varphi_2(t)$, ... $\varphi_m(t)$.

There are, in fact, two distinct problems that need to be solved. These are called the detection problem and the location problem. The task of the detection procedure (which solves the detection problem) is to identify an interval in which it is certain that at least one zero crossing of a switch function occurs. With this interval as its input, the location procedure then has the task of locating the time of the leftmost of these crossings; this constitutes the solution of the location problem.

To correctly deal with known discontinuities, an CTDS simulation model should incorporate, in some form or another, the equivalent of the pseudocode provided below. Step 1 in this code corresponds to the detection procedure and steps 2 and 3 correspond to the location procedure. This code needs to be executed at the completion of each successive time step over the course of the underlying solution procedure. For definiteness, let's assume that the current solution step has moved the solution from $t = t_a$ to $t = t_b$:

1. For each i in the range 1 through m, determine if φ_i signals the occurrence of event i and if so, place i in $\check{\mathbf{I}}$.
2. For each $i \varepsilon \check{\mathbf{I}}$, determine t_i^* such that $\varphi_i\left(t_i^*\right) = 0$ and place t_i^* in $\check{\mathbf{T}}$.
3. Determine t^{**}, the least value in $\check{\mathbf{T}}$.
4. Restart the solution process at $t = t_a$ and solve to t^{**}.
5. Carry out the changes required at the event(s) occurring at t^{**}.
6. Continue the solution process from t^{**}.

Correct and robust implementation of the pseudocode outlined above is not a trivial undertaking because the resolution of both the detection problem and the location problem requires considerable care. Various approximations are typically accepted, but these can introduce substantial error and/or numerical misbehaviour.

Consider, for example, the bouncing ball model; in this case, $m = 1$ because there is only one state event that needs to be monitored; namely, $\varphi_1(t) = y_1(t)$. It is reasonable to conjecture that in the neighbourhood of an event time, t^*, $\varphi_1(t)$ would have the form shown in Fig. 9.4 where we assume that t_a and t_b are adjacent solution points resulting from a fixed step-size solution process. The signal for the occurrence of the state event (the bounce) could be taken simply to be the

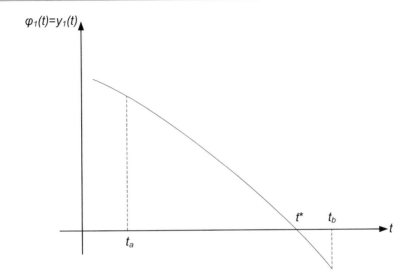

Fig. 9.4 Locating the state event for the bouncing ball

observation that $\varphi_1(t_a)$ and $\varphi_1(t_b)$ have opposite algebraic signs. Having thus established that a state event has occurred, we now need to identify, t^*, the time of its occurrence. A gross (but very convenient) assumption is simply to take $t^* = t_b$. Since the solution process is currently at t_b, it is very straightforward to modify the values of horizontal velocity (x_2) and vertical velocity (y_2) to reflect the changes required by the state event. As a final and entirely artificial change to reflect the intended reality, y_1 can also be set to zero.

The procedure outlined above significantly compromises the accuracy of the solution for the ball's trajectory and hence the accuracy of the results obtained for the underlying modelling and simulation project. But this is not to say that the results are unacceptable. There was no 'accuracy specification' included with the project description, and possibly some latitude is permitted. Note, in fact, that the experiments with the bouncing ball carried out in Annex A4 are undertaken with these same rough approximations.

The approach taken above in handling the location problem (namely, setting $t^* = t_b$) is certainly primitive. In the case where the switch function $\varphi_1(t)$ can safely be assumed to have the form shown in Fig. 9.4 (i.e. a single crossing between t_a and t_b), a relatively simple bisection procedure can be used to solve the location problem in a more credible manner. The idea is simply to half the length of the interval that is known to contain the point of zero crossing on each of a sequence of iterations. This sequence ends either when the interval length is reduced to a sufficiently small size or until the value of φ_1 at the midpoint of the current interval is sufficiently close to zero. A specification of this bisection procedure based on the latter termination criterion is given below.

$$t_c = (t_a + t_b) / 2$$
while $(|\varphi_1(t_c)| > \epsilon)$
    if $(\varphi_1(t_a) * \varphi_1(t_c) < 0)$ $t_b = t_c$
    else $t_a = t_c$
    $t_c = (t_a + t_b) / 2$
endwhile
$t^* = t_c$

Here, ϵ is a parameter that controls the accuracy of the final result that is generated. It should also be appreciated that each evaluation of φ_1 (at time t_c) requires that the underlying solution procedure solve the model equations from time t_a to time t_c. Computational overhead has clearly increased!

Note also that in general, there is no assurance that there is only a single zero crossing in the interval identified by the detection procedure. For example, the behaviour of a switch function (but not the one we have been discussing for the bouncing ball) might have the form shown in Fig. 9.5. Because there are multiple zeros in the interval, the bisection method outlined above would be an inappropriate choice for the location procedure.

The situation in handling discontinuities acquires a different (but none the less challenging) perspective when a variable step-size procedure is used as the equation solving tool.

Some interesting investigations of this challenging numerical problem in handling CTDS models with discontinuities can be found in [1, 3, 6, 14] (a variety of interesting example problems is likewise provided). A comprehensive discussion can be found in Cellier and Kofman [2].

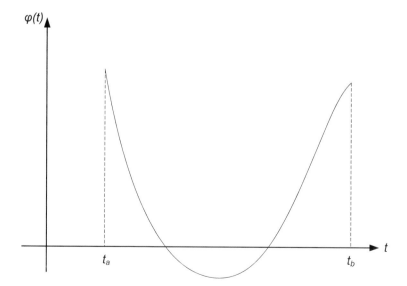

Fig. 9.5 A switch function with multiple crossings

9.4.4   Concluding Remarks

The main purpose of the discussion in Sect. 9.4 has been to demonstrate that the numerical tools required to carry out the simulation phase of a modelling and simulation project in the CTDS domain need to be used with some degree of caution. There are potential pitfalls and these are not always made clear to the users of the many simulation software products that are available in the marketplace. What may appear to be 'interesting' dynamic behaviour in a CTDS model may simply be the reflection of numerical anomalies. Mechanisms to detect such anomalies and bring them to the attention of the user are rarely provided. Thus it is important for the user to be alert and to have reasonable background knowledge and insight in order to be able to assess curious behaviour that may arise. Unfortunately, there are very few guaranteed checks that can be applied to reveal the existence of problems. Nevertheless, one simple option that is always worth considering is the use of an alternate solution method whenever there is some reason to suspect that the numerical solution process is being compromised. Large inconsistencies in the results obtained provide a reasonable signal of underlying difficulty.

It is appropriate finally to stress also that robust solution methods for efficiently handling the differential equation of a CTDS conceptual model continue to evolve, especially in a modelling and simulation context. Readers interested in exploring such developments will find relevant topics in the work of Cellier and Kofman [2].

In this regard, it is particularly interesting to note the work described by Kofman and Junco [10] and further elaborated in [2]. Traditional numerical methods for ode's discretize the time axis as the underlying mechanism for driving the solution forward. The work referenced above takes the alternate approach of discretizing the state space. This introduces an entirely new landscape which is, in particular, well suited to a unified treatment (at the computational level) of models that have both DEDS and CTDS components.

Section 9.4.3 provided an overview of the numerical challenges associated with handling discontinuities within the CTDS domain. Discontinuities can be explored in a much broader simulation context and an interesting perspective on this more general view can be found in [13].

9.5   Options and Choices in CTDS Simulation Software

A wide variety of software products/environments are available for carrying out the simulation phase of any modelling and simulation project in the CTDS realm. Some of these are commercial products (e.g. Dymola [4], Modelica, MapleSim, Matlab), while others are in the public domain (e.g. Desire [11]). By and large, each has a relatively distinctive manner for specifying the model that is to be studied and often has, as well, some special capabilities. Such capabilities (e.g. matrix inversion, eigenvalue calculation, discontinuity handling, animation) can be, especially, relevant to a particular project and thus provide a basis for making a selection from among available alternatives.

From the discussion in Sects. 9.2, 9.3 and 9.4, it is reasonable to suggest that a practical requirement for any CTDS simulation product is a solution engine that provides a variety of numerical solution methods. This is especially important when the conceptual model is large (many differential equations) and/or complex (i.e. derivative function evaluation is time consuming) because in such cases, solution efficiency can become a matter of concern. The availability of solution method alternatives gives the user the option of making trade-offs between computational overhead and accuracy.

Quite apart from a choice from among solution methods, there are still decisions to be made with respect to embedded parameters. The most fundamental, of course, is the step size, h. In the absence of other guidelines or insights, a 'rule of thumb' often used when the solution method is of fourth order is to assign h the value $10^{-3} |I_0|$ (where $|I_0|$ is the length of the observation interval). In the case where a predictor–corrector method has been selected, the parameter, ϵ, that provides some accuracy control (see Sect. 9.2.2.1) may be available for assignment by the user. When a variable step-size method is selected, several associated parameters typically emerge (e.g. the error tolerance parameters K_1 and K_2 introduced in Sect. 9.3), and these must be assigned meaningful values by the user.

Making prudent value choices for these various embedded parameters is not an easy matter for a novice because very little guidance is available. Fortunately, with 'well-behaved' conceptual models, it is usually a noncritical task. However, with ill-behaved situations these value assignments can have significant impact, and improper assignments may even jeopardize the success of the modelling and simulation project.

9.6   The Safe Ejection Envelope Project Revisited

In Chap. 8, an CTDS conceptual model was formulated to provide the data required to establish the 'safe ejection envelope' for a pilot forced to abandon a disabled fighter aircraft by activating the on-board ejection mechanism. Briefly, the objective of this project is to determine for each of a range of horizontal aircraft velocities, the least altitude at which the ejection mechanism will yield an ejection trajectory that avoids the aircraft's tail assembly by a suitable margin of safety. The conceptual model is given by Eq. (8.19) with initial conditions given by Eq. (8.17).

The envelope we seek is, in fact, a display of (V_a, H^*) pairs where H^* is the least 'safe altitude' associated with the horizontal velocity, V_a. The procedure makes use of the fact that if ejection at a particular altitude is unsafe (i.e. results in a trajectory that does not clear the tail assembly by a sufficient distance), then increasing the altitude will eventually locate a safe value. This is a consequence of the fact that the drag force due to air density diminishes as altitude increases.

A procedure for generating the data required to create a graph of the form shown in Fig. 8.6 is given in Fig. 9.6. This procedure assumes the existence of a verified simulation model based on the conceptual model of Eq. (8.19). Several parameters have been introduced to define the boundaries of the study; these are summarized in Table 9.1.

A simulation program, formulated in MATLAB (see Annex 4), that carries out this task is given in Fig. 9.7. The desired safe ejection envelope generated by this program is given in Fig. 9.8.

$$
\begin{aligned}
&V_a \leftarrow V_{start} \\
&H \leftarrow H_{start} \\
&\text{repeat} \\
&\quad \text{while (miss} < S_f) \\
&\qquad \bullet \quad H \leftarrow H + \Delta_h \\
&\qquad \bullet \quad \text{solve ode's of the model from t=t}_E \text{ to t=t}^* \text{ where} \\
&\qquad\qquad t^* \text{ is first occurrence of } X_p(t^*) \leq V_a(t^*) - BT \\
&\qquad \bullet \quad \text{miss} = Y_p(t^*) - HT \\
&\quad \text{endwhile} \\
&\quad \text{record } (V_a, H) \\
&\quad V_a \leftarrow V_a + \Delta_v \\
&\text{until } (V_a > V_{limit}) \\
&\text{Plot the collected } (V_a, H) \text{ pairs}
\end{aligned}
$$

Fig. 9.6   Generating the envelope data

Table 9.1 Parameters used in the safe ejection envelope study

Parameter	Interpretation	Value
V_{start}	Initial horizontal velocity	100 ft/s
H_{start}	Initial altitude	0 ft
V_{limit}	Largest horizontal velocity	950 ft/s
Δ_h	Increment in altitude	500 ft
Δ_v	Increment in velocity	50 ft/s

```
function safe_eject
%---------------------------------------
% pilot ejection problem
% File safe_eject.m
%---------------------------------------

% Define global constants
  global G M BT HT CD SF thetaR tstart Yr Vr Vy Va H RHO
  G=32.2; %  acceleration due to gravity (ft/sec^2)
  M=7;    %  mass of pilot and seat (slugs)
  BT=30;  %  horizontal displacement of tail section (ft)
  HT=12;  %  vertical height of tail section (ft)
  CD=5; %  drag coefficient (ft-sec^2)
  SF=8;   %  safety factor for avoiding tail (ft)
  thetaD=15;  %  angle of ejection rails (degrees)
  thetaR=thetaD*(pi/180);  % angle of ejection rails
                           %  (radians)
  Yr=4;  %  vertical height of rails (ft)
  Vr=40;  %  seat velocity while on rails (ft/sec)
  Vy=Vr*cos(thetaR);  %  vertical velocity when leaving
                      %  rails ft/sec
  Yr=4;  %  vertical height of rails (ft)
  tstart=Yr/(Vr*cos(thetaR));  %  left hand end of
                               %  observation interval
  Va=100;  %  initial aircraft (horizontal) velocity
           %  (ft/sec)
  H=0;  %  initial aircraft altitude (ft)
%-----------------------------------------------------------
% air density data
  alt =[0,1E+3,2E+3,4E+3,6E+3,1E+4,1.5E+4,2E+4,3E+4, ...
       4E+4 5E+4,6E+4];
  air_den = [2.377E-3,2.308E-3,2.241E-3,2.117E-3, ...
             1.987E-3,1.755E-3,1.497E-3, 1.267E-3, ...
             0.891E-3,0.587E-3,0.364E-3,0.2238E-3];

% create RHO interpolation
  RHO = spline(alt, air_den);

% output data storage vectors
  VaV = []; % ---vector of aircraft velocities
  HV  = []; % ---vector of minimum safe ejection altitudes

  refine = 10;
  options = odeset('Events',@eventSpecs, ...
                   'Refine', refine);
```

Fig. 9.7 MATLAB simulation model for safe ejection envelope

```
%---EXPERIMENT
%-----------------------------------
   while Va <= 1000
       s0 = getInitialState();
       [times, sVal] = ode45(@deriv, [tstart, 3], ...
                               s0, options);
       nt = length(times);
       miss = sVal(nt,2) - HT;
       while miss < SF;
         H = H + 500;
         [times, sVal] = ode45(@deriv, [tstart, 3], ...
                                 s0, options);
         nt = length(times);
         miss = sVal(nt,2) - HT;
       end
       VaV = [VaV,Va];
       HV  = [HV,H];
       Va = Va + 50;
   end
-----------------------------------------
%---OUTPUT
% plot of safe ejection altitude vs aircraft velocity
   plot(VaV, HV);
   ylabel('Safe ejection altitude (ft)');
   xlabel('Aircraft velocity (ft/sec)');
   title ('The safe ejection envelop');
%-----------------------------------------
end
%------------------------------
%---DYNAMIC
%------------------------------
function ds_dt = deriv(t, s)
global G M  CD  H  RHO
%  s(1)=Xp, s(2)=Yp, s(3)= Vx, s(4)=Vy
   V = sqrt(s(3)^2 + s(4)^2);
   HplusYp = H + s(2);
   rho = ppval(RHO,HplusYp);
   PSI = (CD*rho*V)/M;
   ds_dt = [ s(3)          % ds(1)/dt = s(3)
             s(4)          % ds(2)/dt = s(4)
             -PSI*s(3)     % ds(3)/dt = -PSI*s(3)
             -PSI*s(4)-G   % ds(4)/dt = -PSI*s(4)-G];
end
```

Fig. 9.7 (continued)

```
function [value, isterminal, direction] = eventSpecs(t, s)
global BT Va
    value(1) = s(1) - (Va*t - BT); %  pilot/seat over tail
    isterminal(1) = 1;  % stop the integration
    direction(1) = -1;  % negative direction only
    value(2) = s(2);  %  pilot/seat has struck the fuselage
    isterminal(2) = 1;  % stop the integration
    direction(2) = -1;  % negative direction only
end

function sInit = getInitialState()
global  thetaR tstart Yr Vr Vy Va
    Xp=(Va-(Vr*sin(thetaR)))*tstart; %  horizontal
                            % position when leaving rails
    Yp=Yr;         % vertical position when leaving rails
    Vx=Va-Vr*sin(thetaR); %   horizontal velocity when
                          %    leaving rails
 % Vy is vertical velocity when leaving rails
    sInit = [Xp Yp Vx Vy];
end
```

Fig. 9.7 (continued)

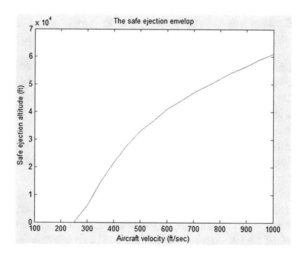

Fig. 9.8 The safe ejection envelope

9.7 Exercises and Projects

9.1. The Foucault pendulum was proposed in 1851 by Léon Foucault as a means of demonstrating the Earth's rotation. Implementations of the Foucault pendulum can be found in science museums throughout the world. The special feature of this pendulum is that the pivot point can turn and consequently the plane in which the swinging bob moves can change in both the x and y directions. In fact, because of the Earth's rotation, the plane of the swing will continuously change. Because of a complex interaction of forces, the rate at which the plane of the swing changes is dependent on the latitude, λ, where the observer is located. For example, at either of the poles $(\lambda = \pm 90°)$, it requires 24 hours for a complete rotation of $360°$ (i.e. angular rate of $2\pi/24$ rad/h), and this period decreases as the observer moves towards the equator $(\lambda = 0)$ where the angular rate is zero. The equations that govern this behaviour (with the assumption that air friction effects can be ignored) are

$$x''(t) - 2\omega \sin(\lambda)y'(t) + K^2 x(t) = 0$$

$$y''(t) + 2\omega \sin(\lambda)x'(t) + K^2 y(t) = 0$$

where ω represents the Earth's rotational velocity $(7.3 \times 10^{-5}$ rad/s) and $K^2 = g/L$ where g is the acceleration due to gravity and L is the pendulum length (necessarily large, e.g. 50 m) and λ is the latitude of the observer:

(a) Formulate a modelling and simulation project based on the conceptual model given above, to determine the angular velocity (radians per hour) of the pendulum's plane of swing for each of the following values of latitude, λ: 5, 10, 15, … 80 and 85° (Hint: observe the graph that results when $x(t)$ is plotted against $y(t)$.)

(b) Determine from a search in the available literature (e.g. the web) what the relation should be and confirm the validity of the results obtained in part (a).

9.2. A bumblebee colony represents an example of a 'stratified population', i.e. one in which the total population is made up of different 'forms' of the same species. Only impregnated females survive the Winter to found a new colony in the Spring. She prepares a simple nest and begins laying eggs at the rate of 12 eggs per day. The life cycle is as follows:

(a) An egg takes 3 days to hatch and what emerges is a larva.
(b) The larva grows for 5 days and then turns into a pupa.
(c) The pupa exists for 14 days and then turns into an adult/worker.
(d) The adult lives for 5 weeks.

Formulate a modelling and simulation project whose goal is to obtain insight into how the population of the colony reacts to the death of the queen bee. Suppose, in particular, that the queen dies after T_0 days. As a result, the population of the colony will eventually diminish to zero. Suppose this happens T_1 days after the death of the queen. The value of T_1 depends on the size of the population at the time T_0 which in turn depends on T_0 itself. Using an appropriate CTDS model, obtain sufficient data to produce a graph of T_1 versus T_0 with T_0 in some suitable range that adequately illustrates the pertinent aspects of the behaviour of interest.

Note that there are four state variables associated with the colony, namely:

(a) $N_e(t)$, the egg population at time t.
(b) $N_r(t)$, the larva population at time t.
(c) $N_p(t)$, the pupal population at time t.
(d) $N_a(t)$, the adult population at time t.

In formulating the model, assume that t has the units of days. The fact that an egg exists for 3 days means that 1/3 of the egg population moves from the egg population to the larva population each day. Similarly, 1/5 of the larva population moves out of the larva population each day. As a consequence, two of the four equations of the conceptual model are

$$N'_e(t) = 12 - \frac{1}{3}N_e(t)$$

$$N'_r(t) = \frac{1}{3}N_e(t) - \frac{1}{5}N_r(t)$$

9.3. In this project, we consider the motion of two masses moving horizontally on frictionless surfaces as shown in Fig. 9.9. Mass m_1 is a block that rolls (without friction) on a horizontal surface, and mass m_2 is a wheel that rolls on top of mass m_1 (again without friction). Each of these masses is individually connected with a spring to a vertical wall. The spring that connects m_1 has a spring constant of k_1, and the spring that connects m_2 has a spring constant of k_2. We assume that up until $t = 0$, this system has been resting in an equilibrium state. The lengths of the two springs are such that at equilibrium the wheel rests at the midpoint of the block whose width is $2w$. We take $m_1 = m_2 = 5$ kg, $k_1 = k_2 = 15$ N/m and $w = 1.6$ m.

If we let $x_1(t)$ and $x_2(t)$ represent the horizontal positions of the two masses relative to their respective equilibrium positions, then the CTDS conceptual model for the system is

$$0.5m_2x''_2(t) - (m_1 + 0.5m_2)x''_1(t) = k_1x_1(t)$$
$$0.5m_2x''_2(t) - 1.5m_2x''_1(t) = k_2x_2(t)$$

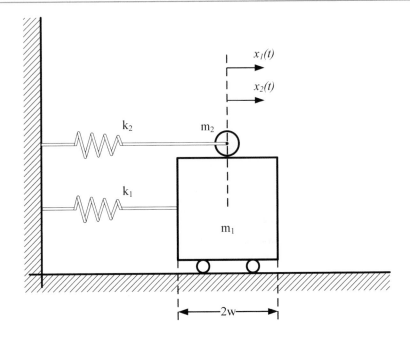

Fig. 9.9 Rolling masses

At $t = 0$, the block is moved to the right by a distance $\alpha = 1.5$ m and then released (the wheel on the other hand remains at its equilibrium position); thus, $x_1(0) = \alpha$, $x_2(0) = 0$, $x_1'(t) = 0$, $x_2'(t) = 0$. The goal of this modelling and simulation project is to gain insight into the circumstances that cause the wheel to fall off the surface of the block:

(a) Determine if the wheel will fall-off the block for the parameter values and the initial conditions that are given.
(b) It is reasonable to assume that there are regions in the (positive) $k_1 - k_2$ plane for which the ball will fall-off the block and conversely regions where the wheel will not fall-off the block (with the given values for the various parameters). Carry out experiments to determine these regions.
(c) Determine how the regions found in part (b) are affected by changes in the value of α.

9.4.   In this study, a proposed system for halting an aircraft that might otherwise overrun the runway during its landing manoeuvre is to be investigated. The system has particular utility in the context of an aircraft carrier. The configuration of the 'upper half' of the system is shown in Fig. 9.10. The complete system is symmetric about the centre line; i.e. an identical configuration to that shown in Fig. 9.10 exists below the centre line.

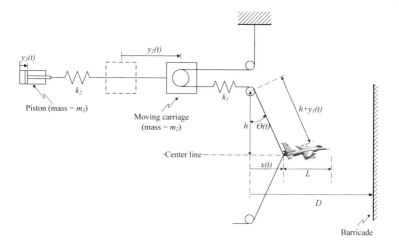

Fig. 9.10 Schematic representation of aircraft arresting mechanism

The 'springs' shown as k_1 and k_2 are fictitious. They are intended to represent the elastic properties of the steel cables which are the connecting members. In particular, this means that these 'springs' cannot be compressed. If, for example, y_2 becomes less than y_3, the cable connecting the piston and the moving carriage simply goes limp.

An appropriate analysis of the elements of the system yields the following conceptual model:

$$m_1\ddot{x}(t) = -2f_1(t)\sin\theta(t)$$

$$\sin\theta(t) = \frac{x(t)}{\sqrt{h^2 + x^2(t)}}$$

$$y_1(t) = \sqrt{h^2 + x^2(t)} - h$$

$$f_1(t) = \begin{cases} k_1(y_1(t) - 2y_2(t)) & \text{for } y_1 > 2y_2 \\ 0 & \text{otherwise} \end{cases}$$

$$m_2\ddot{y}_2(t) = 2f_1(t) - f_2(t)$$

$$f_2(t) = \begin{cases} k_2(y_2(t) - y_3(t)) & \text{for } y_2 > y_3 \\ 0 & \text{otherwise} \end{cases}$$

$$m_3\ddot{y}_3(t) = f_2(t) - f_d(t)$$

Table 9.2 Drag coefficient of the piston

y_3 (meters)	k_3 (Newtons/(m/sec^2))
0	1720
9	1340
18	1100
37	1480
46	1480
55	1480
64	1720
73	1960
82	2500
86	3000
90	3650
93	4650
95	5400
100	7800

Table 9.3 Summary of constants

Constant	Value
m_1	25,000 kg
m_2	1300 kg
m_3	350 kg
k_1	115,000 N/m
k_2	430,000 N/m
h	30 m
L	15 m
D	300 m

The force $f_d(t)$ is a consequence of the 'shock absorber effect' of the piston, which is moving through a cylinder filled with water. Its value is dependent on the square of the velocity $\dot{y}_3(t)$; i.e. $f_d(t) = k_3\,\dot{y}_3^2(t)$. The drag coefficient, k_3, furthermore is dependent on y_3 and its value, as established from experimental data, is given in Table 9.2. The values of the various constants in the model are summarized in Table 9.3.

The specific system to be investigated has a relatively solid barricade located D = 300 m from the contact point (x = 0) which will bring the aircraft to a full stop provided it is not travelling faster than 5 m/s when it strikes the barricade. There are two specific issues that need to be investigated. The first is to determine V^*, where V^* is the largest initial velocity of the aircraft such that its velocity, when the front of the aircraft strikes the barrier, will not exceed 5 m/s. In addition, it is of interest to obtain some insight into the relationship between this maximum initial velocity and the mass of the aircraft. For this purpose, it is required to obtain a graph of V^* versus aircraft mass (m_1) for m_1 in the range 20,000–30,000 kg.

References

1. Birta LG, Ören TI, Kettenis DL (1985) A robust procedure for discontinuity handling in continuous system simulation. Trans Soc Comput Simul 2:189–205
2. Cellier FE, Kofman E (2006) Continuous system simulation. Springer, New York
3. Ellison D (1981) Efficient automatic integration of ordinary differential equations with discontinuities. Math Comput Simul 23:12–20
4. Elmquist H (2004) Dymola—dynamic modeling language, user's manual, Version 5.3, DynaSim AB, Research Park Ideon, Lund
5. Gear CW (1971) Numerical initial value problems in ordinary differential equations. Prentice-Hall, Englewood Cliffs
6. Gear CW, Osterby O (1984) Solving ordinary differential equations with discontinuities. ACM TOMS 10:23–44
7. Hairer E, Wanner G (1996) Solving ordinary differential equations II: stiff and differential-algebraic problems, 2nd edn. Springer, Berlin
8. Iserles A (1996) A first course in the numerical analysis of differential equations. Cambridge University Press, Cambridge
9. Kincaid D, Cheng W (2002) Numerical analysis: mathematics of scientific computing, 3rd edn. Brooks/Cole, Pacific Grove
10. Kofman E, Junco S (2001) Quantized state systems: a DEVS approach for continuous system simulation. Trans SCS 18(3):123–132
11. Korn, GA (2013) Advanced dynamic-system simulation: model replication and Monte Carlo studies, 2nd edn. Wiley
12. Lambert JD (1991) Numerical methods for ordinary differential equations. Wiley, London, p 13
13. Ören TI (2012) Evolution of the discontinuity concept in modeling and simulation: from original idea to model switching, switchable understanding, and beyond. Simulation, Trans Soc Model Simul Int 88(9):1072–1079
14. Shampine LF, Gladwell I, Brankin RW (1991) Reliable solutions of special event location problems for ODEs. ACM Trans Math Softw 17:11–25
15. Watts HA (1984) Step-size control in ordinary differential equation solvers. Trans Soc Comput Simul 1:15–25

The determination of the best value for one or more parameters embedded in a simulation model is a frequently occurring objective of a modelling and simulation study. The solution of this problem has, in recent years, become a significant area of study within the modelling and simulation literature and is generally referred to as 'Simulation Optimization'. Our intent in Part IV of this book is to provide an overview of the underlying challenges that characterize this emerging domain of study together with some of the approaches that have been explored for dealing with them. Appropriate references for further reading are provided in the chapters that follow.

It is interesting to note that in some cases this optimization aspect may simply be a preliminary requirement in the development of a model that is to be subsequently used in the simulation study. In other cases, it may constitute the main aspect of the project goal. We refer to these two alternatives as the *model refinement problem* and the *strategy formulation problem*, respectively.

As an example of the model refinement problem, consider a situation where there exists a general model of how a particular drug that is required in the treatment of some illness, dissipates through the human body. However, before the model can be used it must be adapted ('calibrated') to the particular patient undergoing treatment. In other words, the values for certain parameters within the model have to be established so that it 'best fits' the patient. This could be achieved by finding values for those parameters that minimize the difference between certain of the model's output variables and clinical data obtained from the patient. Once optimized in this sense, the model is available for use by the physician to assist in establishing a proper continuing dosage level of the drug for the patient in question.

As an example of the strategy formulation problem, consider a model of a chemical process which has been developed using known properties of the chemical kinetics that are involved in the process. Suppose one of the model's outputs represents the cost of production over the time period defined by the observation interval. A goal of a modelling and simulation project might be to determine a minimum value for this cost output by the optimum selection of parameters embedded within an operating policy.

Another example of the strategy formulation problem can be developed around the Port Problem examined in Chap. 4. Suppose that for a variety of reasons, the level of activity at this port is expected to increase substantially in the near future. For example, suppose that the average inter-arrival time for tankers is expected to decrease from 8 to 3 hours. In order to ensure that turnaround time for the tankers remains acceptable, port management is considering some combination of increasing the number of berths and increasing the number of tugs. There is a capital cost associated with each of these options and there is a fixed amount of investment dollars available to pay for these capital costs. The underlying optimization problem is to determine how best to allocate the available capital investment dollars between an increase in the number of berths and the acquisition of additional tugs. The criterion for the decision is the minimization of the average tanker turnaround time at the port.

The common element in each of the above examples is the need find a 'best value' (either least or greatest) for a criterion (cost) function, whose value is dependent upon the behaviour of the simulation model of a dynamic system. That behaviour is, in turn, dependent on parameters embedded in the simulation model. The search for the best criterion function value is carried out by finding those values of the parameters which yield behaviour that minimizes (or maximizes) the value of the criterion function.

One might reasonably wonder at this point how 'optimization' impacts the traditional view of a modelling and simulation study. In our preliminary considerations in Sect. 2.3.2, our view of the goal of a modeling and simulation study was taken to be the evaluation of a set of possible solutions to a stated problem formulated within the context of a particular dynamic system that is being studied. Typically, this set of possible solutions is specified by the project stakeholders in terms of values for a parameter vector, $\mathbf{p}$ (of dimension m) that is embedded in the conceptual model of the system. For convenience, we use Ω to denote that set of possible solutions that are to be investigated (i.e. the search space). When the size of Ω is small, the merit of each $\mathbf{p} \in \Omega$ can be evaluated using the output from the simulation model. A decision regarding the best solution alternative can then be made.

Consider now the case where the number of possible solutions of interest becomes very large (possibly infinite). There are two consequences. First, it is no longer feasible to explicitly list the corresponding set of parameter vector values to be evaluated and hence the set Ω can only be implicitly specified; in other words, it is necessarily a specification of that region in m-space that coincides with the solution candidates of interest. The second consequence is that because the values of $\mathbf{p}$ that are implicitly defined within the search space is so large, the examination of the model's response for each $\mathbf{p} \in \Omega$ becomes impractical and usually impossible.

An optimization approach provides a strategy for dealing with such a circumstance. A prerequisite, however, is the formulation of a suitable criterion function J, defined on the model's output variables, whose extreme value (either minimum or maximum) coincides with the solution to the problem underlying the modeling and simulation study. In effect J, provides the basis for evaluating the efficacy of the values of $\mathbf{p}$ within Ω.

The optimization process that is used moves through the space Ω in search of a parameter value $\mathbf{p}^*$, which minimizes/maximizes the criterion function. That value provides the solution to the underlying problem. In effect then, the optimization process automates the exploration of an implicitly defined search space (that may be arbitrarily large) in search of a problem solution.

The perspective outlined above is, however, overly simplistic because significant challenges do emerge! For example, what confidence is there that the criterion function that is being used realistically captures all the solution requirements of the problem originators? Or, what confidence is there that the chosen optimization procedure is sufficiently robust to deal with irregularities that may exist in the search space? Clearly, these are crucial matters and some elaboration is provided in the discussions of Part IV of this book.

There are several aspects of any simulation optimization study that have a significant impact on the concepts and tools that have relevance to dealing with it. These can be identified as follows:

(a) Does the underlying simulation model fall into the CTDS or the DEDS category? In the latter case, considerable care has to be taken to properly accommodate the randomness that is typically inherent in the model's behaviour.
(b) Are the allowable values of the components of the parameter vector $\mathbf{p}$ constrained to the integer domain or is the search space a subset of $\Re^m$? Or possibly there is a combination of these two possibilities.
(c) Are the constraints that are incorporated into the problem statement simple range constraints on the individual components of the parameter vector, $\mathbf{p}$; e.g. $\alpha_j \leq p_j \leq \beta_j$ for $1 \leq j \leq m$ where the α_j and β_j are given constants, or are there functional constraints of the form $\Phi(\mathbf{y}, \mathbf{p}) \geq \mathbf{0}$ where Φ is r-dimensional and $\mathbf{y}$ is vector of variables from the simulation model.

Inasmuch as classical optimization theory provides the foundation for this relatively new facet of modelling and simulation, we begin in Chap. 10 with an overview of some important notions from that area. In Chap. 11, we examine some key aspects of simulation optimization within the CTDS context. In Chap. 12, the special challenges that exist within the DEDS domain when optimization is undertaken, are briefly reviewed and a case study is presented.

Optimization Overview

10

10.1 Introduction

Classical optimization studies are concerned with determining the value of a parameter vector, $\mathbf{p}$, (of dimension m) that yields the least value for a given scalar, real-valued criterion function $J(\mathbf{p})$[1] that is explicitly dependent on the components of $\mathbf{p}$; e.g.,

$$J(\mathbf{p}) = 10\left(p_2 - p_1^2\right)^2 + (1 - p_1)^2$$

In the sequel, we use $\mathbf{p}^*$ to denote that particular value of the parameter vector that a search process seeks to discover. The simplest case occurs when $\mathbf{p} \, \varepsilon \, \Re^m$; i.e., all real values are allowed for each of the m components of $\mathbf{p}$. In other words, the search space $\Omega = \Re^m$. This is referred to as the unconstrained problem.

The alternate, more challenging case, occurs when the permissible values of the components of $\mathbf{p}$ are constrained. In this case, (called the constrained problem) a functional relation of the form $\mathbf{\Phi}(\mathbf{p}) \geq \mathbf{0}$ is associated with the problem. Here, $\mathbf{\Phi}$ is a vector function of dimension q and the implication is that only those values of $\mathbf{p}$ that satisfy $\mathbf{\Phi}(\mathbf{p}) \geq \mathbf{0}$ are admissible solution candidates. A somewhat more straightforward format for constraint specification are range constraints on individual components of $\mathbf{p}$; e.g., $\alpha_j \leq p_j \leq \beta_j$ for $1 \leq j \leq m$ where the α_j and β_j are given constants.

As might be expected, the existence of restrictions on the permitted values for $\mathbf{p}$ introduces additional complexity upon the optimization task. One approach that can be effectively used is called the *penalty function method*. Here, the constraints are manipulated into a special form (a penalty function) which is appended to the criterion function to produce an 'augmented' criterion function whose basic feature is that it penalizes violation of the constraints. The minimization of this augmented

---

[1] Note that the search for the largest value of $J(\mathbf{p})$ (a maximization objective) can be undertaken by searching for the least value of $-J(\mathbf{p})$.

© Springer Nature Switzerland AG 2019
L. G. Birta and G. Arbez, *Modelling and Simulation*, Simulation Foundations,
Methods and Applications, https://doi.org/10.1007/978-3-030-18869-6_10

Fig. 10.1 A response
surface for a criterion function
dependent on two parameters

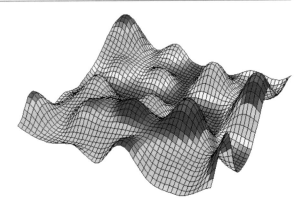

criterion function is, therefore, undertaken without the burden of having to explicitly restrict the search space. The constrained problem is thus transformed into an unconstrained problem (more correctly, there is a requirement for the solution of a *sequence* of unconstrained problems whereby the weight of penalty function violations are increased). In other words, this approach allows the constrained problem to be treated with the same numerical tools as the unconstrained problem. Elaboration of this approach, as well as other methods for handling the constrained optimization problem, can be found in [3, 6, 9].

A frequently occurring situation in regard to constraints occurs when some, or possibly all, components of **p** are restricted to have only integer values. This clearly has a significant impact upon the search space but, as well, eliminates the direct applicability of many optimization procedures.

As might be expected, the difficulty of the minimization task is very much dependent on the geometric nature of the criterion function $J(\mathbf{p})$ as viewed within the search space. In particular, there is a very serious issue of multiple local minima. Most minimization procedures are unable to detect the existence of such 'false' minima and consequently may converge upon such a point, thereby yielding an erroneous result. Another geometric feature that is poorly accommodated by most procedures is the existence of a 'long', gently sloping valley. Such a situation can cause premature termination of a minimization procedure and the presentation of an inferior result. Unfortunately, these difficult circumstances are not uncommon.

By way of illustration, we show in Fig. 10.1 a representative criterion function that is dependent on two parameters.[2] The multiplicity of local extreme values and the existence of sloping valleys are apparent.

10.2 Methods for Unconstrained Minimization

A wide range of methods for dealing with the unconstrained function minimization problem has been developed. A comprehensive review of these is well beyond the scope of our interest in this textbook. Our intent here is simply to provide an

---

[2]Figure 10.1 has been taken from Pinter [15] with the permission of the author.

introduction to some of the basic ideas upon which these methods are based. It is strongly recommended that the reader who needs to carry out an optimization study probe deeper into the topics that are introduced in the discussion that follows. Relevant information can be found in the numerous textbooks that deal specifically with numerical optimization; e.g., [4, 6, 12, 14].

There is a variety of ways for categorizing the relatively large number of available function minimization methods. Perhaps the most fundamental is whether or not gradient information is required by the procedure. Methods not requiring gradient information are often referred to as direct search methods. Because their basis of operation is primarily based on intuitive notions, direct search methods are sometimes referred to as heuristic methods. In the two sections that follow, we outline a representative member of both the gradient-dependent and the direct search categories.

10.2.1 Gradient-Dependent Methods

We begin with a brief review of the notion of the gradient; specifically, the gradient of the criterion function, $J = J(\mathbf{p})$. Inasmuch as $\mathbf{p}$ is a vector of dimension m, the gradient of J is likewise a vector of dimension m. The kth component of this vector is the partial derivative of J with respect to p_k; i.e., with respect to the kth component of $\mathbf{p}$. The gradient of $J(\mathbf{p})$ is denoted by $\mathbf{J_p}(\mathbf{p})$. Suppose, for example, that $J(\mathbf{p}) = 10(p_2 - p_1^2)^2 + (1 - p_1)^2$. Then,

$$\mathbf{J_p}(\mathbf{p}) = \begin{bmatrix} \frac{\partial J}{\partial p_1} \\ \frac{\partial J}{dp_2} \end{bmatrix} = \begin{bmatrix} -(40\,p_1\,(p_2 - p_1^2) + 2\,(1 - p_1)) \\ 20\,(p_2 - p_1^2) \end{bmatrix}$$

The gradient vector is especially relevant in function minimization for two reasons:

(a) If $\bar{\mathbf{p}}$ is a point in m-space, then the negative gradient vector evaluated at $\bar{\mathbf{p}}$ has the property that it points in the direction of greatest decrease in the function J. In other words, for suitably small but fixed ε, $J(\bar{\mathbf{p}} + \varepsilon\,\mathbf{v})$ is smallest when $\mathbf{v}$ is chosen to be $-\mathbf{J_p}(\bar{\mathbf{p}})$.
(b) When J is continuously differentiable, a necessary (but not sufficient) condition for $\mathbf{p}^*$ to be a local minimum for the function $J(\mathbf{p})$ is that $\mathbf{J_p}(\mathbf{p}^*) = \mathbf{0}$.

A concept that has played an important role in the development of numerical minimization procedures that are gradient dependent is that of *conjugate directions*. The concept is formulated in terms of an assumed symmetric positive definite square matrix, $\mathbf{A}$, of dimension m. Specifically, a set of η ($\eta \leq m$) non-zero m-vectors (or equivalently, 'directions') $\mathbf{r}^0$, $\mathbf{r}^1$, ..., $\mathbf{r}^{\eta-1}$ is A-conjugate if $(\mathbf{r}^j)^T \mathbf{A}\,\mathbf{r}^k = 0$ for $j \neq k$ and $j, k = 0, 1, ..., (\eta - 1)$.

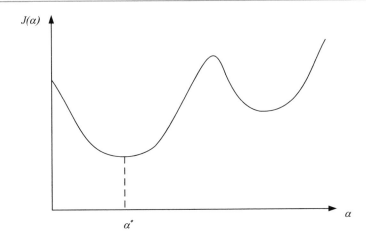

Fig. 10.2 Illustration of the line (linear) search problem

A-conjugate directions have a variety of interesting properties which includes the feature that any such collection of directions is linearly independent. There is one property that is especially relevant from the point of view of function minimization but before outlining it, the notion of a linear or line search needs to be introduced.

Suppose $\bar{\mathbf{p}}$ is a given point in m-space and $\bar{\mathbf{r}}$ is a given m-vector (direction). For any positive value of the scalar α, the m-vector $(\bar{\mathbf{p}} + \alpha\,\bar{\mathbf{r}})$ can be regarded as a point in m-space reached by moving a distance of α away from $\bar{\mathbf{p}}$ in the direction $\bar{\mathbf{r}}$. Suppose now that J is a given scalar valued criterion function whose value depends on the m-vector $\mathbf{p}$; i.e., $J = J(\mathbf{p})$ and suppose we substitute $(\bar{\mathbf{p}} + \alpha\,\bar{\mathbf{r}})$ for $\mathbf{p}$. Because both $\bar{\mathbf{p}}$ and $\bar{\mathbf{r}}$ are fixed, J becomes simply a function of the scalar α and consequently, we can write $J = J(\alpha)$. Furthermore, it is reasonable to assume that there is a value of α (which we denote by $\alpha*$) that yields a minimum value for $J(\alpha)$. Finding the value of $\alpha*$ is called the line (or linear) search problem. In effect, the line search problem corresponds to locating a minimum of J in a specific plane (or 'slice') of the parameter space. This is illustrated in Fig. 10.2. Note the possible existence of multiple local minima.

The following result is the essential property of conjugate directions from the point of view of function minimization.

The CD Lemma

Let

- $J(\mathbf{p}) = \frac{1}{2}\mathbf{p}^T\mathbf{A}\,\mathbf{p} + \mathbf{b}^T\mathbf{p} + c$ (a general quadratic function) with $\mathbf{A}$ symmetric and positive definite and $\mathbf{p}$ an m-vector
- $\mathbf{p}^0$ be a given initial point
- the m-vectors $\mathbf{r}^0, \mathbf{r}^1, \ldots, \mathbf{r}^{\eta-1}$, $(\eta \leq m)$ be a set of A-conjugate directions

- the kth point, $\mathbf{p}^k$ in the sequence $\mathbf{p}^1, \mathbf{p}^2, \ldots, \mathbf{p}^\eta$ be generated by carrying out a line search from $\mathbf{p}^{k-1}$ along $\mathbf{r}^{k-1}$; i.e., $\mathbf{p}^k = \mathbf{p}^{k-1} + \alpha^* \mathbf{r}^{k-1}$ where $J(\mathbf{p}^k) = \min\limits_{\alpha} J(\mathbf{p}^{k-1} + \alpha \, \mathbf{r}^{k-1})$.

Then:

1. $\mathbf{J_p}(\mathbf{p}^k)$ has the property that $(\mathbf{J_p}(\mathbf{p}^k))^T \, \mathbf{r}^j = 0$ for $j = 0, 1, \ldots, (k-1)$.
2. The same point $\mathbf{p}^k$ is reached independently of the order in which the directions $\mathbf{r}^j$ are used in the sequence of line searches.

The criterion function, $J(\mathbf{p})$, considered in the CD Lemma is clearly very specialized. Nevertheless, it is significant because any criterion function has a quadratic approximation in a sufficiently small neighbourhood around its minimum. Consequently, any implications flowing from this Lemma are relevant in such a neighbourhood.

There is, in fact, one especially important consequence of the Lemma; namely, when $\eta = m$, there must exist an index $K \leq m$ such that $\mathbf{p}^K = \mathbf{p}^*$, the minimizing argument of J. This follows from the linear independence of the m-vectors $\mathbf{r}^0, \mathbf{r}^1, \ldots, \mathbf{r}^{m-1}$ and property 1. More specifically, property 1 states that the gradient of J at $\mathbf{p}^K$ (i.e., $\mathbf{J_p}(\mathbf{p}^K)$) is orthogonal to each of a set of m linearly independent m-vectors which, in turn, implies that $\mathbf{J_p}(\mathbf{p}^K)$ must be zero (the zero m-vector is the only one that can be simultaneously orthogonal to each of a set of m linearly independent m-vectors). Because of the assumed special structure of J, the condition $\mathbf{J_p}(\mathbf{p}^K) = 0$ is both necessary and sufficient for $\mathbf{p}^K = \mathbf{p}^*$, the minimizing argument. Note that the case where $K = m$ is a 'worst' case; for certain choices of the initial point $\mathbf{p}^0$, it can occur that $K < m$. In other words, the minimizing argument of J will be located in at most m-steps of the procedure.

The fundamental prerequisite for implementing any function minimization method that is based on conjugate directions is, of course, the availability of the necessary set of directions. Furthermore, it must be borne in mind that any such approach is, at least in principle, relevant only to the minimization of a quadratic function since the directions are, after all, 'A-conjugate' where $\mathbf{A}$ is the matrix that defines (at least in part) the specific quadratic function of interest. Thus, the whole undertaking may appear somewhat pointless since the minimum of a quadratic function can easily be obtained without the need for a numerical search process. (For the quadratic criterion function assumed in the CD Lemma above, the minimizing argument is given by $\mathbf{p}^* = -\mathbf{A}^{-1}\mathbf{b}$.)

The escape from this apparent dilemma is via the observation made earlier that any criterion function has a quadratic approximation in some suitably small region around its minimizing argument, $\mathbf{p}^*$. Thus if a minimization process can move into this region, then the properties of the conjugate directions will result in rapid convergence upon $\mathbf{p}^*$. But it needs to be appreciated that in the general case, the specific quadratic function is never known hence any practical conjugate directions method needs to internally generate directions that will ultimately be A-conjugate even though there is no knowledge of the characterizing matrix, $\mathbf{A}$. While this may appear to be a formidable task, numerous such procedures have been developed. The family of conjugate gradient methods is included among these procedures.

The original function minimization procedure in the conjugate gradient family was proposed by Fletcher and Reeves [7]. The kth step in the procedure ($k \geq 1$) begins with the current estimate of the minimizing argument, $\mathbf{p}^{k-1}$, and a search direction, $\mathbf{r}^{k-1}$. There are two tasks carried out during the step. The first generates a new estimate for the minimizing argument denoted by $\mathbf{p}^k$, where

$$\mathbf{p}^k = \mathbf{p}^{k-1} + \alpha^* \mathbf{r}^{k-1} \text{ and } J(\mathbf{p}^k) = \min_{\alpha} J(\mathbf{p}^{k-1} + \alpha \mathbf{r}^{k-1}).$$

In other words, $\mathbf{p}^k$ is the result of a line search from $\mathbf{p}^{k-1}$ in the direction $\mathbf{r}^{k-1}$.

The second task carried out on the kth step is the generation of a new search direction, denoted by $\mathbf{r}^k$, where

$$\mathbf{r}^k = -\mathbf{J}_{\mathbf{p}}(\mathbf{p}^k) + \beta_{k-1} \mathbf{r}^{k-1} \text{ with } \quad \beta_{k-1} = \frac{\left\| \mathbf{J}_{\mathbf{p}}(\mathbf{p}^k) \right\|}{\left\| \mathbf{J}_{\mathbf{p}}(\mathbf{p}^{k-1}) \right\|}$$

In the above, for an m-vector, $\mathbf{v}$, we use $\|\mathbf{v}\|$ to represent the square of the Euclidean length of $\mathbf{v}$ which is given by $\mathbf{v}^{\mathrm{T}} \mathbf{v}$.

For the first step in this procedure, (i.e., when $k = 1$), $\mathbf{p}^0$ is an initial 'best' estimate of the minimizing argument and $\mathbf{r}^0 = -\mathbf{J}_{\mathbf{p}}(\mathbf{p}^0)$. The sequence of steps ends when some predefined termination criterion is satisfied (e.g., a point $\mathbf{p}^k$ is located at which the length of the gradient vector; i.e., sqrt($\|\mathbf{J}_{\mathbf{p}}(\mathbf{p}^k)\|$, is sufficiently small).

The significant feature of this procedure is that when the criterion function, $J(\mathbf{p})$, is quadratic then the search directions, $\mathbf{r}^0$, $\mathbf{r}^1$, ..., $\mathbf{r}^k$ that are generated are A-conjugate. Consequently, it follows from the CD Lemma that the minimizing argument of J will be located in at most m-steps (or m line searches).

A number of variations on this original procedure have been proposed. Several of these have suggested alternate values for β_{k-1} while others have tried to better accommodate the reality of non-quadratic criterion functions. For example, Polack and Ribière [16] have proposed:

$$\beta_{k-1} = \frac{(\mathbf{J}_{\mathbf{p}}(\mathbf{p}^k))^{\mathrm{T}}(\mathbf{J}_{\mathbf{p}}(\mathbf{p}^k) - \mathbf{J}_{\mathbf{p}}(\mathbf{p}^{k-1}))}{\left\| \mathbf{J}_{\mathbf{p}}(\mathbf{p}^{k-1}) \right\|}$$

while Sorenson [20] recommends:

$$\beta_{k-1} = \frac{(\mathbf{J}_{\mathbf{p}}(\mathbf{p}^k))^{\mathrm{T}}(\mathbf{J}_{\mathbf{p}}(\mathbf{p}^k) - \mathbf{J}_{\mathbf{p}}(\mathbf{p}^{k-1}))}{(\mathbf{p}^{k-1})^{\mathrm{T}}(\mathbf{J}_{\mathbf{p}}(\mathbf{p}^k) - \mathbf{J}_{\mathbf{p}}(\mathbf{p}^{k-1}))}$$

It's perhaps worth observing that if β_{k-1} is set to zero, then the procedure outlined above becomes the classic steepest descent process. The practical performance of that approach, however, is poor and its selection is not recommended, especially in view of the far superior alternatives that are available.

Suggestions have also been made for 're-starting' the conjugate gradient procedure in some cyclic fashion; in other words, abandoning the collection of search directions that have been generated and re-initiating the procedure (which usually means choosing the negative gradient as the search direction). The procedure's m-step property, when applied to a quadratic function, suggests that after a cycle of m-steps (or line searches), the procedure could be re-initialized. Although the natural choice for the restart direction is the negative gradient, Beale [2] has shown that the finite termination property on the quadratic function can be maintained even when the first search direction is not the negative gradient. Based on this observation, a restart strategy that incorporates a novel specification for the search directions was proposed. The approach suggested by Beale was further developed by Powell [17].

The line search problem is one which, on first glance, appears deceptively simple to resolve (see Fig. 10.2). After all, there is only a single parameter, α, that needs to be considered and it is usually known that the minimizing value of α is positive. There is even an easily established orthogonality condition that α^* must satisfy; namely,

$$\left(\mathbf{J}_{\mathbf{p}}\left(\mathbf{p}^{k-1} + \alpha^* \mathbf{r}^{k-1}\right)\right)^{\mathrm{T}} \mathbf{r}^{k-1} = 0$$

Nevertheless, obtaining an accurate solution to the problem can be a challenging numerical task. Note also that there is an implicit requirement for efficiency because a line search problem needs to be solved on each step of the procedure and indeed, the solution of these 'sub-problems' consumes a substantial part of the computational effort in solving the underlying criterion function minimization problem.

A variety of approaches can be considered for solving the line search problem. The first that usually comes to mind is a polynomial fitting process. For example, by evaluating $J(\alpha)$ at three 'test' values of α, it is possible to obtain a quadratic approximation for J whose minimum can be readily determined. That value can be taken as an approximation (albeit rather crude) for α^*. Various refinements of this approach are clearly possible; e.g., obtaining a new quadratic approximation using 'test points' that are in the region of the previous minimum or incorporating a higher order polynomial (possibly cubic).

If it can be assumed that there is available a known interval $\hat{I}$ which contains α^* and that $J(\alpha)$ is unimodal in $\hat{I}^3$ then an interval reduction technique can be used. This involves the judicious placement of points in a sequence of intervals of decreasing length where decisions to discard portions of each interval in the

---

[3]Within the present context, this implies that while α is in $\hat{I}$, $J(\alpha)$ always increases as α moves to the right from α^* and likewise $J(\alpha)$ always increases as α moves to the left from α^*.

sequence are made on the basis of the relative size of $J(\alpha)$ at the selected points. The decisions that are made ensure that the retained interval segment contains α^*. The process ends when the current interval length is sufficiently small and then its midpoint is typically chosen to be α^*. Arguments based on maximizing the usefulness of each evaluation of J give rise to the placement of points in a manner that is related either to the golden section ratio or to the Fibonacci sequence. A discussion of the underlying ideas can be found in [5].

The significance and practical value of carrying out exact line searches is a topic that has received considerable attention in the optimization literature [1, 8]. It can, for example, be easily shown that when the line search is not exact the Fletcher–Reeves formula could generate a search direction, $\mathbf{r}^k$, that is not a descent direction. A variety of conditions have been proposed for terminating the line search when a sufficient decrease has occurred in the value of the criterion function (e.g., the Wolfe conditions [21]). Many of these are outlined in [13].

10.2.2   Heuristic Methods

The Nelder-Mead Simplex method first appeared in the optimization literature in 1965 [11] and has continued to be of practical value and of theoretical interest [10, 19]. Because it is a heuristic method, one of its important features is the absence of any need for gradient information. In a modelling and simulation context, this is especially significant.

The process begins with the specification of a regular simplex which is defined in terms of $(m + 1)$ points in m-space (recall that our parameter vector $\mathbf{p}$ is a vector of m dimensions). When $m = 2$, the simplex is a triangle. The defining points for the initial simplex are part of the initialization procedure. Generally, a (priming) point, $\mathbf{p}^0$, which represents a 'best' solution estimate is prescribed; the remaining m points of the initial simplex are generated by a simple procedure that uses the priming point.

The minimization procedure consists of a sequence of operations referred to as reflection, expansion and contraction. Each step begins with a reflection operation which is then followed by either an expansion operation or a contraction operation. These operations produce a sequence of simplexes that change shape and move through the m-dimensional parameter space until (hopefully) they encompass, and then contact upon, the minimizing argument, $\mathbf{p}^*$.

Let $\{\mathbf{p}^0, \mathbf{p}^1, \mathbf{p}^2, ..., \mathbf{p}^m\}$ be the vertices of the current simplex. Let $\mathbf{p}^L$ be the vertex that yields the largest value for J, $\mathbf{p}^G$ be the vertex that yields the next largest value for J and $\mathbf{p}^S$ be the vertex that yields the smallest value for J. Correspondingly, let $J_L = J(\mathbf{p}^L)$, $J_G = J(\mathbf{p}^G)$ and $J_S = J(\mathbf{p}^S)$. The centroid of the simplex with $\mathbf{p}^L$ excluded is given by

$$\mathbf{p}^C = \frac{1}{m}\left[\left(\sum_{k=0}^{m} \mathbf{p}^k\right) - \mathbf{p}^L\right]$$

A reflection step (Fig. 10.3a) is carried out by 'reflecting' the worst point $\mathbf{p}^L$ about the centroid, to produce a new point $\mathbf{p}^R$ where

$$\mathbf{p}^R = \mathbf{p}^C + \alpha\left(\mathbf{p}^C - \mathbf{p}^L\right)$$

Here, α is one of three user assigned parameters associated with the procedure; the requirement is that $\alpha > 1$ and it is typically chosen to be 1. One of three possible actions now take place depending on the value of $J_R = J(\mathbf{p}^R)$. These are as follows:

1. If $J_G > J_R > J_S$, then $\mathbf{p}^R$ replaces $\mathbf{p}^L$ and the step is completed.
2. If $J_R < J_S$, then a new 'least point' has been uncovered and it is possible that further movement in the same direction could be advantageous. Consequently, an expansion step (Fig. 10.3b) is carried out to produce $\mathbf{p}^E$ where:

$$\mathbf{p}^E = \mathbf{p}^C + \gamma\left(\mathbf{p}^R - \mathbf{p}^C\right) \quad (\gamma > 1 \text{ and is typically } 2)$$

   If $J_E = J(\mathbf{p}^E) < J_R$, then $\mathbf{p}^L$ is replaced with $\mathbf{p}^E$; otherwise $\mathbf{p}^L$ is replaced with $\mathbf{p}^R$. In either case, the step is completed.
3. If $J_R > J_G$, then a contraction step is made to produce the point $\mathbf{p}^D$ where

$$\mathbf{p}^D = \mathbf{p}^C + \beta\left(\tilde{\mathbf{p}} - \mathbf{p}^C\right) \quad (0 < \beta < 1 \text{ and is typically } 0.5)$$

   Here, $\tilde{\mathbf{p}}$ is either $\mathbf{p}^R$ or $\mathbf{p}^L$ depending on whether J_R is smaller or larger than J_L (see Fig. 10.3c, d). If $J_D = J(\mathbf{p}^D) < \min(J_R, J_L)$, then the step ends. Otherwise, the simplex is shrunk about $\mathbf{p}^S$ by halving the distances of all vertices from this point and then the step ends.

Either of two conditions can terminate the search procedure. One is based on the relative position of the vertices of the current simplex; i.e., if they are sufficiently closely clustered then $\mathbf{p}^S$ can be taken as a reasonable approximation of the minimizing argument, $\mathbf{p}^*$. Alternately, the termination can be based on the variation among the values of the criterion function J at the vertices of the simplex. If these values are all within a prescribed tolerance, then again $\mathbf{p}^S$ can be taken as a reasonable approximation of the minimizing argument, $\mathbf{p}^*$.

Fig. 10.3 (a) Reflection step,
(b) expansion step,
(c) contraction step ($J_L < J_R$),
(d) contraction step ($J_R < J_L$)

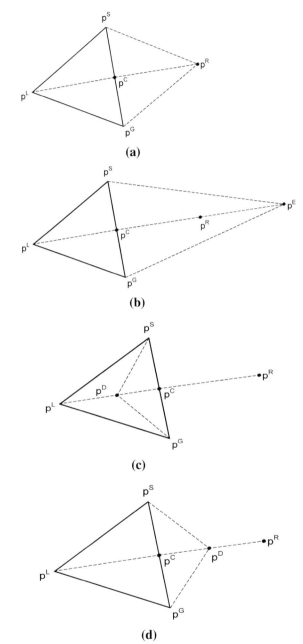

10.3   Exercises and Projects

10.1   The general quadratic function of dimension m can be written as

$$J(\mathbf{p}) = 0.5\mathbf{p}^T\mathbf{A}\mathbf{p} + \boldsymbol{b}^T\mathbf{p} + c$$

where $\mathbf{p}$ is an m-vector, $\mathbf{A}$ is an $m \times m$ positive definite symmetric matrix and $\boldsymbol{b}$ is a m-vector. Consider the point $\mathbf{p}^0$ and a search direction $\mathbf{r}$. Show that if α^* solves the line search problem in the direction $\mathbf{r}$ from the point $\mathbf{p}^0$, i.e., α^* has the property that

$$J(\mathbf{p}^0 + \alpha^*\mathbf{r}) = \min_\alpha J(\mathbf{p}^0 + \alpha\mathbf{r})$$

then

$$\alpha^* = -\frac{r^T\mathbf{J_p}(\mathbf{p}^0)}{\mathbf{r}^T\mathbf{A}\mathbf{r}}$$

10.2.   Develop a computer program that implements an efficient line search procedure which is based on the Golden Section Search. Details about this approach can be found in [9] or [18] or at the Wikipedia site: http://en. wikipedia.org/wiki/Golden_section_search. Test your program on a variety of quadratic functions by comparing your results with the analytic result given in Problem 10.1.

10.3   Probably, the most intuitively appealing approach for locating the minimizing argument of a criterion function $J(\mathbf{p})$ is a succession of line searches along the coordinate axes. This implies that the searches are along the directions $\mathbf{e}_1, \mathbf{e}_2, \ldots, \mathbf{e}_m$ where $\mathbf{e}_k$ is the kth column of the $m \times m$ identity matrix ($\mathbf{e}_k$ is an m-vector whose entries are all 0 except for the entry in the kth position which is 1). One notable feature of this procedure (usually called the univariate search) is that it does not require gradient information. It can be viewed as a series of iterations where each iteration begins at the point $\mathbf{p}^0$ and ends at the point $\mathbf{p}^m$, which is the point that is reached after a sequence of m line searches along the coordinate axes. The procedure is illustrated in Fig. 10.4 for the two-dimensional case.

  The procedure for the univariate search (shown below) can be written in the following way: Choose a value for the termination parameter, ε, and an initial estimate, $\hat{\mathbf{p}}$, for the minimizing argument of J and set $\mathbf{p}^m = \hat{\mathbf{p}}$ (the value of $\mathbf{p}^m$ upon termination of the repeat/until loop is the accepted estimate for the minimizing argument).

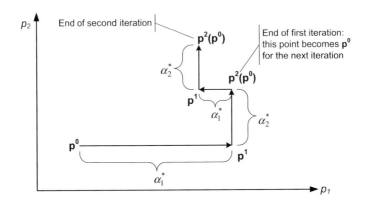

Fig. 10.4 The univariate search in two dimensions ($m = 2$)

> **repeat**
> $\quad k = 0$
> $\quad \mathbf{p}^{\circ} \leftarrow \mathbf{p}^{m}$
> $\quad$**while** $(k < m)$
> $\qquad k \leftarrow k+1$
> $\qquad$ Find $\overset{*}{\alpha}_{k}$ such that $J(\mathbf{p}^{k-1} + \overset{*}{\alpha}_{k}\, \mathbf{e}^{k}) = \min_{\alpha} J(\mathbf{p}^{k-1} + \alpha \mathbf{e}^{k})$
> $\qquad \mathbf{p}^{k} \leftarrow \mathbf{p}^{k-1} + \overset{*}{\alpha}_{k}\, \mathbf{e}^{k}$
> $\quad$**endwhile**
> $\quad J_{\max} \leftarrow \max(|J(\mathbf{p}^{\circ})|, |J(\mathbf{p}^{m})|))$
> **until** $\left| \dfrac{J(\mathbf{p}^{0}) - J(\mathbf{p}^{m})}{J_{\max}} \right| < \varepsilon$

Show that the univariate procedure will converge to the minimizing argument of the general quadratic function given in Problem 10.1 if exact line searches are carried out.

Hint: Consider what must be true if the procedure makes no progress on some particular iteration and then use the fact that the only m-vector that can be simultaneously orthogonal to m orthogonal m-vectors is the zero vector.

10.4 In many situations, the performance of the univariate procedure outlined in Problem 10.3 can be significantly improved by incorporating a slight modification. This simply involves an additional line search in the direction $\mathbf{s} = (\mathbf{p}^{m} - \mathbf{p}^{0})$.

This modified procedure (which we call the extended univariate search) is illustrated in Fig. 10.5.

(a) Modify the procedure given in Problem 10.3 so that it represents the extended univariate search as described above.

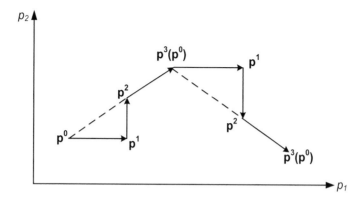

Fig. 10.5 The extended univariate search in two dimensions ($m = 2$)

(b) Formulate an argument that demonstrates that the extended univariate search will also locate the minimizing argument of the general quadratic function.

10.5 (a) Write a program that implements the univariate search procedure as presented in Problem 10.3. Incorporate the line search program that was developed in Problem 10.2.

(b) Test the effectiveness of the program using the following two test problems:

$$J(\mathbf{p}) = 100\left(p_2 - p_1^2\right)^2 + (1 - p_1)^2$$
$$J(\mathbf{p}) = (p_1 + 10p_2)^2 + 5(p_3 - p_4)^2 + (p_2 - 2p_3)^4 + 10(p_1 - p_4)^4$$

Use initial estimates $(0, 1)$ and $(1, 0)$ for the first test problem and initial estimates $(1, 0, 1, 0)$ and $(1, 0, 0, 1)$ for the second test problem. The minimum value of the criterion function for both test problems is zero. The termination parameter, ε, should be set to a value no larger than 10^{-5}.

10.6 Repeat Problem 10.5 for the case of the extended univariate search.

References

1. Al-Baali M (1985) Descent property and global convergence of the Fletcher-Reeves method with inexact line search. IMA J Numer Anal 5:121–124
2. Beale EML (1972) A derivation of conjugate gradients. In: Lottsma FA (ed) Numerical methods for non-linear optimization. Academic, London, pp 39–43
3. Bertsekas DP (1996) Constrained optimization and Lagrange multiplier methods. Athena Scientific, Nashua
4. Bonnans JF, Gilbert JC, Lemaréchal C, Sagastizabal CA (2003) Numerical optimization: theoretical and practical aspects. Springer, Berlin

5. Cormen TH, Leisserson CE, Rivest RL (1990) Introduction to algorithms. MIT Press, Cambridge, MA
6. Fletcher R (1987) Practical methods of optimization, 2nd edn. Wiley, New York
7. Fletcher R, Reeves CM (1964) Function minimization by conjugate gradients. Comput J 7:149–154
8. Gilbert J, Nocedal J (1992) Global convergence properties of conjugate gradient methods for optimization. SIAM J Optim 2:21–42
9. Heath MT (2000) Scientific computing, an introductory survey, 2nd edn. McGraw-Hill, New York
10. Lagarias JC, Reeds JA, Wright MH, Wright PE (1998) Convergence properties of the Nelder-Mead simplex method in low dimensions. SIAM J Optim 9:112–147
11. Nelder J, Mead R (1965) A simplex method for function minimization. Comput J 7:308–313
12. Nocedal J, Wright SJ (1999) Numerical optimization. Springer, New York
13. Oretega JM, Rheinboldt WC (1970) Iterative solution of nonlinear equations in several variables. Academic, New York
14. Pedregal P (2004) Introduction to optimization. Springer, New York
15. Pintér JD (2013) LGO—a model development and solver system for global-local nonlinear optimization, User's guide, 2nd edn. Published and distributed by Pintér Consulting Services, Inc., Halifax. http://www.pinterconsulting.com (First edition: June 1995)
16. Polack E, Ribière G (1969) Note sur la Convergence de Méthodes de Directions Conjuguées. Revue Française d'Informatique et de Recherche Opérationnelle 16:35–43
17. Powell MJD (1978) Restart procedures for the conjugate gradient method. Math Prog 12: 241–254
18. Press WH, Teukolsky SA, Vetterling WT, Flannery BP (1999) Numerical recipes in C: the art of scientific computing, 2nd edn. Cambridge University Press, Cambridge
19. Rykov A (1983) Simplex algorithms for unconstrained optimization. Probl Control Inf Theory 12:195–208
20. Sorenson HW (1969) Comparison of some conjugate directions procedures for function minimization. J Franklin Inst 288:421–441
21. Wolfe P (1969) Convergence conditions for ascent methods. SIAM Rev 11:226–235

Simulation Optimization in the CTDS Domain

11

11.1 Introduction

Incorporation of an optimization phase into a modelling and simulation study within the CTDS domain is, in principle, relatively straightforward. This is especially true when random effects are not present (which is the case we consider here). Its treatment both in general terms and in terms of software support, is not new; e.g. [2, 3].

Nevertheless, relative to classical optimization studies as examined in Chap. 10, there are several distinctive features that need to be recognized. Included here is the special nature of the criterion function whose evaluation now requires the solution of the ode's of the underlying CTDS model. Consequently the computational burden for this evaluation is generally considerably greater than in most classical optimization studies. Furthermore this particular feature of the criterion function undermines the straightforward application of gradient-dependent optimization methods.

The discussion in this chapter explores the various issues outlined above. An application within the optimal control context is provided as an illustration.

11.2 Problem Statement

There are two essential components within the study under consideration. The first is a conceptual model of a CTDS system which incorporates a set of parameters that are represented by an m-vector $\mathbf{p}$. From the discussion in Sect. 9.1.1, that conceptual model can be written as:

$$\dot{\mathbf{x}}(t) = \mathbf{f}(\mathbf{x}(t), t; \mathbf{p}) \tag{11.1}$$

where the dependence on the parameter vector $\mathbf{p}$ is explicitly shown. Associated with this model there necessarily is an initial condition assigned to the N-dimensional state vector $\mathbf{x}$, namely, $\mathbf{x}(t_0) = \mathbf{x}_0$ and, as well, an observation interval

© Springer Nature Switzerland AG 2019
L. G. Birta and G. Arbez, *Modelling and Simulation*, Simulation Foundations,
Methods and Applications, https://doi.org/10.1007/978-3-030-18869-6_11

$I_0 = [t_0, t_f]$. In general it may be that the parameter vector **p** maps onto some (possibly all) components of $\mathbf{x}_0$ and may also include the right boundary of the observation interval, t_f.

The optimization aspect of the study gives rise to the second component of interest; namely, a scalar, real-valued criterion function. The primary dependence of this function is certainly on the state vector $\mathbf{x}(t)$. However the dependence of $\mathbf{x}(t)$ on the parameter vector **p** is of primary interest hence this dependence is emphasized by writing the criterion function simply as $J(\mathbf{p})$. As outlined in Chap. 10, the objective of the study then is to find a value **p*** for **p** which yields an extreme value for J. (As previously indicated, our convention is to assume the requirement of minimization). Thus we seek to find **p*** (the minimizing argument of $J(\mathbf{p})$) such that:

$$J(\mathbf{p}*) \leq J(\mathbf{p}) \text{ for all } \mathbf{p} \in \Phi$$

In general, not all possible m-vectors are permitted candidates for **p*** and consequently the minimization of J could be restricted to a subset of m-dimensional space denoted by Φ. Such restrictions may be explicit; e.g., the first component, p_1, of **p** must be positive. Alternately, the restrictions may be implicitly defined via a collection of functional constraints; e.g., $\varphi_j(\mathbf{x}(t; \mathbf{p})) \geq 0$ for $j = 1, 2, \ldots, c_1$ and $\varphi_j(\mathbf{x}(t; \mathbf{p})) = 0$ for $j = c_1 + 1, c_1 + 2, \ldots, r$. Such a functional constraint would arise, for example, in the case of a manufacturing process where the tensile strength of a plastic material that is being produced is compromised if the rate of cooling at a particular phase of the process is excessive. In such a circumstance only those values of **p** that do not create the unacceptable cooling conditions would be allowed.

11.3 Some Representative Forms for the Criterion Function

The specific form of the criterion function J evolves from the nature of the problem to be solved. The only requirement is that $J(\mathbf{p})$ have a real scalar value for each admissible value of the m-vector **p**. Note however that because the parameter vector **p** is embedded in a conceptual model of some continuous time dynamic system, the evaluation of J, for any given **p**, requires the solution of a set of differential equations. This is, in principle, of no particular consequence for any optimization process; however it can have significant practical consequences in terms of computational overhead.

Some typical forms for the criterion function, J, in the CTDS domain are:

$$J = g(\mathbf{x}(t_f; \mathbf{p})) \qquad \text{(a)}$$

$$J = \sum_{j=1}^{s} g(\mathbf{x}(t_j; \mathbf{p})) \qquad \text{(b)}$$

$$J = \int_{t_0}^{t_f} g(\mathbf{x}(t; \mathbf{p})) dt \qquad \text{(c)}$$

In each of these cases g is some scalar function of the state vector, $\mathbf{x}$. An example where (a) would be an appropriate choice is provided by the bouncing ball problem that was considered earlier (see Sect. 2.2.6). Recall that the task is to find an initial release angle which results in the ball falling through a hole in the ice surface. The release angle represents the parameter (there is only one) and g could be selected to be the distance (absolute value) between the point where the ball strikes the surface and the location of the hole. The implicit assumption that the problem has a solution means that g has a minimum value of zero; i.e., the ball falls through the hole. A successful search for the release angle that minimizes g will therefore provide the solution to the problem.

A criterion function of the form shown in (b) could have relevance to the model refinement problem outlined in the preface to Part IV. The calibration process in question could, for example, be based on the manner in which blood sugar is absorbed following an injection of insulin. In this case, the s time points, $t_j, j = 1, 2, \ldots, s$ that are referenced could be the points in time where blood sugar measurements are taken from the patient and the function, g, could be the absolute value of the difference between the measured data from the patient and the value generated by some particular output variables of the model. Finding values for the set of model parameters that yield a minimum value for J would then correspond to the calibration process.

The criterion function form shown in (c) maps directly onto a classic control system design problem. The feedback controller for a continuous time dynamic system (e.g., an aircraft autopilot) typically has several parameters whose values need to be chosen in a way that, in some sense, optimizes system performance. A frequently used performance measure is "integral-square-error"; i.e., the integral of the square of the deviation between a desired system output and the output that actually occurs when the system has a prescribed input (typically a step function). Assuming that a conceptual model is available for the system and its controller, then the goal of finding best values for the controller parameters would be based on using the model in the minimization of a criterion function of the form shown in (c). In this case $g(t) = (y(t) - \hat{y}(t))^2$ where $y(t)$ is the model's output of interest (some function of the state vector, $\mathbf{x}(t)$) and $\hat{y}(t)$ is the desired value for $y(t)$.

11.4   Using Gradient Dependent Methods

As previously noted, the criterion function J depends on the state vector $\mathbf{x}(t)$ which is, in turn, dependent on the parameter vector $\mathbf{p}$ that is embedded in the conceptual model. This circumstance precludes an explicit evaluation of the gradient vector $\mathbf{J_p}$ via a straightforward differentiation operation. Therefore in order to use any of the gradient dependent optimization methods that are often very effective, an alternate gradient evaluation approach is required.

Recall that the kth component of the gradient vector $\mathbf{J_p}$, evaluated at the specific point $\mathbf{p} = \omega$, is the partial derivative of J with respect to p_k evaluated at $\mathbf{p} = \omega$. However, from first principles we have:

$$\left.\frac{\partial J}{\partial p_k}\right|_{\mathbf{p}=\omega} = \lim_{\varepsilon \to 0} \frac{J(\omega + \varepsilon \mathbf{e}^k) - J(\omega)}{\varepsilon}$$

where $\mathbf{e}^k$ is the kth column of the $m \times m$ identity matrix. The obvious numerical approximation to this formal definition is:

$$\left.\frac{\partial J}{\partial p_k}\right|_{\mathbf{p}=\omega} \approx \frac{J(\omega + \Delta \mathbf{e}^k) - J(\omega)}{\Delta}$$

where Δ is a suitably small positive scalar. In effect then we have a means for obtaining an approximation for the gradient of the criterion function when an optimization study is undertaken within the CTDS context.

Note however that the approximation for $\mathbf{J_p}(\omega)$ depends on m evaluations of J, each of which requires the solution of the underlying ode's of the CTDS conceptual model. (We assume here that the value of J at the reference point ω; i.e., $J(\omega)$, is already known). This can represent a considerable computational overhead in cases where the conceptual model has computationally intensive requirements.

Selecting the most appropriate value for the perturbation, Δ, requires careful consideration because "small" is a highly ambiguous notion. If, for example, Δ is "too small" then the result obtained can become hopelessly corrupted by numerical noise. Nevertheless, with proper care the approach can usually be sufficiently accurate to enable an effective implementation of most optimization procedures requiring gradient information.

We note nevertheless that one particular case that would merit special caution in this respect is the circumstance where a discontinuity is known to exist in the time trajectory of the conceptual model. As pointed out in the discussion in Sect. 9.4.3 dealing with such models has inherent numerical difficulties and the errors introduced could undermine the success of the gradient approximation outlined above.

11.5   An Application in Optimal Control

Typically an optimal control problem involves the determination of the time trajectory for one or more control inputs to a continuous time dynamic system in a manner that minimizes a prescribed performance function. This problem appears, at first glance, to be beyond the scope of our interest in his chapter because the determination of optimal time trajectories was never a part of the intended considerations.

There is however, a substantial body of literature relating to the solution of this generic problem available (see, for example [7, 9, 10, 12]) and among the important results that have emerged is the Pontriagin Minimum Principle [11]. This, in particular, provides a basis for transforming the optimal control problem into a boundary value problem which can then be reformulated as a function minimization problem of the type that coincides with the focus of Part IV of this book. In this section we illustrate this process with a straightforward example.

Our concern in this example is with the control of a first order irreversible exothermic chemical reaction carried out in a stirred tank reactor. Control of the process is achieved by injecting coolant through a valve into a cooling coil inserted into the reactor. The conceptual model is based on the characterizing the perturbations around a steady state condition. It is relatively simple but highly nonlinear. The model is given in Eq. (11.2):

$$x_1'(t) = -(1 + 2x_1(t)) + R(t) - S(t)$$
$$x_2'(t) = 1 - x_2(t) - R(t) \tag{11.2}$$

where:

$$R(t) = 0.5 + (x_2(t) + 0.5)e^{y(t)}$$
$$y(t) = \frac{25x_1(t)}{x_1(t) + 2}$$
$$S(t) = u(t)(x_1(t) + 0.25)$$

Here $x_1(t)$ and $x_2(t)$ represent deviations from steady-state temperature and concentration respectively and $u(t)$ is the control input. We take $x_1(t_0) = 0.05$ and $x_2(t_0) = 0$ and for convenience we assume that $t_0 = 0$. The objective is to rapidly return the reactor to steady-state conditions ($x_1 = x_2 = 0$) at time $t = t_f = 1$ while at the same time avoiding excessive usage of coolant. Choosing $u(t)$ to minimize the following performance function, reflects these objectives:

$$P = \int_0^1 \left(x_1^2(t) + x_2^2(t) + 0.1u^2(t) \right) dt$$

The application of the Pontriagin Minimum Principle gives rise to an auxiliary set of differential equations; namely,

$$v_1'(t) = v_1(t)(2 + u(t)) - Q(t)(v_1(t) - v_2(t)) - 2x_1(t)$$
$$v_2'(t) = v_2(t) - (v_1(t) - v_1(t))e^{y(t)} - 2x_2(t) \tag{11.3}$$

where:

$$Q(t) = \frac{50(x_2(t) + 0.5)e^{y(t)}}{(x_1(t) + 2)^2}$$

and

$$u(t) = 5v_1(t)(x_1(t) + 0.25)$$

The solution to Eqs. (11.2) and (11.3) which corresponds to $v_1(1) = v_2(1) = 0$ provides the necessary conditions for the optimality of $u(t)$.

The difficulty that arises here is that initial conditions are given for $x_1(t)$ and $x_2(t)$ (i.e., conditions at $t = 0$) while the conditions on $v_1(t)$ and $v_2(t)$ are specified at $t = 1$. In other words there is a need to solve a two point boundary value problem. Such problems have been extensively studied in the numerical mathematics literature and a variety of methods are available. One approach is to recast the problem as a criterion function minimization problem within the class considered in this chapter.

In this reformulation, the CTDS model of interest is the group of four equations given by Eqs. (11.2) and (11.3). We assume the set of *initial* conditions:

$$x_1(0) = 0.05, x_2(0) = 0, v_1(0) = p_1, v_2(0) = p_2$$

where p_1 and p_2 are parameters. The values we seek for p_1 and p_2 are those which yield a minimum value for the criterion function:

$$J(p_1, p_2) = v_1^2(1) + v_2^2(1)$$

Then, provided that the minimization process yields a minimum value of zero for J, the value of $u(t)$ which results will be the solution to the original optimal control problem.

The approach which is illustrated in this example has general applicability to a wide range of optimal control problems and is, at least in principle, equally applicable to boundary value problems in general.

11.6   Dealing with Computational Overhead

We begin this section by noting again that with simulation optimization studies in the CTDS context, each evaluation of the criterion function requires the solution of a set of ode's (i.e., the CTDS conceptual model). There are two consequences that are implicit in this requirement:

(a) Each evaluation of the criterion function introduces a computational overhead which can be considerable. This arises from the requirement by the ode solving process for repeated evaluations of the derivative function, $\mathbf{f}()$ (see Eq. (11.1)), as the solution moves across the observation interval.

(b) All computed values of the criterion function are corrupted by numerical "noise" because of inherent error involved in solving the ode's of the conceptual model. That error can generally be managed by appropriate specification for a "solution accuracy parameter" embedded in the solution method used for solving the underlying ode's. In effect then, the "shape" of the criterion function, as "observed" by the minimization procedure is likewise dependent on the value assigned to the accuracy parameter. Note also that shape variability implies variability of both the minimizing argument and the minimum value of the criterion function.

What then qualifies as a solution accuracy parameter for an ode solving method? There are two cases to consider. The first relates to a fixed step size approach and the second to a variable step size alternative. In the first case, solution accuracy can generally be managed by appropriate specification of the value for the fixed step size, h, that is used to advance the current solution point (t_n, x_n) to the next solution point (t_{n+1}, x_{n+1}) (where $t_{n+1} = t_n + h$). The implicit assumption here is that reducing the value of h improves solution accuracy. In this case therefore the solution accuracy parameter is the step size, h, used in the ode solving method.

An equivalent situation applies to variable step size methods. In that case however, step size varies as the solution unfolds across the observation interval. Management of the step size value is controlled by an embedded error control criterion that alters the step size as necessary in order to maintain a solution error estimate within specified bounds. The error control criterion has a user specified parameter that establishes the error bounds in question. The net result is that step size varies automatically over the course of the observation interval to maintain a specified error level (see discussion in Sect. 9.3).

The main point above is that increased solution accuracy of an ode solution method is typically associated with reduced step size for both fixed step size and variable step size methods. The discussion that follows focuses on the fixed step size case but parallel conclusions can be formulated for the variable step size case.

As noted earlier (see (a) above) the computational overhead in obtaining a value for the criterion function originates with the repeated evaluations of the derivative function, $\mathbf{f}()$ required by the ode solving procedure. A practical measure therefore of that computational overhead is provided by a count of the number of evaluations of $\mathbf{f}()$ that are carried out over the course of the observation interval. That number increases with the number of solution values that are generated over that interval. Recall however that the separation of adjacent solution values, as viewed along the time axis, is provided by the value of the step size, h. In other words, as h decreases the computational overhead increases. But we have indicated that reducing the value of h will generally yield a higher accuracy solution. Thus, not surprisingly, it follows that the computational overhead in evaluating the criterion function increases as the accuracy requirement placed on the solution of the underlying ode's is elevated; in other words, as the value of h used in solving the ode's is reduced. Furthermore it needs to be emphasized here that the solution to the minimization problem will require many (possibly hundreds) of criterion function evaluations.

The above discussion provides the basis for formulating a strategy (first proposed in [5]) for reducing the computational overhead in simulation optimization studies in the CTDS context. The underlying idea is to move into the region of the minimizing argument of the criterion function J, by solving a series of intermediate minimization problems each with a successively smaller value of the step size used in the ode solving process and each starting at the minimizing argument located by its predecessor. With a judicious selection of step size values, the total number of evaluations of the derivative function, $\mathbf{f}()$, will (hopefully) be less than the number that would be required with a direct approach; i.e., a single optimization using a predetermined "small" step size in the ode solving method being used. In other words a reduction in computational overhead in solving the minimization problem will be achieved.

We begin with three assumptions:

- For any particular value chosen for h (within some realistic bounds), the optimization procedure being used will succeed in delivering a least value for the criterion function (i.e., one that is consistent with the shape that corresponds to the chosen value of h). We denote the computation cost of that solution by $C(h)$.
- We assume that $C(h)$ is (within limits) a monotonically decreasing function of h.
- A "least value" of h has been determined which is considered to provide an acceptable level of confidence in the quality of the final result returned by the optimization procedure bring used. We denote that least value by H. In other words, the minimized value of J obtained when using H in the ode solution process can be regarded as the "problem solution" (namely J^*) along with the associated value of the parameter vector, $\mathbf{p}$, which we denote by $\mathbf{p}^*$.

The strategy is as follows:

1. Initialization: choose a value for the initial estimate ($\mathbf{p}_{int}$) of the minimizing argument of J; choose also a value for H, a value for Δ (a step size decrement) and a value for N (the number of refinement steps). Let $h = H + N\,\Delta$.
2. Carry out the minimization procedure starting from $\mathbf{p}_{init}$ using h as the step size in the ode solving process. Let $\mathbf{p}_{fin}$ be the minimizing argument returned by the minimization procedure and let J_{fin} be the corresponding criterion function value.
3. Let $h = h - \Delta$
4. If $h \geq$ H, let $\mathbf{p}_{init} = \mathbf{p}_{fin}$ and repeat from step 2.
5. Else $\mathbf{p}^* = \mathbf{p}_{fin}$ and $J^* = J_{fin}$.

As previously noted, the above outline of the strategy is restrictive inasmuch as it deals only with the fixed step size alternative in the ode solving process. A presentation of the more general variable step size case as well as several examples of the effectiveness of the approach, can be found in [5].

Implicit in the computationally oriented discussions thus far in this book, is the assumption that the computational task involved in simulation experiments is being carried out by a single processor. The distribution of the computational workload over multiple processors in order to reduce the time overhead in carrying out simulation experiments has, for the most part, remained an elusive objective. There is, however, one computational area where some progress in this regard has been achieved. This is in the context of the solution of ordinary differential equations, in other words the CTDS context. In particular investigations with "block predictor-corrector methods" have shown that potential gains using multiple processors are possible. Interested readers can find relevant studies in [1, 4, 6, 8, 13, 14].

11.7 Exercises and Projects

11.1 The bouncing ball project was introduced in Chap. 2 (Sect. 2.2.6). The goal is to find a release angle, θ_0, that results in the ball's trajectory entering the hole in the ice surface. This task can be formulated as a line search problem in the following way. Consider the criterion function $J(\theta_0) = (H - x_k)^2$ where H is the location of the hole and x_k is the ball's horizontal position when the kth collision with the ice surface occurs. J has a minimum value of zero when $x_k = H$, i.e. when the ball falls through the hole. Since the criterion function, J, depends only on a scalar parameter (namely, θ_0), the minimization problem is one-dimensional (hence a line search problem).

The solution requirements also stipulate that there must be at least one bounce before the ball passes through the hole, i.e. $k > 1$. This can be handled (somewhat inelegantly) by first finding a value for θ_0 for which the second and third bounces straddle the hole. This value can then be used as the starting point for the line search process.

Embed a syntactically compatible version of the program developed in Problem 10.2 into the MATLAB simulation model for the bouncing ball given in Sect. A4.9.2 of Annex 4 and make appropriate changes to the program so that it finds a suitable value for θ_0.

11.2 Develop an MATLAB simulation program to solve the optimal control problem that is outlined in Sect. 11.5. Use a syntactically compatible version of the extended univariate search program developed for Problem 10.3 to carry out the criterion function minimization. Use (2, 2) as the initial estimate of the minimizing argument of the criterion function.

References

1. Abou-Rabia O, Birta LG, Chen M (1989) A comparative evaluation of the BPC and PPC methods for the parallel solution of ode's. Trans Soc Comput Simul 6:265–290
2. Birta LG (1977) A parameter optimization module for CSSL based simulation software. Simulation 28:113–121
3. Birta LG (1984) Optimization in simulation studies. In: Simulation and model-based methodologies: an integrative view. NATO ASI series, vol 10. Springer, pp. 451–473
4. Birta LG, Abou-Rabia O, (1987). Parallel block predictor-corrector methods for ODE's. IEEE Trans. Comput C-36(3):299-311
5. Birta LG, Deo U (1986) A sequential refinement approach for parameter optimization in continuous dynamic models. Math Comput Simul 28:25–39
6. Birta LG, Yang M (1994) Some step size adjustment procedures for parallel ODE solvers. Trans Soc Comput Simul 12:303–324
7. Cassel KW, (2013) Variational methods with applications in science and engineering. Cambridge University Press
8. Franklin MA (1978) Parallel solution of ODE's. IEEE Trans Comput C-27:413–427
9. Geering HP (2007) Optimal control with engineering applications. Springer
10. Lewis FL, Vrabie D, Syrmos VL (2012) Optimal control, 3rd edn. Wiley, New York
11. Ross IM, (2015) A primer on Pontriagin's principle in optimal control, 2nd edn. Collegiate Publishers
12. Sethi PS, Thompson GL (2000) Optimal control theory: applications to management science and economics. Springer, New York
13. Shampine LF, Watts HA (1969) Block implicit one-step methods. Math-Comput 23:731–740
14. Worland PB (1976) Parallel methods for numerical solution of ordinary differential equations. IEEE Trans Comput C-25(10):1045–1048

Simulation Optimization in the DEDS Domain

12

12.1 Introduction

In this chapter we examine the simulation optimization problem within the DEDS context. The problem has special challenges and has, in recent years, received increasing attention in the modelling and simulation literature. However, an in-depth coverage of the topic is beyond the scope of this book. Nevertheless a brief overview of the "procedure landscape" is provided together with key references for readers interested in exploring this important topic in more detail. It should also be pointed out that almost all contemporary commercial simulation software environments for DEDS incorporate optimization procedures that can be invoked to carry out optimization studies [28].

We begin in Sect. 12.2 with a statement of the simulation optimization problem within the DEDS domain from which its distinctive features are clearly apparent. In Sect. 12.3 an outline of the range of solution approaches that have been discussed in the literature is presented. In Sect. 12.4 we present a case study that includes optimization results based on simple adaptations to the Nelder-Mead simplex method [25] as presented in Chap. 10, an optimization approach which was originally developed for classical (non-simulation) optimization.

12.2 Problem Statement

Consistent with our earlier discussions, the optimization objective relates to a criterion function $J(\mathbf{p})$ where the m dimensional parameter vector,[1] $\mathbf{p}$, is embedded within the specification of the model's behavior; i.e., within the conceptual model. The feature that is fundamental in the simulation optimization problem in the DEDS

---

[1]It is not uncommon to encounter terminology variations such as "decision variables" or "factors" in place of our usage of "parameters" which we use to maintain consistency with our earlier discussions.

© Springer Nature Switzerland AG 2019
L. G. Birta and G. Arbez, *Modelling and Simulation*, Simulation Foundations,
Methods and Applications, https://doi.org/10.1007/978-3-030-18869-6_12

domain emerges from the random effects that typically occur in various contexts within the SUI (and hence within the simulation model). These imply that $J(\mathbf{p})$ is itself a random variable.

In effect then, the value of the criterion function resulting from a single execution of the simulation model with any particular value of the parameter vector, $\mathbf{p}$, is not meaningful. Consequently the decision making required in any optimization procedure cannot be based on a single evaluation of the criterion function at solution estimates that emerge during the search process. Instead, the basis for this decision making needs to be based on an estimate of the mean value of the probability distribution of $J(\mathbf{p})$ where $\mathbf{p}$ is any particular value of the parameter vector that may be of interest. In other words, there is a requirement for replications based on proper management of the underlying random features within the model, consistent with specifications provided in the conceptual model. If we denote by $J_k(\mathbf{p})$ the criterion function value obtained on the kth replication in a series of n such runs, then a meaningful value for the criterion function at the parameter vector value, $\mathbf{p}$, is given by $\bar{J}(\mathbf{p}, n)$ where:

$$\bar{J}(\mathbf{p}, n) = E(J(\mathbf{p})) = \frac{1}{n} \sum_{k=1}^{n} J_k(\mathbf{p})$$

Here $\bar{J}(\mathbf{p}, n)$ represents an estimate of the mean value (namely a "point estimate" see Sect 7.2.1) of the probability distribution of the random variable $J(\mathbf{p})$ and $E[\ ]$ is the expectation operator.

The need to accommodate constraints is, furthermore, usually an essential feature of the optimization process. These can take a variety of forms. One of the simplest is a set of range constrains on some members of the parameter vector; e.g.

$$l_k \leq p_k \leq u_k \quad \text{for } k \, \varepsilon \, I_R$$

where l_k and u_k are constants and I_R is a set of indices in the interval $[1, m]$ that identify those components of the parameter vector, $\mathbf{p}$, that are subject to a range constraint. In addition, it may be that the constraints on the components of the parameter vector are of a functional nature possibly including interrelationships among the components themselves; e.g.,

$$\varphi_j(\mathbf{p}) \geq 0 \text{ for } j = 1, 2, \ldots, c_1 \text{ and } \varphi_j(\mathbf{p}) = 0 \text{ for } j = c_1 + 1, c_1 + 2, \ldots, c_2.$$

Constraints may also involve output variables of the simulation model; e.g., the vector $\mathbf{y}$. Because of the randomness inherent in these output variables such constraints must be written as $E[\varphi_j(\mathbf{y}, \mathbf{p})] \geq 0$ for $j = c_2 + 1, c_2 + 2, \ldots, r$.

Finally there is the fundamental matter of the nature of the search space; is it a subspace of R^m (the m-dimension space of real numbers) or are some, or possibly all, components of **p** constrained to have only integer values? Many of the currently available optimization methods have restrictions in this regard.

12.3   Overview of Search Strategies

There is a variety of ways for categorizing the various search methods that have been suggested in the literature for dealing with the simulation optimization problem within the DEDS context. The presentation below provides a relatively superficial perspective of this optimization landscape. It is by no means fully comprehensive but rather is intended only to provide some basic insights. Interested readers can explore details in the references that are provided.

Direct Search Methods

Two of the original direct search approaches have their origins in the non-linear programming literature; these are the pattern search of Hook and Jeeves [18, 29] and the simplex method of Nelder and Mead [25]. The key feature of both these methods is that they are "gradient independent" which makes them easily applicable to situations where gradient information about the criterion function cannot be easily/reliably obtained (e.g. the DEDS domain). An outline of the Nelder-Mead simplex method has been provided in Chap. 10. Extensions to better adapt the method to the simulation optimization context can be found in Barton and Ivy [3] and Chang [6]. A rudimentary adaptation of the method is used in handling the case study that follows in Sect. 12.4.

Emulation Strategies

This class of search heuristics emerges from an intent to emulate physical processes, either those observed in nature or those that are inherent in man-made processes. The dominant member from the first case are evolutionary/genetic algorithms. In these, the search for the best parameter vector begins with an initial "population" of points in the solution space. This population is then repeatedly updated using notions of "mutation" and "selection" and "recombination" borrowed from evolutionary biology [2].

Particle swarm optimization is an approach that seeks to emulate the manner in which groups of fish, birds, ants etc. work together in their quest for food. Within the optimization context this notion translates into a search procedure in which a collection of "particles" (points in the search space together with their respective criterion function values) move in an organized manner toward the minimizing argument. Each particle has a defined "neighborhood" which includes other particles. The underlying strategy is to ensure that each particle's movement in the search space is biased toward that neighbor whose criterion function value is best. The procedure is an example of a decentralized control and tends to avoid undesirable convergence to a local minimum [20, 27].

Another member of this class is simulated annealing which is a probabilistic method based on the emulation of a cooling strategy for a heated alloy that is designed to decrease the likelihood of defects in the resulting structure. With this analogy in mind, the underlying annealing equation is used to formulate new candidate points in the parameter search space. Each candidate point is evaluated relative to the current solution by a comparison of criterion function values. The new candidate point may or may not become the current solution even if it has a superior criterion function value. That decision is made on the basis of probabilities that favor one or other of the possible outcomes. These probabilities change as the procedure moves forward. The probability distributions at play here emulate the best strategy for lowering of temperature in the annealing process. This provides the procedure with the means to escape local optima [7, 8, 17].

Tabu Search

Tabu search is a metaheuristic procedure that is intended to operate in concert with a predefined local search method with the specific intent of ensuring the underlying method avoids premature termination at inappropriate points in solution space; e.g., a suboptimum solution such as a local minimum. This is achieved by the incorporation of "memory structures" that reflect past behaviour and strategies to exploit this information in a constructive manner. This includes identification of regions that ought not to be revisited (because they are "tabu"). The guidance functions that are provided are flexible and adaptable to specific features of the optimization problem being solved. The approach has its origins in dealing with combinatorial problems [14–16].

Response Surface Methodology (RSM)

This broad area is sometimes referred to as metamodeling inasmuch as the underlying intent is to formulate models (generally polynomial) of the surface being searched using mathematical and statistical techniques. Data values obtained from the simulation model provide the basis for these procedures. The methodology depends to a large extent on notions from the established area of study referred to as the 'Design of Experiments' (Sect. 7.5). The models that evolve are effectively surrogates for the criterion function whose minimization (maximization) is desired. An important development that has emerged for dealing with the optimization requirement when data values are corrupted by random effects are "gradient estimators" that can provide guidance to search procedures used in the RSM methodology [10, 19, 23, 24].

We conclude this brief overview with several references that provide comprehensive and insightful perspectives of the simulation optimization landscape within the DEDS domain. The inherent diversity that prevails is reflected in these presentations. The reference lists that are provided are extensive and provide ample opportunity for readers to explore this important subdomain of the modelling and simulation enterprise [1, 4, 5, 9, 11–13, 21, 22, 26].

12.4   A Case Study (ACME Manufacturing)

12.4.1   Problem Outline

The case study we consider in this section is concerned with a manufacturing process that produces widgets using an automated production line composed of a sequence of four workstations, each of which carries out a specific task.[2] That task can be carried out by any one of several identical processing machines that operate within the workstation. Each partially completed widget moves from one workstation to its downstream "neighbor" for further processing (except for last workstation in the sequence which produces the completed widget). There are, furthermore, temporary storage buffers between workstations to hold partially completed widgets that cannot enter the downstream workstation because all of its machines are busy. Note that it can occur that a machine in a workstation which has finished its task cannot transfer the partially completed widget to the buffer because the buffer is full (all available spaces are occupied). In that circumstance the part remains in the machine. In other words, that machine becomes idle until buffer space becomes available.

The specific configuration we consider is shown in Fig. 12.1. There are four workstations and three buffers areas. A large inventory of "blank parts" is available to feed the machines in the first workstation. The completed widgets are produced at the fourth workstation and are transferred to a large shipping area.

Apart from the possible failure of one of the machines, the production line operates 24 hours a day, 7 days a week. The company's income is determined by the number of widgets that are produced. That number is dependent on the number of machines in each of the workstations and the number of positions in each of the storage buffers (i.e., buffer size) that are used between workstations. There is however a cost associated with operating the machines in the workstations and maintaining the storage positions in the buffers. Consequently the company's profit from this production line corresponds to the income minus the operating cost of the machines and maintenance costs of the storage locations. In more formal terms, the profit, over a particular time interval, I, can be expressed as:

$$\text{Profit} = \Psi N - \left(C_m \sum_{i=0}^{3} m_i + C_s \sum_{i=0}^{2} s_i \right)$$

where:

N is the number of parts produced over the time interval I
Ψ is the income (dollars) per part produced (Ψ = \$200)
C_m is the cost (dollars) of operating a machine over the time interval I (for I = 30 days, C_m = \$25,000)

---

[2]This example is based on the presentation in Law [21], Sect. 12.5.1.

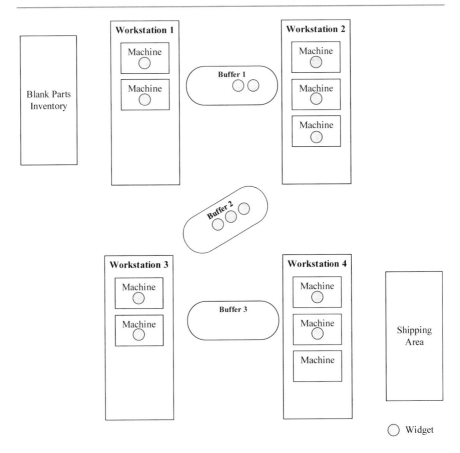

Fig. 12.1 Widget production line

C_s is the maintenance cost (dollars) of a buffer storage position over the time interval I (for $I = 30$ days, $C_s = \$1,000$)

m_j is the number of machines in workstation j, with $j = 1, 2, 3, 4$

s_j is the number of storage positions in buffer j, with $j = 1, 2, 3$.

Note that the variable N is, in fact, a random variable because of the randomness associated with the processing time of parts at the various workstations. Processing times of the machines at the four workstations are exponentially distributed with mean values (in hours) provided in Table 12.1.

The optimization problem we consider is the maximization of profit over a 30 day period. The parameters that are available for adjustment are the number of machines in each of the 4 workstations (i.e., m_j, $j = 1, 2, 3, 4$) and the number of storage positions in each of the buffers (i.e., s_j, $j = 1, 2, 3$). There are, however, size constraints that need to be respected; namely, $1 \leq m_j \leq 3$ and $1 \leq s_j \leq 10$, for each j.

Table 12.1 Mean
processing times

Workstation	Mean processing time (hours)
1	0.33333
2	0.5
3	0.2
4	0.25

In keeping with our standard format, we introduce the parameter vector $\mathbf{p}$ and let $p_j = m_j$ for $j = 1, 2, 3, 4$ and $p_{j+4} = s_j$ for $j = 1, 2, 3$. The criterion function to be *minimized* is:

$$\overline{J}(\mathbf{p}) = -E\left[\Psi N - \left(C_m \sum_{j=1}^{4} p_j + C_s \sum_{j=1}^{3} p_{j+4}\right)\right] \qquad (12.1)$$

where E[ ] is the expectation operator and n has the value 20.

12.4.2   Experimentation

In this section we outline two experimental approaches undertaken to "solve" the simulation optimization problem presented above. In the first an optimization procedure was used (an adaption of the Nelder-Mead simplex method as outlined in Sect. 10.2.2). In the second, a search over all possible combination of the 7 allowable parameter values (i.e., values assigned to the components of the parameter vector $\mathbf{p}$) was carried out; i.e., an exhaustive search over the set of 81,000 ($3^4 \times 10^3$) parameter vectors in the set we designate as Θ_E.

We begin by indicating two particular features governing the determination of a value for the criterion function (Eq. 12.1) for any particular value of the parameter vector, $\mathbf{p}$:

(i) By way of obtaining a meaningful (steady state) value for $\overline{J}(\mathbf{p}, n)$, a warmup period of 20 days was added to the 30 day (720 hours) length of each simulation run.
(ii) To accommodate the expectation operator, the average value following 20 replications was consistently used.

12.4.2.1   Optimization
As previously pointed out, the development of the Nelder-Mead Simplex method was originally formulated for classical optimization studies involving real valued analytic functions. Many extensions/refinements of the procedure to deal with the special requirements of optimization in the DEDS domain have however, been presented in the literature; e.g., [3, 6, 11].

Our application of the Nelder-Mead method for the optimization task incorporated two basic modifications to accommodate requirements of the problem under consideration. Recall (from Chap. 10) that, for the case of an m-dimensional parameter vector, the method progresses from iteration to iteration by moving a simplex (an $(m + 1)$ dimensional structure in m-space) through the parameter search space. The movement of the simplex is based on assessment of the criterion function values at the $(m + 1)$ vertices of the simplex. In general one new vertex of the simplex is replaced on each iteration according a set of underlying rules which endeavor to ensure that the simplex progresses in a direction that will ultimately include the optimizing parameter vector as one of the simplex vertices. However, the coordinates of the simplex vertices are allowed to assume arbitrary real values which is clearly incompatible with the current problem in which only integer values are meaningful.

In order to deal with integer valued constraints on the parameter values, the following simple adaptation was incorporated into the standard procedure. Note that the intent of the approach is not to alter the parameter vector value $\mathbf{p}$ generated by the simplex procedure but rather simply to obtain a criterion function value that is consistent with the integer valued constraints.

Let $\mathbf{p}^{\mathrm{I}}$ be the m-vector obtained by rounding each coordinate of $\mathbf{p}$ to the nearest integer value. Evaluate $\bar{J}(\mathbf{p}^{\mathrm{I}}, n)$ and set $\bar{J}(\mathbf{p}, n) = \bar{J}(\mathbf{p}^{\mathrm{I}}, n)$.

Some safeguards however are necessary because with this approach there is a possibility that the simplex will degenerate; i.e., the $(m + 1)$ vertices may no longer be distinct points in m-space. If this circumstance is detected, then the simplex procedure is re-initialization.

Accommodation of the range constraints that are specified on the components of the parameter vector is handled in a similar way. One change is however required; namely, in the formulation of $\mathbf{p}^{\mathrm{I}}$, a coordinate that violates its prescribed range constraint is replaced by the constraint boundary. Consider, for example, p_1 (the first coordinate of $\mathbf{p}$) whose value must not exceed 3. Suppose that the value assigned to p_1 by the procedure is 3.7, then the first coordinate of $\mathbf{p}^{\mathrm{I}}$ is instead assigned the value 3.

Numerous applications of the optimization procedure were carried out and a sample of the results obtained is provided in Table 12.2. Each row provides pertinent details of one such application of the procedure. The rank value that is shown has been obtained from a ranking of all criterion function values obtained in the exhaustive search. The highest "Final $(-J)$" value in the Table (hence the one with lowest rank) is \$589,140 which is the best value as obtained in the exhaustive search as outlined in Sect. 12.4.2.2. The value in the feKnt column is the number of simulation runs required to locate the "Final $(-J)$" value starting from the initial parameter vector indicated.

Table 12.2 A summary of some optimization experiments

Initial **p** vector	Initial $(-J)$ ($)	Rank	Final **p** vector	Final $(-J)$ ($)	Rank	# parts	feKnt
2 2 2 2 7 8 5	353,090	13,823	3 3 2 2 8 10 4	589,140	1	4306	560
2 2 2 2 6 6 6	353,550	13,711	3 3 2 2 7 8 5	589,020	3	4295	560
2 2 2 2 7 7 7	352,080	14,070	3 3 2 2 8 8 6	588,850	11	4304	560
2 2 2 2 4 4 4	352,620	13,927	3 3 2 2 7 7 5	588,530	29	4288	760
2 3 1 2 5 5 5	455,320	7,938	3 3 2 2 7 7 7	588,370	36	4297	480
2 2 2 2 5 5 5	353,720	13,674	3 3 2 2 6 7 4	588,040	61	4275	560
2 2 2 2 2 2 2	341,700	16,306	3 3 2 2 5 5 3	582,230	389	4226	389
2 2 2 2 3 3 3	348,850	14,752	3 3 2 2 6 4 4	581,340	412	4227	320
1 2 3 1 3 3 3	220,930	43,854	3 3 3 2 6 5 5	566,810	921	4289	1040
1 3 3 1 1 8 9	200,000	48,800	3 3 3 2 4 7 10	557,550	940	4268	940

It is reasonable to conclude that the apparent "erratic behaviour" that appears to be reflected in the results shown in Table 12.2 is due, in part, to what is most certainly an extremely irregular response surface for the criterion function. Contributing as well is the rather rudimentary manner that the Nelder-Mead procedure was modified to accommodate the integer values of the search space and the range constraints specified for the parameter values.

12.4.2.2   Exhaustive Search
We begin by noting that the "best" result uncovered by the exhaustive search is a maximum profit of $589,140 which results from the parameter vector:

$$\mathbf{p}^* = (3, 3, 2, 2, 8, 10, 4).$$

Various interesting insights about the response surface, $\bar{J}(\mathbf{p}, n)$, can be obtained from an analysis of the results obtained from the exhaustive search. For example, there exist regions of the search space where values of the parameter vector yield profit values that are relatively close to the maximum value. Of the 81,000 possible choices for $\mathbf{p} \in \Theta_E$, 4464 (0.55%) of these yield a profit that is within 10% of the maximum. In view of the inherent statistical uncertainties, all of these cases could reasonably be regarded as having equivalent merit.

In Fig. 12.2 this relative insensitivity is examined in more detail. Each increment along the horizontal axis corresponds to a 1% displacement interval from the maximum profit; e.g., the interval between 3 and 4 refers to a profit displacement that is between 3% and 4% from the maximum profit. The vertical height of that segment (533) provides the number of parameter vectors in Θ_E that yield a profit that falls in that range (see Table 12.3).

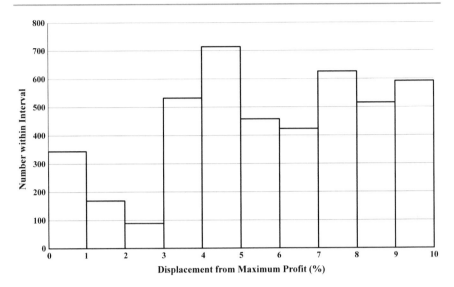

Fig. 12.2 Selected results from the exhaustive search

Table 12.3 Elaboration of Fig. 12.2

Interval (% of max profit)	Number from Θ_E within interval	Percent relative to size of Θ_E (i.e., 81,000) (%)
[0, 1]	344	0.425
[1, 2]	169	0.209
[2, 3]	89	0.110
[3, 4]	533	0.658
[4, 5]	714	0.881
[5, 6]	458	0.565
[6, 7]	423	0.522
[7, 8]	626	0.773
[8, 9]	516	0.637
[9, 10]	592	0.731

References

1. Amaran S, Sahinidis NV, Sharda B, Bury SJ (2014) Simulation optimization: a review of algorithms and applications. 4OR-Q J Oper Res 12:301–333
2. Back T, Schwefel H-P (1993) An overview of evolutionary algorithms for parameter optimization. Evol Comput 1:1–23
3. Barton RR, Ivy JS Jr (1996) Nelder-Mead simplex modifications for simulation optimization. Manag Sci 42:954–973
4. Bhatnager S, Kowshik HJ (2005) A discrete parameter stochastic approximation algorithm for simulation optimization. Simulation 81:757–772

5. Buchholz P (2009) Optimization of stochastic discrete event models and algorithms for optimization logistics. In: Dagstuhl seminar proceedings, vol 09261, Dagstuhl, Germany
6. Chang K-H (2012) Stochastic Nelder-Mead simplex method—a new globally convergent direct search method for simulation optimization. Eur J Oper Res 220:684–694
7. Delahaye D, Chaimatanan S. Mongeau M (2018) Simulated annealing: from basics to applications. In: Gendreau M, Potvin J-Y (eds) Handbook of metaheuristcs, 3rd edn. Springer, pp 1–35
8. Eglese RW (1990) Simulated annealing: a tool for operational research. Eur J Oper Res 46:271–281
9. Fu MC (2002) Optimization for simulation: theory versus practice. INFORMS J Comput 14:192–215
10. Fu MC (2006) Stochastic gradient estimation. In: Henderson S Nelson B (eds) Handbook of OR/MS: simulation. Elsevier
11. Fu MC, Glower FW, April J, (2005) Simulation optimization: a review, new developments, and applications. In: Proceedings of 2005 winter simulation conference, pp 83–95
12. Fu MC (ed) (2015) Handbook of simulation optimization. Springer
13. Fu MC, Henderson SG (2017) History of seeking better solutions, AKA simulation optimization. In: Proceedings of 2017 winter simulation conference, pp 131–157
14. Gendreau M, Potvin J-Y (eds) (2018) Tabu search. In: Handbook of metaheuristics, 3rd edn, Springer, pp 37–55
15. Glover F (1989) Tabu search—Part 1. ORSA J Comput 1:190–206
16. Glover F (1990) Tabu search—Part 2. ORSA J Comput 2:4–32
17. Henderson D, Jacobson SH, Johnson AW (2006) The theory and practice of simulated annealing. In: Grover F, Kochenberger G (eds) Handbook of metaheuristics
18. Hook R, Jeeves TA (1961) 'Direct Search' solution of numerical and statistical problems. J ACM 8:212–229
19. Joshi S, Sherali HD, Tew JD (1998) An enhanced response surface methodology (RSM) algorithm using gradient deflection and second-order search strategies. Comput Oper Res 25:531–541
20. Kennedy J, Eberhart R (1995). Particle swarm optimization. In: Proceedings of the IEEE international conference on neural networks, pp 1942–1948
21. Law AM (2015) Simulation modeling and analysis, 5th edn. McGraw Hill, New York
22. Law M, McComas MG (2002) Simulation-based optimization. In: Proceedings of 2002 winter simulation conference, pp 41–44
23. McAllister CD, Altuntas B, Frank M, Potoradi J (2001) Implementation of response surface methodology using variance reduction techniques in semiconductor manufacturing. In: Proceedings of 2001 winter simulation conference, pp 1225–1230
24. Neddermeijer HG, van Oortmarssen GJ, Piersma N, Dekker R (2000) A framework for response surface methodology for simulation optimization models. In: Proceedings of 2000 winter simulation conference, pp 129–136
25. Nelder J, Mead R (1965) A simplex method for function minimization. Comput J 7:308–313
26. Olafason S, Kim J (2002) Simulation optimization. In: Proceedings of 2002 winter simulation conference, pp 79–84
27. Poli R, Kennedy J, Blackwell T (2007) Particle swarm optimization: an overview. Swarm Intell 1:33–57
28. Swain JJ (2015) Simulation software survey: simulated worlds. OR/MS Today 42:36–49
29. Torczon VJ (1997) On the convergence of pattern search algorithms. SIAM J Optim 7:1–25

Annex 1: ABCmod Applications in M&S Projects

This annex presents seven modelling and simulation (M&S) project examples that are referenced at various points throughout the book. For each example, an ABCmod conceptual model is presented which includes project documentation. That documentation is organized into following sections.

1. Problem Description: Provides a Problem Statement that describes the problem to be solved and also provides the SUI (System Under Investigation) Details.
2. Project Goal and Achievement: This section includes a clear statement of the project goal and how it is to be achieved.
3. Conceptual Model: This section has two subsections.

   a. The first subsection provides the high-level conceptual model which captures the features of the ABCmod conceptual model at a level of detail that is adequate for both communication and comprehension.
   b. The second subsection provides the detailed conceptual model which provides the detail needed for simulation model development.

The first two examples are two versions of the Kojo's Kitchen M&S project as introduced in Sect. 4.4. Section 5.3 transforms Version 1 into a Three-Phase world view simulation model using the Java programming language, the ABSmod/J simulation model for Version 1 is presented in Sect. 6.4.1 and output analysis for this project is discussed in Sect. 7.2.3. Version 2 of Kojo's Kitchen is used to illustrate output analysis of multiple alternatives and is discussed in Sect. 7.4.2. A slightly modified Version 2 is also used to demonstrate the application of the DoE methodology in Sect. 7.5.2.1.

The next three examples are three versions of the Port project. Each of these projects illustrates various features of ABCmod as presented in Sect. 4.3. An approach for transforming the ABCmod conceptual model for Port Version 1 into a GPSS Process-Oriented world view simulation model is presented in Sect. 5.4. It also uses Version 2 to demonstrate how to integrate interruptions into a GPSS simulation model. Output analysis for Version 1 is presented in Sect. 7.3.3. This example is also used to illustrate the use of common random numbers in Sect. 7.4.1. Parts of the ABSmod/J simulation model for Port Version 2 are presented in Sect. 6.4.2.

The Conveyor project is introduced in Sect. 4.4. Parts of the ABSmod/J simulation model for the Conveyor Project are presented in Sect. 6.4.3. That discussion illustrates

© Springer Nature Switzerland AG 2019
L. G. Birta and G. Arbez, *Modelling and Simulation*, Simulation Foundations,
Methods and Applications, https://doi.org/10.1007/978-3-030-18869-6

the use of entity categories with *scope* = Many[*N*]. The last example, 'ACME Man-ufacturing' is used to illustrate the value of the DoE methodology in supporting heuristic experimentation in Sects. 7.5.2.2 and 7.5.2.3, and is discussed in Chap. 12 in the context of carrying out an optimization study within the DEDS M&S context.

A1.1 Kojo's Kitchen (Version 1)

A1.1.1 Problem Description

A1.1.1.1 Problem Statement
Kojo's Kitchen is one of the fast-food outlets in the food court of a shopping mall. Kojo's manager is very happy with business, but has been receiving complaints from customers about long waiting times. He is interested in exploring staffing options that will reduce the number of such complaints.

A1.1.1.2 SUI Details
Customers line up at Kojo's counter where, currently, two employees work throughout the day preparing two types of product, namely, sandwiches and sushi. A schematic view is shown in Fig. A1.1. Most customers purchase one type of product. Kojo's Kitchen is open between 10h00 and 21h00 every day. Two rush hour periods occur during the business day, one between 11h30 and 13h30, and the other between 17h00 and 19h00. At 21h00, the line is closed and any customers remaining in the line are served.

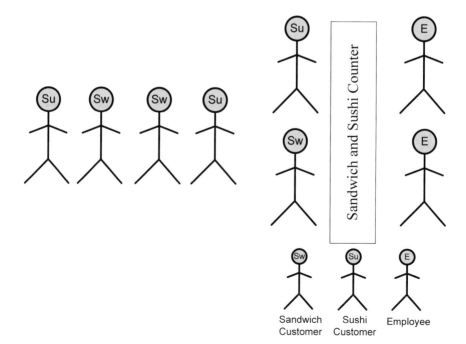

Fig. A1.1 Kojo's Kitchen fast-food outlet

A1.1.2 Project Goal and Achievement

The project goal is to evaluate how the adding of a third employee behind the counter during busy times (between 11h30 and 13h30 and between 17h00 and 19h00) reduces the percentage of customers waiting in line longer than 5 minutes. We compare two cases as specified by the parameter introduced for this purpose. Its assigned values are given below:

RG.Counter.addEmp: Set to 0 for base case and 1 for alternate case; in other words, the value of the parameter gives the number of employees added during the busy periods.

Output

propLongWait: The proportion of customers who wait more than 5 minutes in line before being served at the counter.

Study: Bounded Horizon Study.

Observation Interval (time units are minutes): From $t = 0$ (10h00) to the time after $t = 660$ (21h00) when all customers have left the delicatessen.

A1.1.3 Conceptual Model

A1.1.3.1 High-Level Conceptual Model

Simplification

We consider only two types of customer; one type purchases only sandwiches and the other type purchases only sushi products.

Structural View

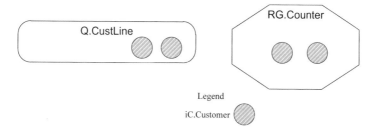

Entity Categories:

- CustLine: An entity category with *role* = Queue and *scope* = Unary that represents the line of customers in front of the counter.
- Counter: An entity category with *role* = Resource Group and *scope* = Unary that represents the counter where customers are being served.
- Customer: An entity category with *role* = Consumer and *scope* = Transient that represents the collection of customers in the model. Customer preference (sandwich/suishi) is indicated by using the Customer attribute *uType*.

Behavioural View

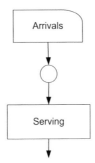

Actions:

- Arrivals: This scheduled action generates the input entity stream of customers.
- StaffChange: The change in the number of employees behind the counter. This Action is not shown in the behavioural diagram since it is not part of the customer life cycle.

Activities:

- Serving: This activity represents the serving of a customer.

Input

Exogenous Input (Entity Stream)			
Variable Name	**Description**	**Domain Sequence**	**Range Sequence**
uC	This input entity stream represents the arriving customers	RVP.DuC()	1 customer arrives at each arrival.
Endogenous Input (independent)			
Variable Name	**Description**	**Domain Sequence**	**Range Sequence**
RG.Counter.uNumEmp	This input variable represents the number of employees at the counter serving customers.	< 0, *(10h00)* 90, *(11h30 start of busy period)* 210, *(13h30 end of busy period)* 420, *(17h00, start of busy period)* 540> *(19h00 end of busy period)*	< 2, 2+RG.Counter. addEmp, 2, 2+RG.Counter. addEmp, 2 >
Endogenous Input (Semi-independent)			
Variable Name	**Description**	**Value(s)**	
uSrvTm	Service time of a customer.	RVP.uSrvTm()	
iC.Customer.uType	Defines the type of an arriving customer.	RVP.uCustomerType()	

A1.1.3.2 Detailed Conceptual Model

Structural Components

Parameters		
Name	**Description**	**Value**
RG.Counter.addEmp	Number of additional employees serving at the counter during the busy periods.	0 (for the base case) or 1 (for the alternate scenario).

Consumer Transient: Customer	
Customer entities represent customers who purchase items at Kojo's Kitchen.	
Attributes	**Description**
startWaitTime	Time stamp that holds the value of time when the customer enters the customer line.
uType	The assigned value indicates the type of customer. Two values are possible: 'W' for a sandwich customer or 'U' for a sushi customer.

Resource/Group Unary: Counter	
This resource group represents the counter where customers receive service.	
Attributes	**Description**
list	The group of the Customer entities that are being served.
n	Number of entities in list (necessarily n ≤ uNumEmp).
uNumEmp	Input variable that defines the number of employees at the counter available to service customers.
addEmp	Parameter that defines the number of additional employees serving at the counter during the busy periods.

QueueUnary: CustLine	
The line of customers waiting for service at the counter.	
Attributes	**Description**
list	A FIFO list of the customers entities waiting in line for service.
n	The number of Customer entities in list.

Behavioural Components

Initialisation

Action: Initialise	
TimeSequence	< 0 >
Event SCS	RG.Counter.n ← 0 Q.CustLine.n ← 0 SSOV.numServed ← 0 SSOV.numLongWait ← 0

Output

OUTPUTS	
Simple Scalar Output Variables (SSOV's)	
Name	**Description**
numServed	Number of customers served.
numLongWait	Number of customers that waited in line longer than 5 minutes.
propLongWait	numLongWait/numServed.

Input Constructs

Random Variate Procedures		
Name	**Description**	**Data Model**
DuC()	Provides the values of the arrival times of customers. No arrivals occur after closing (i.e. for t ≥ 660).	Exponential(MEAN) where MEAN varies over time as shown below: 	**Period (min)**	**MEAN (min)**
---	---
0 to 90	10
90 to 210	1.75
210 to 420	9
420 to 540	2.25
540 to 660	6	
uCustomerType()	Provides the type of the arriving customer. Returns either 'W' (sandwich customer) or 'U' (sushi customer).	Proportion of sandwich customers: PROPW=65% Proportion of sushi customers: PROPU=35%
uSrvTm(type)	Provides a value for the service time of a customer according to the value of type.	UNIFORM(MIN, MAX) | type	MIN	MAX
W	3 min	5 min
U	5 min	8 min	

Action: Arrivals	
The Input Entity Stream of arriving customers.	
TimeSequence	RVP.DuC()
Event SCS	iC.Customer ← SP.Derive(Customer) iC.Customer.uType ← RVP.CustomerType() iC.Customer.startWaitTime ← t SP.InsertQue(Q.CustLine, iC.Customer)

Action: StaffChange	
Manages the value of the input variable RG.Counter.uNumEmp, that is, schedules additional employee during busy times for the alternate case.	
TimeSequence	<0, 90, 210, 420, 540 >

Event SCS	IF(t=0) THEN
	      RG.Counter.uNumEmp ← 2
	ELSE IF(t=90) THEN
	      RG.Counter.uNumEmp +← RG.Counter.addEmp
	ELSE IF(t=210) THEN
	      RG.Counter.uNumEmp ← 2
	ELSE IF(t=420) THEN
	      RG.Counter.uNumEmp +← RG.Counter.addEmp
	ELSE IF(t=540) THEN
	      RG.Counter.uNumEmp ← 2
	ENDIF

Behavioural Constructs

Activity: Serving	
Service for a customer.	
Precondition	(RG.Counter.n < RG.Counter.uNumEmp) AND (Q.CustLine.n ≠ 0)

Event SCS	iC.Customer ← SP.RemoveQue(Q.CustLine)
	IF(t – iC.Customer.StartWaitTime > 5)
	THEN
	    SSOV.numLongWait +← 1
	ENDIF
	SSOV.numServed +← 1
	SSOV.propLongWait ← SSOV.numLongWait/SSOV.numServed
	SP.InsertGroup(RG.Counter, iC.Customer)

Duration	RVP.uSrvTm(iC.Customer.uType)
Event SCS	SP.RemoveGroup(RG.Counter, iC.Customer) SP.Leave(iC.Customer)

A1.2 Kojo's Kitchen (Version 2)

A1.2.1 Problem Description

A1.2.1.1 Problem Statement

Kojo's Kitchen is one of the fast-food outlets in the food court of a shopping mall. Kojo's manager is very happy with business, but has been receiving complaints from customers about long waiting times. He is interested in exploring staffing options that will reduce the number of such complaints.

A1.2.1.2 SUI Details

Customers line up at Kojo's counter where, currently, two employees work throughout the day preparing two types of product, namely, sandwiches and sushi. A schematic view is shown in Fig. A1.2. Most customers purchase one type of product. Kojo's Kitchen is open between 10h00 and 21h00 every day. Two rush hour periods occur during the business day, one between 11h30 and 13h30, and the other between 17h00 and 19h00. At 21h00, the line is closed and any customers remaining in the line are served.

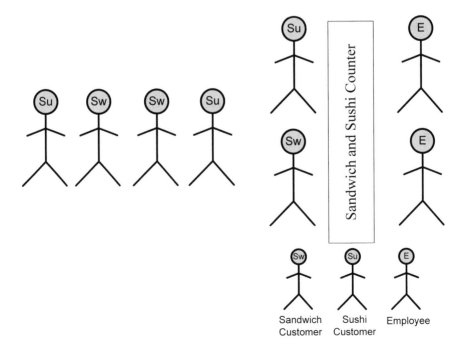

Fig. A1.2 Kojo's Kitchen fast-food outlet

A1.2.2 Project Goal and Achievement

The project goal is to evaluate how each of three different staffing options impacts the percentage of customers who wait in line longer than 5 minutes for service over the course of a business day. We compare three alternate cases to the base case. In each of these alternate cases, a third employee is added during the two busy periods. The option of reducing the number of employees during the non-busy periods is also investigated. A parameter vector of length 5 (*RG.Counter.schedule*) is introduced for this purpose. Its specification is provided below.

RG.Counter.schedule: Five values are assigned to this parameter vector of length 5. Each provides a specification for the number of employees in each of the following five work shifts: 10h00 to 11h30, 11h30 to 13h30, 13h30 to 17h00, 17h00 to 19h00 and 19h00 to 21h00, respectively. These values are shown in Table A1.1. The last column of Table A1.1 provides the number of person-hours that the schedule represents which provides the employee cost of that schedule. Thus Case 2 increases the schedule cost (relative to the base case), Case 3 reduces the schedule cost while Case 4 maintains about the same schedule cost.

Table A1.1 Employee schedules to be evaluated

Case	Parameter RG.Counter.schedule	Person-hours
Case 1 (base case)	<2, 2, 2, 2, 2>	22
Case 2	<2, 3, 2, 3, 2>	26
Case 3	<1, 3, 1, 3, 1>	19
Case 4	<1, 3, 2, 3, 1>	22.5

Output

propLongWait: The proportion of customers who wait in line more than 5 minutes in line before being served at the counter.

Study: Bounded Horizon Study.

Observation Interval (time units are minutes): From $t = 0$ (10h00) to the time after $t = 660$ (21h00) when all customers have left the delicatessen.

A1.2.3 Conceptual Model

A1.2.3.1 High-Level Conceptual Model

Simplifications

We consider only two types of customer: one type purchases only sandwiches and the other type purchases only sushi products.

Structural View

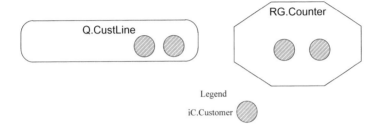

Entity Categories:

- CustLine: An entity category with *role* = Queue and *scope* = Unary that represents the line of customers in front of the counter.
- Counter: An entity category with *role* = Resource Group and *scope* = Unary that represents the counter where customers are being served.
- Customer: An entity category with *role* = Consumer and *scope* = Transient that represents the collection of customers in the model. Customer preference (sandwich/suishi) is indicated by using the Customer attribute *uType*.

Behavioural View

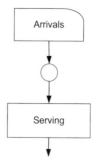

Actions:

- **Arrivals**: This scheduled action generates the input entity stream of customers.
- **StaffChange**: This scheduled action carries out the changes in the number of employees behind the counter. It is not shown in the behavioural diagram because it is not part of the customer life cycle.

Activities:

- **Serving**: This activity represents the serving of a customer.

Input

Exogenous Input (Entity Stream)			
Variable Name	**Description**	**Domain Sequence**	**Range Sequence**
uC	This input entity stream represents the arriving customers	RVP.DuC()	1 customer arrives at each arrival.
Endogenous Input (independent)			
Variable Name	**Description**	**Domain Sequence**	**Range Sequence**
RG.Counter.uNumEmp	This input variable represents the number of employees at the counter serving customers.	< 0, *(10h00)* 90, *(11h30 start of busy period)* 210, *(13h30 end of busy period)* 420, *(17h00, start of busy period)* 540> *(19h00 end of busy period)*	RG.Counter.schedule
Endogenous Input (Semi-independent)			
Variable Name	**Description**	**Value(s)**	
uSrvTm	Service time of a customer.	RVP.uSrvTm()	
iC.Customer.uType	Defines the type of an arriving customer.	RVP.uCustomerType()	

A1.2.3.2 Detailed Conceptual Model

Structural Components

Parameters		
Name	**Description**	**Value**
RG.Counter. schedule	Each value of this parameter is a vector with 5 integer values. The values in the vector represent an employee schedule for a day, in other words the number of employees during each of the following time intervals : 10h00 to 11h30 , 11h30 to 13h30 , 13h30 to 17h00 , 17h00 to 19h00, 19h00 to 21h00. In effect each vector provides the range sequence that corresponds to the domain sequence <0, 90, 210, 420, 540>. We use constants S1, S2, S3, S4 and S5 to index the members of this vector.	One of the four following vectors: $< 2, 2, 2, 2, 2 >$ $< 2, 3, 2, 3, 2 >$ $< 1, 3, 1, 3, 1 >$ $< 1, 3, 2, 3, 1 >$

Consumer Transient: Customer	
Customer entities represent customers who purchase items at Kojo's Kitchen.	
Attributes	**Description**
startWaitTime	Time stamp that holds the value of time when the customer enters the customer line.
uType	The assigned value indicates the type of customer. Two values are possible: 'W' for a sandwich customer or 'U' for a sushi customer.

Resource Group Unary: Counter	
This resource group represents the counter where customers receive service.	
Attributes	**Description**
list	The group of the Customer entities that are being served.
n	Number of entities in list (necessarily $n \leq$ RG.Counter.empNum).
uNumEmp	Input variable that defines the number of employees at the counter available to service customers.
schedule	Parameter that defines the employee schedule.

Queue Unary: CustLine	
The line of customers waiting for service at the counter.	
Attributes	**Description**
list	A FIFO list of the customers entities waiting in line for service.
n	The number of Customer entities in list.

Behavioural Components

Initialisation

Action: Initialise	
TimeSequence	< 0 >
Event SCS	RG.Counter.n ← 0
	Q.CustLine.n ← 0
	SSOV.numServed ← 0
	SSOV.numLongWait ← 0

Output

OUTPUTS	
Simple Scalar Output Variables (SSOV's)	
Name	**Description**
numServed	Number of customers served.
numLongWait	Number of customers that waited in line longer than 5 minutes.
propLongWait	numLongWait/numServed.

Input Constructs

Random Variate Procedures		
Name	**Description**	**Data Model**
DuC()	Provides the values of the arrival times of customers. No arrivals occur after closing (i.e. for t ≥ 660).	Exponential(MEAN) where MEAN varies over time as shown below:

Period (min)	MEAN (min)
0 to 90	10
90 to 210	1.75
210 to 420	9
420 to 540	2.25
540 to 660	6

uCustomerType()	Provides the type of the arriving customer. Returns either 'W' (sandwich customer) or 'U' (sushi customer).	Proportion of sandwich customers: PROPW=65% Proportion of sushi customers: PROPU=35%
uSrvTm(type)	Provides a value for the service time of a customer according to the value of type.	UNIFORM(MIN, MAX)

type	MIN	MAX
W	3 min	5 min
U	5 min	8 min

Action: Arrivals	
The Input Entity Stream of arriving customers.	
TimeSequence	RVP.DuC()
Event SCS	iC.Customer ← SP.Derive(Customer)
	iC.Customer.uType ← RVP.CustomerType()
	iC.Customer.startWaitTime ← t
	SP.InsertQue(Q.CustLine, iC.Customer)

Action: StaffChange	
Manages the value of the input variable RG.Counter.uNumEmp based on the values in the parameter RG.Counter.schedule.	
TimeSequence	<0, 90, 210, 420, 540 >
Event SCS	IF(t=0) THEN
	RG.Counter.uNumEmp ← RG.Counter.schedule[S1]
	ELSE IF(t=90) THEN
	RG.Counter.uNumEmp ← RG.Counter.schedule[S2]
	ELSE IF(t=210) THEN
	RG.Counter.uNumEmp ← RG.Counter.schedule[S3]
	ELSE IF(t=420) THEN
	RG.Counter.uNumEmp ← RG.Counter.schedule[S4]
	ELSE IF(t=540) THEN
	RG.Counter.uNumEmp ← RG.Counter.schedule[S5]
	ENDIF

Behavioural Constructs

Activity: Serving	
Service for a customer.	
Precondition	(RG.Counter.n < RB.Counter.uNumEmp) AND
	(Q.CustLine.n ≠ 0)
Event SCS	iC.Customer ← SP.RemoveQue(Q.CustLine)
	IF(t – iC.Customer.StartWaitTime > 5)
	THEN
	SSOV.numLongWait +← 1
	ENDIF
	SSOV.numServed +← 1
	SSOV.propLongWait ← SSOV.numLongWait/SSOV.numServed
	SP.InsertGroup(RG.Counter, iC.Customer)
Duration	RVP.uSrvTm(iC.Customer.uType)
Event SCS	SP.RemoveGroup(RG.Counter, iC.Customer)
	SP.Leave(iC.Customer)

A1.3 Port Project (Version 1)

A1.3.1 Problem Description

A1.3.1.1 Problem Statement

Tankers of various sizes are loaded with crude oil at a port that currently has berth facilities for loading three tankers simultaneously. Empty tankers arrive in the harbour entrance where they often wait for the necessary assistance from a tug (only one is available) to move them into a berth where the loading takes place. After the loading operation is complete, a tanker again requires assistance from a tug to move from the berth back to the harbour entrance. The port authority wishes to address the increasing number of complaints from tanker captains about excessive turnaround times. It is considering the construction of a fourth berth and requires some insight into the likely impact of such an expansion.

A1.3.1.2 SUI Details

The general SUI configuration is shown in Fig. A1.3.

A harbour master oversees the operation of the port and is in radio contact with the tugboat captain and with each tanker captain. The harbour master ensures that tankers are serviced on a first-come first-served basis by maintaining a list of tankers requiring berthing and a list of tankers ready for deberthing.

Processing of Tankers

1. An arriving tanker's captain communicates his/her arrival at the harbour entrance to the harbour master who then adds the tanker's name to the end of the list of tankers needing to be berthed.
2. Each tanker must wait until all previously arrived tankers have been berthed by the tug. When the tug next becomes available for berthing, it is requested, by the harbour master, to berth the tanker at the top of the berthing list.

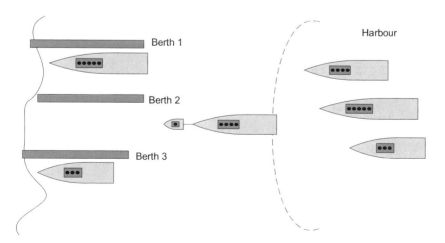

Fig. A1.3 Schematic view of the port's operation

3. The loading procedure starts as soon as the berthing operation is complete at which time the tug is released to carry out its next task.

4. When the loading of a tanker is completed, the tanker's captain contacts the harbour master to request the services of the tug to move it back to the harbour entrance (the 'deberthing' operation). The harbour master adds the tanker's name to the end of a second list of tankers, namely, those waiting for deberthing.

5. Each loaded tanker must wait until all previously loaded tankers have been deberthed (i.e., those that precede it on the deberthing list) before it is allocated the tug by the harbour master. The tug moves the tanker from the berth back to the harbour entrance and the tanker leaves the port. The tug proceeds to its next task.

Tug Operation

When the tug finishes a berthing operation, the harbour master requests the tug to deberth the tanker at the top of the deberthing list. However, if the deberthing list is empty but the berthing list is not empty and a berth is available, then the tug is requested by the harbour master to travel to the harbour entrance and begin berthing the tanker at the top of the berthing list. Otherwise, the tug remains idle in the berth area.

When the tug finishes a deberthing operation and the berthing list is not empty (i.e., there are waiting tankers in the harbour), the harbour master assigns the tanker at the top of the berthing list to the tug for a berthing operation. If the list is empty but there is at least one tanker in the berths, the tug begins the journey back to the berth area without any tanker in tow (i.e., 'empty').

If, at the end of a deberthing operation, the berthing list is empty and there are no occupied berths, the tug remains idle at the harbour entrance waiting for the arrival of a tanker.

Because the port is in a geographical area where storms commonly occur, there is a defined procedure that the tug must follow in order to minimize storm damage. Specifically, when a storm occurs, both the tug and any tanker that is being towed batten down hatches and drop anchor. When the storm is over, the tug resumes the operation that was underway before the storm occurred.

A1.3.2 Project Goal and Achievement

Turnaround time has five components which are (a) waiting time at the harbour, (b) travel time from harbour to berths, (c) loading time, (d) waiting time in the berths and (e) travel time from berths to harbour. The addition of a fourth berth has no effect on travel times or loading times but will impact the waiting times. The project goal is to evaluate how the addition of a fourth berth will affect the waiting time of tankers and the usage of the available berths. We compare two cases as specified by the parameter introduced for this purpose. Its assigned values are given below:

RG.Berths.numBerths: The number of berths for loading tankers. The two values of interest are 3 (base case) and 4 (alternate case).

Output

avgWaitTime: This output variable provides the average waiting time of tankers. Its reduction indicates a reduction of turn-around time for tankers. Waiting time for a tanker includes any time interval during which the tanker is waiting to be serviced by the tug, i.e., the periods when the tanker is waiting to be towed to berths or waiting to be towed back to the harbour entrance. In the ideal case, these tasks would be initiated immediately and waiting time would be zero.

avgOccBerths: This output variable provides the average number of berths that are occupied. This value is a measure of berth usage.

Study: Steady-state study.

Observation Interval: Time units are hours.

A1.3.3 Conceptual Model

A1.3.3.1 High-Level Conceptual Model

Assumptions

- The time required for the tug to travel from the harbour to the berths or from the berths to the harbour with no tanker in tow is a constant.
- The times required for berthing a tanker and for deberthing a tanker are both constants that are independent of tanker size.

Simplifications

- The occurrence of storm conditions is not considered.
- The tankers fall into three size categories: SMALL, MEDIUM and LARGE.
- It is not necessary to model the harbour master, because his/her directives are captured in the behaviour constructs of the model.
- The berths are not actually modelled as individual entities but rather as a resource that can simultaneously service several tanker entities; namely, 3 or 4.

Structural View

The structural diagram shows the entities used to represent the tankers, tug, berths and lists used by the harbour master.

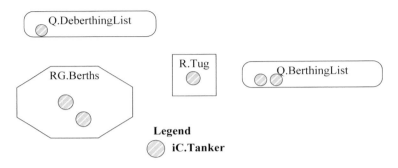

Entity Categories

- BerthingList: A single queue entity that represents the berthing list of empty tankers waiting to be berthed by the tug.
- DeberthingList: A single queue entity that represents the list of tankers loaded with oil ready for deberthing.
- Tug: A single resource entity that berths and deberths the tankers as it travels between the harbour and berths.
- Berths: A single resource group entity that represents the set of berths used for loading the tankers with oil. The attribute numBerths of this entity is a parameter.
- Tanker: A consumer entity category representing the tankers that arrive empty at the port and are then loaded with oil. Because of the transient existence of tankers within the model, this entity category has *scope* = Transient.

Notes:

- Except for the Tanker entity category, all other categories have *scope* = Unary.
- Any particular Tanker entity can necessarily be referenced simultaneously by both the *RG.Berths* and the *Q.DeberthingList* entities.

Behavioural View

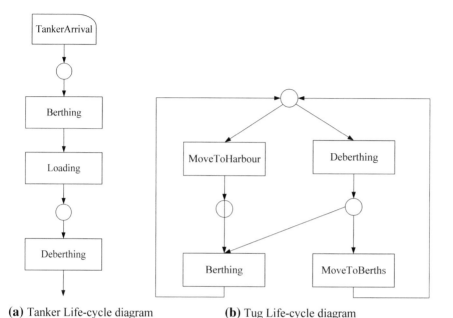

(**a**) Tanker Life-cycle diagram   (**b**) Tug Life-cycle diagram

Actions:

TankerArrival: This scheduled action generates the input entity stream of tankers. With each arrival, tanker attribute values are assigned and the tanker entity is placed into the Q.BerthingList entity.

Activities:

Berthing: This activity represents the Tug entity towing an empty Tanker entity from the harbour to the berths.

Deberthing: This activity represents the Tug entity towing a loaded Tanker entity from the berths to the harbour (at which point the Tanker entity leaves the model).

MoveToBerths: This activity represents the Tug entity moving to the berths area with no Tanker entity in tow.

MoveToHarbour: This activity represents the Tug entity moving to the harbour with no Tanker entity in tow.

Loading: This activity represents the loading of a Tanker. This is a sequel activity and is initiated at the completion of the Deberthing activity.

Input

<table>
<tr><th colspan="4">Exogenous Input (Entity Stream)</th></tr>
<tr><th>Variable Name</th><th>Description</th><th>Domain Sequence</th><th>Range Sequence</th></tr>
<tr><td>uTnkr(t)</td><td>This input entity stream represents the arriving tankers.</td><td>RVP.DuTnkr()</td><td>N/A; 1 tanker arrives at each arrival time.</td></tr>
<tr><th colspan="4">Endogenous Input (Semi-Independent)</th></tr>
<tr><th>Variable Name</th><th colspan="2">Description</th><th>Value(s)</th></tr>
<tr><td>iC.Tanker.uSize</td><td colspan="2">The size of the arriving tanker.</td><td>RVP.uTankerSize()</td></tr>
<tr><td>uLoadTime</td><td colspan="2">Loading time for a tanker. The loading time depends on the size of the tanker.</td><td>RVP.uLoadTime(size)</td></tr>
</table>

A1.3.3.2 Detailed Conceptual Model

Structural Components

<table>
<tr><th colspan="3">Constants</th></tr>
<tr><th>Name</th><th>Description</th><th>Value</th></tr>
<tr><td>BERTH_TIME</td><td>Time required for the berthing operation.</td><td>1 hour</td></tr>
<tr><td>DEBERTH_TIME</td><td>Time required for the deberthing operation.</td><td>2 hours</td></tr>
<tr><td>EMPTY_TIME</td><td>Harbour to berth (and berth to harbour) travel time for tug when travelling without a tanker in tow.</td><td>0.25 hour</td></tr>
<tr><th colspan="3">Parameters</th></tr>
<tr><th>Name</th><th>Description</th><th>Values</th></tr>
<tr><td>RG.Berths.numBerths</td><td>Number of berths.</td><td>3 and 4</td></tr>
</table>

<table>
<tr><th colspan="2">Consumer Transient: Tanker</th></tr>
<tr><td colspan="2">Consumer entities that represent the tankers that arrive in the port, load and leave.</td></tr>
<tr><th>Attributes</th><th>Description</th></tr>
<tr><td>uSize</td><td>This input variable represents the size of the tanker (value is one of SMALL, MEDIUM, LARGE) as assigned via RVP.uTankerSize().</td></tr>
<tr><td>startWait</td><td>A timestamp used to determine waiting times.</td></tr>
<tr><td>totalWait</td><td>Accumulated time waiting for a tanker entity.</td></tr>
</table>

<table>
<tr><th colspan="2">Resource Group Unary: Berths</th></tr>
<tr><td colspan="2">A resource group entity that represents the collection of berths used for loading tankers.</td></tr>
<tr><th>Attributes</th><th>Description</th></tr>
<tr><td>list</td><td>A FIFO list of the Tanker entities that occupy berths.</td></tr>
<tr><td>n</td><td>The number of entities in list (maximum value is numBerths).</td></tr>
<tr><td>numBerths</td><td>The number of berths (this is a parameter).</td></tr>
</table>

Queue Unary: BerthingList	
A queue entity that represents the harbour master's berthing list of tankers in the harbour waiting to be assisted into a berth by the tug.	
Attributes	**Description**
list	A FIFO list of the Tanker entities in the harbour master's berthing list.
n	The number of entities in list.

Queue Unary: DeberthingList	
A queue entity that represents the harbour master's list of tankers that have completed loading and are waiting in a berth for deberthing assistance from the tug.	
Attributes	**Description**
list	The FIFO list of Tanker entities in the harbour master's deberthing list.
n	The number of entities in list.

Resource Unary: Tug	
A resource entity that represents the tugboat needed to berth tankers that arrive in the harbour and to deberth tankers that have finished loading.	
Attributes	**Description**
status	Indicates the status of the tug as specified by one of the following values: BUSY – travelling with or without a tanker in tow, PAUSE_H – stopped in the harbour ready for the next task, PAUSE_B – stopped in the berth area ready for the next task. The status attribute is initialised to PAUSE_H, that is, the Tug starts in the harbour.

Behavioural Components

Initialisation

Action: Initialise	
TimeSequence	< 0 >
Event SCS	R.Tug.status ← PAUSE_H
	RG.Berths.n ← 0
	Q.BerthingList.n ← 0
	Q.DeberthingList.n ← 0

Output

OUTPUTS	
Trajectory Sequences	
Name	**Description**
TRJ[RG.Berths.n]	This trajectory sequence records the values of the attribute RG.Berths.n which represents the number of occupied berths.
Sample Sequences	
Name	**Description**
PHI[iC.Tanker.totalWait]	Each value in PHI[iC.Tanker.totalWait] is of the form (t_k, y_k) where y_k is the value of the attribute totalWait for some tanker entity and t_k is the time when the tug begins deberthing that tanker.

Derived Scalar Output Variables (DSOV's)			
Name	**Description**	**Output Sequence Name**	**Operation**
avgOccBerths	Average number of occupied berths.	TRJ[RG.Berths.n]	AVG
avgWaitTime	Average time that the tankers spend waiting for the tug.	PHI[iC.Tanker.totalWait]	AVG

User-Defined Procedures

None required.

Input Constructs

Random Variate Procedures		
Name	**Description**	**Data Model**
uLoadTime(size)	The returned value is the loading time for a tanker entity. It is assigned according to the tanker's size; i.e., SMALL, MEDIUM or LARGE).	Each of the data models is UNIFORM(Min, Max) where Min and Max take on the following values (in hours) according to the tanker size.
		<table><tr><td>size</td><td>Min</td><td>Max</td></tr><tr><td>SMALL</td><td>8</td><td>10</td></tr><tr><td>MEDIUM</td><td>10</td><td>14</td></tr><tr><td>LARGE</td><td>15</td><td>21</td></tr></table>
uTankerSize()	Returns the 'size' of an arriving tanker entity; value is one of SMALL, MEDIUM or LARGE.	The percentage of SMALL, MEDIUM and LARGE arriving tankers is given in the following table.
		<table><tr><td>Size</td><td>Percent</td></tr><tr><td>SMALL</td><td>25</td></tr><tr><td>MEDIUM</td><td>25</td></tr><tr><td>LARGE</td><td>50</td></tr></table>
DuTnkr()	Provides the values of the arrival times of tanker entities.	First arrival at t=0 and interarrival given by: EXPONENTIAL(AVGARR) where AVGARR = 8 hours.

Action: TankerArrivals
Provides the input entity stream of arriving tanker entities; includes the associated derive operation and attribute value assignment.

TimeSequence	RVP.DuTnkr()
Event SCS	iC.Tanker ← SP.Derive(Tanker) iC.Tanker.uSize ← RVP.uTankerSize() iC.Tanker.startWait ← t iC.Tanker.totalWait ← 0 SP.InsertQue(Q.BerthingList, iC.Tanker)

Behavioural Constructs

Activity: Deberthing
This activity represents the Tug entity towing a loaded Tanker entity from the berths to the harbour (at which point the Tanker entity leaves the model).

Precondition	(R.Tug.status = PAUSE_B) AND (Q.DeberthingList.n ≠ 0)
Event SCS	iC.Tanker ← SP.RemoveQue(Q.DeberthingList)
	iC.Tanker.totalWait +← (t – iC.Tanker.startWait) *(Update totalWait)*
	(Update sample sequence)
	SP.Put(PHI[iC.Tanker.totalWait], iC.Tanker.totalWait)
	(Remove Tanker from Berths)
	SP.RemoveGrp(RG.Berths, iC.Tanker)
	R.Tug.status ← BUSY
Duration	DEBERTH_TIME
Event SCS	R.Tug.status ← PAUSE_H
	SP.Leave(iC.Tanker)

Activity: Berthing
This activity represents the Tug entity towing an empty tanker entity from the harbour to the berths. Note that the availability of an empty berth is implicit in the fact that the tug is located in the harbour.

Precondition	(R.Tug.status = PAUSE_H) AND
	(Q.BerthingList.n > 0)
Event SCS	R.Tug.status ← BUSY
	iC.Tanker ← SP.RemoveQue(Q.BerthingList)
	iC.Tanker.totalWait +← (t – iC.Tanker.startWait)
Duration	BERTH_TIME
Event SCS	SP.InsertGrp(RG.Berths, iC.Tanker)
	SP.StartSequel(Loading, iC.Tanker)
	R.Tug.status ← PAUSE_B

Activity: MoveToHarbour
This activity represents the Tug entity moving to the harbour area with no Tanker entity in tow.

Precondition	(R.Tug.status = PAUSE_B) AND
	(Q.DeberthingList.n = 0) AND
	(Q.BerthingList.n > 0) AND
	(RG.Berths.n < RG.Berths.numBerths)
Event SCS	R.Tug.status ← BUSY
Duration	EMPTY_TIME
Event SCS	R.Tug.status ← PAUSE_H

Activity: MoveToBerths	
This activity represents the Tug entity moving to the berths area with no Tanker entity in tow.	
Precondition	(R.Tug.status = PAUSE_H) AND
	(Q.BerthingList.n = 0) AND (RG.Berths.n > 0)
Event SCS	R.Tug.status ← BUSY
Duration	EMPTY_TIME
Event SCS	R.Tug.status ← PAUSE_B

Activity: Loading	
This activity represents the loading of a Tanker. This is a sequel activity and is initiated at the completion of the Berthing activity	
Causal	(iC.Tanker)
Event SCS	
Duration	RVP.uLoadTime(iC.Tanker.uSize)
Event SCS	iC.Tanker.startWait ← t
	SP.InsertQue(Q.DeberthingList, iC.Tanker)

A1.4 Port Project (Version 2)

A1.4.1 Problem Description

A1.4.1.1 Problem Statement

Tankers of various sizes are loaded with crude oil at a port that currently has berth facilities for loading three tankers simultaneously. Empty tankers arrive in the harbour entrance where they often wait for the necessary assistance from a tug (only one is available) to move them into a berth where the loading takes place. After the loading operation is complete, a tanker again requires assistance from a tug to move from the berth back to the harbour entrance. The port authority wishes to address the increasing number of complaints from tanker captains about excessive turnaround times. It is considering the construction of a fourth berth and requires some insight into the likely impact of such an expansion.

A1.4.1.2 SUI Details

The general SUI configuration is shownin Fig. A1.4.

A harbour master oversees the operation of the port and is in radio contact with the tug boat captain and with each tanker captain. The harbour master ensures that tankers are serviced on a first-come first-served basis by maintaining a list of tankers requiring berthing and a list of tankers ready for deberthing.

Processing of Tankers

1. An arriving tanker's captain communicates his/her arrival at the harbour entrance to the harbour master who then adds the tanker's name to the end of the list of tankers needing to be berthed.
2. Each tanker must wait until all previously arrived tankers have been berthed by the tug. When the tug next becomes available for berthing, it is requested by the harbour master to berth the tanker at the top of the berthing list.

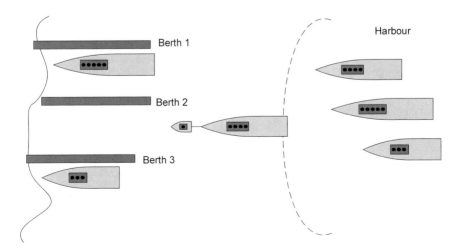

Fig. A1.4 Schematic view of the port's operation

3. The loading procedure starts as soon as the berthing operation is complete at which time the tug is released to carry out its next task.
4. When the loading of a tanker is completed, the tanker's captain contacts the harbour master to request the services of the tug to move it back to the harbour entrance (the "deberthing" operation). The harbour master adds the tanker's name to the end of a second list of tankers; namely, those waiting for deberthing.
5. Each loaded tanker must wait until all previously loaded tankers have been deberthed (i.e., those that precede it on the deberthing list) before it is allocated the tug by the harbour master. The tug moves the tanker from the berth back to the harbour entrance and the tanker leaves the port. The tug proceeds to its next task.

Tug Operation

When the tug finishes a berthing operation, the harbour master requests the tug to deberth the tanker at the top of the deberthing list. However, if the deberthing list is empty but the berthing list is not empty and a berth is available, then the tug is requested by the harbour master to travel to the harbour entrance and begin berthing the tanker at the top of the berthing list. Otherwise, the tug remains idle in the berth area.

When the tug finishes a deberthing operation and the berthing list is not empty (i.e., there are waiting tankers in the harbour), the harbour master assigns the tanker at the top of the berthing list to the tug for a berthing operation. If the list is empty but there is at least one tanker in the berths, the tug begins the journey back to the berth area without any tanker in tow (i.e., 'empty'). If a tanker arrives in the harbour and the tug has not yet completed 70% of its journey back to the berths or if there is no tanker waiting to be deberthed, the harbour master will instruct the tug to return to the harbour entrance to start berthing the newly arrived tanker. If, at the end of a deberthing operation, the berthing list is empty and there are no occupied berths, the tug remains idle at the harbour entrance waiting for the arrival of a tanker.

Because the port is in a geographical area where storms commonly occur, there is a defined procedure that the tug must follow in order to minimize storm damage. Specifically, when a storm occurs, both the tug and any tanker that is being towed batten down hatches and drop anchor. When the storm is over, the tug resumes the operation that was underway before the storm occurred. After a storm, the "70% rule" is nullified.

A1.4.2 Project Goal and Achievement

Turnaround time has five components which are (a) waiting time at the harbour, (b) travel time from harbour to berths, (c) loading time, (d) waiting time in the berths and (e) travel time from berths to harbour. The addition of a fourth berth has no effect on travel times or loading times but will impact the waiting times. The project goal is to evaluate how the addition of a fourth berth will affect the waiting time of tankers and the usage of the available berths. We compare two cases as specified by the parameter introduced for this purpose. Its assigned values are given below:

RG.Berths.numBerths: The number of berths for loading tankers. The two values of interest are 3 (base case) and 4 (alternate case).

Output

avgWaitTime: This output variable provides the average waiting time of tankers. Its reduction indicates a reduction of turn-around time for tankers. Waiting time for a tanker includes any time interval during which the tanker is waiting to be serviced by the tug, i.e., the periods when the tanker is waiting to be towed to berths or waiting to be towed back to the harbour entrance. In the ideal case, these tasks would be initiated immediately and waiting time would be zero.

avgOccBerths: This output variable provides the average number of berths that are occupied. This value is a measure of berth usage.

Study: Steady state study.

Observation Interval: Time units are hours.

A1.4.3 Conceptual Model

A1.4.3.1 High-Level Conceptual Model

Assumptions

- The time required for the tug to travel from the harbour to the berths or from the berths to the harbour with no tanker in tow, is a constant.
- The times required for berthing a tanker and for deberthing a tanker are both constants that are independent of tanker size.

Simplifications

- The occurrence of storm conditions is not considered.
- The tankers fall into three size categories: SMALL, MEDIUM and LARGE.
- It is not necessary to model the harbour master, because his/her directives are captured in the behaviour constructs of the model.
- The berths are not actually modelled as individual entities but rather as a resource that can simultaneously service several tanker entities; namely, 3 or 4.

Structural View

The structural diagram shows the entities used to represent the tankers, tug, berths and lists used by the harbour master.

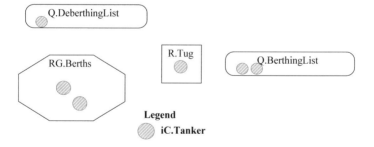

Entity Categories:

- BerthingList: A single queue entity that represents the berthing list of empty tankers waiting to be berthed by the tug.
- DeberthingList: A single queue entity that represents the list of tankers loaded with oil ready for deberthing.
- Tug: A single resource entity that berths and deberths the tankers as it travels between the harbour and berths.
- Berths: A single resource group entity that represents the set of berths used for loading the tankers with oil. The attribute numBerths of this entity is a parameter.

- Tanker: A consumer entity category representing the tankers that arrive empty at the port and are then loaded with oil. Because of the transient existence of tankers within the model this entity category has *scope* = Transient.

Notes:

- Except for the Tanker entity category, all other categories have *scope* = Unary.
- Any particular Tanker entity can necessarily be referenced simultaneously by both the *RG.Berths* and the *Q.DeberthingList* entities.

Behavioural View

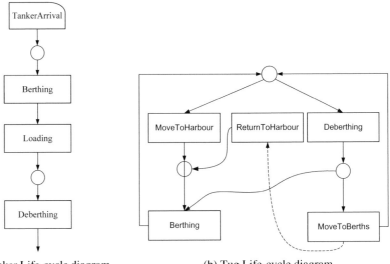

(a) Tanker Life-cycle diagram   (b) Tug Life-cycle diagram

Actions:

- **TankerArrival**: This scheduled action generates the input entity stream of tankers. With each arrival, tanker attribute values are assigned and the tanker entity is placed into the Q.BerthingList entity.

Activities:

- **Berthing**: This activity represents the Tug entity towing an empty Tanker entity from the harbour to the berths.
- **Deberthing**: This activity represents the Tug entity towing a loaded Tanker entity from the berths to the harbour (at which point the Tanker entity leaves the model).
- **MoveToBerths**: This activity represents the Tug entity moving to the berths area with no Tanker entity in tow and may suffer an interruption if a Tanker arrives in the habour before the Tug reaches the Berths.

- **ReturnToHarbour**: This (sequel) activity is initiated when the tug, while moving to the berth area without a tanker in tow, is required to return to the harbour because an empty tanker has arrived.
- **MoveToHarbour**: This activity represents the Tug entity moving to the harbour with no Tanker entity in tow.
- **Loading**: This activity represents the loading of a Tanker. This is a sequel activity and is initiated at the completion of the Deberthing activity.

Input

Exogenous Input (Entity Stream)			
Variable Name	**Description**	**Domain Sequence**	**Range Sequence**
uTnkr(t)	This input entity stream represents the arriving tankers.	RVP.DuTnkr()	N/A; 1 tanker arrives at each arrival time.
Endogenous Input (Semi-Independent)			
Variable Name	**Description**		**Value(s)**
iC.Tanker.uSize	The size of the arriving tanker.		RVP.uTankerSize()
uLoadTime	Loading time for a tanker. The loading time depends on the size of the tanker.		RVP.uLoadTime(size)

A1.4.3.2 Detailed Conceptual Model

Structural Components

Constants		
Name	**Description**	**Value**
BERTH_TIME	Time required for the berthing operation.	1 hour
DEBERTH_TIME	Time required for the deberthing operation.	2 hours
EMPTY_TIME	Harbour to berth (and berth to harbour) travel time for tug when travelling without a tanker in tow.	0.25 hour
Parameters		
Name	**Description**	**Values**
RG.Berths.numBerths	Number of berths.	3 and 4

Consumer Transient: Tanker
Consumer entities that represent the tankers that arrive in the port, load and leave.

Attributes	Description
uSize	This input variable represents the size of the tanker (value is one of SMALL, MEDIUM, LARGE) as assigned via RVP.uTankerSize().
startWait	A timestamp used to determine waiting times.
totalWait	Accumulated time waiting for a tanker entity.

Resource Group Unary: Berths
This resource group entity structure represents the collection of berths used for loading tankers.

Attributes	Description
list	A FIFO list of the Tanker entities that occupy berths.
n	The number of entities in list (maximum value is numBerths).
numBerths	The number of berths (this is a parameter).

Queue Unary: BerthingList
This queue entity structure represents the harbour master's berthing list of tankers in the harbour waiting to be assisted into a berth by the tug.

Attributes	Description
list	A FIFO list of the Tanker entities in the harbour master's berthing list.
n	The number of entries in list.

Queue Unary: DeberthingList
A queue entity that represents the harbour master's list of tankers that have completed loading and are waiting in a berth for deberthing assistance from the tug.

Attributes	Description
list	The FIFO list of Tanker entities in the harbour master's deberthing list.
n	The number of entries in list.

Resource Unary: Tug
A resource entity that represents the tugboat needed to berth tankers that arrive in the harbour and to deberth tankers that have finished loading.

Attributes	Description
status	Indicates the status of the tug as specified by one of the following values: BUSY—travelling with or without a tanker in tow, PAUSE_H—stopped in the harbour ready for the next task, PAUSE_B—stopped in the berth area ready for the next task. The status attribute is initialised to PAUSE_H, that is, the Tug starts in the harbour.
startTime	Time stamp indicating the time when the tug leaves the harbour to travel back to the berth area.

Behavioural Components

Initialisation

Action: Initialise	
TimeSequence	< 0 >
Event SCS	R.Tug.status ← PAUSE_H RG.Berths.n ← 0 Q.BerthingList.n ← 0 Q.DeberthingList.n ← 0

Output

OUTPUTS			
Trajectory Sequences			
Name	**Description**		
TRJ[RG.Berths.n]	This trajectory sequence records the values of the attribute RG.Berths.n which represents the number of occupied berths.		
Sample Sequences			
Name	**Description**		
PHI[iC.Tanker.totalWait]	Each value in PHI[iC.Tanker.totalWait] is of the form (t_k, y_k) where y_k is the value of the attribute totalWait for some tanker entity and t_k is the time when the tug begins deberthing that tanker.		
Derived Scalar Output Variables (DSOV's)			
Name	**Description**	**Output Sequence Name**	**Operation**
avgOccBerths	Average number of occupied berths.	TRJ[RG.Berths.n]	AVG
avgWaitTime	Average time that the tankers spend waiting for the tug.	PHI[iC.Tanker.totalWait]	AVG

User-Defined Procedures

None required.

Input Constructs

Random Variate Procedures		
Name	**Description**	**Data Model**
uLoadTime(size)	The returned value is the loading time for a tanker entity. It is assigned according to the tanker's size; i.e., SMALL, MEDIUM or LARGE).	Each of the data models is UNIFORM(Min, Max) where Min and Max take on the following values (in hours) according to the tanker size. <table><tr><td>size</td><td>Min</td><td>Max</td></tr><tr><td>SMALL</td><td>8</td><td>10</td></tr><tr><td>MEDIUM</td><td>10</td><td>14</td></tr><tr><td>LARGE</td><td>15</td><td>21</td></tr></table>
uTankerSize()	Returns the 'size' of an arriving tanker entity; value is one of SMALL, MEDIUM or LARGE.	The percentage of SMALL, MEDIUM and LARGE arriving tankers is given in the following table. <table><tr><td>Size</td><td>Percent</td></tr><tr><td>SMALL</td><td>25</td></tr><tr><td>MEDIUM</td><td>25</td></tr><tr><td>LARGE</td><td>50</td></tr></table>
DuTnkr()	Provides the values of the arrival times of tanker entities.	First arrival at t=0 and interarrival given by: EXPONENTIAL(AVGARR) where AVGARR = 8 hours.

Action: Tanker Arrivals	
Provides the input entity stream of arriving tanker entities; includes the associated derive operation and attribute value assignment.	
TimeSequence	RVP.DuTnkr()
Event SCS	iC.Tanker ← SP.Derive(Tanker) iC.Tanker.uSize ← RVP.uTankerSize() iC.Tanker.startWait ←t iC.Tanker.totalWait ← 0 SP.InsertQue(Q.BerthingList, iC.Tanker)

Behavioural Constructs

Activity: Deberthing	
This activity represents the Tug entity towing a loaded Tanker entity from the berths to the harbour (at which point the Tanker entity leaves the model).	
Precondition	(R.Tug.status = PAUSE_B) AND (Q.DeberthingList.n ≠ 0)

Event SCS	iC.Tanker ← SP.RemoveQue(Q.DeberthingList)
	iC.Tanker.totalWait +← (t – iC.Tanker.startWait) *(Update totalWait)*
	    (Update sample sequence)
	SP.Put(PHI[iC.Tanker.totalWait], iC.Tanker.totalWait)
	    (Remove Tanker from Berths)
	SP.RemoveGrp(RG.Berths, iC.Tanker)
	R.Tug.status ← BUSY

Duration	DEBERTH_TIME
Event SCS	R.Tug.status ← PAUSE_H SP.Leave(iC.Tanker)

Activity: Berthing	
This activity represents the Tug entity towing an empty tanker entity from the harbour to the berths. Note that the availability of an empty berth is implicit in the fact that the tug is located in the harbour.	
Precondition	(R.Tug.status = PAUSE_H) AND (Q.BerthingList.n > 0)
Event SCS	R.Tug.status ← BUSY iC.Tanker ← SP.RemoveQue(Q.BerthingList) iC.Tanker.totalWait +← (t – iC.Tanker.startWait)
Duration	BERTH_TIME
Event SCS	SP.InsertGrp(RG.Berths, iC.Tanker) SP.StartSequel(Loading, iC.Tanker) R.Tug.status ← PAUSE_B

Activity: MoveToHarbour	
This activity represents the Tug entity moving to the harbour area with no Tanker entity in tow.	
Precondition	(R.Tug.status = PAUSE_B) AND (Q.DeberthingList.n = 0) AND (Q.BerthingList.n > 0) AND (RG.Berths.n < RG.Berths.numBerths)
Event SCS	R.Tug.status ← BUSY
Duration	EMPTY_TIME
Event SCS	R.Tug.status ← PAUSE_H

Activity: MoveToBerths	
This activity represents the Tug entity moving to the berths area with no Tanker entity in tow.	
Precondition	(R.Tug.status = PAUSE_H) AND (Q.BerthingList.n = 0) AND (RG.Berths.n > 0)
Event SCS	R.Tug.status ← BUSY R.Tug.startTime ← t
Duration	EMPTY_TIME

Interruption
    Precondition
	(Q.BerthingList.n > 0) AND
	(((t-R.Tug.startTime) < 0.7 * EMTPY_TIME)
	    OR (Q.DeberthingList.n = 0))

Event SCS	SP.StartSequel(ReturnToHarbour) SP.Terminate()
Event SCS	R.Tug.status ← PAUSE_B

Activity: ReturnToHarbour	
This (sequel) activity is initiated when the tug, while moving to the berth area without a tanker in tow, is required to return to the harbor because an empty tanker has arrived.	
Causal	( )
Event SCS	
Duration	(t – R.Tug.startTime)
Event SCS	R.Tug.status ← PAUSE_H

Activity: Loading	
This activity represents the loading of a Tanker. This is a sequel activity and is initiated at the completion of the Deberthing activity	
Causal	(iC.Tanker)
Event SCS	
Duration	RVP.uLoadTime(iC.Tanker.uSize)
Event SCS	iC.Tanker.startWait ← t SP.InsertQue(Q.DeberthingList, iC.Tanker)

A1.5 Port Project (Version 3)

A1.5.1 Problem Description

A1.5.1.1 Problem Statement

Tankers of various sizes are loaded with crude oil at a port that currently has berth facilities for loading three tankers simultaneously. Empty tankers arrive in the harbour entrance where they often wait for the necessary assistance from a tug (only one is available) to move them into a berth where the loading takes place. After the loading operation is complete, a tanker again requires assistance from a tug to move from the berth back to the harbour entrance. The port authority wishes to address the increasing number of complaints from tanker captains about excessive turnaround times. It is considering the construction of a fourth berth and requires some insight into the likely impact of such an expansion.

A1.5.1.2 SUI Details

The general SUI configuration is shown in Fig. A1.5.

A harbour master oversees the operation of the port and is in radio contact with the tug boat captain and with each tanker captain. The harbour master ensures that tankers are serviced on a first-come first-served basis by maintaining a list of tankers requiring berthing and a list of tankers ready for deberthing.

Processing of Tankers

1. An arriving tanker's captain communicates his/her arrival at the harbour entrance to the harbour master who then adds the tanker's name to the end of the list of tankers needing to be berthed.

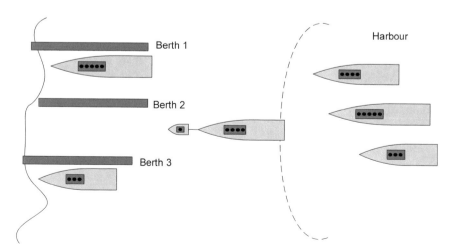

Fig. A1.5 Schematic view of the port's operation

2. Each tanker must wait until all previously arrived tankers have been berthed by the tug. When the tug next becomes available for berthing, it is requested by the harbour master to berth the tanker at the top of the berthing list.
3. The loading procedure starts as soon as the berthing operation is complete at which time the tug is released to carry out its next task.
4. When the loading of a tanker is completed, the tanker's captain contacts the harbour master to request the services of the tug to move it back to the harbour entrance (the "deberthing" operation). The harbour master adds the tanker's name to the end of a second list of tankers; namely, those waiting for deberthing.
5. Each loaded tanker must wait until all previously loaded tankers have been deberthed (i.e., those that precede it on the deberthing list) before it is allocated the tug by the harbour master. The tug moves the tanker from the berth back to the harbour entrance and the tanker leaves the port. The tug proceeds to its next task.

Tug Operation

When the tug finishes a berthing operation, the harbour master requests the tug to deberth the tanker at the top of the deberthing list. However, if the deberthing list is empty but the berthing list is not empty and a berth is available, then the tug is requested by the harbour master to travel to the harbour entrance and begin berthing the tanker at the top of the berthing list. Otherwise, the tug remains idle in the berth area.

When the tug finishes a deberthing operation and the berthing list is not empty (i.e., there are waiting tankers in the harbour), the harbour master assigns the tanker at the top of the berthing list to the tug for a berthing operation. If the list is empty but there is at least one tanker in the berths the tug begins the journey back to the berth area without any tanker in tow (i.e., "empty"). If a tanker arrives in the harbour and the tug has not yet completed 70% of its journey back to the berths or if there is no tanker waiting to be deberthed, the harbour master will instruct the tug to return to the harbour entrance to start berthing the newly arrived tanker.

If, at the end of a deberthing operation, the berthing list is empty and there are no occupied berths, the tug remains idle at the harbour entrance waiting for the arrival of a tanker.

Because the port is in a geographical area where storms commonly occur, there is a defined procedure that the tug must follow in order to minimize storm damage. Specifically, when a storm occurs, both the tug and any tanker that is being towed batten down hatches and drop anchor. When the storm is over, the tug resumes the operation that was underway before the storm occurred. After a storm, the "70% rule" is nullified.

A1.5.2 Project Goal and Achievement

Turnaround time has five components which are (a) waiting time at the harbour, (b) travel time from harbour to berths, (c) loading time, (d) waiting time in the berths and (e) travel time from berths to harbour. The addition of a fourth berth has no effect on travel times or loading times but will impact the waiting times. The project goal is to evaluate how the addition of a fourth berth will affect the waiting time of tankers and the usage of the available berths. We compare two cases as specified by the parameter introduced for this purpose. Its assigned values are given below:

RG.Berths.numBerths: The number of berths for loading tankers. The two values of interest are 3 (base case) and 4 (alternate case).

Output

avgWaitTime: This output variable provides the average waiting time of tankers. Its reduction indicates a reduction of turn-around time for tankers. Waiting time for a tanker includes any time interval during which the tanker is waiting to be serviced by the tug; i.e., the periods when the tanker is waiting to be towed to berths or waiting to be towed back to the harbour entrance. In the ideal case, these tasks would be initiated immediately and waiting time would be zero.

avgOccBerths: This output variable provides the average number of berths that are occupied. This value is a measure of berth usage.

Study: Steady state study.

Observation Interval: Time units are hours.

A1.5.3 Conceptual Model

A1.5.3.1 High-Level Conceptual Model

Assumptions

- The time required for the tug to travel from the harbour to the berths or from the berths to the harbour with no tanker in tow is a constant.
- The times required for berthing a tanker and for deberthing a tanker are both constants that are independent of tanker size.

Simplifications

- The tankers fall into three size categories: SMALL, MEDIUM and LARGE.
- It is not necessary to model the harbour master, because his/her directives are captured in the behaviour constructs of the model.
- The berths are not actually modelled as individual entities but rather as a resource that can simultaneously service several tanker entities; namely, 3 or 4.

Structural View

The structural diagram shows the entities used to represent the tankers, tug, berths and lists used by the harbour master.

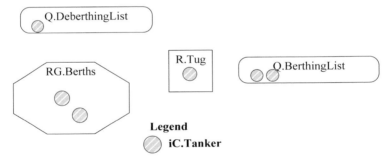

Entity Categories

- BerthingList: A single queue entity that represents the berthing list of empty tankers waiting to be berthed by the tug.
- DeberthingList: A single queue entity that represents the list of tankers loaded with oil ready for deberthing.
- Tug: A single resource entity that berths and deberths the tankers as it travels between the harbour and berths.
- Berths: A single resource group entity that represents the set of berths used for loading the tankers with oil. The attribute numBerths of this entity is a parameter.
- Tanker: A consumer entity category representing the tankers that arrive empty at the port and are then loaded with oil. Because of the transient existence of tankers within the model, this entity category has *scope* = Transient.

Notes:

- Except for the Tanker entity category, all other categories have *scope* = Unary.
- Any particular Tanker entity can necessarily be referenced simultaneously by both the *RG.Berths* and the *Q.DeberthingList* entities.

Behavioural View

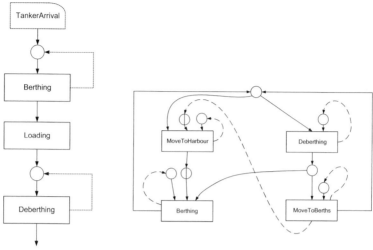

(a) Tanker Life-cycle diagram   (b) Tug Life-cycle diagram

Actions:

TankerArrival: This scheduled action generates the input entity stream of tankers. With each arrival, tanker attribute values are assigned and the tanker entity is placed into the Q.BerthingList entity.

Activities:

Berthing: This activity represents the Tug entity towing an empty Tanker entity from the harbour to the berths.

Deberthing: This activity represents the Tug entity towing a loaded Tanker entity from the berths to the harbour (at which point the Tanker entity leaves the model).

MoveToBerths: This activity represents the Tug entity moving to the berths area with no Tanker entity in tow and may suffer an interruption if a Tanker arrives in the habour before the Tug reaches the Berths.

MoveToHarbour: This activity represents the Tug entity moving to the harbour with no Tanker entity in tow.

Loading: This activity represents the loading of a Tanker. This is a sequel activity and is initiated at the completion of the Deberthing activity.

Storm: This scheduled activity provides the means for implementing the policy of the port authority with respect to the tug's operation when a storm occurs. This activity updates the value of the exogenous input variable uStorm which assumes the value TRUE when the storm is present and FALSE otherwise.

The Berthing, Deberthing, MoveToBerths and MoveToHarbour can all be interrupted by a storm when the tug must drop anchor and wait for the storm to finish. The MoveToBerths activity can also be interrupted by a tanker that arrives before the activity terminates.

Input

Exogenous Input (Entity Stream)			
Variable Name	Description	Domain Sequence	Range Sequence
uTnkr	This input entity stream represents the arriving tankers.	RVP.DuTnkr()	N/A; 1 tanker arrives at each arrival time.
Exogenous Input (Environment Variable)			
Variable Name	Description	Domain Sequence Procedure	Range Sequence Procedure
uStormT	The domain sequence of this sample discrete variable models the times at which the storms start. Its range gives the duration of the storm.	RVP.DuStormTiming()	RVP.RuStormTiming()
Exogenous Input (Environment Variable)			
Variable Name	Description	Domain Sequence	Range Sequence
uStorm	The input variable that indicates the presence of a storm in the port.	Sequence $< t_1, t_2, t_3, \ldots t_n >$ t_i is given the values of $CS_D[uStormT]$ for i odd t_i is given the values in the sequence $< CS_D[uStormT(t_k)] + CS_R[uStormT(t_k)]$: k = 1, 2, 3, …>, for i even.	FALSE at $t_0 = 0$ TRUE for t_i, i odd FALSE for t_i, i even. The value alternates between TRUE and FALSE.

Endogenous Input (Semi-Independent)		
Variable Name	Description	Value(s)
iC.Tanker.uSize	The size of the arriving tanker.	RVP.uTankerSize()
uLoadTime	Loading time for a tanker. The loading time depends on the size of the tanker.	RVP.uLoadTime(size)

A1.5.3.2 Detailed Conceptual Model

Structural Components

Constants		
Name	**Description**	**Value**
BERTH_TIME	Time required for the berthing operation.	1 hour
DEBERTH_TIME	Time required for the deberthing operation.	2 hours
EMPTY_TIME	Harbour to berth (and berth to harbour) travel time for tug when travelling without a tanker in tow.	0.25 hour

Parameters		
Name	**Description**	**Values**
RG.Berths.numBerths	Number of berths.	3 and 4

Consumer Transient: Tanker	
Consumer entities that represent the tankers that arrive in the port, load and leave.	
Attributes	**Description**
uSize	This input variable represents the size of the tanker (value is one of SMALL, MEDIUM, LARGE) as assigned via RVP.uTankerSize().
startWait	A timestamp used to determine waiting times.
totalWait	Accumulated time waiting for a tanker entity.

Resource Group Unary: Berths	
A resource group entity that represents the collection of berths used for loading tankers.	
Attributes	**Description**
list	A FIFO list of the Tanker entities that occupy berths.
n	The number of entities in list (maximum value is numBerths).
numBerths	The number of berths (this is a parameter).

Queue Unary: BerthingList	
A queue entity that represents the harbour master's berthing list of tankers in the harbour waiting to be assisted into a berth by the tug.	
Attributes	**Description**
list	A FIFO list of the Tanker entities in the harbour master's berthing list.
n	The number of entities in list.

Queue Unary: DeberthingList	
A queue entity that represents the harbour master's list of tankers that have completed loading and are waiting in a berth for deberthing assistance from the tug.	
Attributes	**Description**
list	The FIFO list of Tanker entities in the harbour master's deberthing list.
n	The number of entities in list.

Resource Unary: Tug	
A resource entity that represents the tugboat needed to berth tankers that arrive in the harbour and to deberth tankers that have finished loading.	
Attributes	**Description**
status	Indicates the status of the tug as specified by one of the following values: BERTHING – berthing a tanker, DEBERTHING – deberthing a tanker, TO_HARBOUR – going to the harbour with no tanker in tow, TO_BERTHS – going to berth area with no tanker in tow PAUSE_H – in the harbour ready for the next task, PAUSE_B – in the berth area ready for the next task. The status attribute is initialised to PAUSE_H, that is, the Tug starts in the harbour.
startTime	Time stamp indicating the time the tug begins any one of the following activities: Berthing, Deberthing, MoveToBerths, MoveToHarbour. All these activities may be interrupted.
tnkr	Reference to a tanker in tow (the Tanker entity must be referenced when activities are interrupted by storms).
travelTime	The time required to complete the task currently being carried out by the tug.
anchored	Set to TRUE when the tug was stopped (i.e. is anchored) by the storm and FALSE otherwise. This attribute is used to re-initiate an activity that is interrupted by a storm.
startMoveToBerths	Use to record the time a MoveToBerths activity is first started before any interruptons by a storm may have occurred. This attribute and the next 2 attributes are required to deal with 2 different interruptions possible with the MoveToBerths activity.
returnToHarbour	Set to TRUE when the tug was interrupted by the arrival of a Tanker in the habour and it has not reached the Berths yet (see MoveToBerths interruptions).
timeReturnHarbour	Set to time required to return to harbour after MoveToBerths has been interrupted and 0 when MoveToBerths has not been interrupted.

Behavioural Components

Initialisation

Action: Initialise	
TimeSequence	< 0 >
Event SCS	R.Tug.status ← PAUSE_B R.Tug.anchored ← FALSE R.Tug.returnToHarbour ← FALSE uStorm ← FALSE RG.Berths.n ← 0 Q.BerthingList.n ← 0 Q.DeberthingList.n ← 0

Output

OUTPUTS			
Trajectory Sequences			
Name	**Description**		
TRJ[RG.Berths.n]	This trajectory sequence records the values of the attribute RG.Berths.n which represents the number of occupied berths.		
Sample Sequences			
Name	**Description**		
PHI[iC.Tanker.totalWait]	Each value in PHI[iC.Tanker.totalWait] is of the form (t_k, y_k) where y_k is the value of the attribute totalWait for some tanker entity and t_k is the time when the tug begins deberthing that tanker.		
Derived Scalar OutputVariables (DSOV's)			
Name	**Description**	**Output Sequence Name**	**Operation**
avgOccBerths	Average number of occupied berths.	TRJ[RG.Berths.n]	AVG
avgWaitTime	Average time that the tankers spend waiting for the tug.	PHI[iC.Tanker.totalWait]	AVG

User-Defined Procedures
None required.

Input Constructs

Random Variate Procedures		
Name	**Description**	**Data Model**
uLoadTime(size)	The returned value is the loading time for a tanker entity. It is assigned according to the tanker's size; i.e., SMALL, MEDIUM or LARGE).	Each of the data models is UNIFORM(Min, Max) where Min and Max take on the following values (in hours) according to the tanker size. <table><tr><td>size</td><td>Min</td><td>Max</td></tr><tr><td>SMALL</td><td>8</td><td>10</td></tr><tr><td>MEDIUM</td><td>10</td><td>14</td></tr><tr><td>LARGE</td><td>15</td><td>21</td></tr></table>
uTankerSize()	Returns the 'size' of an arriving tanker entity; value is one of SMALL, MEDIUM or LARGE.	The percentage of SMALL, MEDIUM and LARGE arriving tankers is given in the following table. <table><tr><td>Size</td><td>Percent</td></tr><tr><td>SMALL</td><td>25</td></tr><tr><td>MEDIUM</td><td>25</td></tr><tr><td>LARGE</td><td>50</td></tr></table>
DuTnkr()	Provides the values of the arrival times of tanker entities.	First arrival at t=0 and interarrival given by: EXPONENTIAL(AVGARR) where AVGARR = 8 hours.
DuStormTiming()	The return value gives the times at which storms start.	The time values from the sequence of inter-arrivals times are determined from the inter-arrival times with: EXP(LAMBDA, MIN_STORM_ARR) where LAMBDA = 1/(8 hours) and MIN_STORM_ARR = 20 hours (location parameter for the exponential distribution). The start of the first storm is defined by an inter-arrival time from t=0. The mean of the exponential distribution is 8 + 20 = 28 hours.
RuStormTiming()	The procedure provides values for the duration of the storm.	UNIFORM(MIN_SDUR, MAX_SDUR) Where MIN_SDUR = 2 hours, MAX_SDUR = 6 hours

Action: TankerArrivals	
Provides the input entity stream of arriving tanker entities.	
TimeSequence	RVP.DuTnkr()
Event SCS	iC.Tanker ← SP.Derive(Tanker)
	iC.Tanker.uSize ← RVP.uTankerSize()
	iC.Tanker.startWait ← t
	iC.Tanker.totalWait ← 0
	SP.InsertQue(Q.BerthingList, iC.Tanker)

Activity: Storm	
This activity specifies storm behaviour. The input variable *uStorm* is initialised to FALSE by the Initialise action.	
TimeSequence	RVP.DuStormTiming()
Event SCS	uStorm ← TRUE
Duration	RVP.RuStormTiming()
Event SCS	uStorm ← FALSE

Behavioural Constructs

Activity: Loading	
This activity represents the loading of a Tanker. This is a sequel activity and is initiated at the completion of the Deberthing activity.	
Causal	(iC.Tanker)
Event SCS	
Duration	RVP.uLoadTime(iC.Tanker.uSize)
Event SCS	iC.Tanker.startWait ← t
	SP.InsertQue(Q.DeberthingList, iC.Tanker)

Activity: Deberthing	
This activity represents the Tug entity towing a loaded Tanker entity from the berths to the harbour (at which point the Tanker entity leaves the model).	
Precondition	(uStorm = FALSE) AND (((R.Tug.status = PAUSE_B) AND (Q.DeberthingList.n ≠ 0)) OR ((R.Tug.anchored = TRUE) AND (R.Tug.status = DEBERTHING)))

Event SCS	IF(R.Tug.status = PAUSE_B)
	    iC.Tanker ← SP.RemoveQue(Q.DeberthingList)
	    R.Tug.tnkr ← iC.Tanker
	    SP.RemoveGrp(RG.Berths, iC.Tanker)
	    R.Tug.status ← DEBERTHING
	    R.Tug.travelTime ← DEBERTH_TIME
	ELSE
	    R.Tug.anchored ← FALSE
	    iC.Tanker ← R.Tug.tnkr
	ENDIF
	iC.Tanker.totalWait +← (t – iC.Tanker.startWait) *(Update total Wait)*
	    (Update sample sequence)
	SP.Put(PHI[iC.Tanker.totalWait], iC.Tanker.totalWait)
	R.Tug.startTime ← t

Duration	R.Tug.travelTime
Interruption Precondition	uStorm = TRUE
Event	R.Tug.travelTime –← t – R.Tug.startTime R.Tug.anchored ← TRUE R.Tug.tnkr.startWait ← t SP.Terminate()
Event SCS	R.Tug.status ← PAUSE_H SP.Leave(iC.Tanker)

Activity: Berthing	
This activity represents the Tug entity towing an empty Tanker entity from the harbour to the berths. Note that the availability of an empty berth is implicit in the fact that the tug is located in the harbour.	

Precondition	(uStorm = FALSE) AND
	(
	   ((R.Tug.status = PAUSE_H) AND (Q.BerthingList.n > 0))
	   OR
	   ((R.Tug.anchored = TRUE) AND (R.Tug.status = BERTHING))
	)

Event SCS	IF(R.Tug.status = PAUSE_H)
	   R.Tug.status ← BERTHING
	   iC.Tanker ← SP.RemoveQue(Q.BerthingList)
	   R.Tug.tnkr ← iC.Tanker
	   R.Tug.travelTime ← BERTH_TIME
	ELSE
	   R.Tug.anchored ← FALSE
	   iC.Tanker ← R.Tug.tnkr
	ENDIF
	iC.Tanker.totalWait +← (t – iC.Tanker.startWait)
	R.Tug.startTime ← t

Duration	R.Tug.travelTime
Interruption Precondition	uStorm = TRUE
Event SCS	R.Tug.travelTime –← t – R.Tug.startTime R.Tug. anchored ← TRUE iC.Tanker.startWait ← t SP.Terminate()
Event SCS	SP.InsertGrp(RG.Berths, iC.Tanker) SP.StartSequel(Loading, iC.Tanker) R.Tug.status ← PAUSE_B

Activity: MoveToHarbour	
This activity represents the Tug entity moving to the harbour area with no Tanker entity in tow. This activity can be intiated after an interruption: either an interruption of the MoveToBerths activity (while moving to the berth area without a tanker in tow, is required to return to the harbor because an empty tanker has arrived) or an interruption of the MoveToHarbour activity by a storm.	

Precondition	(uStorm = FALSE) AND
	(
	  ((R.Tug.status = PAUSE_B) AND
	    (Q.DeberthingList.n = 0) AND
	    (Q.BerthingList.n > 0) AND
	    (RG.Berths.n < RG.Berths.numBerths))
	  OR
	  ((R.Tug.anchored = TRUE) AND
	    (R.Tug.Status = TO_HARBOUR))
	  OR
	  (R.Tug.returnToHarbour = TRUE)
	)

Event SCS	IF(R.Tug.status = PAUSE_B)
	  R.Tug.status ← TO_HARBOUR
	  R.Tug.travelTime ← EMPTY_TIME
	ELSE IF(R.Tug.anchored = TRUE)
	  R.Tug.anchored = FALSE
	ELSE (Must be returning to harbour)
	  R.Tug.returnToHarbour = FALSE
	ENDIF
	R.Tug.startTime ← t

Duration	R.Tug.travelTime
Interruption Precondition	uStorm = TRUE
Event SCS	R.Tug.travelTime −← t – R.Tug.startTime R.Tug.anchored ← TRUE SP.Terminate()
Event SCS	R.Tug.status ← PAUSE_H

Activity: MoveToBerths	
This activity represents the Tug entity moving to the berths area with no Tanker entity in tow.	

Precondition	(uStorm = FALSE) AND
	(
	    ((R.Tug.status = PAUSE_H) AND
	    (Q.BerthingList.n = 0) AND
	    (RG.Berths.n > 0))
	    OR
	    ((R.Tug.anchored = TRUE) AND
	    (R.Tug.status = TO_BERTHS))
	)

Event SCS	IF(R.Tug.status = PAUSE_H)
	    R.Tug.status ← TO_BERTHS
	    R.Tug.travelTime ← EMPTY_TIME
	    R.Tug.startMoveToBerths ← t *(Time at which the*
	                     activity started before any interruptions)
	    R.Tug. timeReturnHarbour ← 0
	ELSE
	    R.Tug.anchored = FALSE
	ENDIF
	R.Tug.startTime ← t *(Time at which the*
	        MoveToBerths activity started or isresumed)

Duration	R.Tug.travelTime

Interruption
    Precondition
	(Q.BerthingList.n > 0) AND
	    (Note that the first comparison will not be true after an
	interruption by a storm, because the duration of storms exceeds
	EMPTY_TIME))
	((((t-R.Tug.startMoveToBerths) < 0.7 * EMTPY_TIME) OR
	(Q.DeberthingList.n = 0))

Event SCS	IF (R.Tug.timeReturnHarbour = 0) *(Activity has not been*
	                    interrupted by storm)
	    R.Tug.travelTime ← (t – R.Tug.startTime)
	ELSE *(Activity has been interrupted by a storm, i.e.*
	                tug is currently anchored)
	    R.Tug.travelTime ← R.Tug. timeReturnHarbour
	ENDIF
	R.Tug.status ← TO_HARBOUR
	R.Tug. returnToHarbour ← TRUE
	SP.Terminate()

Interruption Precondition	uStorm = TRUE
Event SCS	R.Tug.travelTime –← t – R.Tug.startTime *(In case there is a need to return to harbour)* R.Tug. timeReturnHarbour ← t – R.Tug.startTime R.Tug.anchored ← TRUE SP.Terminate()
Event SCS	R.Tug.status ← PAUSE_B

A1.6 Conveyor Project

A1.6.1 Problem Description

A1.6.1.1 Problem Statement

In a manufacturing plant, management is interested in automating the loading/unloading of components into/out of three machines. Two conveyors unload components from the first machine and move the components into two other machines. These conveyors have limited capacity which can result in downtime of the first machine when one of the conveyors becomes full.

Because the cost of the conveyors increases as their capacity increases, management wishes to keep the capacity as small as possible while maintaining the downtime of the first machine below a threshold of 10%.

A1.6.1.2 SUI Details

A department in a manufacturing plant processes two streams of partially completed components (A and B) using three different machines. The plant operates 24 hours/day, 7 days/week.

Components and Plant Resources

Components: The department in the manufacturing plant process two types of components called A and B.

Machines: Three machines are used to process the components. Machine 1 processes both types of components, while machine 2 processes only type A components and machine 3 processes only type B components. Figure A1.6 shows the layout of the three machines used to process the components.

Conveyors: Power roller conveyors are being added to automate the loading/unloading of components and moving components to and from the

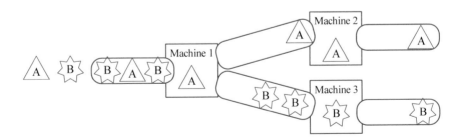

Fig. A1.6 Plant Layout

machines to improve throughput. The conveyors leading to machine 1 and leading away from machines 2 and 3 are of fixed size (they each have a capacity of three parts). Conveyor 2 between machine 1 and machine 2 and conveyor 3 between machine 1 and machine 3 have a minimum capacity of 3 parts and a maximum capacity of 10 parts.

Employees: Three employees are responsible for loading components onto and off of conveyors.

Processing Components

The components are first processed at machine 1 and upon completion each component type flows onto a separate path so that A components are processed at machine 2 and B components are processed at machine 3.

Conveyors for augmenting throughput will move components between machines and load/unload components to/from machines automatically. Employees load components onto the conveyor leading to machine 1 and remove completed components from conveyors leading away from machines 2 and 3.

When either one of conveyor 2 or 3 is full, a finished component in machine 1 destined for that conveyor will remain in machine 1 until space becomes available in on the conveyor. This prevents further processing in machine 1 and slows down the processing procedure.

A1.6.2 Project Goal and Achievement

The goal of the simulation project is to determine the smallest capacity of conveyors 2 and 3 such that the downtime for machine 1 is less than 10% of the duration of the observation interval, a value designated as an acceptable maximum downtime. The parameters introduced for this purpose are as follows:

Q.Conveyor[M2].capacity: The capacity of conveyor 2. This capacity can vary between 3 and 10.
Q.Conveyor[M3].capacity: The capacity of the conveyor 3. This capacity can vary between 3 and 10.

The following heuristic solution strategy is used.

1. Initially, set the capacity of the conveyors 2 and 3 to 3 components.
2. Run an experiment with the simulation model to determine the percentage of time each conveyor is full and the percentage of time machine 1 is down.
3. The capacity of the conveyor with the largest percentage of time it is full, is increased to accommodate one more component.
4. Repeats steps 2 and 3 until the downtime of machine 1 less than 10% of the duration of the observation interval.

Output

percentTimeConvM2Full, percentTimeConvM3Full: Percentage of time conveyor 2 is full, and percentage of time conveyor 3 is full, respectively. This output provides the basis for increasing the capacity of each conveyor between experiments as outlined in the strategy.

percentTimeDown: Percent of time machine 1 is down: Downtime is measured as the percentage of time machine 1 contains a finished component that cannot be moved out of the machine. Note that when machine 1 has no component it is not considered to be down.

Study: Steady-state study.

Observation Interval: Time units are minutes.

A1.6.3 Conceptual Model

A1.6.3.1 High-Level Conceptual Model

Simplifications

- Employees are not modelled because the focus of the study is on the capacity of the conveyors 2 and 3.
- The conveyor that brings components to machine 1 can be regarded as having an unlimited capacity, because arriving components are loaded into machine 1 when space is available.
- Conveyors that remove components from machines 2 and 3 are not modelled. This is because employees immediately remove components from these conveyors and therefore they never reach their respective capacities. In effect when either machines 2 or 3 finishes with a component, that component can be regarded as leaving the SUI.
- The time for conveyors to move components into the machines is considered negligible.

Structural View

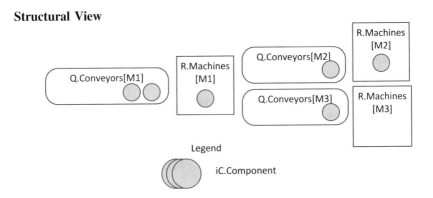

Legend

iC.Component

Entity Categories

1. iC.Component: This category (with *scope* = Transient) represents the components that are processed. The value of the attribute *type* identifies the component type (either *A* or *B*).
2. R.Machines: The machines that process the components. We use M1, M2 and M3 to represent the member identifiers of the category which has *scope* = Many[*N*] where *N* = 3. Thus the entities representing the machines are R.Machines[M1], R.Machines[M2] and R.Machines[M3].
3. Q.Conveyors: A set of three queues which represent the three conveyors that feed each of the machines. We use M1, M2, and M3 to identify the members of this category which has *scope* = Many[*N*] where *N* = 3. These identifiers also associate each conveyor with a specific machine. The queue Q.Conveyors[M1] represents the conveyor leading to machine M1 and effectively has an unlimited capacity, given that an employee continuously feeds the conveyor. The conveyors, Q.Conveyors[M2] and Q.Conveyors[M3], represent the conveyors that move components from R.Machines[M1] to R.Machines[M2] and to R.Machines[M3], respectively. The capacities of these queues are limited and defined by parameters *Q.Conveyors[M2].capacity* and *Q.Conveyors[M3].capacity*, respectively.

Behavioural View

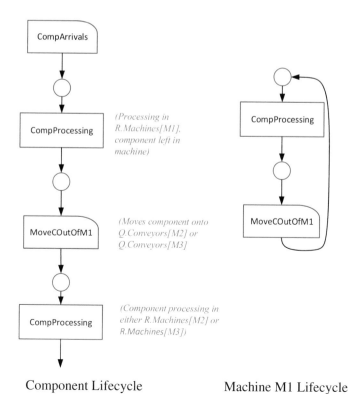

Component Lifecycle   Machine M1 Lifecycle

Scheduled Action:

CompArrivals: This scheduled action generates the input entity stream of components.

Conditional Action:

MoveCOutOfM1: When the processing of a component is completed in machine M1, this conditional action moves it either to conveyor M2 or M3 depending on whether the component type is A or B, respectively. The move, however, is conditional on there being available space on the destination conveyor. If space is unavailable, then the component remains in machine M1.

Activity Constructs:

CompProcessing: Carries out the processing of components at any of the three machines. In the case of component processing in machine M1, this activity does not move components out of that machine. The conditional action MoveCOutOfM1 performs this action when space is available on the appropriate conveyors. For machines M2 and M3, the component leaves the system once processing is completed (shown in the component life cycle diagram).

Input

Exogenous Input (Entity Streams)			
Variable	**Description**	**Domain Sequence**	**Range Sequence**
uCArr	Input entity stream variable of components.	RVP.DuArr()	1 component arrives at each arrival time.
Endogenous Input (Entity Streams)			
Variable	**Description**	**Values**	
iC.Component.uType	The type of component that is set upon arrival of the component.	RVP.uCompType()	
uProcTime	Processing times in machine M1, M2 and M3. The time is dependent on the machine. For machine M1, processing time is also depended on the type of component; i.e., A or B.	RVP.uProcTime(machineId, type)	

A1.6.3.2 Detailed Conceptual Model

Structural Components

Constants		
Name	**Description**	**Value**
M1, M2, M3	Identifier for the members of the entity categories R.Machines and Q.Conveyors which have *scope* = Many[3]. These identifiers serve to associate Q.Conveyors entities with a R.Machines entities.	0, 1, 2
Parameters		
Name	**Description**	**Value**
Q.Conveyors[M2].capacity	The maximum number of components that the Q.Conveyors[M2] entity can hold.	3 to 10
Q.Conveyors[M3].capacity	The maximum number of components that the Q.Conveyors[M3] entity can hold.	3 to 10.

Consumer Transient: Components	
Components that are processed.	
Attributes	**Description**
uType	Assigned value indicates the type of the component (either A or B)

Resource Many[3]: Machines	
The three machines in the model. Their respective identifiers are the constants M1, M2 and M3.	
Attributes	**Description**
busy	Set to TRUE when the machine is processing a component and FALSE otherwise.
component	References a Component entity that is being processed and is set to NOCOMP to indicate that there is no component present in the machine.

Queue Many[3]: Conveyors	
The three conveyors in the model. Their respective identifiers are M1, M2 and M3.	
Attributes	**Description**
n	The number of Component entities in any particular Conveyor entity.
list	The list of Component entities in any particular Conveyor entitiy.
capacity	The capacity of the conveyor. The attribute is used for R.Conveyors[M2] and R.Conveyors[M3]. This attribute is not relevant to R.Conveyors[M1].

Behavioural Components

Initialisation

Action: Initialise	
TimeSequence	< 0 >

Event SCS	FOR id in M1 to M3
	  R.Machines[id].busy ← FALSE
	  R.Machines[id].component ← NOCOMP
	  Q.Conveyors[id].N ← 0
	ENDFOR

Output

OUTPUTS			
Trajectory Sequences			
Name	**Description**		
TRJ[m1Down]	The value of the output variable m1Down indicates when machine M1 is down. It is has the value 1 when R.Machines[M1].busy is FALSE and R.Machines[M1].component is not NOCOMP and the value 0 otherwise.		
TRJ[timeConv2Full]	The value of the output variable timeConv2Full indicates when Q.Conveyors[M2] is full. It has the value 1 when Q.Conveyors[M2].n = Q.Conveyors[M2].capacity and the value 0 otherwise.		
TRJ[timeConv3Full]	The value of the output variable timeConv3Full indicates when Q.Conveyors[M3] is full. It has the value 1 when Q.Conveyors[M3].n = Q.Conveyors[M3].capacity and the value 0 otherwise.		
Derived Scalar Output Variables (DSOV's)			
Name	**Description**	**Data Sequence Name**	**Operator**
percentTimeDown	Percentage of time the machine M1 is down.	TRJ[m1Down]	Average
percentTimeC2Full	Percentage of time the conveyor M2 is full.	TRJ[timeConv2Full]	Average
percentTimeC3Full	Percentage of time the conveyor M3 is full.	TRJ[timeConv3Full]	Average

Input Constructs

Action: CompArrivals	
Arrival of a component.	
TimeSequence	RVP.DuCArr()
Event	iC.Component ← SP.Derive(Component)
	iC.Component.type ← RVP.uCompType()
	SP.InsertQue(Q.Conveyors[M1], iC.Component)

Random Variate Procedures		
Name	**Description**	**Data Model**
DuCArr()	Returns the next arrival time for a component, based on the assumption that t an arrival has occurred at current time t.	t + Exponential(MEAN) where MEAN = 7.0 minutes
uCompType()	Returns the component type (namely A or B).	Returns the value A, 55% of the time. Returns the value B, 45% of the time.

Behavioural Constructs

Activity: CompProcessing	
Processing components at any of the machines. In the case of machines M2 and M3, the component leaves the system after processing. In the case of machine M1, the component is left in the machine for removal by the conditional action.	
Precondition	UDP.MachineReadyForProcessing() ≠ NONE
Event	id ← UDP.MachineReadyForProcessing() R.Machines[id].busy ← TRUE R.Machines[id].component ← Q.RemoveQue(Q.Conveyors[id])
Duration	RVP.uProcTime(id, R.Machines[id].component.uType)

Event	R.Machines[id].busy ← FALSE
	IF(id ≠ M1) THEN *(Conditional action will remove component from*
	                              machine M1)
	      R.Machines[id].component ← NOCOMP
	      SP.Leave(R.Machines[id].component)
	ENDIF

Embedded User-Defined Procedures

Name	Description
MachineReadyForProcessing()	Returns the identifier, *id*, of a member of the R.Machines category under the following conditions: 1) Machine is not busy (R.Machines[id].busy = FALSE) 2) There is no component in the machine (R.Machines[id].component = NOCOMP) 3) There exists a component in the conveyor feeding the machine (Q.Conveyors[id].n ≠ 0) If no machine is ready for component processing, NONE is returned.

Embedded Random Variate Procedures

Name	Description	Data Model
uProcTime (machineId, type)	Provides the processing times for machines M1, M2 and M3; *machineId* has one of the id values M1, M2, or M3. The processing time for M1 is dependent on the component type given by *type* whose value is either A or B.	Exponential(MEAN) where the value of MEAN depends on the procedure parameters as follows | *machineId*	*type*	**MEAN**
M1	*A*	2.1
M1	*B*	4.2
M2	*N/A*	9.4
M3	*N/A*	10.5 | Where *N/A* means that the value of type is disregarded.

Action: MoveCOutOfM1	
Moves a component out of M1 when processing is complete and space is available on conveyor that is to receive the component.	
Precondition	UDP.ConveyorReadyForComp() ≠ NONE
Event	qid ← UDP. ConveyorReadyForComp() SP.InsertQue(Q.Conveyors[qid], R.Machines[M1].component) R.Machines[M1].component ← NOCOMP

User-Defined Procedures	
Name	**Description**

ConveyorReadyForComp()	Returns the identifier of the member in the Q.Conveyors category into which the component from R.Machine[M1] needs to be moved. The following conditions must be met to move a component:
	1. Machine is not busy (R.Machines[M1].busy is FALSE)
	2. Machine contains a component (R.Machines[M1].component ≠ NOCOMP).
	3. Space is available for on the conveyor for the component, **either**:
	    a) Component in the machine is type A and Conveyor M2 has space (R.Machines[M1].component.uType = A AND Q.Conveyors[M2].n < Q.Conveyors[M2].capacity)
	    OR
	    b) Component in the machine is type B and Conveyor M3 has space (R.Machines[M1].component.uType = B AND Q.Conveyors[M3].n < Q.Conveyors[M3].capacity
	If 3(a) is true, return M2, if 3(b) is true, return M3, otherwise return NONE (no component can be moved).

A1.7 ACME Manufacturing

A1.7.1 Problem Description

A1.7.1.1 Problem Statement

ACME Manufacturing produces widgets using an automated production line composed of a series of workstations, each of which carries out a specific task. That task can be carried out by any one of several identical processing machines that operate within each workstation. Each partially completed widget moves from one workstation to its 'downstream' neighbour for further processing. This transition occurs via a buffer.

The company's income is determined by the number of widgets that are produced. That number depends on the number of machines in each of the workstations and the size of the buffers that are used between workstations. However, there is a cost associated with the machines in the workstations and with the locations in the buffers.

The company wishes to determine the number of machines in each workstation and the size (number of locations) in each of the buffers such that profit (income minus the costs associated with the machines and buffer locations) is maximized.

A1.7.1.2 SUI Details

Production Line Layout

As shown in Fig. A1.7, the constituents of the automated production line are as follows:

(1) **Workstations**: There are four workstations each of which has a collection of identical machines. Because of space limitations, the number of machines in any workstation cannot exceed 3.

(2) **Machines**: At each workstation, a partially completed widget undergoes some processing (with random processing time) by one of the machines in that workstation. The cost of operating a machine (in any one of the workstations) for 30 days is $25,000.

(3) **Buffers**: After processing by a machine, a partially completed widget is passed on to a buffer where it is held until there is an available machine in the downstream workstation. For practical reasons, the size of each buffer cannot exceed 10 positions. The maintenance cost of each position in a buffer is $1000 for a 30 days period.

(4) **Blank Parts Inventory**: A large inventory of blank parts is available to feed the first workstation in the production line.

(5) **Shipping Area**: The widgets produced in the last workstation are stored in a widget inventory.

(6) **Widget**: The circles in Fig. A1.7 represent widgets in intermediate states of completion. Each widget that is produced provides an income of $200.

Widget Production

To produce a widget, a part must be processed by one of the machines in each of the four workstations (numbered 1–4) that are arranged in a linear sequence. Parts are transferred from a machine in one workstation to a machine in the next using buffers (numbered 1–3). This transfer to a buffer happens as soon as the machine processing is complete, provided space in the buffer is available. When buffer K is full, a partially completed widget whose processing by a machine in workstation K has been completed, must remain in that machine until space becomes available in buffer K. The production of a widget involves the following steps:

1. Blank parts enter workstation 1 as soon as a machine is available in that workstation. Thus, all machines in workstation 1 will always be occupied by a part. When a machine has finished processing a part, and space is available in buffer 1, the part is moved to the buffer and another blank part is obtained from the blank parts inventory for processing. If there is no room in buffer 1, then the

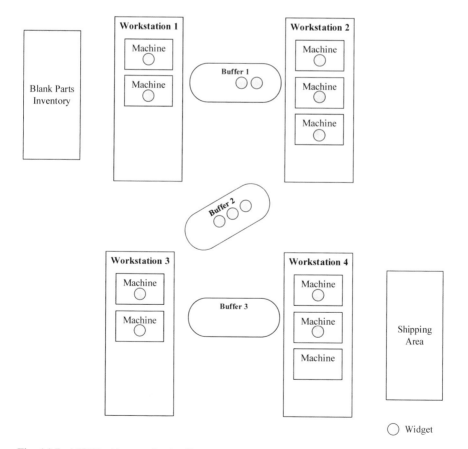

Fig. A1.7 ACME widget production line

completed part is left in the machine until a space becomes available in buffer 1. Thus, a machine in workstation 1 can either be busy (processing a part) or blocked (waiting for space in buffer 1).

2. For workstations 2 and 3, a part enters the workstation when a machine is available and there is a part in buffers 1 and 2, respectively (source buffers). When part processing is completed, it is moved to buffers 2 and 3, respectively (destination buffers) if space is available. If there is no space in the destination buffer, then the completed part remains in the machine until space becomes available in destination buffer. The machines in workstations 2 and 3 can either be idle (no part available in the source buffer), busy (processing a part) or blocked (waiting for space in the destination buffer).

3. A partially completed widget in buffer 3 enters workstation 4 when a machine is available in that workstation. A completed widget is produced when a machine in workstation 4 completes its processing. The completed widget is immediately moved from the machine to the shipping area. The machines in workstation 4 can either be idle (no part available in buffer 3) or busy (processing a part).

Widgets are produced 24 hours/day, 7 days/week.

A1.7.2 Project Goal and Achievement

The project goal is to determine the number of machines in each of the four workstations and the size (number of locations) in each of the three buffers which maximizes profit (income minus the cost of machines and buffer locations). In this regard, we consider a seven-dimensional space, Ω, with coordinates $\mathbf{p} = (p_1, p_2, p_3, \ldots, p_7)$ with the following interpretation:

(a) p_k, $k = 1, 2, 3, 4$ corresponds to the number of machines in workstation k with $k = 1, 2, 3, 4$.

(b) p_{k+4}, $k = 1, 2, 3$ corresponds to the number of spaces in the buffer area k with $k = 1, 2, 3$.

A search space of interest, called Ω^*, is defined to be that subspace of Ω for which:

(a) p_k is constrained to the integer interval $[1, 3]$ for $k = 1, 2, 3, 4$.

(b) p_{k+4} is constrained to the integer interval $[1, 10]$ for $k = 1, 2, 3$.

The goal of the project is to determine, via an optimization strategy, the seven dimensional vector $\mathbf{p}^*$ with the parameter search space Ω^* which yields a maximum value for profit where profit is specified in Eq. (A1.1):

$$\text{profit} = \psi N - \left(C_m \sum_{j=0}^{3} p_j + C_s \sum_{j=0}^{2} p_{j+4} \right) \tag{A1.1}$$

where

N is the number of widgets produced over the time interval $I = 30$ days,

$\psi = \$200$ is the income (dollars) per widget produced,

$C_m = \$25,000$ is the cost (dollars) of operating a machine over the time interval I, $C_s = \$1000$ is the maintenance cost of a buffer storage position over the time interval I.

Output

profit: The value of profit as specified in Eq. (A1.1).

Study: Steady-State Study.

Observation interval: Time units are hours.

A1.7.3 Conceptual Model

A1.7.3.1 High-Level Conceptual Model

Simplification

We consider the inventory of blank parts to be unlimited.

Structural View

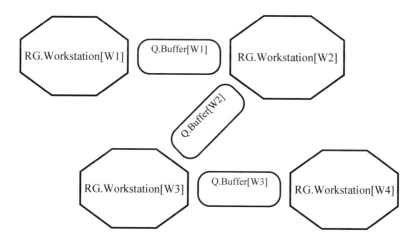

Entity Categories:

- RG.WorkStation: The four workstations where parts are processed are collectively referenced by this entity category which has *role* = Resource Group and *scope* = Many[4].
- Q.Buffer: The three buffer areas used to transfer parts between workstations are collectively referenced by this entity category which has *role* = Queue and *scope* = Many[3].

Note that entity categories are not required for either the partially completed widgets as they flow through the manufacturing process or for the individual machines within the workstations.

Behavioural View

The life-cyle diagram for the widget is shown in the following behavioural diagram.

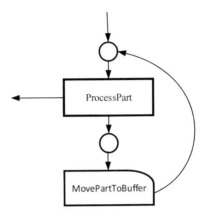

Actions

MovePartToBuffer: When a machine in any one of the first three workstations completes its processing task, the part needs to be moved into a buffer that provides the transfer mechanism to the downstream workstation. This conditional action carries out the transfers when appropriate conditions exist.

Activities

ProcessPart: This conditional activity carries out the processing of a part by a machine in any one of the four workstations. In the case of the last workstation, the fabrication of the widget is completed and that completion is duly registered.

Input

Endogenous Input (Semi-Independent)		
Variable Name	**Description**	**Value(s)**
ProcessingTime(wid)	Time required for a machine within the workstation whose member identification is wid to process a part.	RVP.uProcessingTime(wid)

A1.7.3.2 Detailed Conceptual Model

Structural Components

Constants and Parameters

Parameters		
Name	**Description**	**Value**
M	A vector of length 4 whose entries provide the number of machines within each of the four members of the Workstation category	$<m_1, m_2, m_3, m_4>$ where $1 \le m_i \le 3$.
S	A vector of length 3 whose entries provide the number of spaces in each of the three members of the Buffer category.	$<s_1, s_2, s_3>$ where $1 \le s_i \le 10$.

Entity Structures

Resource Group Many[4]: Workstation	
A set of workstations where widgets are processed. Member identifiers are W1, W2, W3, and W4.	
Attributes	**Description**
n	The number of occupied machines within the Workstation.
numMachines	Number of machines within the Workstation.
numFinished	Number of machines within the Workstation that have a finished widget.

Queue Many[3]: Buffer	
A set of buffers used to transfer parts between workstations. Identifiers are W1, W2, and W3.	
Attributes	**Description**
n	Number of widgets in the Buffer.
size	Number of spaces in the Buffer.

Behavioural Components
Initialisation

Action: Initialise	
TimeSequence	< 0 >

Event SCS	FOR id = W1 to W3
	    Q.Buffers[id].n ← 0
	    Q.Buffers[id].size ← S[id]
	ENDFOR
	FOR id = W1 to W4
	    RG.Workstations[id].n ← 0
	    RG.Workstations[id].numMachines ← M[id]
	    RG.Workstations[id].numFinished ← 0
	ENDFOR
	SSOV.Np = 0

Output

OUTPUTS	
Simple Scalar Output Variables (SSOV's)	
Name	**Description**
Np	Number of widgets produced over the 30 day observation interval

Input Constructs

Random Variate Procedures		
Name	**Description**	**Data Model**
uProcessingTime(wid)	Gives the time to process a part in that member of the Workstation category whose identifier is wid.	Exponentially distributed with one of the following means according to the value of wid: W1: 1/3 hour W2: 0.5 hours W3: 0.2 hours W4: 0.25 hours

Action: MovePartToBuffer	
Moves a widget from a workstation into a buffer.	
Precondition	UDP.CanMovePart() ≠ NONE
Event SCS	id ← UDP.CanMovePart() RG.Workstation[id].n –← 1 RG.Workstation[id].numFinished –← 1 Q.Buffer[id].n +← 1
Embedded User Defined Procedure	
Name	**Description**
CanMovePart()	Returns the identifier, id, of a member of the Buffer category if it is true that: 1) The buffer has free space, (i.e., Q.Buffer[id].n < Q.Buffer[id].size) and 2) A machine within the workstation that feeds the buffer contains a finished part, (i.e., RG.Workstation[id].numFinished > 0). If there is no such member of the Buffer category, the value NONE is returned.

Behavioural Constructs

Activity: ProcessPart	
This activity can be instantiated for any machine in any of the workstations. It represents the processing of a widget by one of the machines in any particular workstation.	
Precondition	UDP.CanProcessPart() ≠ NONE
Event SCS	id ← UDP.CanProcessPart() IF id ≠ W1 Q.Buffer[id – 1].n –← 1 ENDIF RG.Workstation[id].n +← 1
Duration	RVP.ProcessingTime(id)

Event SCS	IF(id = W4)
	 RG.Workstation[id].n -← 1
	 SSOV.Np +← 1
	 ELSE
	 RG.Workstation[id].numFinished +← 1
	 ENDIF

Embedded User Defined Procedure	
Name	**Description**
canProcessPart()	Returns the identifier, id, of a member of the Workstation category if it is true that: 1) The workstation has a free machine, (i.e., RG.Workstation[id].n < RG.Workstation[id].numMachines). and 2) The buffer that feeds the workstation contains a part, (i.e., Q.Buffer[id-1].n ≠ 0). For R.Workstation[W1], this condition is always TRUE. If there is no such member of the Workstation category, the value NONE is returned.

Annex 2: Probability and Statistics Primer

A2.1 Motivation

The exploration of a number of important issues relating to the development of DEDS-based models and to the correct interpretation of the results acquired from experiments with simulation programs developed for such models (i.e., 'simulation experiments') requires some familiarity with the fundamental notions of probability theory. Our purpose in this Annex is to briefly review these so that these discussions (e.g., in Chaps. 3, 4 and 7) will have a reasonable foundation. Our presentation is relatively informal; more extensive and mathematically comprehensive treatments can be found in references such as [1–5].

A2.2 Random Experiments and Sample Spaces

We begin with the basic notion of a *random experiment*, which is an experiment whose outcome is uncertain or unpredictable. Although unpredictable, it is nevertheless assumed that the outcome of such an experiment will fall within some known set of possible outcomes called the *sample space*. A sample space that consists of a finite number, or an infinite sequence, of members is called a discrete sample space. A sample space may also be an interval or a collection of disjoint intervals from the real line, in which case the sample space is said to be continuous.

Consider, for example, the random experiment that corresponds to rolling two dice: one red and the other green. If we regard the outcome of this experiment to be the pair of integers (n_1, n_2) which corresponds to the number of dots on the red and green dice, respectively, then the sample space is the set

$$\widehat{S} = \{(1,1), (1,2), , \ldots, (6,6)\}$$

which is discrete and has 36 members. An alternate representation of $\widehat{S}$ which provides visual features that are helpful in later examples is shown below:

© Springer Nature Switzerland AG 2019
L. G. Birta and G. Arbez, *Modelling and Simulation*, Simulation Foundations,
Methods and Applications, https://doi.org/10.1007/978-3-030-18869-6

$$\widehat{S} = \begin{Bmatrix} (1,1) & (1,2) & (1,3) & (1,4) & (1,5) & (1,6) \\ (2,1) & (2,2) & (2,3) & (2,4) & (2,5) & (2,6) \\ (3,1) & (3,2) & (3,3) & (3,4) & (3,5) & (3,6) \\ (4,1) & (4,2) & (4,3) & (4,4) & (4,5) & (4,6) \\ (5,1) & (5,2) & (5,3) & (5,4) & (5,5) & (5,6) \\ (6,1) & (6,2) & (6,3) & (6,4) & (6,5) & (6,6) \end{Bmatrix} \qquad (A2.1)$$

Or alternately, consider the experiment of flipping a single coin. Here, there are only two members in the sample space, and these are usually designated as 'heads' and 'tails' (the possibility of the coin landing on its edge is excluded). In other words, the discrete sample space in this case is

$$\widehat{S} = \{\text{heads}, \text{tails}\}$$

A continuous sample space can be illustrated by considering a manufacturing process that produces ball bearings that have a nominal diameter of 1 cm. Because of inherent imprecision at the various manufacturing stages, the actual diameters of the ball bearings have random variations. A basic screening device ensures that all bearings that are not in the range [0.9, 1.1] cm are automatically rejected. Before packaging the bearings in the 'accepted stream', the diameter of each one is measured as part of a quality control procedure. The sample space for this experiment (the result obtained from the diameter measurements) is, at least in principal, the interval [0.9, 1.1] (The qualifier 'at least in principal' is intended to acknowledge the finite precision of any measuring device.).

The notions of probability theory are, of course, equally applicable to random experiments with either discrete or continuous sample spaces. However, acquiring an appreciation for the ideas involved is considerably simpler within the framework of discrete sample spaces. Consequently, the presentation that follows first explores the fundamentals exclusively in the simpler framework of discrete sample spaces. Extension of these notions to the case of continuous sample spaces is presented later.

A2.3 Discrete Sample Spaces

A2.3.1 Events

An *event* is any subset of a sample space. The individual elements that populate a discrete sample space are therefore valid events, but they are somewhat special and hence are usually called *simple events*. A total of 2^n events can be associated with a sample space having n elements. Included here is the entire sample space as well as the empty set.

Within the context of a random experiment, an event, $\widetilde{E}$, is said to occur if the outcome of the experiment is a member of $\widetilde{E}$. For example, in the rolling dice experiment, we might designate an event, $\widetilde{E}$, as

$$\widetilde{E} = \{(1,6), (2,5), (3,4), (4,3), (5,2), (6,1)\}$$

This set has been constructed by selecting all the possible experiment outcomes for which the sum of the number of dots displayed on the two dice is 7. Thus, this event $\widetilde{E}$ occurs if a 7 is rolled.

As we have pointed out, a large number of events can be associated with a random experiment. It is of some interest to observe that predefined events can be combined to yield 'new' events. There are two basic ways of carrying out such a combination, namely, intersection and union:

(a) The intersection of two events E_1 and E_2 is denoted by $E_1 \cap E_2$ and represents the event that corresponds to those experiment outcomes (simple events) that are common to both E_1 and E_2. Clearly, $E_1 \cap E_2 = E_2 \cap E_1$. Note that it is possible that $E_1 \cap E_2$ is the empty set, ϕ.
(b) The union of two events E_1 and E_2 is denoted by $E_1 \cup E_2$ and represents the event that corresponds to those experiment outcomes (simple events) that are either in E_1 or in E_2 or in both. Clearly, $E_1 \cup E_2 = E_2 \cup E_1$.

A2.3.2 Assigning Probabilities

A fundamental concern in probability theory is with the characterization of the likelihood of the occurrence of events where the notion of an 'event' is interpreted in the special sense outlined above. However, a prerequisite for any such exploration is the assignment of a 'valid' set of probabilities to each of the simple events in the sample space under consideration. Each such assigned value is intended to reflect the likelihood that the simple event in question will be the outcome of the random experiment.

Consider then a sample space, S, with n members (i.e., n simple events):

$$S = \{s_1, s_2, s_3, \ldots, s_n\}$$

The probability assigned to the simple event, s_k, is a non-negative real number denoted by $p(s_k)$. An assignment of probabilities to the members of S is *valid* if

$$\sum_{k=1}^{n} p(s_k) = 1 \qquad (A2.2)$$

Notice that because $p(s_k) \geq 0$ for each k, the validity condition of Eq. (A2.2) naturally implies the constraint that $p(s_k) \leq 1$ for each k.

The assignment of a valid set of probabilities to the simple events in a sample space has no fundamental and/or formal basis. However, the chosen values generally do have a logically or psychologically satisfying basis. Often they reflect some idealized circumstance (such as an 'ideal' coin or a pair of "unbiased" dice). In any case, once a valid set of probabilities has been assigned, it is often referred to as a 'probability model' for the random phenomenon under consideration.

To illustrate, let's return to the rolling dice example. A valid assignment of probabilities to each of the 36 possible outcomes is 1/36. This particular assignment corresponds to an assumption that both the red and green dice are 'unbiased' and consequently each of the 36 possible outcomes is equally likely. This is a particular case of the general result that if a random experiment has N possible outcomes and each is equally likely, then assigning a probability of $1/N$ to each of the possible outcomes (i.e., to each of the simple events in the sample space) is both a valid and a logically satisfying assignment of probabilities.

Suppose E is an event associated with some random experiment. The notation $P[E]$ represents the probability that the outcome of the experiment will be the event E. In particular, note that for any simple event, $s_k, P[s_k] = p(s_k)$. Furthermore, if E_a is the event

$$E_a = \{s_1, s_2, \ldots, s_m\}$$

where each s_j is a simple event, then

$$P[E_a] = \sum_{i=1}^{m} p(s_i) \tag{A2.3}$$

Suppose in the rolling dice example, we consider the event:

$$E = \{(1, 6), (2, 5), (3, 4), (4, 3), (5, 2), (6, 1)\}$$

As pointed out earlier, this event E occurs if (and only if) a 7 is rolled. From Eq. (A2.3), it follows that $P[E] = 6/36 = 1/6$; in other words, the probability of rolling a 7 is 1/6.

A fundamental result that relates the union and intersection of two events E_1 and E_2 is the following:

$$P[E_1 \cup E_2] = P[E_1] + P[E_2] - P[E_1 \cap E_2] \tag{A2.4}$$

A2.3.3 Conditional Probability and Independent Events

Recall again that a large number of events can be associated with a random experiment. The notion of 'conditional probability' is concerned with the following question: what impact can knowledge that an event, say E_2, has occurred have on the probability that another event, say E_1, has occurred? As an extreme case (in the context of the rolling dice example), suppose E_1 is the event 'the sum of the dots showing is odd' and E_2 is the event 'the sum of the dots showing is seven'. Clearly, if it is known that E_2 has occurred, then the probability of E_1 is 1 (i.e., certainty). Without the knowledge that E_2 has occurred, $P[E_1] = 0.5$ (this can be easily verified by examining the sample space $\widehat{S}$ given in Eq. (A2.1)).

In general, if E_1 and E_2 are two events associated with a random experiment, then the probability that the event E_1 has occurred given that event E_2 has occurred (hence $P[E_2] \neq 0$) is called the conditional probability of E_1 given E_2. Such a conditional probability is written as $P[E_1|E_2]$. The following is a fundamental result relating to conditional probabilities:

$$P[E_1|E_2] = P[E_1 \cap E_2]/P[E_2] \tag{A2.5}$$

Notice that if the events E_1 and E_2 have no simple events in common [i.e., $E_1 \cap E_2 = \phi$ (the empty set)], then $P[E_1|E_2] = 0$ because $P[\phi] = 0$. This result is entirely to be expected because if E_2 has occurred and E_1 shares no simple events with E_2, then the occurrence of E_1 is not possible. Notice also that if E_1 and E_2 are mutually exclusive ($E_1 \cap E_2 = \phi$), then there is a suggestion of a 'lack of independence' (i.e., dependence) in the sense that if one occurs, then the simultaneous occurrence of the other is excluded. This notion of independence is explored further in the discussion that follows.

From Eq. (A2.5), it necessarily follows that the conditional probability of E_2 given E_1 is

$$P[E_2|E_1] = P[E_1 \cap E_2]/P[E_1]$$

where we have incorporated the fact that $E_2 \cap E_1 = E_1 \cap E_2$ Combining this result with Eq. (A2.5) gives

$$P[E_1 \cap E_2] = P[E_1] * P[E_2|E_1] = P[E_2] * P[E_1|E_2] \tag{A2.6}$$

Earlier, we defined E_1 and E_2 to be two specific events within the context of the rolling dice example (E_1 is the event 'the sum of the dots showing is odd' and E_2 is the event 'the sum of the dots showing is seven'). We already know $P[E_1] = 0.5$ and $p[E_2] = 1/6$. Also $E_1 \cap E_2 = E_2$ (because E_2 is a subset of E_1); hence $P[E_1 \cap E_2] = p[E_2]$. Thus it follows from Eq. (A2.1) that

$$P[E_1|E_2] = P[E_2]/P[E_2] = 1$$

which confirms the result from our earlier intuitive arguments.

Consider, on the other hand, the conditional probability $P[E_2|E_1]$, namely, the probability of E_2 given E_1. From Eq. (A2.5), we have

$$P[E_2|E_1] = P[E_2 \cap E_1]/P[E_1] = (1/6)/0.5 = 1/3$$

In other words, the probability that the outcome of the roll is 7 is doubled if it is already known that the outcome of the roll has an odd sum.

To illustrate another possibility, consider the case where we alter the definition of E_2 to be the event that the red dice shows an even number. Now

$$E_1 \cap E_2 = \{(2,1), (2,3), (2,5), (4,1), (4,3), (4,5), (6,1), (6,3), (6,5)\}$$

and $P[E_1 \cap E_2] = 1/4$ Then from Eq. (A2.5), we have

$$P[E_1|E_2] = P[E_1 \cap E_2]/P[E_2] = 0.25/0.5 = 0.5$$

Here, the information that event E_2 has occurred (red dice shows an even number) does *not* alter the probability of the occurrence of E_1 (the sum of the dots is odd).

A particularly important relationship that can exist between two events is that of independence. The event E_1 is said to be independent of the event E_2 if

$$P[E_1|E_2] = P[E_1] \tag{A2.7}$$

In other words, knowledge of event E_2 does not change the probability of event E_1. This is the case that is illustrated in the previous example.

From Eq. (A2.6), it follows that $P[E_1]$ in Eq. (A2.7) can be replaced with $P[E_2] * P[E_1|E_2]/P[E_2|E_1]$ which then gives the result that

$$P[E_2|E_1] = P[E_2]$$

In other words, if event E_1 is independent of event E_2, then E_2 is independent of E_1.

Note now that if the independence condition $P[E_1|E_2] = P[E_1]$ is substituted in Eq. (A2.5), then we get the following equivalent condition for independence:

$$P[E_1 \cap E_2] = P[E_1] * P[E_2] \tag{A2.8}$$

which, in particular, implies that $P[E_1 \cap E_2] \neq 0$ is a necessary condition for the events E_1 and E_2 to be independent.

A2.3.4 Random Variables

Suppose that S is a sample space. A *random variable*, X, is a function which assigns a real value to each $s \in S$ (This is sometimes equivalently expressed by saying that X is a mapping from the sample space, S, to some particular set of the real numbers.). In more formal terms, a random variable is a function whose domain is a sample space and whose range is a set of real numbers. If Ψ represents the range, then this assertion, in conventional mathematical notation, is written as

$$X : S \rightarrow \Psi \tag{A2.9}$$

The feature that needs to be emphasized here is that a random variable is not, in fact, a variable but rather is a function! In spite of this apparently inappropriate terminology, the usage has become standard. It is often useful to review this definition when related concepts become cloudy.

It is interesting to observe that if $X : S \rightarrow \Psi$ is a random variable and g is a real-valued function whose domain includes the range of X, i.e.,

$$g : \Phi \rightarrow R$$

where $\Psi \subseteq \Phi$ then $X^* = g(X)$ is also a random variable because for each $s \in S, X^*(s)$ has the numerical value given by $g(X(s))$. In other words, X^* is likewise a mapping from the sample space S to the real numbers.

The notation of Eq. (A2.9) conveys the important information that X maps each member of S onto a value in Ψ. However, it does not reveal the rule that governs how the assignment takes place. More specifically, if $s \in S$, we need to know the rule for determining the value in Ψ that corresponds to s (this value is denoted by $X(s)$).

To illustrate, let's continue with the earlier example of the pair of rolled dice and introduce a random variable Y whose value is the sum of the dots showing on the two dice. In this case, the domain set is $\widehat{S}$ as given in Eq. (A2.1) and the range set, Ψ_y, is

$$\Psi_y = \{2, 3, 4, 5, \ldots, 11, 12\}$$

The definition for Y is

$$Y(s) = n_1 + n_2 \text{ when } s = (n_1, n_2)$$

The random variable Y is one of many possible random variables that can be identified for this random experiment of throwing two (distinct) dice. Consider, for example, Z whose value is the larger of the two numbers that show on the pair of dice. The range set in this case is

$$\Psi_z = \{1, 2, \ldots, 6\}$$

and the definition for Z is

$$Z(s) = \max(n_1, n_2) \text{ when } s = (n_1, n_2)$$

Both random variables Y and Z have the same domain (namely, $\widehat{S}$ as given in Eq. (A2.1)) but different range sets. Note also that in both of these cases, the size of the range set is smaller than the size of the domain set, implying that several values in the sample space get mapped onto the same value in the range set (in other words, both Y and Z are 'many-to-one' mappings). This is frequently the case but certainly is not essential. In the discussion that follows, we will frequently make reference to the 'possible values' which a random variable can have; this will be reference to the range set for that random variable.

Consider the random variable $X : S \rightarrow \Psi$. Suppose we use # to represent any one of the relational operators $\{=, <, >, \leq, \leq \}$. Observe now that for each $x \in \Psi$, the set

$$\{s \in S : X(s) \# x\}$$

is, in fact, an event. The conventional simplified notation for this event is $(X \# x)$, and the probability of its occurrence is denoted by $P[X \# x]$. With any specific choice for the operator #, $P[X \# x]$ can be viewed as a function of x. Two particular choices have special relevance:

1. Choose # to be '='. This yields $F(x) = P[X = x]$ which is generally called the probability mass function (or simply the probability function) for the random variable X.
2. Choose # to be '$\leq$'. This yields $F(x) = P[X \leq x]$ which is generally called the cumulative distribution function (or simply the distribution function) for the random variable X.

The cumulative distribution function, F, for the random variable, X, has several important properties that are intuitively apparent:

- $0 \leq F(x) \leq 1, \ -\infty < x < \infty$
- If $x_1 < x_2$, then $F(x_1) \leq F(x_2)$ i.e., $F(x)$ is non-decreasing.
- $\lim\limits_{x \to -\infty} F(x) \to 0$ and $\lim\limits_{x \to \infty} F(x) \to 1$.

Earlier, we introduced the random variable $Y : \widehat{S} \rightarrow \Psi_y$ to represent the sum of the dots showing on the dice in our dice-rolling example. The probability mass function for Y is shown in Fig. A2.1a, and the cumulative distribution function is shown in Fig. A2.1b.

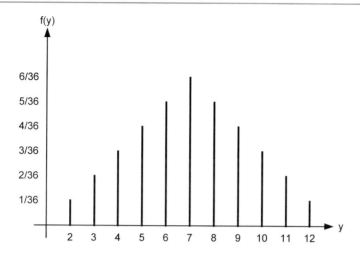

Fig. A2.1a Probability mass function for the random variable Y

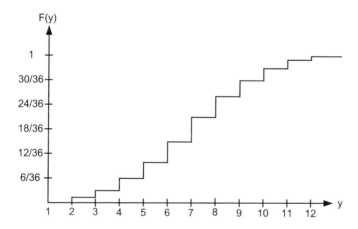

Fig. A2.1b Cumulative distribution function for the random variable Y

A2.3.5 Expected Value, Variance and Covariance

Let X be a random variable with possible values x_1, x_2, ..., x_n and probability mass function $f(x)$. Thus the possible values occur with probabilities $f(x_1)$, $f(x_2)$, ..., $f(x_n)$, respectively. The *expected value* (or *mean*) of X is denoted by $E[X]$ and has the value:

$$E[X] = \sum_{k=1}^{n} x_k f(x_k) \qquad (A2.10)$$

In other words, expected value of X is the weighted sum of its possible values where each value is weighted by the probability of its occurrence.

Consider again the random variable Y that represents the sum of the dots showing on the pair of rolled dice. From the information presented in Fig. A2.1a, it can be easily verified that $E[Y] = 7$.

Suppose X is a random variable with possible values x_1, x_2, ..., x_n and probability mass function $f(x)$. Let the random variable Y be specified as $Y = g(X)$; then:

$$E[Y] = E[g(X)] = \sum_{k=1}^{n} g(x_k) f(x_k) \qquad (A2.11)$$

Using (A2.11), it can be easily demonstrated that

$$E[aX + b] = aE[X] + b \qquad (A2.12)$$

The expected value of a random variable, X, is one of the parameters of its probability mass function, and it serves to reflect where it is 'located' or 'centred' in the sense that it provides a rough indication of the middle of the likely values that X can assume. Another property (or parameter) that is generally of interest is a measure of the spread or dispersion of the probability mass function. For this purpose, we might consider evaluating the expected value of the deviation of X from its mean, namely, $E[X - \mu_x]$ where we have introduced μ_x as an alternate symbol for $E[X]$ to simplify notation. However, from (A2.12) it follows (taking $a = 1$ and $b = -E[X]$) that

$$E[X - \mu_x] = E[X] - E[X] = 0$$

which shows that this particular measure is not useful.

The measure of dispersion that is generally used is called the variance of X which is denoted by $\mathrm{Var}[X]$. Its definition is

$$\mathrm{Var}[X] = E\left[(X - \mu_x)^2\right] = \sum_{k=1}^{n} (x_k - \mu_x)^2 f(x_k) \qquad (A2.13)$$

where, as earlier, we assume that x_1, x_2, ..., x_n are the possible values for the random variable X and its probability mass function is $f(x)$. A related notion is the *standard deviation* of X which is normally denoted by σ_x and defined as

$$\sigma_x = \sqrt{\text{Var}[X]}$$

(hence $\text{Var}[X] = \sigma_x^2$).

Two properties relating to the variance of X are especially noteworthy and can be easily demonstrated:

- $\text{Var}[X] = E[X^2] - \mu_x^2$
- $\text{Var}[aX + b] = a^2 \text{Var}[X]$ where a and b are arbitrary constants.

The second of these (with $a = 1$) reinforces the dispersion feature of the variance notion inasmuch as it demonstrates (in a formal mathematical sense) that the addition of a constant to a random variable does not alter the dispersion of its distribution, a result which is entirely consistent with intuition.

It is sometimes convenient to 'normalize' a random variable, X, by creating an associated variable which we denote by Z, where

$$Z = (X - \mu_x)/\sigma_x$$

The random variable Z is called the standardized random variable corresponding to X. It can be demonstrated that $E[Z] = 0$ and $\text{Var}[Z] = 1$.

Covariance is a notion that relates to a pair of random variables X and Y and provides a measure of their mutual dependence. The covariance of the random variables X and Y is denoted by $\text{Cov}[X, Y]$ and is, by definition,

$$\text{Cov}[X, Y] = E[(X - \mu_x)(Y - \mu_y)]$$

where $\mu_x = E[X]$ and $\mu_y = E[Y]$. An alternate expression can be obtained by expanding the right-hand side and using the distributive property of the expected value operator. This yields

$$\text{Cov}[X, Y] = E[XY] - E[X]E[Y] \tag{A2.14}$$

The covariance relationship is especially useful because it makes possible a convenient expression for the variance of the sum of two random variables. We have that

$$
\begin{aligned}
\text{Var}[X + Y] &= E\left[(X + Y - \mu_x - \mu_y)^2\right] \\
&= E\left[(X - \mu_X)^2 + (Y - \mu_y)^2 + 2(X - \mu_x)(Y - \mu_y)\right] \tag{A2.15} \\
&= \text{Var}[X] + \text{Var}[Y] + 2\,\text{Cov}[X, Y]
\end{aligned}
$$

On the basis of Eqs. (A2.14) and (A2.15), it is now possible to show that if the random variables X and Y are independent, then the variance of their sum is equal to the sum of their variances. Note that independence implies that

$$P[X = x_j, Y = y_k] = P[X = x_j]P[Y = y_k]$$
$$\text{for all } x_j \text{ and } y_k.$$

(A2.16)

Observe that the desired result follows directly from Eq. (A2.15) if $\text{Cov}[X, Y] = 0$ when X and Y are independent. To show that this is indeed the case, we only need to show (from Eq. (A2.14)) that $E[XY] = E[X]E[Y]$ when X and Y are independent:

$$
\begin{aligned}
E[XY] &= \sum_{j=1}^{n}\sum_{k=1}^{m} x_j y_k P[X = x_j, Y = y_k] \\
&= \sum_{j=1}^{n}\sum_{k=1}^{m} x_j y_k P[X = x_j]P[Y = y_k] \\
&= \sum_{j=1}^{n} x_j P[X = x_j] \sum_{k=1}^{m} y_k P[Y = y_k] \\
&= E[X]E[Y]
\end{aligned}
$$

(A2.17)

where the above development has used Eq. (A2.16). Thus, we have demonstrated that if the random variables X and Y are independent, then

$$\text{Var}[X + Y] = \text{Var}[X] + \text{Var}[Y]$$

(A2.18)

In general, the 'strength' of the relationship between the random variables X and Y is measured by a dimensionless quantity $\rho[X, Y]$ called the correlation coefficient where

$$\rho[X, Y] = \frac{\text{Cov}[X, Y]}{\sqrt{\text{Var}[X]\text{Var}[Y]}}$$

(A2.19)

When X and Y are independent, then $\rho[X, Y] = 0$. It can be shown that in the general case, the value of $\rho[X, Y]$ lies in the interval $[-1, +1]$. When the correlation coefficient $\rho[X, Y]$ is positive, a positive correlation exists between X and Y which implies that if a large value is observed for one of the variables, then the observed value for the other will likely also be large. Negative correlation indicates that when an observation of one of the variables is large, the observation for the other will tend to be small. The value of the correlation coefficient reflects the level of this dependency. Values close to $+1$ or -1 indicate high levels of correlation.

A2.3.6 Some Discrete Distribution Functions

A2.3.6.1 Bernoulli Distribution

A random variable, X, has a Bernoulli distribution with parameter p if it has the two possible values 1 and 0 and $X = 1$ with probability p and $X = 0$ with probability (necessarily) of $(1 - p)$. X could, for example, characterize a random experiment that has two possible outcomes, e.g., 'success' or 'failure' (such binary-valued experiments are often called Bernoulli trials). The probability mass function for X is

$$f(1) = P[X = 1] = p$$
$$f(0) = P[X = 0] = (1 - p)$$

The mean value of X is $E[X] = f(1) + 0\,f(0) = p$ and the variance of X is $\mathrm{Var}[X] = p(1 - p)^2 + (1 - p)\,(0 - p)^2 = p\,(1 - p)$.

A2.3.6.2 Binomial Distribution

The binomial distribution can be viewed as a generalization of the Bernoulli distribution. More specifically, if X is a random variable that represents the number of 'successes' in n consecutive Bernoulli trials, then X has a binomial distribution. In other words, the binomial distribution relates to the probability of obtaining x successes in n independent trials of an experiment for which p is the probability of success in a single trial. If the random variable X has a binomial distribution, then its probability mass function is

$$f(x) = \frac{n!}{x!(n - x)!}p^x(1 - p)^{n-x}$$

The mean value of X is $E[X] = n\,p$ and the variance of X is $\mathrm{Var}[X] = n\,p\,(1 - p)$.

A2.3.6.3 Poisson Distribution

The Poisson distribution is a limiting case of a binomial distribution. It results when p is set to μ/n and $n \to \infty$. The probability mass function is

$$f(x) = \frac{\mu^x e^{-\mu}}{x!}$$

If X has a Poisson distribution, then the mean and variance of X are equal and $E[X] = \mu = \mathrm{Var}[X]$. The mean value, μ, is sometimes called the rate parameter of the Poisson distribution.

The Poisson distribution is commonly used as a model for arrival rates, e.g., messages into a communication network or vehicles into a service station. Suppose, for example, that the random variable, X, represents the number of vehicles arriving at a service station per half-hour. If X has a Poisson distribution with mean of 12

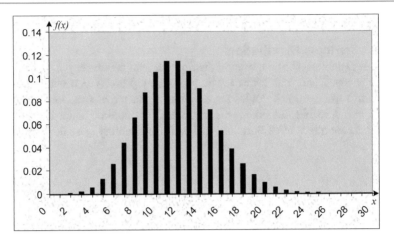

Fig. A2.2 Example of a probability mass function for the Poisson distribution ($\mu = 12$)

(i.e., the rate parameter $\mu = 12$ implying a mean arrival rate of 12 cars per half-hour), then the probability of six cars arriving in the 'next' half-hour interval is

$$
\begin{aligned}
f(6) &= \frac{\mu^6 e^{-\mu}}{6!} \\
&= \frac{12^6 e^{-12}}{6!} \\
&= 0.0255
\end{aligned}
$$

The probability mass function for this example is shown in Fig. A2.2.

A2.4 Continuous Sample Spaces

A2.4.1 Background

Continuous sample spaces are typically associated with experiments that involve the measurement of physical attributes, e.g., weight, length, temperature and time (where the latter may arguably not fall in the category of a physical attribute). In reality, however, no such measurement can be carried out with unbounded precision; hence, the notion of a continuous sample space is one of theoretical rather than practical relevance. Nevertheless, it is a concept that has credible modelling power in many practical situations.

As an example, consider a circumstance where the manufacturer of a particular consumer product needs to characterize the 'time to failure' of the product in order to determine a realistic warranty to offer purchasers. The sample space, S, associated with experiments carried out to measure the time to failure is the positive real line (i.e., $S = R^+$), and a random variable X introduced to represent this property would have the specification $X : S \rightarrow R^+$. Because the domain of X is a segment of the real line, X is a continuous random variable. Note also that $X(s) = s$ for all $s \in S$. In other words, the random variable in this case is the identity function. Such situations are frequently encountered with continuous random variables.

A2.4.2 Continuous Random Variables

The distinguishing feature of a continuous random variable is that it can assume an uncountably infinite number of values. Because of this, it is not possible to enumerate the probabilities of these values as is done in the case of discrete random variables. In fact, we are logically obliged to conclude that the probability that X will assume any particular value that lies within its domain is zero. But, on the other hand, the probability that it will assume some value in its domain is 1 (certainty).

As consequence of this, it is preferable to begin the characterization of a continuous random variable, $X : S \rightarrow R$ (where R is the real line and S is a collection of disjoint intervals from R), by first noting the properties of its cumulative distribution function $F(x) = P[X \leq x]$. These are as follows:

- $F(x)$ is continuous.
- The derivative of $F(x)$, i.e., $F'(x)$, exists except possibly at a finite set of points.
- $F'(x)$ is piecewise continuous.

Recall that a discrete random variable has a probability mass function. Its analog in the case of a continuous random variable, X, is a *probability density function*, $f(x)$ which has the following properties:

- $f(x) \geq 0$.
- $f(x) = 0$ if x is not a member of the range set of X.
- $\int_{-\infty}^{\infty} f(x)\, dx = 1$.
- $F(x) = \int_{-\infty}^{x} f(t)\, dt$.

A2.4.3 Expected Value and Variance

As in the case of a discrete random variable, a continuous random variable, X, likewise has an expected value and a variance, and these reflect characteristics of the probability density function for X. Not surprisingly, the specifications for these parallel those given in Eqs. (A2.10) and (A2.13) but with the summation operation replaced with an integration operation:

$$E[X] = \int_{-\infty}^{\infty} xf(x)dx$$

$$\text{Var}[X] = \int_{-\infty}^{\infty} (x - E[X])^2 dx$$

As in the discrete case, $\sigma_x = \sqrt{\text{Var}[X]}$ is called the standard deviation of X, and μ_x is used as an alternate symbol for $E[X]$. It can be demonstrated that

$$\text{Var}[X] = \int_{-\infty}^{\infty} x^2 f(x)\,dx - \mu_x^2 = E[X^2] - \mu_x^2 \qquad \text{(A2.20)}$$

If g is a real-valued continuous function whose domain includes the range of X and the continuous random variable Y is specified as $Y = g(X)$, then

$$E[Y] = \int_{-\infty}^{\infty} g(x)f(x)\,dx$$

from which it follows that

$$E[aX + b] = a\,E[X] + b$$

It can likewise be shown that

$$\text{Var}[aX + b] = a^2 \text{Var}[X].$$

A2.4.4 Some Continuous Distribution Functions

A2.4.4.1 The Uniform Distribution

A random variable, X, is uniformly distributed if its probability density function has the form:

$$f(x) = \begin{cases} \frac{1}{(b-a)} & \text{for } a \le x \le b \\ 0 & \text{otherwise} \end{cases}$$

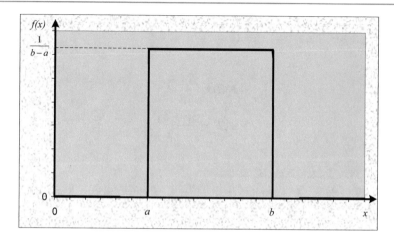

Fig. A2.3a Probability density function for the uniform distribution

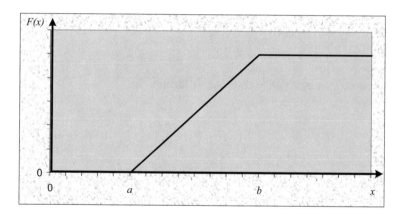

Fig. A2.3b Cumulative distribution function for the uniform distribution

It then follows that the cumulative distribution function is

$$
F(x) = \begin{cases} 0 & \text{for} \quad x < a \\ \frac{x-a}{b-a} & \text{for} \quad a \leq x \leq b \\ 1 & \text{for} \quad x > b \end{cases}
$$

The probability distribution function and the cumulative distribution function for the uniform distribution are shown in Fig. A2.3a and b, respectively.

It can be readily shown that the mean and variance of the uniformly distributed random variable X are

$$E[X] = \frac{a+b}{2}$$

$$\mathrm{Var}[X] = \frac{(b-a)^2}{12}$$

A2.4.4.2 The Triangular Distribution

The triangular distribution is especially useful for modelling random phenomena in circumstances where very little information is available. If the random variable X has a triangular distribution, then the probability density function (pdf) for X has the form shown in Fig. A2.4.

The specifications for the pdf given in Fig. A2.4 are as follows:

$$f(x) = \begin{cases} 0 & \text{for} \quad x < a \\ \frac{2(x-a)}{(b-a)(c-a)} & \text{for} \quad a \le x \le c \\ \frac{2(b-x)}{(b-a)(b-c)} & \text{for} \quad c < x \le b \\ 0 & \text{for} \quad x > b \end{cases}$$

The corresponding cumulative distribution function is:

$$F(x) = \begin{cases} 0 & \text{for} \quad x < a \\ \frac{(x-a)^2}{(b-a)(c-a)} & \text{for} \quad a \le x \le c \\ 1 - \frac{(b-x)^2}{(b-a)(b-c)} & \text{for} \quad c < x \le b \\ 1 & \text{for} \quad x > b \end{cases}$$

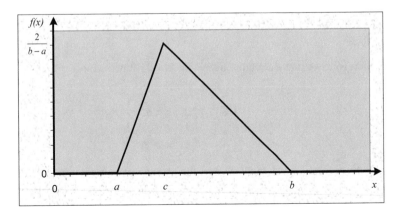

Fig. A2.4 Probability density function for the triangular distribution

The mean and the variance of X are:

$$E[X] = \frac{a+b+c}{3}$$

$$\text{Var}[X] = \frac{a^2 + b^2 + c^2 - ab - ac - bc}{18}$$

A statistical parameter that is especially relevant to the triangular distribution is the *mode*. For X a continuous distribution, the mode is the value of x where the maximum value of the pdf occurs. When X has a triangular distribution as shown in Fig. A2.4, the mode equals c.

A2.4.4.3 The Exponential Distribution

A random variable, X, is exponentially distributed if its probability density function has the form:

$$f(x) = \begin{cases} \lambda e^{-\lambda x} & \text{for } x \geq 0 \\ 0 & \text{otherwise} \end{cases} \qquad (A2.21a)$$

It then follows that the cumulative distribution function is

$$F(x) = \begin{cases} \displaystyle\int_{-\infty}^{x} f(t)\, dt = \lambda \int_{0}^{x} e^{-\lambda t}\, dt = 1 - e^{-\lambda x} & \text{for } x \geq 0 \\ 0 & \text{otherwise} \end{cases} \qquad (A2.21b)$$

The probability density function and the cumulative density function for the exponential distribution are shown in Fig. A2.5a and b, respectively, for the cases where $\lambda = 0.5$, 1.0 and 2.0.

It can be readily shown that the mean and variance of the exponentially distributed random variable X are:

$$E[X] = 1/\lambda$$

$$\text{Var}[X] = (1/\lambda)^2 = (E[X])^2$$

One of the most significant properties of the exponential distribution is that it is 'memoryless' (It is the only distribution with this property.). This notion corresponds to the following relationship:

$$P[X > s + t \mid X > s] = P[X > t] \qquad (A2.22)$$

The implication here is well illustrated in terms of the life of many electronic components. Suppose, for example, that the random variable X represents length of time that a memory chip in a cell phone will function before failing. If X has an exponential distribution, then (from Eq. (A2.22)) the probability of it being functional at time $(s + t)$ given that it is functional at time s is the same as the

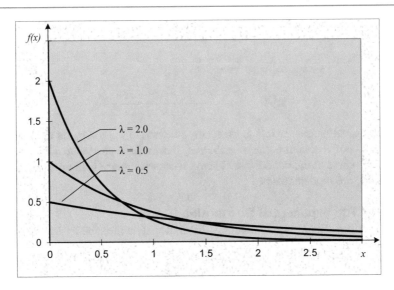

Fig. A2.5a Probability density function for the exponential distribution

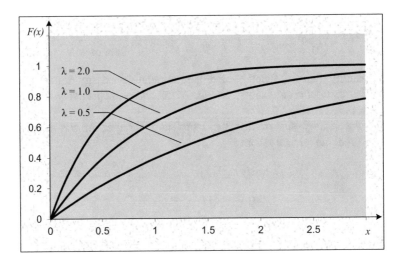

Fig. A2.5b Cumulative distribution function for the exponential distribution

probability of it being functional at time t. In other words, the chip does not 'remember' that it has already been in use for s time units. If, for example, X has a mean of 4 (years), i.e., $\lambda = 0.25$, then the probability of the cell phone remaining functional for 2 years is:

$$P[\text{remaining life} > 2\,\text{years}] = 1 - F(2)$$
$$= 1 - (1 - e^{-0.5})$$
$$= e^{-0.5}$$
$$= 0.604$$

and this is independent of how many years the phone has already been in use.

The following important relationship exists between the Poisson distribution and the exponential distribution:

Suppose that arrivals have a Poisson distribution with mean μ, then the corresponding inter-arrival times have an exponential distribution with mean $1/\mu$.

A2.4.4.4 The Gamma Distribution

A random variable, X, has a gamma distribution if its probability density function has the form:

$$f(x) = \begin{cases} \frac{\lambda^{\alpha} x^{\alpha-1} \exp(-\lambda x)}{\Gamma(\alpha)} & \text{for } x \ge 0 \\ 0 & \text{otherwise} \end{cases}$$

where α and λ are positive parameters. The function $\Gamma(\alpha)$ is called the *gamma function* and has the following definition:

$$\Gamma(\alpha) = \int_0^{\infty} x^{\alpha-1} \exp(-x)\,dx$$

It can be shown that

$$\Gamma(\alpha+1) = \alpha\,\Gamma(\alpha)$$

A general analytic expression for the cumulative distribution, $F(x)$, cannot be formulated. Values can only be obtained by numerical integration of $f(x)$.

If the random variable, X has a gamma distribution with parameters α and λ, then

$$E[X] = \frac{\alpha}{\lambda}$$
$$\text{Var}[X] = \frac{\alpha}{\lambda^2} = \frac{(E[X])^2}{\alpha} \tag{A2.23}$$

Notice that if the mean is held constant and α increases, then the variance approaches zero.

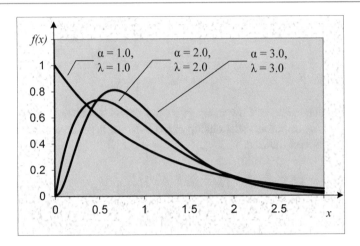

Fig. A2.6a Probability density function for the gamma distribution

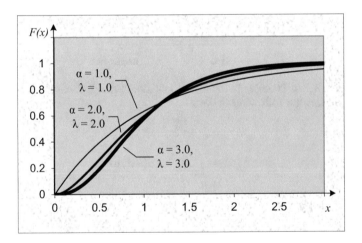

Fig. A2.6b Cumulative distribution function for the gamma distribution

The probability density function and the cumulative density function for the gamma distribution are shown in Fig. A2.6a and b, respectively, for the cases where $\alpha = \lambda = 1$, $\alpha = \lambda = 2$ and $\alpha = \lambda = 3$.

A2.4.4.5 The Erlang Distribution

The Erlang distribution is a specialization of the gamma distribution which corresponds to the case where the parameter α is restricted to (positive) integer values, i.e., $\alpha = n > 0$. In this case, it can be shown that:

$$\Gamma(\alpha) = \Gamma(n) = (n-1)!$$

To facilitate examples of the application of this distribution, it's convenient to replace the parameter λ of the gamma distribution with $n\omega$ and to rewrite the probability density function as:

$$f(x) = \begin{cases} \dfrac{n\omega(n\omega x)^{n-1}\exp(-n\omega x)}{(n-1)!} & \text{for } x > 0 \\ 0 & \text{otherwise} \end{cases}$$

The probability density function of the Erlang distribution as given above can be explicitly integrated, and the cumulative distribution function, $F(x)$, can be written as:

$$F(x) = \begin{cases} 1 - \displaystyle\sum_{i=0}^{n} \dfrac{(n\omega x)^{i}\exp(-n\omega x)}{i!} & \text{for } x > 0 \\ 0 & \text{otherwise} \end{cases}$$

If the random variable X has an Erlang distribution with parameters n and ω, then:

$$E[X] = \frac{1}{\omega}$$

$$\mathrm{Var}[X] = \frac{1}{n\,\omega^2} = \frac{(E[X])^2}{n}$$

It can be demonstrated that if:

$$X = \sum_{i=1}^{n} X_i$$

and the random variables X_i are independent and exponentially distributed with mean $(1/n\omega)$, then X has an Erlang distribution with parameters n and ω.

A2.4.4.6 The Chi-Square Distribution

The chi-square distribution is another special case of the gamma distribution. It can be obtained by setting $\alpha = (m/2)$ and $\lambda = 0.5$. This results in:

$$f(x) = \begin{cases} \dfrac{(0.5x)^{\frac{m}{2}-1}\exp(-0.5x)}{2\Gamma\left(\frac{m}{2}\right)} & \text{for } x > 0 \\ 0 & \text{otherwise} \end{cases}$$

The parameter m is called the 'degrees of freedom' of the distribution. It follows directly from Eq. (A2.23) that if X has a chi-square distribution, then:

$$E[X] = m$$

$$\mathrm{Var}[X] = 2m$$

It can be demonstrated that if

$$X = \sum_{i=1}^{m} X_i^2$$

and the random variables X_i are independent and have a standard normal distribution (see below), then X has a chi-square distribution with m degrees of freedom.

A2.4.4.7 The Normal Distribution

A random variable, X, has a normal distribution with mean μ and variance σ^2 if its probability density function can be written as:

$$f(x) = \frac{1}{\sqrt{2\pi}\sigma} \exp\left(\frac{-(x-\mu)^2}{2\sigma^2}\right); \quad -\infty < x < \infty$$

This function has the classic 'bell shape' which is symmetric about the mean μ. The maximum value of f occurs at $x = \mu$ and $f(\mu)\frac{1}{\sqrt{2\pi}\sigma}$. It is sometimes convenient to use the notation $X \sim N[\mu, \sigma^2]$ to indicate that X is a normally distributed random variable with mean μ and variance σ^2.

The probability density function and the cumulative density function for the normal distribution are shown in Fig. A2.7a and b, respectively, for the cases where $\mu = 0$ and $\sigma = 1.0$, 2.0 and 4.0. Note, however, that a closed-form analytic expression for the cumulative distribution function is not possible and consequently the data for the construction of Fig. A2.7 were necessarily obtained using numerical approximation for the underlying integration operation.

Suppose Y is a random variable that is linearly related to the normally distributed random variable, $X \sim N[\mu, \sigma^2]$, e.g., $Y = aX + b$ where a and b are constants. Then Y likewise has a normal distribution with $E[Y] = (a\mu + b)$ and $Var[Y] = a^2\sigma^2$, i.e., $Y \sim N[a\mu + b, a^2\sigma^2]$.

The above feature provides a convenient way of 'normalizing' the normal distribution. In particular, by choosing $a = (1/\sigma)$ and $b = -\mu/\sigma$, then the normally distributed random variable

$$Z = aX + b = \frac{X - \mu}{\sigma}$$

has zero mean and unit variance, i.e., $Z \sim N[0,1]$. A random variable such as Z is said to have a *standard normal distribution*.

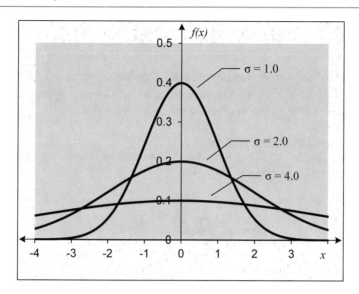

Fig. A2.7a Probability density function for the normal distribution

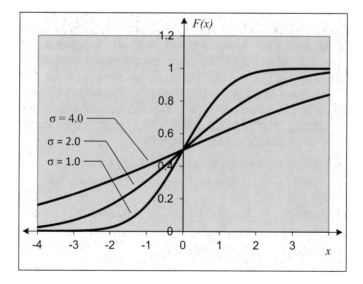

Fig. A2.7b Cumulative distribution function for the normal distribution

The following feature of a collection of normally distributed random variables has an important practical relevance. Let X be a normally distributed random variable with mean μ and variance σ^2 and let $X_1, X_2, \ldots, X_n$ be random variables that correspond to a sequence of n independent samples of X (hence the X_k are identically distributed). The random variable

$$\overline{X}(n) = \frac{1}{n}\sum_{k=1}^{n} X_k$$

which is called the sample mean is normally distributed with mean μ and variance σ^2/n, i.e., $\overline{X}(n) = N[\mu, \sigma^2/n]$. This result can be viewed as a special case of the Central Limit Theorem that is given in the following section.

A2.4.4.8 The Beta Distribution

A random variable, X, has a beta distribution with parameters α_1 and α_2 if its probability density function is given by

$$f(x) = \begin{cases} \frac{x^{\alpha_1-1}(1-x)^{\alpha_2-1}}{B(\alpha_1,\alpha_2)} & \text{for } 0<x<1 \\ 0 & \text{otherwise} \end{cases}$$

where $\alpha_1 > 0$, $\alpha_2 > 0$ and

$$B(\alpha_1, \alpha_2) = \frac{\Gamma(\alpha_1)\Gamma(\alpha_2)}{\Gamma(\alpha_1 + \alpha_2)}.$$

A closed-form representation for the corresponding cumulative distribution does not exist. The probability density function is shown for several values of the parameters α_1 and α_2 in Fig. A2.8a (the corresponding cumulative distribution functions are shown in Fig. A2.8b). The figure illustrates, to some degree, one of the interesting features of the beta pdf, namely, the wide range of 'shapes' which it can assume as its parameters are varied. This can be useful in dealing with the problem of selecting appropriate data models for random phenomena as discussed in Chap. 3. Some general properties are as follows:

(a) If $\alpha_1 = \alpha_2$, then the beta pdf is symmetric about 0.5.
(b) If $\alpha_1 > 1$ and $\alpha_2 \leq 1$ or $\alpha_1 = 1$ and $\alpha_2 < 1$, then the beta pdf is strictly increasing.
(c) If $\alpha_1 < 1$ and $\alpha_2 \geq 1$ or $\alpha_1 = 1$ and $\alpha_2 > 1$, then the beta pdf is strictly decreasing.

Several specific shapes that can be achieved are:

(a) Constant (i.e., the uniform distribution) when $\alpha_1 = \alpha_2 = 1$.
(b) Uniformly rising to the right (special case of the triangular distribution where $a = 0$, $b = c = 1$) when $\alpha_1 = 2$ and $\alpha_2 = 1$.
(c) Uniformly falling to the right (special case of the triangular distribution where $a = c = 0$ and $b = 1$) when $\alpha_1 = 1$ and $\alpha_2 = 2$.
(d) Parabolic when $\alpha_1 = 2 = \alpha_2$.
(e) U-shaped when $\alpha_1 < 1$ and $\alpha_2 < 1$ (e.g., $\alpha_1 = 0.5 = \alpha_2$).

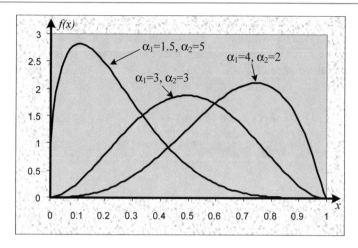

Fig. A2.8a Probability density function for the beta distribution

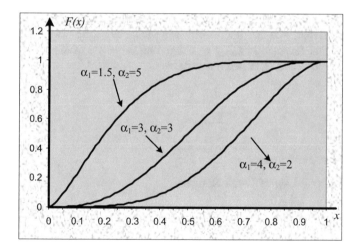

Fig. A2.8b Cumulative distribution function for the beta distribution

An 'extension' of the beta distribution as given above can be obtained by letting $Y = a + (b - a)X$. The distribution of random variable Y maintains the shape of the distribution of X, but its domain extends over the interval (a, b) (rather than $(0, 1)$). The mean and variance of Y are

$$E[Y] = a + (b - a)\frac{\alpha_1}{\alpha_1 + \alpha_2}$$
$$\mathrm{Var}[Y] = (b - a)^2 \frac{\alpha_1 \alpha_2}{(\alpha_1 + \alpha_2)^2 (\alpha_1 + \alpha_2 + 1)}$$

A2.5 Some Theorems

Two results in probability theory stand out as being especially significant and valuable. These are the Strong Law of Large Numbers and the Central Limit Theorem and ; they are stated below (without proof). Because of their wide relevance, they are frequently referenced in the discussions in this book.

The results in question are all formulated in terms of a sequence, $X_1, X_2, \ldots, X_n$, of random variables that are independent and identically distributed.[1] Suppose, for definiteness, that for each k, $E[X_k] = \mu$ and $\mathrm{Var}[X_k] = \sigma^2$, where we assume that both μ and σ are finite. The random variable

$$\overline{X}(n) = \frac{1}{n}\sum_{k=1}^{n} X_k \tag{A2.24}$$

is called the sample mean.

Theorem (Strong Law of Large Numbers)
With probability 1,

$$\lim_{n \to \infty} \overline{X}(n) = \mu$$

Various intuitive interpretations are possible. One is simply that the average of a sequence of independent, identically distributed random variables converges to the common mean as the sequence becomes larger.

---

[1]Typically, within the context of our consideration in this book, this collection of random variables corresponds to a sequence of n independent samples of a population.

Theorem (Central Limit Theorem)

The sample mean random variable $\overline{X}(n)$ is normally distributed with mean μ and variance σ^2/n as $n \to \infty$. Consequently, the random variable

$$Z = \frac{\overline{X}(n) - \mu}{(\sigma/\sqrt{n})}$$

has a distribution that approaches the standard normal distribution as $n \to \infty$.

Typically, the random variables X_k are samples from an underlying population X. The Central Limit Theorem states that regardless of what distribution X has, the distribution of the sample mean, $\overline{X}(n)$, will tend towards a normal distribution as $n \to \infty$. Furthermore, $E[\overline{X}(n)] \to \mu$ and $\mathrm{Var}[\overline{X}(n)] \to \sigma^2/n$.

While the Central Limit Theorem has exceptional theoretical importance, it does give rise to the very natural practical question of how large n needs to be in order to achieve a 'satisfactory' approximation. While no specific result is available, it is generally true that if the parent distribution is symmetric and short-tailed, then approximate normality is reached with smaller sample sizes than if the parent distribution is skewed and/or long-tailed. The value of 30 is a widely recommended as a satisfactory value for n. There is convincing experimental evidence that when $n > 50$, the shape of the parent distribution has minimal impact.

A2.6 The Sample Mean as an Estimator

Let X_1, X_2, ..., X_n be random variables that correspond to a sequence of n independent samples of a population whose mean (μ) and variance (σ^2) are unknown. Note that the X_k are therefore identically distributed and have the mean and variance of the underlying population, namely, μ and σ^2, respectively. The random variable

$$\overline{X}(n) = \frac{1}{n}\sum_{k=1}^{n} X_k$$

is called a sample mean. Notice that

$$E[\overline{X}(n)] = E\left[\frac{1}{n}\sum_{k=1}^{n} X_k\right] = \frac{1}{n}\sum_{k=1}^{n} E[X_k] = \mu \tag{A2.25}$$

Because the expected value of $\overline{X}(n)$ yields the value of the distribution parameter μ of the underlying distribution, $\overline{X}(n)$ is said to be an (unbiased) estimator of μ (it is unbiased because $E\left[\overline{X}(n)\right]$ is not displaced from μ by a non-zero constant). The interesting and important observation here is that the value of a parameter of a distribution (μ in this case) can be obtained in a somewhat indirect fashion.

The variance of $\overline{X}(n)$ provides insight into its 'quality' as an estimator of μ because $\mathrm{Var}\left[\overline{X}(n)\right]$ is a measure of deviation of $\overline{X}(n)$ from the mean, μ. In other words, the estimator's quality is inversely proportional to the value of $E\left[\left(\overline{X}(n) - \mu\right)^2\right] = \mathrm{Var}\left[\overline{X}(n)\right]$. The determination of $\mathrm{Var}\left[\overline{X}(n)\right]$ is straightforward:

$$
\begin{aligned}
\mathrm{Var}\left[\overline{X}(n)\right] &= \mathrm{Var}\left[\frac{1}{n}\sum_{k=1}^{n} X_k\right] \\
&= \frac{1}{n^2}\sum_{k=1}^{n} \mathrm{Var}[X_k] \\
&= \frac{\sigma^2}{n}
\end{aligned}
\tag{A2.26}
$$

where we have used the relation

$$
\mathrm{Var}\left[\sum_{k=1}^{n} c_k X_k\right] = \sum_{k=1}^{n} c_k^2 \mathrm{Var}[X_k]
$$

whose validity is restricted to the case where the X_k are independent. Note that the implication of Eq. (A2.26) is that the quality of our estimator $\overline{X}(n)$ improves with increasing n.

From a practical point of view, the result of Eq. (A2.26) is not particularly useful because, in general, σ^2, like μ, is not known. To some extent, this can be remedied by introducing the random variable

$$
\overline{S}^2(n) = \frac{\displaystyle\sum_{k=1}^{n} \left(X_k - \overline{X}(n)\right)^2}{n-1}
$$

which is called the sample variance. It can now be demonstrated that

$$
E\left[\overline{S}^2(n)\right] = \sigma^2
\tag{A2.27}
$$

In other words, $E\left[\overline{S}^2(n)\right]$ is an (unbiased) estimator of σ^2. This demonstration begins by first noting that

$$
\begin{aligned}
(n-1)E\left[\overline{S}^2(n)\right] &= E\left[\sum_{k=1}^{n} X_k^2\right] - nE\left[\overline{X}^2(n)\right] \\
&= n\left(E\left[X_1^2\right] - E\left[\overline{X}^2(n)\right]\right)
\end{aligned}
\tag{A2.28}
$$

because

$$
\sum_{k=1}^{n}(X_k - \overline{X}(n))^2 = \sum_{k=1}^{n} X_k^2 - n\overline{X}^2(n)
$$

and because the X_k all have the same distribution. From the general result of Eq. (A2.20), we can write

$$
E\left[X_1^2\right] = \mathrm{Var}[X_1] + (E[X_1])^2
$$

and similarly that

$$
E\left[\overline{X}^2(n)\right] = \mathrm{Var}\left[\overline{X}(n)\right] + \left(E\left[\overline{X}(n)\right]\right)^2
$$

Substitution into Eq. (A2.28) together with the observations that $E[X_1] = \mu$, $\mathrm{Var}[X_1] = \sigma^2$, $E\left[\overline{X}(n)\right] = \mu$ and $\mathrm{Var}\left[\overline{X}(n)\right] = \sigma^2/n$ yields the desired result of Eq. (A2.27).

A2.7 Interval Estimation

In Sect. A2.6, we explored the problem of obtaining a 'point' (single value) estimate of the mean of an unknown distribution. In this section, our goal is to identify an interval in which the unknown mean is likely to fall together with an indication of the 'confidence' that can be associated with the result. We begin the discussion in the context of the standard normal distribution whose density function, $\phi(z)$, is described in Sect. A2.4.4.

Choose a value a in the range (0, 1). Then there necessarily exists a value of z (which we denote by z_a) such that the area under $\phi(z)$ that is to the right of z_a is equal to a (see Fig. 2.9).

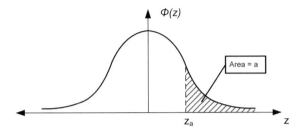

Fig. A2.9 A standard normal density function (perspective 1)

Table A2.1 Standard normal distribution data

$a = 1 - \Phi(z_a) = P[Z > z_a]$	$\Phi(z_a)$	z_a
0.05	0.95	1.64
0.025	0.975	1.96
0.01	0.99	2.33

Observe furthermore that this area corresponds to $P[Z > z_a] = 1 - P[Z \leq z_a] = 1 - \Phi(z_a)$, where $\Phi(z)$ is the cumulative distribution function of the standard normal distribution. In other words, we have

$$a = \int_{z_a}^{\infty} \phi(z)\mathrm{d}z = 1 - \phi(z_a) = P[Z > z_a]$$

Using tabulated values for $\Phi(z)$, a summary of the form shown in Table A2.1 can be developed. Thus it follows that $z_{0.05} = 1.64$, $z_{0.025} = 1.96$ and $z_{0.01} = 2.33$.

We consider now a slight variation of the above discussion. Again we choose a value of a in the range $(0, 1)$. As earlier, there necessarily is a value of z (which we again denote by z_a) such that (see Fig. A2.10):

$$C = P[-z_a \leq Z \leq z_a] = (1 - 2a) \tag{A2.29}$$

Our interest here is to find a value for z_a given a value for C. However, we know that $a = (1 - C)/2$ and that

$$P[Z > z_a] = a = 1 - \Phi(z_a)$$

Consequently, Table A2.2 can be derived.

It then follows that:

$$\begin{aligned}
C &= P[-z_{0.05} \leq Z \leq z_{0.05}] = 0.90, && \text{with } z_{0.05} = 1.64 \\
C &= P[-z_{0.025} \leq Z \leq z_{0.025}] = 0.95, && \text{with } z_{0.025} = 1.96 \\
C &= P[-z_{0.01} \leq Z \leq z_{0.01}] = 0.98, && \text{with } z_{0.01} = 2.33
\end{aligned} \tag{A2.30}$$

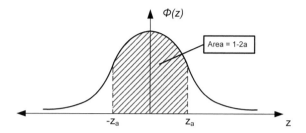

Fig. A2.10 Standard normal density function (perspective 2)

Table A2.2 Standard normal distribution data (extended)

C	$a = \frac{1-C}{2}$	$\Phi(z_a)$	z_a
0.90	0.05	0.95	1.64
0.95	0.025	0.975	1.96
0.98	0.01	0.99	2.33

The results shown in Eq. (A2.30) have important practical interpretations. For example, the first can be interpreted to imply that a random sample from a standard normal distribution will fall, with 90% confidence (100 C%), in the range $[-1.64, 1.64]$ (The 0.9 probability is interpreted here as an indicator of 'confidence level'.).

We know from the Central Limit Theorem that the distribution of the random variable:

$$Z(n) = \frac{\overline{X}(n) - \mu}{(\sigma/\sqrt{n})} \qquad (A2.31)$$

approaches the standard normal distribution as n becomes large. Here, μ and σ^2 are the mean and the variance of a parent distribution, and our task is to formulate an interval estimate for μ. To the extent that we can assume that n is sufficiently large, $Z(n)$ could be substituted in Eq. (A2.30). Instead, however, we substitute the right-hand side of Eq. (A2.31), and with some straightforward manipulation, the first result of Eq. (A2.30) becomes:

$$C = P\left[\overline{X}(n) - z_{0.05}\frac{\sigma}{\sqrt{n}} \leq \mu \leq \overline{X}(n) + z_{0.05}\frac{\sigma}{\sqrt{n}}\right] = 0.9 \qquad (A2.32)$$
$$\text{with } z_{0.05} = 1.64$$

Unfortunately, Eq. (A2.32) has little practical value because the bounds on the unknown mean, μ, depend on σ which likewise is unknown. Recall, however, that earlier it was demonstrated that $E\left[\overline{S}^2(n)\right] = \sigma^2$ (see Eq. (A2.27)). The implication here is that for large n the random variable $\overline{S}^2(n)$ can be used as an estimator for σ^2. This in effect provides the basis for replacing σ in Eq. (A2.32) with a random variable for which an observation (i.e., value) can be established. In effect, then, we can restate our earlier illustrative conclusion of Eq. (A2.32) in the following way:

$$C = P\left[\overline{X}(n) - z_{0.05}\frac{\overline{S}(n)}{\sqrt{n}} \leq \mu \leq \overline{X}(n) + z_{0.05}\frac{\overline{S}(n)}{\sqrt{n}}\right] = 0.9 \tag{A2.33}$$

$$\text{with } z_{0.05} = 1.64$$

Suppose now that observed values of $\bar{x}$ and $\bar{s}$ have been obtained for $\overline{X}(n)$ and $\overline{S}(n)$, respectively (with the use of suitably large values for n). A practical interpretation for the result of Eq. (A2.33) is that with *approximately* 90% confidence, we can assume that the unknown mean, μ, lies in the interval:

$$\left[\bar{x} - z_{0.05}\frac{\bar{s}}{\sqrt{n}}, \quad \bar{x} + z_{0.05}\frac{\bar{s}}{\sqrt{n}}\right] \tag{A2.34}$$

$$\text{with } z_{0.05} = 1.64$$

Or alternately, the above can be interpreted to mean that if a large number of independent observations for $\bar{x}$ and $\bar{s}$ are obtained, then, in approximately 90% of the cases, the mean will fall in the interval given by Eq. (A2.34). The qualifier 'approximately' is required for two reasons: the use of the estimator $\overline{S}(n)$ and the uncertainty about the adequacy of n.

The approximation of the interval estimate given above which arises because of the uncertainty introduced by reliance on n being sufficiently large can be remedied in one special case. This is the special case where the random variables, X_k, that are the constituents of $\overline{X}(n)$ are themselves normally distributed. Earlier we pointed out (see Sect. A2.4.4.3) that under such circumstances $\overline{X}(n) = N[\mu, \sigma^2/n.]$, i.e., for *any* n, the sample mean is normally distributed with mean equal to the mean of the underlying distribution and with variance that is reduced by a factor of n from the variance of the underlying distribution. Under these circumstances, the analysis can be recast in the context of the *Student's t-distribution*. This distribution is similar in shape to the standard normal distribution but is less peaked and has a longer tail (see Fig. A2.11).

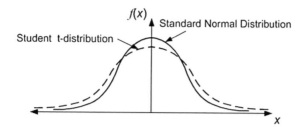

Fig. A2.11 Illustration of the relationship between the Student's t-distribution and the standard normal distribution

Table A2.3 Some representative values for $t_{n-1,0.05}$ (C = 0.9)

n	$t_{n-1,0.05}$
10	1.83
20	1.73
30	1.70
∞	1.64

More specifically, the random variable

$$T(n) = \frac{\overline{X}(n) - \mu}{\left(\overline{S}(n)/\sqrt{n}\right)} \tag{A2.35}$$

has a *t-distribution* with $(n-1)$ *degrees of freedom*. The feature that is especially significant here is that this distribution accommodates the approximation introduced by replacing σ by $\overline{S}(n)$. The distribution is, in fact, a family of distributions that are indexed by the value of n. Consequently, the transition from our earlier analysis to the framework of the *t*-distribution requires that the 'anchor point' z_a acquire a second index to identify the degrees of freedom that relate to the circumstances under consideration. To emphasize this transition to the *t*-distribution, we also relabel the anchor point to get $t_{n-1,a}$. Equation (A2.33) can be now be rewritten as

$$C = P\left[\overline{X}(n) - t_{n-1,0.05}\frac{\overline{S}(n)}{\sqrt{n}} \leq \mu \leq \overline{X}(n) + t_{n-1,0.05}\frac{\overline{S}(n)}{\sqrt{n}}\right] = 0.9 \tag{A2.36}$$

The value of $t_{n-1,0.05}$ is determined from tabulated data relating to the *t*-distribution, but a prerequisite is the specification of a value for n. Some representative values for $t_{n-1,0.05}$ are given in Table A2.3.

It should be observed that as n becomes large, the value of $t_{n-1,a}$ approaches the value z_a for any value of a (see last row of Table A2.3). A more complete section of the Student's *t*-distribution is given in Table A2.4 where the limiting case is again given in the last row.

Table A2.4 A section of the student's t-distribution

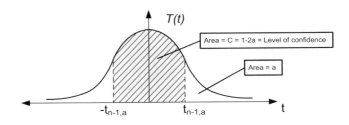

	a						
	0.2	0.15	0.1	0.05	0.025	0.005	0.0025
				C			
(n-1)	0.6	0.7	0.8	0.9	0.95	0.99	0.995
1	1.376382	1.962611	3.077684	6.313752	12.7062	63.65674	127.3213
2	1.06066	1.386207	1.885618	2.919986	4.302653	9.924843	14.08905
3	0.978472	1.249778	1.637744	2.353363	3.182446	5.840909	7.453319
4	0.940965	1.189567	1.533206	2.131847	2.776445	4.604095	5.597568
5	0.919544	1.155767	1.475884	2.015048	2.570582	4.032143	4.773341
6	0.905703	1.134157	1.439756	1.94318	2.446912	3.707428	4.316827
7	0.89603	1.119159	1.414924	1.894579	2.364624	3.499483	4.029337
8	0.88889	1.108145	1.396815	1.859548	2.306004	3.355387	3.832519
9	0.883404	1.099716	1.383029	1.833113	2.262157	3.249836	3.689662
10	0.879058	1.093058	1.372184	1.812461	2.228139	3.169273	3.581406
11	0.87553	1.087666	1.36343	1.795885	2.200985	3.105807	3.496614
12	0.872609	1.083211	1.356217	1.782288	2.178813	3.05454	3.428444
13	0.870152	1.079469	1.350171	1.770933	2.160369	3.012276	3.372468
14	0.868055	1.07628	1.34503	1.76131	2.144787	2.976843	3.325696
15	0.866245	1.073531	1.340606	1.75305	2.13145	2.946713	3.286039
16	0.864667	1.071137	1.336757	1.745884	2.119905	2.920782	3.251993
17	0.863279	1.069033	1.333379	1.739607	2.109816	2.898231	3.22245
18	0.862049	1.06717	1.330391	1.734064	2.100922	2.87844	3.196574
19	0.860951	1.065507	1.327728	1.729133	2.093024	2.860935	3.173725
20	0.859964	1.064016	1.325341	1.724718	2.085963	2.84534	3.153401
21	0.859074	1.06267	1.323188	1.720743	2.079614	2.83136	3.135206
22	0.858266	1.061449	1.321237	1.717144	2.073873	2.818756	3.118824
23	0.85753	1.060337	1.31946	1.713872	2.068658	2.807336	3.103997
24	0.856855	1.059319	1.317836	1.710882	2.063899	2.796939	3.090514
25	0.856236	1.058384	1.316345	1.708141	2.059539	2.787436	3.078199
26	0.855665	1.057523	1.314972	1.705618	2.055529	2.778715	3.066909
27	0.855137	1.056727	1.313703	1.703288	2.05183	2.770683	3.05652
28	0.854647	1.055989	1.312527	1.701131	2.048407	2.763262	3.046929
29	0.854192	1.055302	1.311434	1.699127	2.04523	2.756386	3.038047
30	0.853767	1.054662	1.310415	1.697261	2.042272	2.749996	3.029798
40	0.8507	1.050046	1.303077	1.683851	2.021075	2.704459	2.971171
50	0.848869	1.047295	1.298714	1.675905	2.008559	2.677793	2.936964
75	0.84644	1.043649	1.292941	1.665425	1.992102	2.642983	2.89245
100	0.84523	1.041836	1.290075	1.660234	1.983971	2.625891	2.870652
∞	0.841621	1.036433	1.281552	1.644854	1.959964	2.575829	2.807034

Generated using Microsoft Excel functions TINV and NORMINV

A practical implication of Eq. (A2.36) (which parallels that of Eq. (A2.34)) is that if a large number of independent observations for $\bar{x}$ and $\bar{s}$ are obtained, then, in 90% of the cases, the unknown mean, μ, will fall in the interval given by Eq. (A2.37):

$$\left[\bar{x} - t_{n-1,0.05} \frac{\bar{s}}{\sqrt{n}}, \quad \bar{x} + t_{n-1,0.05} \frac{\bar{s}}{\sqrt{n}} \right]$$

(A2.37)

$$\text{with } t_{n-1,0.05} = 1.83 \quad \text{when } n = 10$$

Notice furthermore that there is no need here for the qualifier 'approximately' because of our assumption that the unknown distribution is normal.

It needs to be emphasized that the discussion above with respect to the application of the Student's t-distribution was founded on the assumption of a normal distribution for the underlying population being investigated. This is clearly an idealized case. Nevertheless, the t-distribution is the tool generally used to establish interval estimates for population means even when the normality assumption is not applicable. This naturally means that the results obtained are approximate and this should be clearly indicated.

A2.8 Stochastic Processes

A discrete stochastic process is a finite sequence of random variables defined on the same population, e.g., $X = (X_1, X_2, X_3, \ldots, X_n)$. Often the indices have a correspondence with a set of discrete points in time, i.e., index i corresponds to t_i and $t_i < t_j$ for $i < j$. The X_i's are associated with some measurable feature that is of interest, and in general each may have its own distinct distribution. Consider, for example, the customer waiting times in a queue in front of a store cashier. A random variable W_i can be used to represent the time waited by the customer whose service begins at time t_i for $i = 1, 2, \ldots, n$. Thus we can associate the discrete stochastic process $(W_1, W_2, W_3, \ldots, W_n)$ with this activity.

The following properties of a discrete stochastic process $X = (X_1, X_2, X_3, \ldots, X_n)$ are of particular relevance:

$$\text{mean: } \mu_i = E[X_i] \quad (\text{for } i = 1, 2, \ldots, n)$$

$$\text{variance: } \sigma_i^2 = E[(X_i - \mu_i)^2] \quad (\text{for } i = 1, 2, \ldots, n)$$

$$\text{covariance: } Cov[X_i, X_j] = E[(X_i - \mu_i)(X_j - \mu_j)]$$

$$= E[X_i X_j] - \mu_i \mu_j \quad (\text{for } i, j = 1, 2, \ldots, n)$$

$$\text{correlation: } \rho_{i,j} = \frac{Cov[X_i, X_j]}{\sigma_i \sigma_j} \quad (\text{for } i, j = 1, 2, \ldots, n)$$

Note that for all i, $\text{Cov}[X_i, X_i] = \sigma_i^2$ and $\rho_{i,i} = 1$.

A discrete stochastic process is *stationary* when the following invariance properties hold:

$$\mu_i = \mu \quad \text{for all } i$$
$$\sigma_i^2 = \sigma^2 \quad \text{for all } i$$
$$\rho_{i,i+k} = \frac{\text{Cov}[X_i, X_{i+k}]}{\sigma^2} \quad \text{for all } i$$

Note that when a discrete stochastic process is stationary, covariance (and hence correlation) between two members of the sequence is dependent only on their separation (or lag, k) and is independent of the specific indices involved.

We define a *homogeneous stochastic process* to be a stationary discrete stochastic process for which the constituent random variables are independent and identically distributed (in other words, the stochastic process becomes a sequence of independent, identically distributed random variables defined on the same population). Recall (see Eq. (A2.17)) that independence implies that $E[X_iX_j] = \mu_i\,\mu_j$ for $i \neq j$ and consequently $\text{Cov}[X_i, X_j] = 0$ for $i \neq j$. The properties of special interest become

$$\mu_i = \mu \quad \text{for all } i$$
$$\sigma_i^2 = \sigma^2 \quad \text{for all } i$$
$$\text{Cov}[X_i, X_j] = \begin{cases} \sigma^2 & \text{for } i = j \\ 0 & \text{for } i \neq j \end{cases}$$
$$\rho_{i,j} = \begin{cases} 1 & \text{for } i = j \\ 0 & \text{for } i \neq j \end{cases}$$

An observation of a discrete stochastic process, $X = (X_1, X_2, \ldots, X_n)$, yields a sequence of values $x = (x_1, x_2, \ldots, x_n)$ where each x_i is obtained as an observed value of X_i. Such a sequence of values is often called a *time series*. The possibility of a recurring relationship between values in this sequence (e.g., every seventh value is consistently almost zero) is an important statistical consideration. The notion of *serial correlation* addresses this issue and a measure called *autocorrelation* is used as a means of quantification.

The underlying idea is to determine a set of *sample autocovariance* values where each such value corresponds to a different 'alignment' of the values in x, i.e.,

$$c_k = \frac{1}{(n-k)} \sum_{i=1}^{n-k} (x_i - \bar{x}(n))(x_{i+k} - \bar{x}(n))$$

where $\bar{x}(n) = \sum_{i=1}^{n} x_i$ is the sample mean. The measure c_k is called the *sample autocovariance for lag k*. In principle, c_k has a meaningful definition for $k = 1, 2, \ldots, (n-1)$, but in practice, the largest k should be significantly smaller than n to ensure that a reasonable number of terms appear in the summation. An associated measure is the *sample autocorrelation for lag k* which is defined as

$$\hat{\rho}(k) = \frac{c_k}{s^2(n)}$$

where $s^2(n) = \frac{1}{(n-1)} \sum_{i=1}^{n} (x_i - \bar{x}(n))^2$ is an estimator for the sample variance. A non-zero value for $\hat{\rho}(k)$ implies that the stochastic process, X, cannot be regarded as homogeneous because there is dependence among the constituent random variables.

It is generally accepted that a discrete stochastic process that is stationary has steady-state behaviour. Generally, real-life stochastic processes are nonstationary. Often, however, a nonstationary stochastic process may move from one steady-state behaviour to another steady-state behaviour where the underlying distribution parameters have changed. In other words, the nonstationary process does have 'piecewise' stationarity. During the transition, the process has transient behaviour. Note that a discrete stochastic process can also start with transient behaviour that eventually leads to steady-state behaviour.

Our discussion thus far has been restricted to discrete stochastic processes. Not surprisingly, there exists as well an alternate category of *continuous* stochastic processes. Each member of this category is a function whose domain is an interval of the real line. Because of the continuous nature of this domain, the stochastic process in question could be viewed as an uncountably infinite sequence of random variables.

A special class of continuous stochastic process evolves when the function in question is piecewise constant (see Fig. A2.12). The function, $L(t)$, in Fig. A2.12 could, for example, represent the length of a queue. Although $L(t)$ has a value for every t in the observation interval, there are only a finite number of 'relevant' time points in the domain set, namely, those at which L changes value. Because the domain set for this special class of continuous stochastic process can be interpreted as being finite, members of this class can be represented as a finite collection of ordered random variables. In this regard, they share an important common feature with discrete stochastic processes. This similarity, however, does not extend to the manner in which statistical parameters are evaluated.

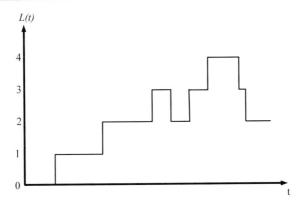

Fig. A2.12 Piecewise constant stochastic process

References

1. Borovkov AA (2013) Probability theory. Springer Science & Business Media, Springer-Verlag, London
2. Florescu I (2014) Probability and stochastic processes. Wiley
3. Karlin S, Taylor HE (2012) A first course in stochastic processes. Academic Press
4. Mendenhall W, Beaver R, Beaver B (2019) Introduction to probability and statistics, 15th edn. Duxbury Press
5. Spiegelhalter D (2019) The art of statistics: learning from data. Pelican Books

Annex 3: GPSS Primer

A3.1 Introduction

This primer provides an introduction to GPSS (General-Purpose Simulation System) and thus provides background material for Chap. 5 (Sect. 5.4) for readers unfamiliar with this software. GPSS is a general process-oriented simulation software environment. A student version of GPSS World, a current GPSS system developed by Minuteman Software, is freely available at http://www.minutemansoftware.com/product.htm. The reader is encouraged to obtain a copy of this version and use it with this primer. A number of examples will be presented and problems suggested allowing the reader to gain practical experience with creating process-oriented simulation models.

GPSS World is a Microsoft Windows application designed to run on the various Windows operating systems. The student version of the software includes a complete reference manual. As well, a 20-lesson tutorial developed by Minuteman is available from their website.

The objective of this primer is to describe how GPSS functions and to highlight several important subtleties. It is by no means a replacement for the GPSS World Reference Manual made available by Minuteman. The authors hope that this primer will enable the student to understand the basic operation of GPSS which can then be extended with the details provided by Minuteman documentation.

All GPSS program elements are referred to as entities. Note that in this text we adopt the convention of capitalizing the first letter of GPSS entity terms. For example, the GPSS Transaction entity will be referred to as a GPSS Transaction, or simply as a Transaction.

GPSS carries out simulation tasks by processing GPSS *Transaction* entities (a data structure that consists of a list of attributes). These entities are viewed as flowing through a simulation model. The programmer's view of this movement consists of the Transaction flowing through a sequence of *Block* entities. In reality, the Transaction actually moves through a number of underlying data structures (that are classified as *structural entities*). Each GPSS Block in fact represents a GPSS function triggered by Transactions that traverse, or try to traverse, the Block.

© Springer Nature Switzerland AG 2019   491
L. G. Birta and G. Arbez, *Modelling and Simulation*, Simulation Foundations,
Methods and Applications, https://doi.org/10.1007/978-3-030-18869-6

A GPSS Block corresponds to a GPSS programming statement. Figure A3.2 in the next section provides a simple example.

Consider the following example of a Transaction moving through the SEIZE Block (shown in Fig. A3.2). The SEIZE Block is used to gain the ownership of a Facility structural entity (more on the Facility entity later). When a Transaction tries to enter the SEIZE Block, it triggers a function that determines if the Transaction is allowed ownership. If the Transaction is allowed ownership (i.e., the Facility is not currently owned by another Transaction and the Facility is available), it is allowed into the SEIZE Block. Otherwise, it cannot enter the Block and stays in the Current Block until the Facility is freed and/or made available (actually the Transaction is placed on a list in the Facility). The relationship between the movement through the SEIZE Block and gaining ownership of the Facility will be clarified in the next few sections.

A GPSS segment is composed of a sequence of Blocks. It provides a specification of how Transaction entities flow through the simulation model beginning with their arrival in the simulation model until their exit from the simulation model (note that is it permissible for Transactions to never exit the model). A GPSS program generally contains multiple segments that can interact with each other as will be shown later. A segment generally corresponds to a process.

This primer introduces GPSS by exploring how Transactions flow among the GPSS data structures or more specifically among lists that are called chains in GPSS. When a GPSS program is compiled and executed, its statements (Blocks) will reference structural entities that contain these chains. The structural entities are often places of interaction between processes where Transactions from different processes compete with each other.

The flow of Transactions from Block to Block is the most common perspective taken in describing GPSS. In reality, however, GPSS is a list processor that moves Transactions from chain to chain. This list processing is, of course, automatic. The simulation modeller typically focuses on the functions provided by Blocks but will benefit considerably from an understanding of the background list processing.

A3.1.1 GPSS Transaction Flow

As suggested above, the most fundamental entities in a GPSS program are the 'Transaction' entities and Block entities. Recall that Transactions flow through Blocks to invoke their functions. These functions include list processing that move the Transactions between chains that are part of structural entities. For example, one of the chains associated with a Facility entity is a chain called the delay chain. When a Transaction is denied entry into the SEIZE Block, the Transaction is moved into the delay chain, creating a queue of Transactions waiting to gain ownership of the Facility. Let's introduce a simple example to explore how GPSS list processing is related to the flow of Transactions within a GPSS simulation model.

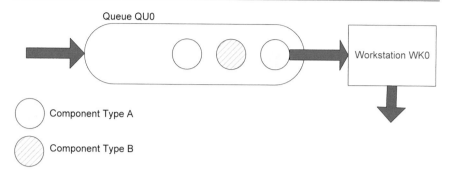

Fig. A3.1 Manufacturing system

We consider a simple manufacturing process where a workstation, WK0, works on parts one at a time. These parts can be of two different types (called A and B). Figure A3.1 illustrates the system. Transactions can be used to represent parts that are serviced by the workstation. The workstation and queue on the other hand would typically be represented as a GPSS Facility entity. Only one Transaction can 'own' the workstation Facility because only one part can be serviced by the workstation at any moment in time. All other Transactions will be placed on the Facility delay chain, thereby forming a queue of components waiting for servicing.

Figure A3.2 shows a corresponding GPSS simulation model. Note that it is organized into two code groups, composed of a sequence of Blocks and called Block segments. Block segments generally start with a GENERATE Block that inserts Transactions to the simulation model and end with a TERMINATE block that removes Transactions from the simulation model. Such a Block segment specifies a process, i.e., a life cycle, for Transactions (in some very special cases, more than one segment is required to represent a GPSS process). Two segments (processes) are included in our simulation model, one for each component type. For the moment, we will focus on the first Block segment that represents the servicing of type A components.

```
*****Component A Process (single segment)
CompA     GENERATE 6,3
          SEIZE WK0
          ADVANCE 10
          RELEASE WK0
          TERMINATE

*****Component B Process (single segment)
CompB     GENERATE 7,4
          SEIZE WK0
          ADVANCE 5
          RELEASE WK0
          TERMINATE
```

Fig. A3.2 GPSS simulation model for the manufacturing system

Each GPSS statement in Fig. A3.2 corresponds to a GPSS Block entity. Section A3.6.2.2 provides details about the use of these blocks. Here we simply present a brief summary of each Block:

- GENERATE: This Block inserts Transaction into the simulation model according to its arguments. For the component A case, Transactions enter the simulation model every 6 ± 3 time units.
- SEIZE: A Transaction enters this Block when it gains ownership of the WK0 Facility.
- ADVANCE: This is used to represent a service time duration. A Transaction remains in this Block for some duration of time as defined by the statement argument(s). In the case of component A, the duration is 10 time units.
- RELEASE: A Transaction releases ownership of the named Facility when it enters this block.
- TERMINATE: Transactions leave the simulation model when they enter this Block.

The functions invoked by the Blocks generally update the status of the simulation model (variables, data structures, etc.). In fact many Blocks do not affect the Transaction that entered the Block but instead update some other part of the simulation model. Structural entities are acted upon by the Blocks. Note, for example, the statement 'SEIZE WK0' which references a Facility entity, named WK0. The Facility entity is a structural entity (see Sect. A3.6.3). In general, any GPSS simulation model can be viewed as some set of functions that act on the program's structural entities such as the Facility entity.

The Transaction itself is a data structure consisting of a data tuple that passes from chain to chain (i.e., list to list) in the simulation model. The data values in the data tuple represent values for *parameters* (attributes). Standard parameters exist for each Transaction and are essential for processing the Transaction. The most common standard parameters include the following:

- Number: An integer that uniquely identifies each Transaction currently within the simulation model.
- Current Block: Indicates the Block entity at which the Transaction is located.
- Next Block: Indicates the Block entity the Transaction will next enter.
- Priority: Defines the priority[2] of the Transaction which is used to order Transactions within GPSS chains.
- Mark time: Initially set to the (simulation) time the Transaction entered the simulation model.

---

[2]In GPSS World, larger values indicate a higher priority. Negative values are valid as priority values.

- BDT (block departure time): Contains a time value which indicates when the Transaction can be processed. This value is present when the Transaction is part of the future events chain, a list used to advance time. It is created when the Transaction enters an ADVANCE Block or when a Transaction is created by a GENERATE Block. The future events chain is described below.

Additional parameters can be defined and created for Transactions in a GPSS program (by using the ASSIGN Block). These additional parameters are specified by the simulation modeller and reflect requirements flowing from the conceptual model.

The passage of a Transaction through a sequence of Blocks is managed with the Current Block and Next Block parameters of that Transaction. These parameters contain numerical identifiers that GPSS assigns to each Block in the simulation model. The GPSS processor consults the value of the Next Block parameter to identify the Block (and the operation) when processing a Transaction. When a Transaction 'enters' a Block, its corresponding identifier is saved in the Current Block parameter, and the Next Block parameter is updated with the identifier of the subsequent Block. If a Transaction is refused entry, then no changes are made to the Block parameters, and the Transaction is typically placed on some chain, such as the Facility delay chain. Some Blocks (see the TRANSFER Block) will update the Next Block parameter of the entering Transaction.

As previously indicated, Transactions are moved between chains that are usually priority-based first-in, first-out lists. When a Transaction is placed on a chain, it is placed ahead of any lower priority Transactions and behind any Transaction with equal or higher priority. Let's examine now the main GPSS chains used to process Transactions.

The current events chain (CEC) and the future events chain (FEC) are the main chains used in the processing of Transactions and in moving simulation time forward. Note that there exists only one CEC and one FEC within any GPSS simulation program. Other chains that are part of GPSS structural entities such as the Facility are nevertheless very important in managing the execution of a GPSS program.

The Transaction at the head of the CEC is the 'active' Transaction. GPSS moves the active Transaction using the Next Block parameter. More specifically, the Block identified by the Next Block parameter provides the functions invoked by the active Transaction. Returning to our SEIZE Block example, an active Transaction whose Next Block parameter refers to a SEIZE Block undergoes the following processing:

- If it can gain ownership of the Facility, it enters the SEIZE Block.

  - Its Current Block parameter is updated to the Block number of the SEIZE Block.
  - The Next Block parameter is updated to the Block number of the Block that would logically follow the SEIZE Block.
  - The Facility is tagged with the Transaction identifier to identify it as the owner.

- *Or*, if it cannot gain ownership of the Facility, it is refused entry into the SEIZE Block.
  - The active Transaction is removed from the CEC and placed on the delay chain of the Facility. The delay chain corresponds to the queue of Transactions waiting to access the Facility.

In the first case (the Transaction is successful in gaining ownership of the Facility), GPSS then moves the Transaction to the Block following the SEIZE Block, namely, the ADVANCE Block in our example. In the second case, the Transaction is removed from the CEC and placed on the Facility delay chain where it can move no further. Now the next Transaction at the head of the CEC becomes the active Transaction, and it will be processed by GPSS using its Next Block parameter.

Consider now the case where all Transactions on the CEC have been processed, i.e., the list is empty. The chain must somehow be replenished. The FEC is the source for replenishing the CEC in this situation (as will be discussed later, the CEC can also be replenished from other sources). Note that the FEC is populated with Transactions flowing from two sources, namely, ADVANCE Blocks and GENERATE Blocks.

Consider the first case. When a Transaction enters an ADVANCE Block, the invoked function will calculate the time when the delay will terminate and assigns this value to the Transaction's BDT parameter. The Transaction is then placed on the FEC according to the value of the BDT parameter, with due regard to the BDT of Transactions that are already on the FEC. Thus, the FEC contains a list of Transactions that are time ordered.

In the second case, the GENERATE Block places the Transaction on the FEC for entry into the simulation model with the BDT set to the time when the Transaction is scheduled to arrive in the simulation model. The Current Block parameter is set to 0 (non-existing Block), and the Next Block parameter is set to the Block number of the GENERATE Block. When a Transaction is moved from the FEC to the CEC and it enters the GENERATE Block, a successor Transaction will be placed on the FEC (ready to enter the GENERATE Block) by the GENERATE Block. This bootstrapping technique ensures a continuous flow of Transactions into the simulation model.

To advance time, GPSS moves the Transaction at the head of the FEC to the CEC and updates the time according to that Transaction's BDT parameter. If more than one Transaction on the FEC has the same BDT value, they are all moved to the CEC. Transactions with the same priority will be moved to the CEC in random order; otherwise, priority conditions are enforced.

The CEC can also be replenished from other chains during the processing of the active Transaction. To see how the CEC could be replenished from the Facility delay chain (see Fig. A3.3), consider the case when the GPSS processor finds a reference to a RELEASE Block in the Next Block parameter of the active Transaction. This Transaction enters the RELEASE Block and relinquishes ownership of the Facility, and the Transaction at the head of the delay chain becomes the new owner of the Facility. The new owner is also moved into the

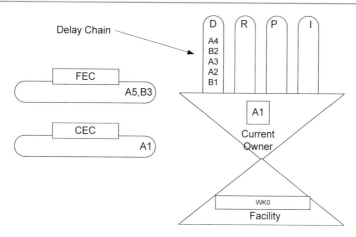

Fig. A3.3 Structure diagram for the manufacturing problem

SEIZE Block by updating its Current Block parameter and is moved from the delay chain to the CEC. This illustrates how the CEC can be replenished from chains other than the FEC.

Figure A3.3 shows the state of the Facility when Transaction A1 is the owner of the Facility and five Transactions are on the delay chain, namely, B1, A2, A3, B2 and A4. Assuming that Transactions A1–A4 represent components of type A and B1 and B2 represent components of type B, it is clear that Transactions from different processes (traversing different Block segments) can be found in the delay chain (more on this in the next section). Note that the owner A1 is part of the CEC and in fact is the active Transaction. Assuming that A1 just entered the SEIZE Block (see Fig. A3.4), it will be moved to the FEC when it enters the ADVANCE Block, that is, the ADVANCE Block function will place the Transaction on the FEC. As a consequence of the bootstrapping process of the two GENERATE Blocks in Fig. A3.2, the FEC will always contain two Transactions (one of type A and one of type B) that are scheduled to enter the simulation model at some future time.

The above view demonstrates that although GPSS simulation models are typically constructed from the perspective of Transactions flowing through some designated sequence of Blocks, the underlying reality is that Transactions are moved among the chains that are implicit in a GPSS simulation model. By keeping this fact in mind, the program developer should be better able to understand the interaction between different processes. In fact, it is this background processing that allows interaction between processes. The advantage of GPSS, and other process-oriented simulation environments, is that the simulation software takes care of managing the interactions among processes. There is no need for the program developer to be concerned with this detail. But structure diagrams are useful for keeping the program developer mindful of this underlying processing when creating Block segments. The next section outlines how the GPSS Blocks are used to implement processes.

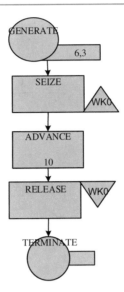

Fig. A3.4 GPSS process for type A components

A3.1.2 Programmer's View of Transaction Flow

A GPSS program specifies how Transactions flow through a simulation model, i.e., through Block entities (and implicitly through the structural entities) of a simulation model. In the manufacturing example in the previous section, Transactions are used to represent parts that are serviced by the workstation WK0. The workstation together with the queue of Transactions waiting for the workstation becomes a GPSS Facility entity. Transactions flow through this Facility, one at a time.

Recall that the GPSS Block segment (a sequence of Blocks) that is traversed by Transactions having a common interpretation is generally a specification of a process. A GPSS simulation program specifies one or more interacting processes formulated using multiple Block segments. This specification includes but is not limited to

- How Transactions enter into the model,
- How they flow among structural entities,
- How their service times are established,
- The decisions for selecting different paths through the Block segment,
- How their existence in the model is terminated.

A GPSS Transaction can represent a consumer entity with *scope* = Transient as presented in Sect. 4.2.1. Examples of such instances are messages flowing in a communications network or customers in a department store. But it is also possible to use Transactions to represent resources that move within a model. Consider a tugboat that moves in harbour, a waiter/waitress in a restaurant or repair personnel in a manufacturing plant.

A GPSS Block segment can be graphically represented, thereby providing a pictorial representation of a process. Figure A3.4 illustrates the Blocks used to specify the process of the type A components for the simple manufacturing system described in the previous section. The arrows indicate paths for the flow of Transactions from Block to Block triggering the functions associated to the Blocks. We consider now the manufacturing problem in more detail in order to examine these concepts more carefully.

The GENERATE Block defines how Transactions typically enter the model, that is, when they first appear on the FEC. In our example, a type A Transaction is generated randomly every 6 ± 3 time units using the uniform distribution (i.e., randomly between 3 and 9 time units). Other distributions are available, for example, GPSS World provides over 20 different probability distributions for this purpose. As pointed out earlier, the operation of the GENERATE Block is based on a bootstrapping approach.

The Transaction exiting the GENERATE Block moves to the SEIZE Block that will let the Transaction enter it if the associated Facility WK0 is free (a Facility is free when it is not already owned by some Transaction). Otherwise, the Transaction remains in the GENERATE Block. Entering the SEIZE Block triggers the function that causes the Transaction to become the owner of the Facility. Subsequent Transactions that try to enter the SEIZE Block will remain in the GENERATE Block until the owner Transaction releases the Facility (by entering the RELEASE Block).

The description in the previous paragraph is a process view of the Transaction flow presented in terms of updating the Block location parameters. As we saw in the previous section, the Transactions that cannot gain access to the Facility are placed on the Facility delay chain.

The owner Transaction then proceeds from the SEIZE Block to the ADVANCE Block where it remains for a specified length of time (10 time units in this case). This duration represents the service time the workstation WK0 requires to service the part (i.e., the 'type A Transaction').

Once the Transaction exits the ADVANCE Block, it will pass through the RELEASE Block, triggering the function that relinquishes the ownership of the WK0 Facility. The Transaction then exits the model by entering the TERMINATE Block. When the Facility ownership is released, a Transaction waiting to enter the SEIZE Block will be enabled to do so. When the Facility is released, the Transaction at the head of the Facility delay chain will become the owner of the Facility and allowed to move into the SEIZE Block. Note that GPSS processor continues to move an active Transaction until it cannot be moved any further (i.e., it comes to rest in a chain when refused entry into a Block or it exits the model). Only then does the processor consider other Transactions first on the CEC, otherwise on the FEC.

Figure A3.4 shows a single Block segment that specifies the processing of type A components. Figure A3.5, on the other hand, shows two Block segments (hence two processes), one for each of the two component types, A and B. Recall that the workstation WK0 can service either component type. This is reflected by the SEIZE

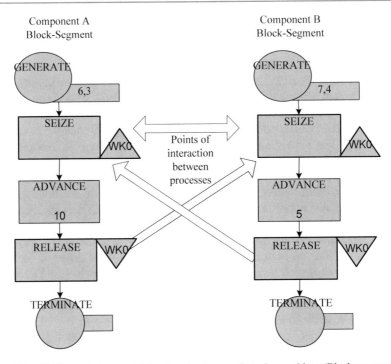

Fig. A3.5 GPSS simulation model for the simple manufacturing problem (Block segments for processing both types of components)

Block in each Block segment that refers to *the same Facility* WK0. This demonstrates that a process represented by one Block segment can affect the operation of another process represented by another Block segment. If a Transaction from either Block segment has seized the WK0 workstation, no other Transaction from either Block segment can seize it. The SEIZE and RELEASE Blocks are points of interaction between each Block segment or, in other words, between each process. This illustrates how the processes for each component type interact with each other. Note also that components of the same type interact in the same way.

The Block segments provide a program developer's view of the Transaction flow. They are used to specify the behaviour of the system. As was seen in the previous section, when a GPSS program is compiled, each Block is numbered, and the data structures associated with GPSS structural entities such as the Facility referenced by the Block entities are created. During execution, Block functions will change the state of these data structures. Each Transaction is tagged both with its current location in the Block segment and the Next Block it should enter.

The remainder of this Annex presents system numeric attribute (SNA), GPSS commands, details about the various GPSS Transaction chains and an overview of available GPSS entities. The SNA plays an important role in accessing the state of entities in the simulation model. Commands are used to declare and initialize various entities and to control the execution of simulations.

A3.2 System Numeric Attributes

The system numeric attributes (SNAs) provide the means for accessing a whole range of entity data in a simulation model, e.g., Transaction parameters, Facility utilization (time Facility is occupied), Facility available (indicates if it is free), system clock, active transaction number and the number of Transactions in a Block.

The SNAs fall into categories. Each category is aligned with a specific entity type. Any particular SNA has relevance only to the entities of the designated entity type. An SNA has two parts. The first is the type of data to be delivered, and the second identifies the specific entity (or sometimes parameter) of the relevant type from which the data is to be acquired. The entity identifier refers the specific entity of interest. Consider the SNA used to provide the number of Transactions currently in a Block entity labelled *CompA*. The first part of the necessary SNA is 'W' (the number of Transactions in a Block entity) and the entity identifier is *CompA*; hence the SNA is 'W$CompA' (the $ serves to indicate a symbolic reference for the entity identifier).

The entity identifier can be explicitly given as a numerical value or a symbolic representation or a reference to a parameter of the active Transaction whose value is the numeric identifier of the entity. This latter mechanism for referencing the entity identifier is called indirect addressing.

The forms of specific SNA class, W, are as follows (note that the characters * serve to indicate indirect addressing):

- *W*22: Refers the number of Transactions in Block 22.
- *W$CompA*: Refers the number of Transactions in Block labelled CompA (the GENERATE block in the example of the previous section).
- *W*5: Refers the number of Transactions in the Block identified by the value of the active Transaction's parameter 5.
- *W*BLKNUM* (*or W*$BLKNUM*): Refers the number of Transactions in the Block identified by the value of the active Transaction's parameter named BLKNUM.

Other examples of SNAs include the following:

P$Name – To reference the value of the parameter labelled *Name* of the active Transaction.

F$WK0 – Gives the value 1 if a Facility labelled WK0 is busy (i.e., owned by a Transaction) and 0 otherwise.

FC$WK0 – Facility capture count that gives the number of times the Facility labelled WK0 has been SEIZEd.

FR$WK0 – Facility utilization that indicates the percentage of time the Facility labelled WK0 was busy.

N$CompA – Block entry count that gives the number of Transactions that have entered the Block labelled CompA.

Note that all the symbolic names (together with the $) in the above examples can be replaced with a numeric value that identifies the entity in question. The following 'atomic' SNAs do not include an entity identifier:

- Parameters of the active Transaction:

  - A1: Assembly set.
  - M1: Transit time which is initially set to AC1 when the transaction enters the simulation (often called the mark time).
  - PR: Priority.
  - XN1: Transaction number.

- AC1: Absolute value of the system clock (measured with respect to the start of the simulation run).
- C1: Elapsed simulation time since the last RESET Command.
- TG1: Remaining termination count. See the TERMINATE Block.

The example in Fig. A3.6 illustrates the use of an SNA with the TEST Block to check the availability of the Facility called *Machine*. The SNA FV$Machine is 1 when the Facility is available in which case the TEST Block allows the Transaction to move to the Block labelled *Continue*. Otherwise, the SNA is 0 and the TEST Block halts further progress of the Transaction (i.e., it will be placed on the retry chain of the Machine Facility).

To demonstrate the use of indirect addressing, we revise the example in Fig. A3.6 by replacing the explicit reference to the Machine Facility with an indirect address as shown in Fig. A3.7.

```
                              TEST E 1,FV$Machine
                Continue   .
                           .
                           .
                           SEIZE Machine
```

Fig. A3.6 Using SNAs

```
                              TEST E 1,FV*Station
                           .
                           .
                           .
                           SEIZE  P$Station
```

Fig. A3.7 Using indirect addressing

We assume here that the Transaction parameter named *Station* contains the number of the *Machine* Facility. The TEST Block uses an SNA with indirect addressing to reference the Facility using the *Station* parameter of the active Transaction. It will only allow Transactions to pass if the SNA is equal (*E*) to 1, that is, the Facility is available. The SEIZE statement references the Facility using the parameter value reference given by the SNA 'P$Station'.

When a Transaction parameter is used to identify entities, a generic Block segment (process) specification can be defined. The Block segment is generic in the sense that explicit references to the entities are not included in the Block segment but rather are provided by the parameters of Transactions traversing the Block segment. This approach can simplify GPSS programs by collapsing a family of identical Block segments into a single (parameterized) Block segment. This is particularly useful, for example, when there is a large number of Facilities (consider modelling a system with many tens or hundreds of Facilities).

A3.3 GPSS Standard Output

Figure A3.8 shows part of GPSS reports generated at the end of a simulation run for the simple manufacturing model. For Facility entities, GPSS provides the following values at the end of a simulation run:

1. ENTRIES – Total number of Transactions that entered the Facility entity.
2. UTIL. – The utilization of the Facility, that is, the percentage of time that the Facility was busy.
3. AVE. TIME – The average time a Transaction owns the Facility entity.
4. AVAIL. – Indicates the availability of the Facility entity, that is, equal to 1 when the Facility is available, 0 otherwise.
5. OWNER – Transaction identifier of the current owner.
6. PEND – Number of Transactions on the pending chain of the Facility entity.
7. INTER – Number of Transactions on the interrupt chain of the Facility entity.
8. RETRY – Number of Transactions on the retry chain of the Facility entity.
9. DELAY – Number of Transactions on the delay chain of the Facility entity.

Note that the ENTRIES, UTIL. and AVE. TIME can be used as DSOV values. Consult the GPSS World Reference Manual for details on available output values for other GPSS entities.

The first part of Fig. A3.8 provides information on the Block entities including the Block Number (LOC), the type of Block (BLOCK TYPE), the number of Transactions that have traversed the Block (ENTRY COUNT), the number of Transactions currently in the Block (CURRENT COUNT) and the number of Transactions on the Block's retry chain (RETRY). The last part of the figure shows the contents of the FEC and provides for each Transaction its Transaction identifier (XN), its priority (PRI), its block departure time (BDT), its assembly set identifier (ASSEM), its Current Block parameter (CURRENT), its Next Block parameter (NEXT) and non-standard parameter values (PARAMETER and VALUE).

```
LABEL            LOC  BLOCK TYPE    ENTRY COUNT  CURRENT COUNT RETRY
COMPA             1   GENERATE        2417              0         0
                  2   SEIZE           2417              0         0
                  3   ADVANCE         2417              1         0
                  4   RELEASE         2416              0         0
                  5   TERMINATE       2416              0         0
COMPB             6   GENERATE        2053              0         0
                  7   SEIZE           2053              0         0
                  8   ADVANCE         2053              0         0
                  9   RELEASE         2053              0         0
                 10   TERMINATE       2053              0         0
                 11   GENERATE           1              0         0
                 12   TERMINATE          1              0         0


FACILITY         ENTRIES  UTIL.   AVE. TIME AVAIL. OWNER PEND INTER RETRY DELAY
WK0               4470    0.906       2.918   1     4470    0    0     0      0


FEC XN    PRI       BDT      ASSEM  CURRENT  NEXT  PARAMETER    VALUE
 4470      0    14401.728    4470      3       4
 4472      0    14402.357    4472      0       6
 4473      0    14403.421    4473      0       1
 4474      0    28800.000    4474      0      11
```

Fig. A3.8 GPSS output for a simulation run of the manufacturing model

A3.4 GPSS Commands

The existence of certain entities needs to be declared to the GPSS processor and also initialized. One of the purposes of GPSS commands is to carry out these tasks, and the following commands are relevant:

- BVARIABLE: Used to declare Boolean Variable entities that implement Boolean expressions. The expression is evaluated during a simulation run using the 'BV' SNA.
- VARIABLE or FVARIABLE: Used to declare Variable entities that implement arithmetic expressions. The expression is evaluated where the 'V' SNA is specified.
- FUNCTION: Used to declare a Function entity. The Function is invoked where the 'FN' SNA is specified.
- MATRIX: Used to declare a Matrix entity.
- QTABLE: Used to declare a Table entity for collecting statistics from a Queue Entity.
- TABLE: Used to declare a Table entity for collecting statistics from other simulation data items.
- STORAGE: Used to declare and configure a Storage entity.
- EQU: Used to assign a value to a label, e.g., for assigning labels to constants.
- INITIAL: Used to initialize a Matrix entity, a SaveValue entity or a Logic Switch entity.
- RMULT: To set the seeds for the random number generators.

Another role for GPSS commands is the control of the simulation execution:

- CLEAR: Completely re-initializes the simulation including, but not limited to, the removal of all Transactions, clearing all data values, resets the system clock and primes the GENERATE Blocks with their first Transactions (See the GPSS World documentation for details.).
- RESET: Resets the statistical accumulators. This Command is useful for clearing collected data after a transient warm-up period.
- START: To start a simulation run.
- STOP: To set or remove a stop condition (breakpoint) at a Block.
- HALT: To halt the execution of the simulation run (The GPSS World hotkey Ctrl–Alt–H can be used to invoke HALT.).
- STEP: To execute a specified number of Blocks (useful for debugging).
- CONTINUE: Used to continue the simulation run following a stop condition.
- SHOW: To display the state of the simulation model. Can be used to examine SNAs or calculate expressions.

Other miscellaneous Commands include the following:

- REPORT: At the end of a simulation run, a standard report is generated. This standard report can be generated before the end of the simulation run with this Command.
- INCLUDE: To include contents from another file into a simulation model. This feature is frequently needed in complex models.

GPSS World provides a GUI interface that can invoke many of the Commands to control the simulation run. Details about this interface can be found in the GPSS World Reference Manual.

A3.5 Transaction Chains

Different types of Transaction chains exist in a GPSS simulation model. The two main Transaction chains used for processing Transactions have already been introduced in Sect. A3.1.1:

- Current events chain: Transactions waiting for processing.
- Future events chain: Transactions waiting for time to advance.

Other chains used by the GPSS processor include user chains, delay chains, retry chains, pending chains and interrupt chains. Note that at any moment in time, a Transaction will be on only one of the following chains:

- Current events chain,
- Future events chain,
- Delay chain,
- Pending chain,
- User chain.

Various other rules about occupying a chain also exist. For example, a Transaction on an interrupt chain cannot be on the future events chain. A Transaction may be on any number of interrupt chains or any number of retry chains at the same time.

(a) *Retry chains*: Blocks such as GATE or TEST perform tests using some logical expression. These expressions typically contain SNAs related to one or more entities. When a Transaction tries to enter such a Block and the test fails, the Transaction is placed on all retry chains of the referenced entities. When in the underlying data structure one of entities changes, the Transactions on the entity's retry chain are moved back onto the CEC. Thus, the GPSS processor will re-evaluate the test to see if the change in the entity has caused the logical expression to become TRUE. If the test is successful, then the Transaction will enter the Block and be removed from all retry chains.

(b) *Delay chains*: Delay chains are used to queue Transactions that have been refused access to a Facility or a Storage entity (See the discussion on the Facility and the Storage entities for details.). These chains use a priority FIFO queuing discipline. User chains must be used for implementing more complex queuing disciplines.

(c) *Pending chain*: Each Facility entity contains a pending chain used to queue Transactions that are waiting to pre-empt a Facility via an interrupt mode PREEMPT Block, i.e., were refused entry into the PREEMPT Block (See the discussion of the Facility entity for further details.).

(d) *Interrupt chain*: Each Facility entity contains an interrupt chain used to queue Transactions that have been pre-empted (See the discussion of the Facility entity for further details.).

(e) *User chains*: The user chain entity is simply a general Transaction chain. Such chains can be used to implement different queuing disciplines. The UNLINK block provides many features for selecting a Transaction from a user chain.

A3.6 GPSS Entities

We first divide all available GPSS entities into four categories and examine each category in turn. An overview of each entity is presented that includes related Commands, Blocks and SNAs where applicable. Further details about the usage of these various entities can be found in the GPSS World Reference Manual.

A3.6.1 GPSS Entity Categories

We separate the GPSS Entities that are divided into the following four categories:

1. Basic category: Transaction entities and Block entities are the two members of this category. They represent the principle building blocks for GPSS simulation models.

2. Structural category: This category is comprised of entities used to represent the structural aspect of the simulation model. In Sect. A3.1, the Facility entity is used to represent a workstation together with the queue leading to it. This category has the following members:

   (a) Facility: Represents a resource together with a queue leading to it (There are also two other queues defined for the Facility that are used in the implementation of pre-emption.).
   (b) Storage: Represents a group together with a queue leading to it. A Transaction occupies space in a Storage entity. The amount of space available and the amount of space a Transaction occupies in the Storage entity are assignable by the program developer.
   (c) User chain: A queue that is a priority ordered list of Transactions.
   (d) Logic Switch: A simple entity with two states: set or reset.
   (e) Numeric Group: A set of numeric values that can be useful for recording events or representing the state of a process.
   (f) Transaction Group: A set of Transactions. This is different from the Storage entity. A Transaction can be a member of one or more Transaction Groups. This membership serves as a means of classifying Transactions.

3. Data and procedures category: This category is comprised of entities that can be used to represent data or be used to develop expressions and functions:

   (a) Matrix: A multidimensional array of data values. Matrices are global in scope and can be accessed (including modification) by both the PLUS procedures and GPSS Blocks. Within GPSS statements, the 'MX' SNA is used to reference a matrix, while PLUS procedures use the 'Matrix' label.
   (b) Save Value: A variable that can take on any value.
   (c) Variable: The variable entity is an arithmetic or logical expression (of arbitrary complexity) that can be evaluated at any time using an SNA reference. There are three types of Variable entities: arithmetic, floating point and Boolean.
   (d) Function: Used to create data functions. For example, this entity is useful in creating discrete probability mass functions.

4. Output category: This category contains entities that can be used to collect data during a simulation run and carry out subsequent statistical analysis:

   (a) Queue: This entity is often used to collect data and provide average values (such as the average resident time) relative to the delay chains for Facility or Storage entities. SNAs for Queue entities include the average queue size, maximum queue size, average queue resident time (with and without entries with zero time in queue) and number of entries with zero time in the queue. The Queue entity can also be used to determine other times such as the time that a Transaction spends in an area in the system (such as the time waiting for and being serviced by some resource, e.g., a Facility) or the total (simulation) time in the model. Note that the Queue entity is not, in fact, a genuine queue (as is a user chain).

(b) Table: This entity provides a means of creating a histogram for data obtained from the simulation run. Tables can be created for Queue entities or for any other data values in the system, such as parameter values.

A3.6.2 Basic Category

A3.6.2.1 Transaction Entity

Recall that Transactions are data tuples that 'move' through a simulation model. Although the programmer views movement of Transactions from Block to Block, GPSS actually moves Transactions from chain to chain (and invokes list processing operations). The state of a Transaction in a simulation is defined by the following:

- Parameters: A set of values referenced by numerical parameter identifiers (symbolic references are also permitted). A set of standard parameters is associated with every Transaction. This set includes the following:

  - Transaction number – This parameter uniquely identifies the Transaction in the simulation model. It is set when the Transaction enters the simulation model.
  - Priority – The priority impacts on the selection of Transactions from the various GPSS chains. Higher priority Transactions will be selected first.
  - Mark time – A timestamp that indicates the time that a Transaction entered the simulation model. This value can be altered by a MARK Block.
  - Assembly set – It is possible to split a Transaction into a number of 'parallel' Transactions that form a set called an assembly set. The SPLIT Block carries out this split operation. Each of resulting Transactions is given its own Transaction number. All Transactions resulting from such an operation have the same value for the assembly set parameter (most other parameters are also duplicated from the parent). This makes it possible to reassemble the split Transactions with the ASSEMBLE Block. When a Transaction enters a simulation model, its assembly set parameter is set to its Transaction number.
  - Current Block – This parameter contains the number of the Block that the Transaction currently occupies.
  - Next Block – This parameter contains the number of the Block that the Transaction wishes to enter.
  - Trace indicator – A flag that enables printing of tracing messages to the console each time a Transaction enters a Block. This flag is set and reset by the TRACE and UNTRACE Blocks, respectively.
  - The state of a Transaction is a reference to the chain on which it is located.

The possible states of a Transaction are as follows:

- Active – The Transaction has reached the head of the CEC and is the one that is currently being processed.
- Suspended – The Transaction is either on the FEC or the CEC and is waiting for processing.
- Passive – The Transaction is resting in a delay chain, pending chain or user chain.

A Transaction can also be in the pre-empted state which is not mutually exclusive with the states listed above. A Transaction is in this state when located in an interrupt chain. The Transaction is in the terminated state when it has left the simulation model.

Related Block Entities

As Transactions move into Blocks, they invoke some processing that impacts the simulation model's state. Some Blocks, however, have a direct impact on the Transactions themselves. A list of these Blocks is given below:

- TRANSFER – The TRANSFER Block, in subroutine mode, assigns the Block number of the TRANSFER Block to one of the Transaction's parameters. This allows the Transaction to return to this point, thereby providing a subroutine function.
- SELECT – The SELECT Block provides a variety of features to select an entity from a set specified by lower and upper entity numbers, such as selecting from a set of Facility entities. The number of the selected entity is saved in a parameter of the active Transaction (which can be used with indirect addressing).
- SPLIT – This Block splits a Transaction into a number of 'replicated' Transactions. These replicated Transactions have the same value for their respective assembly set parameter.
- ASSEMBLE – This Block reassembles the Transactions created by the SPLIT Block. When a specified number of Transactions in the assembly set have been collected, the first Transaction will exit the ASSEMBLE Block, and the rest are destroyed.
- GATHER – This Block collects Transactions created by the SPLIT Block (i.e., from the same assembly set). When a specified number of Transactions from the same assembly set have been collected, they are allowed to exit the Block. Note that this Block is similar to the ASSEMBLE Block but does not destroy Transactions.
- MATCH – This Block works in pairs. That is, two MATCH Blocks are placed in a simulation model and will allow Transactions through only when both Blocks see Transactions from the same assembly set.
- COUNT – This Block carries a counting operation on a set of entities (identified by lower and upper entity numbers) and places the count in one of the active Transaction's parameters.

- INDEX – INDEX has two operands. This Block adds the value of a parameter of the active Transaction identified by the first operand to the value of the second operand and saves the result into parameter 1 of the active Transaction. For example, INDEX 4,23.4 will add 23.4 to the value in parameter 4 and save the results in parameter 1.
- LOOP – This Block decrements the parameter in the active Transaction that is identified by a LOOP operand. If the resulting parameter value is not 0, then the Transaction will be sent to a Block identified by another LOOP operand. Otherwise, the Transaction proceeds to the Block following the LOOP Block.
- TRACE and UNTRACE – These Blocks turn on and off the Transaction's trace indicator.
- ADOPT – This Block will change the assembly set value of the active Transaction.
- GENERATE – This Block generates Transactions.
- TERMINATE – This Block removes Transactions from the simulation model.

Related SNAs

The following SNAs provide information about the active Transaction:

- A1 – This refers to the value of the assembly set number of the active Transaction.
- MB – This SNA can be used to determine if another Transaction from the same assembly set is located at a specific Block with number *EntNum* using the expression MB*EntNum*. The SNA returns a 1 if such a Transaction exists and 0 otherwise.
- MP – This SNA can be used to compute a transit time. It provides a value equal to the current simulation time minus the value in a parameter referenced by the parameter identifier in the SNA (e.g., MP5 yields the value of the current time minus the value found in the 5th parameter of the active Transaction).
- M1 – This SNA returns a value equal to the current simulation time minus the mark time of the active Transaction.
- P – This SNA can be used to reference the value of a parameter (e.g., P*Param*) of the active Transaction.
- PR – This SNA is a reference to the value of the active priority parameter of the active Transaction.
- XN1 – This SNA is a reference to the value of the active Transaction number (i.e., its unique identifier within the program as assigned by the GPSS processor).

A3.6.2.2 Block Entity

The Block entities are the basic programming elements of the GPSS environment. The Block segments (a sequence of Blocks) are used to specify the various processes within a simulation model. The Blocks and the Transactions work together to generate changes in the various structural entities in the simulation model, and these, in turn, give rise to the model's behaviour. Furthermore, it is the through these structural entities that processes interact with each other (see the structural entity category). As was already seen in Sect. A3.1.2, it is possible to represent Block segments using a graphical representation of the Blocks.

Related SNAs

- *N* – The total number of Transactions that have entered a Block.
- *W* – The current number of Transactions occupying a Block.

Several Block entity types will be examined in the following subsections in terms of their relationship with various other GPSS entities. First, however, we list some Blocks that have a variety of specialized roles:

Miscellaneous Blocks

- ADVANCE – This Block provides the fundamental mechanism for representing the passage of time (typically some service time). This is achieved by placing the Transaction entering the Block on the FEC with its BDT parameter set as described in Sect. A3.1.1. Many options exist for defining the time delay.
- BUFFER – This Block allows a Transaction to interrupt its processing and allows other Transactions on the CEC to be processed first. In effect, the active Transaction on the CEC is moved behind other Transactions of equal or greater priority.
- EXECUTE – The EXECUTE operand identifies a Block number. When a Transaction enters this Block, the action associated with the Block identified in the operand is executed.
- PLUS – This Block evaluates an expression that may include PLUS procedures. If a second operand is provided, the result of the evaluation is saved in the active Transaction's parameter as specified by the second operand.

Test and Displacement Blocks

- DISPLACE – This Block allows a Transaction to displace any other Transaction in the simulation model. The displaced Transaction is the one identified by the Transaction number in the first operand of the DISPLACE Block.
- GATE – This Block changes the flow of a Transaction based on testing the state of some entity in the simulation model. It operates in two modes. If the test is 'successful', the Transaction traverses the Block. Otherwise, in the Refuse mode, the GATE will block Transactions until the test becomes successful, and in the Alternate Exit mode, the GATE will send the Transaction to an alternate Block.
- TEST – This Block changes the flow of a Transaction based on a test that compares values, often using expressions with SNAs. Like the GATE Block, it can also operate in either Refuse mode or an Alternate Exit mode.
- TRANSFER – This Block results in the entering Transaction jumping to a new location. The Block basically sets the value for the Transaction's Next Block parameter by evaluating its operands. Many options are available with this Block, e.g., unconditional transfer, picking a location randomly, selecting among a number of Blocks that are nonblocking and using a Function to compute the Block location.

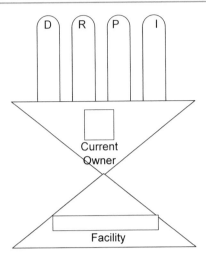

Fig. A3.9 The GPSS Facility

Data Stream Blocks

- OPEN – Opens a 'data stream'. For example, OPEN ('Data.Txt'), 2, ERROR will open the file *Data.Txt* as stream 2. If an error occurs, then the active Transaction is sent to the Block labelled *ERROR*.
- READ – Retrieves a line of text from a data stream identified by an operand.
- WRITE – Writes a line of text to a data stream identified by an operand.
- SEEK – Sets the line position of a data stream identified by an operand.
- CLOSE – Used to close a data stream identified by an operand.

A3.6.3 Structural Category

A3.6.3.1 Facility Entity

The Facility entity represents a resource that can be owned by a single Transaction. Our representation for the Facility as shown in Fig. A3.9 is intended as a snapshot of the Facility during the simulation run. A Facility contains a number of Transaction chains. These support various queuing requirements of waiting Transactions as well as pre-emption and interruption. GPSS maintains statistical information for Facilities. The Facility is always in one of two states, namely, 'busy' when owned by a Transaction or 'idle' otherwise.

In Fig. A3.9, the box labelled 'Facility' is the name assigned to the Facility and the box labelled 'current owner' identifies the Transaction that currently owns the Facility. The following Transaction chains are associated with a Facility:

- Delay chain (D),
- Retry chain (R),

- Pending chain (P) and
- Interrupt chain (I).

The retry chain contains Transactions waiting for a change in the Facility's state (e.g., from 'busy' to 'idle'). The delay, pending and interrupt chains are used to support pre-emption of the Facility. Two modes of pre-emption are available:

1. Interrupt mode: Allows a Transaction to pre-empt the Facility if the current owner of the Facility has not already pre-empted the Transaction. Transaction priority is not relevant in this mode.
2. Priority mode: A Transaction will pre-empt the Facility if its priority is higher than the priority of the current owner.

When a Transaction tries to pre-empt the Facility using the interrupt mode of the PREEMPT Block, it is refused ownership when the Facility is owned by a Transaction that acquired the Facility through pre-emption. In this case, it is placed on the pending chain.

When a Transaction tries to pre-empt the Facility using the Priority mode of the PREEMPT Block, it is refused ownership when the Facility is owned by a Transaction of equal or higher priority. In this case, it is placed on the delay chain.

When the Transaction that owns the Facility is pre-empted, it is placed on the interrupt chain. Transactions may be allowed to move through Blocks while on an interrupt chain, but movement is restricted. For example, such Transactions can never be placed on the FEC (e.g., enter an ADVANCE Block).

A Transaction that owns a Facility must return ownership using the RETURN or RELEASE Blocks, e.g., a Transaction that owns a Facility must not directly exit the simulation model. This applies equally to Transactions that are on a delay chain, a pending chain or an interrupt chain. Pre-empted Transactions, on the other hand, are allowed to exit the simulation model. Certain Blocks such as the PREEMPT and FUNAVAIL Blocks have options that can remove Transactions that are on delay chains, pending chains or interrupt chains (such Transactions are said to be 'in contention').

When a Transaction releases ownership of a Facility, ownership is given to a Transaction from one of the Facility's chains in the following order:

- Pending chain, the Transaction enters the PREEMPT Block and is placed on the CEC.
- Interrupt chain.
- Delay chain, the Transaction enters the SEIZE or PREEMPT Block and is placed on the CEC.

SNAs Related to a Facility's State

- F – Returns 1 if the Facility is busy and 0 otherwise.
- FI – Returns 1 if the Facility is currently pre-empted using interrupt mode pre-emption and 0 otherwise.
- FV – Returns 1 if the Facility is available and 0 otherwise.

SNAs Related to Data Collection

- FC – Returns the number of Transactions that have gained ownership of the Facility (SEIZEd or PREEMPTed).
- FR – Returns the fraction of time the Facility has been busy. It is expressed as parts per thousand using a value between 0 and 1,000.
- FT – Returns the average time that a Transaction owns the Facility.

Related Block Entities

- SEIZE and RELEASE: The SEIZE Block is used by a Transaction to gain ownership of a Facility and RELEASE is then used to release the ownership.
- PREEMPT and RETURN: The PREEMPT Block is used to pre-empt a busy Facility. PREEMPT allows both interrupt mode and priority mode pre-emption. A pre-empting Transaction must enter a RETURN Block to release the ownership of the Facility.
- FAVAIL and FUNAVAIL: The FUNAVAIL Block will place the Facility in the unavailable state, while the FAVAIL Block places the Facility in the available state. The FUNAVAIL Block offers many options to remove Transactions from contention.

A3.6.3.2 Storage Entity

The Storage entity provides the means for allowing a Transaction to occupy 'space'. Such an entity must be declared using the STORAGE command to define its size, i.e., how much space is available in the entity. By default, a Transaction occupies a single unit of space when entering the Storage entity, but this default can be overridden with an ENTER Block operand. There are two chains associated with a Storage entity as shown in our representation in Fig. A3.10, namely, a delay chain (D) and a retry (R) chain. The 'Members' box can be used to record the Transactions that currently occupy space in the Storage Entity. The 'Max' box provides the maximum storage units available and the 'Storage' box gives the name of the Storage entity.

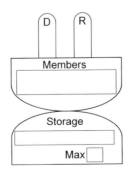

Fig. A3.10 The storage entity

When there is insufficient space to accommodate a storage request from a Transaction, the Transaction is placed on the Storage entity's delay chain in order of priority. When space is freed by the departure of a member Transaction, the delay chain is scanned, and the first Transaction that can be accommodated becomes a member of the Storage entity. The retry chain contains Transactions waiting for a change in the underlying data structure of the Storage entity.

SNAs Related to the Storage Entity's State

- R – Provides the number of storage units available within the Storage entity.
- S – Provides the number of storage units occupied within the Storage entity.
- SE – Returns 1 if the Storage entity is empty and 0 otherwise.
- SF – Returns 1 if the Storage entity is full and 0 otherwise.
- SV – Returns 1 if the Storage entity is available and 0 otherwise.

SNAs Related to Data Collection

- SA – Returns the time-weighted average of storage capacity in use.
- SC – Returns the total number of storage units that have been occupied.
- SR – Returns utilization as the fraction of total possible utilization, expressed in parts per thousand, i.e., using a value between 0 and 1000.
- SM – Returns the maximum storage value used within the entity, i.e., the high watermark.
- ST – Returns the average holding time per unit storage.

Related Block Entities

- ENTER and LEAVE: A Transaction uses the ENTER Block to enter a Storage Entity and the LEAVE Block to exit the Storage entity.
- SAVAIL and SUNAVAIL: The SAVAIL makes the Storage entity available, while the SUNAVAIL makes the Storage entity unavailable. Transactions that are already members of the Storage entity are not affected by the SUNAVAIL Block.

Related Commands

- STORAGE: Must be used to declare the capacity of the Storage entity.

A3.6.3.3 User Chain

A user chain provides a means to queue Transactions. Our representation of the user chain is shown in Fig. A3.11. It offers a priority FIFO queuing discipline. It is possible to alter such a queuing discipline by using the UNLINK Block that offers

Fig. A3.11 The user chain

many options for selecting Transactions for removal based on expressions and Transaction parameter values.

The user chain also contains a link indicator, a sort of 'red' and 'green' light that turns to red when a Transaction encounters an empty user chain with a green light indicator. When the Transaction subsequently enters an UNLINK Block, the red light is turned green if the user chain is empty. If user chain contains Transactions, the one at the head of the chain is removed and the light is left red. This link indicator feature is useful when placing a user chain in front of a Facility (in essence replacing the Facility delay chain).

SNAs Related to the User Chain Entity's State

- CH – Returns the number of Transactions currently in the user chain SNAs related to data collection.
- CA – Returns the average number of Transactions that have been in the user chain.
- CC – Returns the total number of Transactions that have been in the user chain.
- CM – Returns the maximum number of Transactions that have occupied the user chain, i.e., the high watermark.
- CT – Returns the average time that Transactions resided in the user chain.

Related Block Entities

- LINK and UNLINK: When a Transaction enters the LINK Block, it is placed on the user chain (unless the link indicator option is used). A Transaction is removed from a user chain by some other Transaction that enters the UNLINK Block. The UNLINK Block offers many options to select Transactions based on expressions and Transaction parameters.

A3.6.3.4 Logic Switch

A Logic Switch (see our representation in Fig. A3.12) simply assumes one of two states: set or reset. The 'Logic Switch' box contains the name of the entity while the 'State' circle is shaded (Logic Switch is set) or clear (Logic Switch is reset).

SNAs Related to the Logic Switch Entity's State

- LS – Returns 1 if when the Logic Switch is set and 0 otherwise.

Related Block Entities

- LOGIC – This Block can be used to set, reset or invert the Logic Switch state.

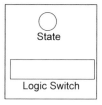

Fig. A3.12 The Logic Switch entity

Related Commands

- INITIAL – Used to initialize a Logic Switch.

A3.6.3.5 Numeric Group

The Numeric Group entity can be useful for describing the state of a process or recording events in a simulation run. Such entities contain a set of integer or real values.

SNAs Related to the Numeric Group's State

- GN – Returns the number of members in the Numeric Group.

Related Block Entities

- JOIN and REMOVE: The JOIN Block adds a value to a Numeric Group, while the REMOVE Block removes a value from a Numeric Group.
- EXAMINE: This Block tests for the existence of specified values in a Numeric Group.

A3.6.3.6 Transaction Group

A Transaction Group is a set of Transactions. Transactions can be members of any number of Transaction Groups. These entities are useful for classifying Transactions.

SNAs Related to the Transaction Group's States

- GT – Returns the number of Transactions in a Transaction Group.

Related Block Entities

- JOIN and REMOVE: The JOIN Block adds a Transaction to a Transaction Group, while the REMOVE Block removes a Transaction from a Transaction Group.
- EXAMINE: This Block tests for the presence of the active Transaction in a Transaction Group.
- SCAN and ALTER: These Blocks can test and or modify the members of a Transaction Group.

A3.6.4 Data and Procedure Category

A3.6.4.1 Matrix Entity

GPSS World supports matrices with up to six dimensions. All six dimensions are available in PLUS procedures, but only two dimensions are accessible when used with SNAs, that is, within Block arguments.

Related SNA

- MX – The expression MX*matA*(m, n) returns the value in row *m* and column *n* of the matrix *matA*.

Related Block Entity

- MSAVEVALUE: This Block can be used to assign a value (values may also be added or subtracted) to an element in the matrix. Only two dimensions of the matrix can be accessed by this Block.

Related Commands

- INITIAL: The command can be used to initialize a matrix element.
- MATRIX: This command declares the existence of a Matrix entity, from 2 to 6 dimensions (in GPSS World). This command must be used before a Matrix entity can be referenced.

A3.6.4.2 SaveValue Entity

SaveValue entities are variables that can assume any value.

Related SNA

- X – Returns the value of the SaveValue entity, e.g., X$VarB.

Related Block Entity

- SAVEVALUE: This Block assigns or modifies the value of a SaveValue entity.

Related Command

- INITIAL: Can be used to initialize the value of a SaveValue entity.

A3.6.4.3 Variable Entity

The Variable entity is associated with an expression. These expressions can contain constants, SNAs and procedure calls and include logic and arithmetic operators. Details can be found in the GPSS World Reference Manual.

Related SNAs

- BV – Returns the result of evaluating a Boolean Variable Entity.
- V – Returns the results of evaluating an arithmetic or floating-point Variable entity.

Related Commands

- VARIABLE: Creates an arithmetic Variable entity.
- FVARIABLE: Creates a floating-point Variable entity.
- BVARIABLE: Creates a Boolean Variable entity.

A3.6.4.4 Function Entity

The Function entity provides a number of options for computing a value. These options include the use of random numbers. For example, it is possible to return a random value according to some discrete probability mass function.

Many times it is more convenient to use a PLUS procedure rather than using a Function entity. Functions are particularly useful for creating list base functions and empirical probability distributions. Details about the use of the Function entity can be found in the GPSS World Reference Manual.

Related SNA

- FN – Returns the result of evaluating the Function entity.

Related Command

- FUNCTION: Used to define a Function

A3.6.5 Output Category

A3.6.5.1 Queue Entity

Queue entities are used for collecting data. It is important not to confuse Queue entities with chains. Transaction queues are formed in GPSS chains, not in Queue entities. As in the case of Transaction Groups, Transactions can become members of one or more Queue entities. GPSS will then compute queuing information based on this membership.

Typical use of Queue entities is to compute statistics related to Facility delay chains (a Transaction becomes a member of the Queue entity just before trying to SEIZE a Facility and releases the membership just after acquiring the ownership of the Facility). This can be done by sandwiching the SEIZE Block between the QUEUE and DEPART Blocks.

SNAs Related to the Queue Entity's State

- Q – Returns the current number of Transactions that are members of the entity SNAs related to data collection.
- QA – The average number of members in the Queue entity.
- QC – The total number of Transactions that entered the entity.
- QM – The maximum number of Transactions that occupied the entity.
- QT – The average time a Transaction spent in the entity.
- QX – The average time a Transaction spent in the entity without including Transactions that spent 0 time in the entity.
- QZ – The number of Transactions that spent 0 time in the entity.

Related Block Entities

- QUEUE and DEPART: When a Transaction enters a QUEUE Block, it becomes a member of the associated Queue entity. When a Transaction enters the DEPART Block, it releases its membership from the associated Queue entity.

Related Command

- QTABLE – Declares a Table entity that can be used to create a histogram of the times that Transactions spend in a Queue entity.

A3.6.5.2 Table Entity

The Table entity is used to accumulate data for creating a histogram.

Related SNAs

- TB – Average of the entries in the Table entity.
- TC – Number of entries in the entity.
- TD – Standard deviation of the entries in the entity.

Related Block Entity

- TABULATE: Used by a Transaction to add entries to a Table entity.

Related Commands

- TABLE – Used to declare a Table entity.
- QTABLE – Declares a Table entity that can be used to create a histogram of the times that Transactions spend in a Queue entity.

Annex 4: MATLAB Primer

A4.1 Introduction

MATLAB is a programming environment that is heavily oriented towards solving problems that have a numerical nature. Included here are problems whose solution requires numerical techniques from areas such as linear algebra, function minimization (optimization), integration, curve fitting and interpolation and the solution of ordinary differential equations (ode's). In other words, it provides a rich collection of software tools for dealing with many of the problems that arise in various disciplines within science and engineering. Furthermore, it provides a significant toolkit for graphical display of the solution information that is generated (both 2D and 3D). Valuable insights often flow from such visual presentations.

The brief overview in this Annex makes no attempt to provide a comprehensive exploration of the capabilities that are available in MATLAB. Our primary concern is with features related to the solution of ode's, and even then we do not explore the full range of available options. In order to set a meaningful basis for that discussion, we begin with an overview of some salient MATLAB features. The reader interested in a more comprehensive and detailed presentation should refer to the many available references (a short list is provided at the end of this Annex).

A4.2 Some Fundamentals

MATLAB's basic data element is a matrix (a two-dimensional array). This has a significant impact upon how data is managed within MATLAB. A noteworthy consequence, for example, is that a variable such as x which is intended as a scalar variable has, in fact, three distinct but equivalent reference modes. Suppose, for example, that it is desired to assign x the value 3. Three referencing alternatives are available, namely, $x = 3$, $x(1) = 3$ and $x(1, 1) = 3$.

A special case of an array is distinguished in MATLAB; this is the case of an array with a single row or a single column. Such a 'one-dimensional' structure is called a vector and the usage here parallels the convention within linear algebra.

© Springer Nature Switzerland AG 2019
L. G. Birta and G. Arbez, *Modelling and Simulation*, Simulation Foundations, Methods and Applications, https://doi.org/10.1007/978-3-030-18869-6

There is, however, one noteworthy difference. The term 'vector', as used in linear algebra, typically refers to a data element having vertical structure as opposed to horizontal structure. In MATLAB, the term is used without any implied 'orientation' for the data element in question. There are nevertheless many circumstances where orientation does matter, and consequently care has to be taken.

Unlike conventional programming languages, variables do not need to be 'typed' (declared). A variable springs into existence when a value is assigned to it, and the type of that variable is inferred from the context at the time of value assignment. A MATLAB variable must begin with a character which can then be followed by an (essentially) arbitrary number of characters (upper or lower case), digits and the underscore character. Variables' names are case sensitive; hence, the variables Good_Morning and good_morning are different.

Inasmuch as MATLAB has a matrix foundation, it is important to be aware of conventions that govern the assignment of values to this key data structure. Consider, for example, the assignment:

$$A = [2, \quad 4, \quad 6, \quad 8; \quad 3, \quad 6, \quad 9, \quad 12; \quad 4, \quad 8, \quad 12, \quad 16];$$

The resulting value assigned to A is the 3×4 array (3 rows, 4 columns):

$$
\begin{array}{cccc}
2 & 4 & 6 & 8 \\
3 & 6 & 9 & 12 \\
4 & 8 & 12 & 16
\end{array}
$$

The 'rule' demonstrated here is that values are assigned row by row and these rows are separated by semicolons. In this example, the entries in a row are separated by commas but blanks are equally valid. Square brackets, i.e., [ ], delimit the structure.

Another option is to replace the semicolon with a carriage return, i.e., a new line. The assignment with this approach is as follows (commas omitted here between row entries):

$$
A = \begin{array}{cccc}
[2 & 4 & 6 & 8 \\
3 & 6 & 9 & 12 \\
4 & 8 & 12 & 16];
\end{array}
$$

It is worth noting that the assignment

$$x = [1; \quad 2; \quad 3; \quad 4; \quad 5; \quad 6; \quad 7; \quad 8; \quad 9; \quad 10];$$

yields a vector value (vertical orientation) for x because the semicolon is used as a row separator.

A reference to any particular element within an existing matrix uses a row/column convention with these indices enclosed in parentheses. For example, $A(3, 2) = 10$ alters the value of the element in the third row/second column of A from 8 to 10.

The colon (:) plays an important role in referencing elements of a matrix. This role is conditioned by whether or not a matrix is already in existence. We give below several examples (here A is the matrix created above):

(a) $B = A(:, 4)$ assigns to B the fourth column of A.
(b) $C = A(3, :)$ assigns to C the third row of A.
(c) $D = A(1:2, 1:3:4)$ assigns to D the first and second rows of A and the first and fourth columns of A^3; in other words, the value of D is

$$
\begin{array}{cc}
2 & 8 \\
3 & 12
\end{array}
$$

If a matrix Q does not yet exist, then a value assignment to an element of Q infers a matrix structure that flows from explicit indexing, for example, $Q(3, 2) = 7$ implies that value in the third row and second column of Q has the value 7. For this assignment to be meaningful, there must be an associated implication that Q is an array with three rows and two columns, and, in fact, Q is created as

$$
\begin{array}{cc}
0 & 0 \\
0 & 0 \\
0 & 7
\end{array}
$$

The convention illustrated above is somewhat more general, e.g., $R(3, 1:4) = [10, 20, 30, 40]$ creates the following value for the matrix R:

$$
\begin{array}{cccc}
0 & 0 & 0 & 0 \\
0 & 0 & 0 & 0 \\
10 & 20 & 30 & 40
\end{array}
$$

The indexing convention can also be used remove an entire row or column from an existing matrix. For example, $R(:, 3) = []$ removes the entire third column from R so that the resulting value for R is

$$
\begin{array}{ccc}
0 & 0 & 0 \\
0 & 0 & 0 \\
10 & 20 & 40
\end{array}
$$

---

[3]In MATLAB, the structure: k:m:n defines a vector whose entries are the integers: k, (k + m), (k + 2m), etc. continuing up to (k + Nm) where (k + Nm) is the largest integer that does not exceed n.

The extreme case of entirely removing the existence of a previously created matrix (such as R above) can be achieved with: clear R.

Specialized conventions for constructing vectors likewise exist. For example, if x1, x2, ..., xn are existing row vectors, not necessarily of the same length, then

$$rv = [x1, x2, x3, \ldots, xn]$$

is a row vector obtained by placing the constituent row vector side by side. Similarly, if y1, y2, ..., yn are existing column vectors, not necessarily of the same length, then

$$cv = [y1; y2; y3; \ldots; yn]$$

is a column vector obtained by 'stacking' the constituent column vectors one below the other.

A4.3 Printing

Acquiring printed output from a computation is an essential requirement. In MATLAB, this important feature is linked to the notion of a string which is any character string enclosed between single quotes. The *disp* command displays the value of a string and its format is

$$disp\,(\mathord{<}string\mathord{>})$$

Thus

```
outText = 'what a fine day';
disp(outText);
```

prints the single line:

what a fine day.

The notion of a string is not, in itself, adequate for the task at hand which is to provide values for variables. There is an associated command that creates a string that includes both the values for named variables and, as well, information about the format to be used in presenting these values. The command, which is used in conjunction with *disp*, is *sprintf*[4] and its format is

```
sprintf(<string including format specification>, ...
           <variable list>)
```

Thus

```
sixty = 60;
cos60 = cos(pi/3);
outInfo = sprintf(                              ...
           'The cosine of %4.0f degrees is %6.2f', ...
           sixty, cos60);
disp(outInfo);
```

prints the line:

```
           The cosine of 60 degrees is 0.50
```

Illustrated above are two occurrences of a format specification whose general form is '%m.nf'. Here, m specifies a field width, and n specifies the number of decimal digits to be provided in the displayed value.

A4.4 Plotting

Plotting can be regarded as a graphical presentation of the relationship between the data within two (or more) vectors. We briefly review some of the fundamental features relating to this task that is provided in MATLAB. Our concern is entirely with the creation of 2D line plots of accumulated data (while 3D renditions are likewise available, they are beyond our considerations here).

The basic MATLAB command is

```
                    plot(x, y);
```

where x and y are the vectors of interest and necessarily have identical length. The result is a graph of the corresponding data pairs (x_k, y_k) where the x_k values are distributed across the horizontal axis, while the y_k relate to the vertical axes. Some

---

[4]This MATLAB function is based on the C standard library function *sprintf*.

Table A4.1 Options for Colour, Line Style and Data Markers for the plot Command

Line colour		Line style		Data markers	
k	black	-	solid line (default)	+	plus sign
r	red	–	dashed line	.	point
b	blue	:	dotted line	*	asterisk
g	green	-.	dash-dot line	o	circle

annotation is always required in order to clarify the contents of the graphical display that results. The following are three relevant MATLAB commands:

```
xlabel('text enclosed in quotes');
        % Annotates the horizontal axis of the plot
ylabel('text enclosed in quotes');
        % Annotates the vertical axis of the plot
title('text enclosed in quotes');
        % Provides a title for the displayed graph
```

The basic plot command given above has many extensions and we explore several of these. We begin by noting that options are available for the manner in which the line for the plot is presented, e.g., it may be coloured, it need not be a solid line, and the data points along the trajectory can be explicitly 'marked'. In fact, we can write the plot command as

$$plot\ (<curve\_specification>);$$

where $<curve_specification> = $ x-values, y-values, 'style-options'. The third entry in this triplet is new. The general format for *style-options* is a sequence of characters representing style options where a somewhat reduced summary of choices is given in Table A4.1.

By way of illustration:

(a) `plot(x, y, 'g')` displays the plot of y versus x with a green solid line.
(b) `plot (x, y, 'r-')` displays the plot of y versus x with a red dashed line.
(c) `plot (x, y, 'b*')` displays the plot of y versus x with a blue solid line with the data points indicated with an *.
(d) `plot(x, y, '+')` displays y versus x as a set of unconnected points marked by +.

Notice that all three components of *style-options* need not be given.

A frequent requirement is for plotting multiple curves within the same plot window. These are referred to as 'overlay plots' in MATLAB. There are two distinct cases. The first is when the data for each of the curves is available when the plot is to be created. The straightforward approach that deals with this case is the invocation of the general form of the plot command, namely,

```
plot(<curve1_specification>, ...
     <curve2_specification>, ...
              .
              .
              .
     <curveN_specification>);
```

We provide below an illustration of overlay plotting which begins with a particular value assignment to the vector x, namely,

$$x = \text{linspace}(0, 1, 201)$$

The MATLAB function linespace(0, 1, 201) used above assigns to x a row vector of length 201 whose entries are evenly spaced beginning at 0 and ending at 1; thus, x = [0, 0.005, 0. 01, 0.015, ..., 0.995, 1]. The script below

```
x = linspace(0, 1, 201);
y1 = sin(2*pi*x);
y2 = cos(2*pi*x);
plot(x,y1, x, y2,'--r');
title('The sine and cosine functions');
xlabel('x');
ylabel('sin(2\pix) and cos(2\pix)');
```

generates the plot shown in Fig. A4.1.

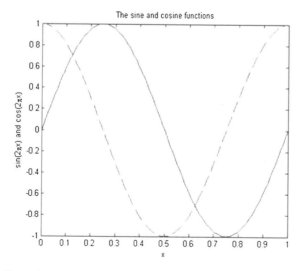

Fig. A4.1 An illustration of overlay plotting

The alternate case of overlay plots occurs when the data for the curves to be displayed is not simultaneously available, e.g., it is generated within a looping construct. In this case, the plots (and any other programming constructs that are needed) are enclosed between a 'hold on' command and a 'hold off' command. An illustration of this option is provided in the discussion of the bouncing ball example in Sect. A4.8.

A4.5 Functions

The notion of a function plays a key role in all programming environments and MATLAB is no exception. This applies to both 'built-in' functions (those inherent to the programming environment) and user-defined functions. The general syntax of the definition line that initiates the construction of a user-defined function in MATLAB is as follows:

```
function [<output variables>] = ...
                name_of_function(<input variables>);
```

As might be anticipated, both the list of input variables and the list of output variables can include any combination of vectors and arrays. An important feature, furthermore, is that the input variable list may include a reference to another function. This is illustrated in the discussion of the bouncing ball example in Sect. A4.8.

As a simple illustration, we return to the script provided earlier for the generation of a display of the sine and cosine functions over the 0 to 2π interval. Our intent now is to enhance the functionality by embedding the program code within a MATLAB function that incorporates a minor generalization. More specifically, the function (which we call sin_cos_plot) will allow the specification of the number of cycles that are to be displayed using an input variable and provide the means to output (return) the values used to create the plot (as previously, only a single cycle is displayed as shown in Fig. A4.1):

```
function [x, y1, y2] = sin_cos_plot(num_cycle)
   num_pts = 1 + num_cycle*200;
   x =  linspace(0, num_cycle, num_pts);
   y1 = sin(2*pi*x);
   y2 = cos(2*pi*x);
   plot(x, y1, x, y2, 'k--');
   title('the sine and cosine functions');
   xlabel('x');
   ylabel('sin(2\pix) and cos(2\pix)');
end
```

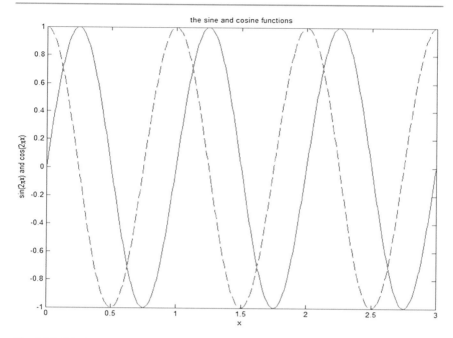

Fig. A4.2 Display resulting from the `sin_cos_plot` function with ncycle = 3

The execution of any function necessarily requires that values be assigned to all input variables. Figure A4.2 shows the display resulting from:

```
ncycle = 3;
[out1, out2] = sin_cos_plot(ncycle);
```

The same result would flow from the single line:

```
[out1, out2] = sin_cos_plot(3);
```

If there is no particular need for the data in the output variables defined for this function, then an equally admissible form is

```
sin_cos_plot(3);
```

A4.6 Interpolation

Often a functional relationship in a CTDS conceptual model is known only in tabular form. As an example recall the pilot ejection problem discussed in both Chaps. 8 and 9 where there is a given tabular relationship between altitude and air density (Table 8.1). This relationship plays a key role in the drag force acting on the ejected pilot/seat. However, the 12 data pairs in this table are insufficient for the requirements of any solution procedure that is used to solve the differential equations of the conceptual model because air density values for altitudes lying between the given data will almost certainly be needed.

The requirement here is for interpolation, i.e., extraction of estimated values based on some polynomial approximation between available data points. The general situation is the following: a functional relation, $y = F(x)$, between a variable x (the independent variable) and a variable y (the dependent variable) is known only as a finite set of data points $\{(x_k, y_k); k = 1, 2, \ldots, N\}$ where $x_{k+1} > x_k$. Interpolation is concerned with providing a means for obtaining an estimate, $y^* = F(x^*)$, for any x^* in the range $[x_1, x_N]$.

The topic of interpolation is well covered in the MATLAB toolbox. We provide here one particular option that is used in our investigation of the pilot ejection problem in Chap. 9. The approach is based on a procedure called cubic spline interpolation and uses available data values to construct a cubic polynomial approximation between adjacent data pairs. This polynomial then provides the basis for generating estimates of required values between the available data.

There are two stages involved which we call specification and usage. The specification stage is concerned with making known to MATLAB the available data values. Suppose the values of the independent variable have been assigned to the vector xx and the corresponding values of the dependent variable have been assigned to the vector yy. Appropriate MATLAB specification syntax is

```
pp = spline(xx, yy);
```

The syntax for the subsequent usage stage (invocation of the interpolation operation) has the form

```
y = ppval(pp, x);
```

which assigns to y the interpolated value corresponding to x.

A4.7 Solving Ordinary Differential Equations

We now focus on our primary interest in this Annex, namely, with the mechanism within MATLAB that relates to the solution of ordinary differential equations (ode's). As in the previous discussion, the intent is not to be all-inclusive in the coverage but rather to provide sufficient information to enable the novice MATLAB user to explore simulation problems in the domain of continuous-time dynamic systems (CTDS).

A basic requirement in MATLAB is that the equations of interest be formulated as a set of first-order ordinary differential equations, e.g.,

$$
\begin{aligned}
s_1'(t) &= f_1(t, s_1(t), s_2(t), \ldots, s_n(t)) \\
s_2'(t) &= f_2(t, s_1(t), s_2(t), \ldots, s_n(t)) \\
&\vdots \\
s_n'(t) &= f_n(t, s_1(t), s_2(t), \ldots, s_n(t))
\end{aligned}
\tag{A4.1}
$$

Here the n variables $s_1(t)$, $s_2(t)$, ..., $s_n(t)$ represent the n state variables of the model and f_k $(t, s_1(t), s_2(t), \ldots, s_n(t))$ is the time derivate of the kth state variable, $s_k(t)$. The presentation in Eq. (A4.1) can be written in a more compact vector format as

$$
\mathbf{s}'(t) = \mathbf{f}(t, \mathbf{s}(t))
\tag{A4.2}
$$

where we refer to $\mathbf{f}$ as the derivative vector. In effect Eq. (A4.1) (or equivalently, Eq. (A4.2)) represents the conceptual model that is of interest.

The importance of the derivative vector in the numerical solution of ode's should be apparent from the discussion in Chap. 9. Therefore, it should not be surprising that there is a MATLAB requirement for a user-defined function that evaluates the derivative vector given the value of time, t, and the state vector, $s(t)$. In principle, the name of this function and indeed the identifier chosen for the state vector are arbitrary. However, in the interests of clarity, we shall adopt several conventions: (a) we shall use $s(t)$ as our standard identifier for the state vector, (b) we shall call the required user-defined function for evaluating the derivative vector *deriv* and (c) we shall use *ds_dt* as the standard output of the function *deriv*, namely, as the value of the derivative vector. Consequently, our standard definition line for the required function is

```
function [ds_dt] = deriv(t, s)
```

By way of illustration, consider

$$s_1'(t) = \left(s_1^2(t) + s_2^2(t)\right)^{0.5} s_2(t)$$

$$s_2'(t) = \left(\left(s_1^2(t) + s_2^2(t)\right)^{0.5} - |s_2(t)|\right) s_1(t)$$

The derivative vector construct (with our chosen format) could be

```
function [ds_dt] = deriv(t,s)
    A = sqrt(s(1)^2 + s(2)^2);
    B = abs(s(2));
    C = A - B;
    ds_dt =  [
                A * s(2)
                C * s(1)
             ]
end
```

MATLAB provides a number of numerical methods for the solution of Eq. (A4.2), and these are summarized in the discussion to follow. Each of these numerical methods is provided as a MATLAB 'built-in' function, and, not surprisingly, one of the required input variables to each one of these MATLAB functions is the name of a user-defined function that provides the specification for the derivative vector.

Because MATLAB provides a wide range of features in respect to obtaining a solution to Eq. (A4.2), there are, correspondingly, a wide range of associated syntactic formats. The simplest is given below:

```
[times, sVal] = <ode solver name> (@deriv, ...
                                    tspan, szero)
```

This format accommodates the basic requirements which are as follows:

- Name of the derivative function construct, namely, `deriv`.
- Specification of the observation interval, namely, `tspan`.
- Specification of the initial value of the state vector, namely, `szero`.
- Specification of the numerical solution procedure to be invoked in order to generate the solution values (i.e., `<ode solver name>`, see discussion below).
- Specification of the data structure that holds the output generated by the solution procedure, namely, `[times, sVal]`.

We note the following:

(a) Typically, `tspan` is a vector with two entries. These specify the start time and the end time of the solution, i.e., the observation interval. In this case, the times at which the solution trajectory is presented are determined by the ode solving procedure. If, on the other hand, this vector has more than two entries, then output trajectory values are provided at those values of time (and only at those values).

(b) szero is the column vector $(s_1(t_0), s_2(t_0), \ldots, s_n(t_0))^T$ where t_0 is the first value in tspan, i.e., tspan(1).

(c) times is a column vector whose entries are the times at which the solution values along the output trajectory that will be computed.

(d) sVal is a matrix with n columns. The values in column k are the computed values of the kth state variable, s_k, at the time values contained in times.

MATLAB provides a suite of ode solvers. This suite does, however, vary depending on the version of MATLAB being used. Some typical solvers are summarized below. The names that appear correspond to the 'place holder' <ode solver name>, as referenced above:

ode45 This ode procedure generates two estimates for the solution value s_{n+1} at time t_{n+1} based on fourth-order and fifth-order Runge–Kutta formulas. The difference between these two estimates provides an error estimate which is used for appropriate step-size adjustment according to the stipulated error tolerance. The procedure is essentially an implementation of the Runge–Kutta–Fehlberg method (hence an explicit method) outlined in Sect. 9.3. The method is not well suited to the solution of stiff equations and is considered to be of 'medium accuracy'.

ode23 This procedure parallels the one used in ode45 except that the two Runge–Kutta formulas used are of second and third order. The method is considered to be of low accuracy.

ode113 This is the most robust and accurate explicit method provided for non-stiff equations. It is a variable order (from 1 to 13) and uses Adam–Bashforth–Moulton formulas.

ode15s This solver is specifically intended for stiff equations. It uses implicit formulas that can vary from first to fifth order.

ode23s This is a low-accuracy solver (implicit) for stiff equations but may provide better stability than ode15s in certain cases.

Details about the MATLAB ode solvers can be found in Ashino et al. [1] and Shampine and Reichelt [10].

A4.7.1 Dealing with Discontinuities

The process for dealing with discontinuities that is provided in MATLAB begins with the specification of a set of possible 'events' that need to be detected during the solution process. Each is formulated as a 'zero crossing' of a given scalar function of the state vector $s(t)$. A key issue relates to whether or not it matters if the crossing takes place from a positive value to a negative value (or vice versa). If the direction of crossing does *not* matter, then the crossing itself constitutes an event. Otherwise, the crossing is an event only if the scalar function passes through zero in a stipulated direction. When an event does occur, the solution process may or may not be halted.

The requirements outlined above are realized by a MATLAB function which we choose to call eventSpecs that has the following format:

```
function [value, isterminal, direction] = ...
                              eventSpecs(t, s);
```

Here, assuming specification of m events numbered from 1 to m:

value is a vector of length m whose kth entry is the value of a scalar function
  $E_k(s, p)$ associated with event k, namely, a function of the state vector $s(t)$ and
  the auxiliary vector of constants p with the understanding that the condition
  given by $E_k(s, p) = 0$ is of interest.
direction is a vector of length m whose entries are either -1, 0 or 1. If the
  kth entry of direction is:

-1, then $E_k(s, p) = 0$ defines a valid event k only if $E_k(s, p)$ crosses zero while
    changing from a positive value to a negative value, i.e., with a negative
    derivative.
$+1$, then $E_k(s, p) = 0$ defines a valid event k only if $E_k(s, p)$ crosses zero while
    changing from a negative value to a positive value, i.e., with a positive
    derivative.
0, then $E_k(s, p) = 0$ defines a valid event k independent of the direction of
    crossing.

isterminal is a vector of length m whose entries are either 1 or 0. If the kth
  entry of isterminal is:

1, then the solution procedure is terminated when the kth event occurs.
0, then the solution procedure is not terminated when the kth event occurs.

The eventSpecs function outlined above must be made known to the ode solver. This is achieved using a fourth input argument to the ode solver, often called options. This argument is created using the MATLAB function called odeset. The input argument list for odeset allows changing a variety of properties of the selected ode solver. Included here are 'RelTol' (the relative error parameter whose default value is 10^{-3}) and 'AbsTol' (the absolute error tolerance whose default value is 10^{-6} for all solution components). Among the arguments of odeset is 'Events' which is assigned the chosen name of the events function (which by out convention we call eventSpecs). A representative assignment is

```
options = odeset('RelTol', 1.e-6, 'AbsTol',   ...
                 [1.e-5],  'Events', @eventSpecs)
```

When the options input references the `'Events'` property (as above), then the invocation format for the ode solver is augmented as follows:

```
[times, sVal, tEvent, sEvent, eIdentifiers] = ...
    <ode solver name>(deriv, tspan, szero, options)
```

Here,

tEvent is a vector of length r that holds the times when events occurred up to and including the end point of the generated solution (often the last value in times).

sEvent is a matrix with r rows and n columns (assuming n state variables) whose kth row provides the value of the state vector, s, at the time t_k that is in the kth entry of tEvent.

eIdentifiers is a vector of length r whose kth entry provides the identifier of the event that is associated with the time t_k that is in the kth entry of tEvent.

A4.8 An Example (The Bouncing Ball)

In Chap. 2 (Sect. 2.2.6), we introduced an example problem in which a boy, standing on an ice surface, would like to throw a ball so that it falls through a hole in the surface after at least one bounce. There are several simplifications of this problem that we consider in the discussion that follows. First, however, we review the pertinent details of the SUI.

The ball is released from the boy's hand with a velocity of V_0 where the velocity vector makes an angle of θ_0 with the horizontal (see Fig. 2.3). At the moment of release, the ball has a vertical displacement of y_0 above the ice surface. The hole is located a horizontal distance of H (metres) from the boy whose horizontal position is at $x = 0$ (i.e., the hole is at $x = H$). There is a wind blowing horizontally.

By the way of a simplification, we assume that the ball's trajectory lies within a plane that passes through the boy's hand and the hole in the ice surface. Furthermore, the direction of the wind is parallel to this plane. Consequently, the ball will never strike the ice surface 'beside' the hole. If it does not enter the hole, then it can only fall in front of it or beyond it.

The ball's motion, up to the first encounter with the ice surface (i.e., the first bounce), can be described by the following CTDS model (i.e., set of ode's):

$$\frac{dx_1(t)}{dt} = x_2(t)$$

$$\frac{dx_2(t)}{dt} = -W(t)/m$$

$$\frac{dy_1(t)}{dt} = y_2(t)$$

$$\frac{dy_2(t)}{dt} = -g$$

Here, the state variables $x_1(t)$ and $x_2(t)$ represent the horizontal displacement and velocity of the ball, and $y_1(t)$ and $y_2(t)$ represent its vertical displacement and velocity. The relevant initial conditions are $x_1(0) = 0$, $x_2(0) = V_0 * \cos(\theta_0)$, $y_1(0) = y_0$, $y_2(0) = V_0 * \sin(\theta_0)$. We choose the following values for the problem parameters: $H = 5$ metres, $m = 0.25$ kg, $y_0 = 1.0$ m, and we take the wind force, $W(t)$, to have a constant value of 0.15 N.

The first simulation model we present postpones the issue of dealing with 'bouncing'. In this initial investigation, our goal is simply to determine which of 31 release angles, $\theta_0 = 15°$, $17°$, $19°$, ..., $73°$, $75°$, will place the ball closest to the hole when it first strikes the ice surface, given that the release velocity is $V_0 = 5$ m/s. In particular, we relax the originally stated objective of finding a value of θ_0 for which the ball drops into the hole (although there is nothing to preclude one of the test values having this consequence!).

The MATLAB simulation model for this investigation is given in Sect. A4.9.1. The ball trajectories generated by this program are shown in Fig. A4.3. The summary for the best release angle is given in Fig. A4.4.

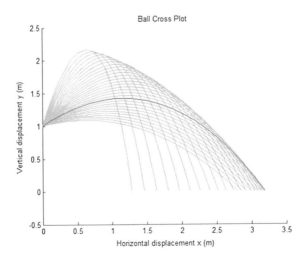

Fig. A4.3 Ball trajectories for the single-bounce case

```
Best release angle is   35.00 degrees

Minimum distance is:    1.80 meters, with impact to
left of hole at time    0.83
```

Fig. A4.4 Summary for the best release angle

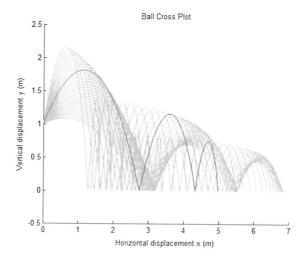

Fig. A4.5 Ball trajectories for the three-bounce case

We now enhance the simulation model given in Sect. A4.9.1 by incorporating a function (called three-Bounce) that restarts the solution procedure at each of the first two bounces and continues until the third bounce (i.e., third contact with the ice surface). In other words, the simulation model incorporates two bounces and stops on the third. This simulation model is given in Sect. A4.9.2. Two separate output displays are provided. The first (Fig. A4.5) shows the ball trajectories for the 31 test cases (i.e., release angles). The second display shown in Fig. A4.7 provides a plot of the output variable dist versus the 31 release angles (Recall that dist is the distance between the location of the hole and the location of the ball at the third contact with the ice surface.).

A solution to the originally stated problem corresponds to the circumstance where dist = 0 at the moment of the third impact with the ice surface. For convenience a horizontal line has been drawn at this value of dist in Fig. A4.7. From the information in Figs. A4.6 and A4.7, the release angle value of 53° provides a very

```
Best release angle is   53.00 degrees

Minimum distance is:    0.02 meters, with impact to
left of hole at time    2.77
```

Fig. A4.6 Summary for the best release angle

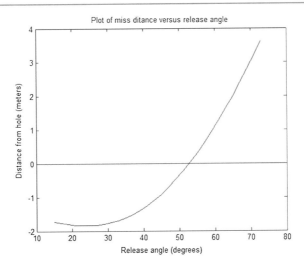

Fig. A4.7 Plot of `dist` versus `thetaD` (release angle)

good approximation of a release angle of interest. It must, however, be acknowledged that this is the special case of two bounces prior to the ball's entry into the hole. Recall that the problem statement simply required that the ball falls into the hole 'after at least one bounce' which leaves considerable latitude. The solution inferred above certainly qualifies!

We note finally that although the solution suggested above (53°) is a reasonable conjecture, the identification of a more precise solution can be easily undertaken. This is based on the observation that the output variable `dist` is clearly a function of `thetaD` (the release angle), and this functional relationship is well behaved (as shown in Fig. A4.7). Note that the function

$$J(\text{thetaD}) = dist^{\wedge}2$$

has a minimum value (of zero) at the value of `thetaD` that we seek. Finding that desired value of `thetaD` can therefore be regarded as a function minimization problem and can be undertaken using the concepts outlined in Chap. 11. Furthermore, the necessary numerical tools are available in MATLAB.

A4.9 MATLAB Program Code

A4.9.1 MATLAB Simulation Model for the Single-Bounce Case

```
function ball1
  %-----------------------------------------
  % Bouncing Ball (One bounce version)
  % File Ball1.m
  %-----------------------------------------
  % Define global constants
  global G W M ALPHA tend
  G = 9.8;      % acceleration due to gravity (m/s^2)
  W = 0.15;     % wind force (Newtons)
  M = 0.25;     % ball mass (kg)
  ALPHA = 0.8;  % energy absorption factor
  tstart = 0;
  tend = 10;    % arbitrarily assigned right hand
                % boundary of solution interval
  % Define local constants
  H = 5;        % location of hole (meters)
  INC = 2;      % increment in release angle (degrees)
  V0 = 5.0;     % initial release velocity (m/s)
  X1 = 0;       % initial horizontal position of the
                % ball
  Y1 = 1.0;     % initial vertical position of the ball
  %-----------------------------------------------------
  % EXPERIMENT
  %-----------------------------------------------------
  thetaD = 15;  % initial release angle (degrees)
  minDist = H;
  refine = 10;
  options = odeset('Events',@eventSpecs, ...
                   'Refine', refine);
  hold on
  for i = 1:30  % check 31 angles
     X2 = V0*cos(thetaD*(pi/180)); % reset horizontal
                                   % velocity (m/sec)
     Y2 = V0*sin(thetaD*(pi/180)); % reset vertical
                                   % velocity (m/sec)
     s0 = [ X1 ; X2 ; Y1 ; Y2 ];   % initialize state
                                   % vector
     [t, sVal] = ode23(@deriv, [tstart tend], ...
                       s0, options);
  % Accumulate data
     nt = length(t);
     x = sVal(:,1);
```

```
      y = sVal(:,3);
      % corresponding values in the vectors x and y
      % are points along the ball's trajectory
      dist = H-x(length(x)); % the last value in x
                             %  gives horizontal
                             %  position at bounce
      plot(x,y,'g');
      if abs(dist) < minDist
          bestTD = thetaD;
          minDist = abs(dist);
          side = sign(dist);
          xbest = x;
          ybest = y;
          tbest = t(nt);
      end
      thetaD = thetaD + INC;
  end
  %  Plot of the bouncing ball's trajectory for best
  & release angle (in red)
  plot(xbest,ybest,'r');
  xlabel('Horizontal displacement x (m)');
  ylabel('Vertical displacement y (m)');
  title('Ball Cross Plot');
  hold off
  bestSummary(bestTD, minDist, side, tbest)
end
%------------------------------------------------------
function bestSummary(angle, distance, side, time)
% this function provides a summary of the features of
& the best release angle
  out1 = sprintf('Best release angle is %6.2f ...
                                 degrees\n', angle);
  disp(out1);
  out2 = sprintf('Minimum distance is: %6.2f ...
              meters, with impact to ',distance);
  disp(out2);
  if side > 0
    out3 = sprintf('left of hole ...
                            at time %6.2f\n', time);
    disp(out3);
  else
    out3 = sprintf('right of hole ...
                            at time %6.2f\n', time);
    disp(out3);
  end
end
```

```
%--------------------------------
% DYNAMIC
%--------------------------------
function ds_dt = deriv(t, s)
  global G W M
  % s(1)=x1, s(2)=x2, s(3)= y1, s(4)=y2
  ds_dt = [
             s(2)    % ds(1)/dt = s(2)
            -W/M     % ds(2)/dt = -W/M
             s(4)    % ds(3)/dt = s(4)
            -G       % ds(4)/dt = -G
         ];
end

function [value,isterminal,direction] = eventSpecs(t,s)
  value = s(3);     % height of ball
  isterminal = 1;   % stop the integration
  direction = -1;   % negative direction only
end
```

A4.9.2 MATLAB Simulation Model for the Three-Bounce Case

```
function ball3
  %------------------------------------------
  % Bouncing Ball (Three bounce version)
  % File ball3.m
  %------------------------------------------
  % Define global constants
   global G W M ALPHA tend
   G = 9.8;      % acceleration due to gravity (m/sec^2)
   W = 0.15;     % wind force (Newtons)
   M = 0.25;     % ball mass (kg)
   ALPHA = 0.8;  % energy absorption factor
   tend = 10;    % arbitrarily assigned right hand
                 %boundary of solution interval
  % Define local constants
   H = 5;        % location of hole (meters)
   INC = 2;      % increment in initial release angle
                 %(degrees)
   V0 = 5.0;     % initial release velocity (m/sec)
   X1 = 0;       %initial horizontal position of the ball
   Y1 = 1.0;     % initial vertical position of the ball
  %-------------------------------------------------------
  % EXPERIMENT
  %-------------------------------------------------------
   thetaD = 15;  % initial release angle (degrees)
   thetaV = [];  % vector of release angle values
   distV  = [];  % vector or miss distances
   minDist = H;
   refine = 10;
   options = odeset('Events',@eventSpecs, 'Refine',...
                                            refine);
   hold on
   for i = 1:30  % check 31 angles
       X2 = V0*cos(thetaD*(pi/180)); % reset horizontal
                                     % velocity (m/sec)
       Y2 = V0*sin(thetaD*(pi/180)); % reset vertical
                                     % velocity (m/sec)
                                     % (meters/sec)
       s0 = [ X1 ; X2 ; Y1 ; Y2 ];   %initialize state
                                     % vector
       [tlast, x, y] = threeBounce(0, s0, options);
```

```
%
    % corresponding values in the vectors x and y are
    % points along the ball's trajectory
    % the last value in x gives horizontal position
    % at the third bounce
        dist = H-x(length(x));
        plot(x,y,'g');
        thetaV = [thetaV; thetaD];
        distV = [distV; dist];
        if abs(dist) < minDist
            bestTD = thetaD;
            minDist = abs(dist);
            side = sign(dist);
            xbest = x;
            ybest = y;
            tbest = tlast;
        end

        thetaD = thetaD + INC;
    end
    % Plot of the bouncing ball's trajectory for best
    % release angle (in red)
    plot(xbest,ybest,'r');
    xlabel('Horizontal displacement x (m)');
    ylabel('Vertical displacement y (m)');
    title('Ball Cross Plot');
    hold off
    bestSummary(bestTD, minDist, side, tbest)
    %
    sprintf('Hit any key to continue ...');
    pause
    % plot of miss distance vs release angle
    plot(thetaV, distV);
    hold on
    plot( [10 80], [0 0], 'k'); % prints black line at
                                % dist = 0
    ylabel('Distance from hole (meters)');
    xlabel('Release angle (degrees)');
    title ('Plot of miss distance versus release ...
                                        angle');
    hold off
end
```

```
%----------------------------------------------------------
function [tstart, x, y] = threeBounce(tstart, s0, ...
                                      options)
% this function bounces the ball three times
   global ALPHA tend
   % Record initial values for x and y
   x=s0(1);
   y=s0(3);
   for i=1:3  % solve up to the third bounce
       [t, sVal] = ode23(@deriv, [tstart tend], ...
                               s0, options);
   % Accumulate data
       nt = length(t);
       x = [x; sVal(2:nt,1)];
       y = [y; sVal(2:nt,3)];
   % reset initial values
       s0(1) = sVal(nt,1);              % x1
       s0(2) = ALPHA * sVal(nt,2);      % x2
       s0(3) = 0.0;                     % y1
       s0(4) = -ALPHA * sVal(nt,4);     % y2
   % set start time for next trajectory segment
       tstart = t(nt); % when i=3, this value is the
                       %time of third bounce
   end
end


%----------------------------------------------------------
function  bestSummary(angle, distance, side, time)
% this function provides a summary of the features of
%the best angle of release
   out1 = sprintf('Best release angle ...
                       is %6.2f degrees\n', angle);
   disp(out1);
   out2 = sprintf('Minimum distance ...
     is: %6.2f meters, with impact to ', distance);
   disp(out2);
   if side > 0
      out3 = sprintf('left of hole ...
                            at time %6.2f\n', time);
      disp(out3);
   else
      out3 = sprintf('right of hole ...
                            at time %6.2f\n', time);
      disp(out3);
   end
end
```

```
%--------------------------------
% DYNAMIC
%--------------------------------
function ds_dt = deriv(t, s)
  global G W M
  % s(1)=x1, s(2)=x2, s(3)= y1, s(4)=y2
    ds_dt = [
          s(2)     % ds(1)/dt = s(2)
         -W/M      % ds(2)/dt = -W/M
          s(4)     % ds(3)/dt = s(4)
         -G        % ds(4)/dt = -G
      ];
end

function [value, isterminal, direction] =
                        eventSpecs(t, s)
  value = s(3);       % height of ball
  isterminal = 1;     % stop the integration
  direction = -1;     % negative direction only
end
```

References

1. Ashino R, Nagase M, Vaillancourt R (2000) Behind and beyond the MATLAB ODE suite. Comput Math Appl 40:491–512
2. Davis TA (2011) MATLAB primer, 8th edn. CRC Press, Boca Raton
3. Driscoll TA (2009) Learning MATLAB. Society for industrial and applied mathematics (SIAM), Philadelphia
4. Hahn B, Valentine D (2019) Essential MATLAB for engineers and scientists. 7th edn, Academic, Boston
5. Hanselman DC, Littelfield BL (2011) Mastering MATLAB 8. Prentice-Hall, Upper Saddle River
6. Higham DJ, Higham NJ (2017) MATLAB guide, 3rd edn. Society for industrial and applied mathematics (SIAM), Philadelphia
7. Moler C (2004) Numerical computing with MATLAB. Society for industrial and applied mathematics (SIAM), Philadelphia
8. Pratap R (2017) Getting started with MATLAB. Oxford University Press, New York
9. Recktenwald GW (2000) Introduction to numerical methods and MATLAB: implementations and applications. Addison-Wesley, New York
10. Shampine LF, Reichelt MW (1997) The MATLAB ODE suite. SIAM J Sci Comput 18:1–22

Index

Printed in the United States
by Baker & Taylor Publisher Services